Advances in Intelligent Systems and Computing

Volume 1022

The series "Advances in Intelligent Systems and Computing" contains publications on theory, applications, and design methods of Intelligent Systems and Intelligent Computing. Virtually all disciplines such as engineering, natural sciences, computer and information science, ICT, economics, business, e-commerce, environment, healthcare, life science are covered. The list of topics spans all the areas of modern intelligent systems and computing such as: computational intelligence, soft computing including neural networks, fuzzy systems, evolutionary computing and the fusion of these paradigms, social intelligence, ambient intelligence, computational neuroscience, artificial life, virtual worlds and society, cognitive science and systems, Perception and Vision, DNA and immune based systems, self-organizing and adaptive systems, e-Learning and teaching, human-centered and human-centric computing, recommender systems, intelligent control, robotics and mechatronics including human-machine teaming, knowledge-based paradigms, learning paradigms, machine ethics, intelligent data analysis, knowledge management, intelligent agents, intelligent decision making and support, intelligent network security, trust management, interactive entertainment, Web intelligence and multimedia.

The publications within "Advances in Intelligent Systems and Computing" are primarily proceedings of important conferences, symposia and congresses. They cover significant recent developments in the field, both of a foundational and applicable character. An important characteristic feature of the series is the short publication time and world-wide distribution. This permits a rapid and broad dissemination of research results.

** Indexing: The books of this series are submitted to ISI Proceedings, EI-Compendex, DBLP, SCOPUS, Google Scholar and Springerlink **

More information about this series at http://www.springer.com/series/11156

Bidyut B. Chaudhuri · Masaki Nakagawa ·
Pritee Khanna · Sanjeev Kumar
Editors

Proceedings of 3rd International Conference on Computer Vision and Image Processing

CVIP 2018, Volume 1

 Springer

Editors
Bidyut B. Chaudhuri
Computer Vision and Pattern
Recognition Unit
Indian Statistical Institute
Kolkata, India

Techno India University
Kolkata, India

Pritee Khanna
Department of Computer Science
Indian Institute of Information
Technology, Design and
Manufacturing, Jabalpur
Jabalpur, Madhya Pradesh, India

Masaki Nakagawa
Department of Advanced Information
Technology and Computer Sciences
Tokyo Institute of Agriculture
and Technology
Koganei, Tokyo, Japan

Sanjeev Kumar
Department of Mathematics
Indian Institute of Technology Roorkee
Roorkee, Uttarakhand, India

ISSN 2194-5357   ISSN 2194-5365   (electronic)
Advances in Intelligent Systems and Computing
ISBN 978-981-32-9087-7   ISBN 978-981-32-9088-4   (eBook)
https://doi.org/10.1007/978-981-32-9088-4

This Springer imprint is published by the registered company Springer Nature Singapore Pte Ltd.
The registered company address is: 152 Beach Road, #21-01/04 Gateway East, Singapore 189721,
Singapore

Preface

The Third International Conference on Computer Vision and Image Processing (CVIP 2018) was organized at PDPM Indian Institute of Information Technology, Design and Manufacturing, Jabalpur (IIITDMJ), during 29 September–1 October 2018. The conference was endorsed by the International Association of Pattern Recognition (IAPR) and co-sponsored by BrahMos, Council of Scientific and Industrial Research (CSIR), Defence Research and Development Organisation (DRDO), Indian Space Research Organisation (ISRO), MathWorks and Science and Engineering Research Board (SERB), India.

The general theme of the conference being Computer Vision and Image Processing, papers on Image Registration, Reconstruction and Retrieval; Object Detection, Recognition, Classification and Clustering; Biometrics, Security, Surveillance and Tracking; Deep Learning Methods and Applications as well as Medical Image Processing and Remote Sensing were submitted from India and abroad. Out of the 206 submitted papers, 81 were accepted for presentation after peer review, making the acceptance rate of 39%. Among those, 47 were oral and 34 posters. Out of 81 papers, 76 papers were presented at the conference. These papers were distributed over eight oral and four poster sessions.

In addition to the contributory paper sessions, the conference included plenary talks of Prof. Masaki Nakagawa (Tokyo Institute of Agriculture and Technology, Japan), Prof. Venu Govindaraju (SUNY, Buffalo, USA) and Prof. Hironobu Fujiyoshi (Chubu University, Japan). Besides, Dr. R. Venkatesh Babu (IISc, Bangalore) and Ms. Ramya Hebbalaguppe (TCS Innovation Labs, New Delhi, India) delivered the invited talks. The talks ranged from historical epochs of Artificial Intelligence research to theoretical and application-based works on Deep Learning. The talks enriched and expanded the knowledge horizon of the researchers attending the conference.

Besides these, a coding challenge supervised classification of bird species from a set of bird images was organized and the works of the respondents were judged by a committee. A pre-conference workshop was organized by Mathworks on Deep Learning for Computer Vision Applications. Like the previous year, the best oral and poster papers were selected by a panel of experts. The last session of the conference unfolded the Best Paper Award, the Best Student Paper Award and the Best Poster Award along with the declaration of the names of the Coding Challenge winners and the venue and organizing institute name of the next CVIP conference (2019).

Overall, the conference was a grand success. Young researchers were immensely benefitted by interacting with academic and industry experts. Like previous years, the proceedings of this conference are also compiled in this edited volume brought out by Springer Nature under their series of *Advances in Intelligent Systems and Computing*.

The success of such an event was due to harmonious contributions of various stakeholders including the members of the international advisory committee, the technical programme committee, the plenary and invited speakers, the local organizing committee, the sponsors, the endorser and the researchers who attended the conference. Our sincere thanks to all of them! Last but not least, warm thanks are due to Springer for printing and publishing this proceedings in such a beautiful form.

Kolkata, India
Koganei, Japan
Jabalpur, India
Roorkee, India

Bidyut B. Chaudhuri
Masaki Nakagawa
Pritee Khanna
Sanjeev Kumar

Contents

About the Editors

Bidyut B. Chaudhuri is currently Pro Vice Chancellor of Techno India University, Salt Lake, Calcutta, India. Previously he was INAE Distinguished Professor at Indian Statistical Institute, Calcutta. He received his B.Sc. (Hons), B.Tech, and M.Tech. degrees from Calcutta University, India, and his Ph.D. from the Indian Institute of Technology Kanpur, in 1980. He did his postdoc work as Leverhulme fellow at Queen's University, UK and acted as a visiting faculty at the Technical University, Hannover, Germany. His main research interests are in Pattern Recognition, Image Processing, Language processing and Machine learning in which he has published 450 research Papers and five books. He is a life fellow of IEEE, IAPR, TWAS as well as fellow of Indian Academies like INAE, INSA, INASc. He has received many awards for his research work. Prof. Chaudhuri is now an Associate Editor of the International Journal of Document Analysis and Recognition (IJDAR), International Journal of Pattern Recognition and Artificial Intelligence (IJPRAI). In the past he worked in such capacity in several other international journals.

Masaki Nakagawa is a Professor of Media Interaction at the Department of Computer and Information Sciences, Tokyo University of Agriculture and Technology, Japan. He graduated from the University of Tokyo in March 1977 and pursued an M.Sc. course in Computer Studies at Essex University, England, sponsored by the Japanese Government. In March 1979, he graduated from the University of Tokyo with an M.Sc. in Physics. In July 1979, he completed his M.Sc. in Computer Studies at Essex University in England, and in December 1988, he completed his Ph.D. at the University of Tokyo. His work chiefly focuses on handwriting recognition and pen-based user interfaces and applications, especially educational applications. Prof. Nakagawa has over 300 publications to his credit.

Pritee Khanna is an Associate Professor and Head of the Computer Science & Engineering Discipline, PDPM Indian Institute of Information Technology, Design and Manufacturing, Jabalpur. Her main areas of interest include Biometrics, Biomedical Image Processing, Image Retrieval and Indexing, Dynamic Gesture

Recognition, and Computer-Aided Product Design. She is a recipient of UGC Fellowship, India and Long Term JSPS Fellowship, Japan. She is a senior member of the IEEE Computer Society and a life member of IAENG. She has published 89 papers in journals and conference proceedings.

Sanjeev Kumar is an Associate Professor of Mathematics at the IIT Roorkee, India. His areas of interest include Computer Vision & Mathematical Imaging, Inverse Problems, and Machine Learning. He completed his Ph.D. in Mathematics at the IIT Roorkee in 2008. He is a member of the IEEE Computer Society and International Association of Pattern Recognition, and a life member of the ACEEE and IACSIT. He has published over 40 papers in journals and conference proceedings.

CARTOONNET: Caricature Recognition of Public Figures

**Pushkar Shukla, Tanu Gupta, Priyanka Singh
and Balasubramanian Raman**

Abstract Recognizing faces in the cartoon domain is a challenging problem since the facial features of cartoon caricatures of the same class vary a lot from each other. The aim of this project is to develop a system for recognizing cartoon caricatures of public figures. The proposed approach is based on the Deep Convolutional Neural Networks (DCNN) for extracting representations. The model is trained on both real and cartoon domain representations of a given public figure, in order to compensate the variations in the same class. The IIIT-CFW (Mishra et al., European conference on computer vision, 2016) [1] dataset, which includes caricatures of public figures, is used for the experiments. It is seen from these experiments that improving the performance of the model can be achieved when it is trained on representations from both real and cartoon images of the given public figure. For a total of 86 different classes, an overall accuracy of 79.65% is achieved with this model.

Keywords Face recognition · Cartoon recognition · Deep Learning · Convolutional Neural Networks

1 Introduction

The rise in the number of pictures and videos has increased the need for different face recognition algorithms. Recognizing cartoon faces is one of these algorithms proposed to be applicable in many sectors, such as education, media, and entertainment. For example, unwanted graphic contents can be removed using this algorithm. Moreover, educational interfaces can be built for children with special needs or machines

P. Shukla (✉)
University of California, Santa Barbara, USA
e-mail: pushkarshukla@umail.ucsb.edu

T. Gupta · P. Singh · B. Raman
Indian Institute of Technology Roorkee, Roorkee, India

B. Raman
e-mail: balarfma@iitr.ac.in

© Springer Nature Singapore Pte Ltd. 2020
B. B. Chaudhuri et al. (eds.), *Proceedings of 3rd International Conference on Computer Vision and Image Processing*, Advances in Intelligent Systems and Computing 1022,
https://doi.org/10.1007/978-981-32-9088-4_1

can be upgraded to have a better understanding in humor and art, with the help of cartoon face recognition algorithm.

Recognizing images in the cartoon domain is challenging since there are many variations. The two caricatures of the same person might be entirely different from each other. Caricatures are usually exaggerated version of faces, where certain features are highly emphasized.

Furthermore, cartoonists have different styles. Two cartoonists may have different perceptions of a human face and may choose to emphasize different aspects of the same face. Therefore, cartoon images may have lots of variations, where the sizes and shapes of various facial features like nose, ears, or eyes differ. These images may also vary in pose and illumination. The differences in the facial features of the caricatures of the same public figure are illustrated in Fig. 1. It can be seen that Obama's oval face and ears, Albert Einstein's hairstyle, Gandhi's and Stephen Hawkins' glasses are some of the distinct features that are exaggerated in their caricatures. Because of these variations, face recognition in the cartoon domain is a challenging problem.

The previous approaches to this problem were focusing mostly on recognizing and distinguishing queries of images or videos as cartoons or non-cartoons. Also, these approaches use low-level descriptors for cartoon recognition, which means there was

Fig. 1 A block diagram representation of the proposed approach. The proposed approach uses both real and cartoon features for training

no use of any sophisticated machine learning strategy. Mostly, smaller datasets of videos or images were used in these experiments and the focus was primarily on recognizing digital cartoons, instead of recognizing caricatures of public figures.

This project's main contribution is in proposing a framework capable of identifying faces in the cartoon domain. Recognizing faces in the cartoon domain is a new problem with no prior experiments. The proposed framework tries to establish uniformity between the representations of the real and cartoon faces. It uses representations that are extracted from both real faces and cartoon faces. Also, the proposed framework uses saliency-based approaches to capture distinct aspects of real and cartoon faces. The extracted representations are then used with the real representations in order to recognize cartoon faces.

The paper has been organized as follows. A review of the related existing work is presented in Sect. 2. The proposed architecture is described in Sect. 3. In Sect. 4, the results of the various experiments are presented in Sect. 4. Conclusions and discussions of the future scope of the work are done in Sect. 5.

2 Related Work

In this section, a review of the existing works on cartoon recognition is presented.

Although, it has been an old problem, there are still few researches on recognizing cartoons from images and videos. A framework for recognizing cartoons from mpeg-2 videos was proposed by Glasberg et al. [2]. This framework used audio and video descriptors to detect and distinguish sequences as cartoons or non-cartoons. An approach focusing on jawline extraction and skin color extraction to detect cartoon faces was proposed by Takamya et al. [3]. A Euclidean distance based similarity measure was applied to retrieve the relevant cartoon image, once the face was detected. A neural network based architecture that identifies a given query as cartoon or non-cartoon was proposed by Humphrey [4].

There has been little research in the cartoon domain, also because of the absence of publicly available datasets. A dataset including caricatures of 100 public figures was proposed in the IIIT-CFW [1] dataset. Generating caricatures from images is also becoming an active area of research since the introduction of generative adversarial networks. Taigman et al. [5] proposed a cross-domain image transfer network that can generate caricatures. Another method was also proposed by Liu et al. [6] to generate cartoons from sketches using conditional generative adversarial networks [7].

3 Proposed Method

A detailed explanation of the framework proposed for cartoon recognition has been provided in this section.

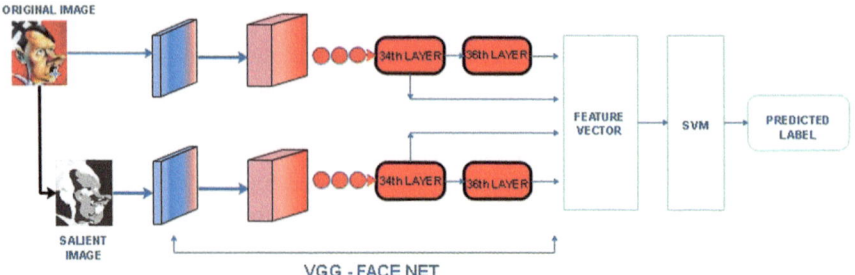

Fig. 2 A block diagram representation of the proposed approach. The proposed approach uses both real and cartoon features for training

3.1 Architecture

While building the proposed architecture, two key intuitions are taken into consideration. First, it is considered that even though caricatures of the same face may not look alike, they certainly resemble the original face in some manner. Therefore, considering representations of the faces for the same class may highly increase the performance of the cartoon face recognition system. Second, many caricatures are exaggerated versions of the original faces, where some facial features are highlighted. Therefore, in the model, the local facial features must also be incorporated (Fig. 2).

The model is trained on representations of the real faces and cartoon faces of the same class in order to account for the similarity between real faces and cartoon faces. For incorporating the local features, salient regions of the face are detected and representations are extracted from both the image and the salient parts of the face.

Because of their superiority in object recognition tasks [8], deep convolutional neural networks are chosen for constructing the framework over traditional hand-crafted features. Then, the extracted salient features are separately fed into different CNNs. The training data is formed of images of real faces and cartoon faces for each class. Representations extracted from different layers of the CNNs for the cartoon and real subspace are concatenated into a single representation vector L:

$$L = \{L1, L2\} \tag{1}$$

where L represents the concatenated vector and L1 and L2 are the two CNN representations. Then, this single representation vector L is fed into a multi-class Support Vector Machine (SVM). The SVM is used for learning these representations and for further classification.

3.2 Detecting Salient Regions

In order to extract the important facial features of real and cartoon faces, a cluster-based saliency method [9] is employed. The framework is proposed for detecting co-salient regions and used three different cues, but in this proposed method, two different saliency-based cues are used for extracting salient regions. The model only uses contrast and spatial cues that are extracted via a cluster-based pipeline for a single image.

Let the ith part of the image be denoted as p_i^j. Then, the jth image lattice is denoted by N_j. Let there be K clusters denoted by $\{C_k\}_{k=1}^K$. These clusters are represented by a set of D-dimensional vectors, where the center of the cluster is denoted by $\{\mu^k\}_{k=1}^K$. The function b: $R^2 \rightarrow \{-1, 1\}$ associates pixel p_i and cluster index $b(p_i^j)$.

Contrast Cues: The contrast cue is calculated for representing the uniqueness of visual features in an image. It is one of the most vital cues for detecting salient parts of the image. The contrast cue for a cluster $w^c(k)$ is defined as follows:

$$w^c(k) = \sum_{i=1:i\neq k}^{K} (n^i \, ||\mu^k - \mu^i||_2 \frac{1}{N}) \tag{2}$$

where N represents the total number of pixels in the image and C^i represents the total pixels of the cluster space. The contrast cue proves to be extremely useful for images where the foreground significantly differs from the background. However, the cue does not perform well if the background is complex.

Spatial Cues: People usually focus more on objects that are close to the center of the image, which is often called "Central Bias Rule". The spatial cue uses this phenomenon to extract salient regions in an image. While doing so, the spatial cue considers their proximity with the center of the image. For a given cluster C^K, the spatial cue is given as follows:

$$w^*(k) = \frac{1}{n^k} \sum_{i=1}^{M} \sum_{j=1}^{N} (N(||z^j - o^j||^2|0, \sigma^2)\delta[b(p^i - c^k)] \tag{3}$$

where the Kronecker delta function is denoted as $\delta()$. $N()$ is the function that is used for calculating the Euclidean distance between the center of the image z^j and the given pixel o^j. The total number of pixels in a cluster is denoted by n^k and it is used as a normalization constant. The spatial cue is especially useful for complex backgrounds since it can suppress the adverse effects of the contrast cue on a complex background. Examples of salient regions extracted by the algorithm for cartoon faces can be seen in Fig. 3.

3.3   Representation Extraction Using Deep CNNs

Pretrained CNN's were preferred over handcrafted CNNs due to the small size of the dataset. The vgg-face architecture was preferred over other pretrained DCNNs primarily because the network has been specifically designed for recognizing faces. Therefore, the representations extracted from the network would be more useful for

Fig. 3 Salient regions extracted from cartoon faces

recognizing cartoon faces than other pretrained networks. DCNNs have had some very interesting applications in the past [10–12], and therefore we employ them for our model.

The vgg-face model is built of 37 different layers comprising 13 convolutional layers, 5 max pooling layers, 3 fully connected layers, and a softmax layer at the end of the network. Representations were extracted from the 34th and the 37th layer of the network. The dimensions of both these representations were 4096*1. These representations were then concatenated to form a vector of 8192*1. Similar representations were also extracted from the salient image. The final feature vector had a dimension of 16384*1. The feature vector was then fed into a support vector machine.

3.4 Training Using Multi-class SVM

A multi-class linear SVM was used for training the extracted representations. A one-vs-all classification strategy was employed for training the SVM. The SVM was trained and tested on a total of 86 different classes. The SVM was trained with a sequential minimal optimization strategy with a kernel offset of 0.

4 Experimental Setup

The section outlines and discusses the outcomes of various experimental scenarios that were considered for testing the performance of the proposed model. A description of the data that was used for testing and training the model has also been provided in this section.

4.1 Dataset Description

The IIIT-CFW [1] database is a publicly available dataset comprising 8298 images of cartoon faces of 100 public figures. The dataset has been annotated in terms of position, pose, age, and expression of the different cartoon faces. Classes with greater than 35 images were considered for experimentation purposes. Hence, the model was trained on a total of 86 classes. The dataset was then split further into training and testing data where 10 images per class were used for testing whereas the remaining images were used for training the model. As per the requirements of the model, real images for each public figure were also needed for training purposes. Further, a dataset was also created comprising actual facial images of these public figures. Each class in the real domain comprised of more than 50 images. Some of these images were previously provided as a part of the IIIT-CFW dataset whereas others were collected from various sources across the Internet. The experiments were performed on a system with Intel(I5) processor with an NVIDIA GEFORCE GTX gpu. MATLAB2017a was used as a software for performing these experiments. A sample of images present in the dataset has been shown in Fig. 4.

4.2 Experimental Scenarios

Two different experimental scenarios were considered to test the performance of the proposed model. The scenarios and their outcomes have been discussed below:

Scenario 1 In the first set of experiments, the focus was on comparing the proposed framework with other state-of-the-art descriptors and classifiers. Table 1 provides a statistical comparison of the performance of different descriptors and classifiers with the proposed framework. The model was also compared with the vgg-face [13] that is a state-of-the-art face recognition algorithm. Recognition accuracy was used as a measure for evaluating the performance of these models. The architecture achieved an overall recognition accuracy of 74.65%. It was seen that not only did the proposed method outperform several handcrafted and deep feature descriptors but also the individual descriptors were not able to achieve a significant recognition accuracy while recognizing cartoon images.

Fig. 4 Sample cartoon faces present in the dataset

Discussion: The following conclusions can be made on the basis of the results that have been presented in Table 1:

– Handcrafted representation extraction methods like HOG [14], LBP [15], etc. performed poorly.
– Pretrained models like Alex-Net [16] did not perform well and were not able to achieve a recognition accuracy greater than 50%.
– The state-of-the-art deep models that were trained for face recognition performed considerably better than the other models.
– Adding salient aspects of the image improved the performance of the model and lead to an increase in the recognition accuracy by 4.

Scenario 2 In the second scenario, we wanted to focus on the impact of adding representations from the real face domain while training the model. Therefore, experiments were carried out to check the efficiency of the framework in the real, cartoon, and mixed domains. Although, the test dataset for the experiments remained the same, different SVMs were trained on representations that were extracted from the real faces and cartoon faces, respectively. The results were then tested against the results

Table 1 A statistical comparison of various handcrafted and deep features with the proposed architecture to recognize cartoon faces

Technique	Accuracy (%)
HOG+SVM	18.37
HOG+Random Forest	12.67
HOG+KNN	13.37
BSIF+SVM	15.46
GIST+Random Forest	26.51
GIST+SVM	25.93
GIST+KNN	19.77
AlexNet (fc7)+SVM	37.67
AlexNet (fc7)+Random Forest	31.98
AlexNet (fc7)+KNN	31.98
Alex NET (fc6+fc7)+Random Forest	32.87
AlexNet (fc6+fc7)+SVM)	49.06
Alex NET (fc6+fc7)+ANN	48.21
Vgg-face	70.69
Proposed approach	74.65

Table 2 Recognition accuracies of different models when trained separately on cartoon images, real images, and a combination of real and cartoon images

	Cartoon (%)	Real (%)	Combined (%)
Fc7	37.67	17.67	39.65
Fc6+Fc7	49.06	25.53	49.06
Proposed model	73.48	46.27	74.65

of the model that were trained from combined representations of real and cartoon faces. These experiments were carried out for the fc7, fc6 layers. It can be clearly inferred from Table 2 that the performance of the model improves when the model was trained on combined representations of cartoon and real faces.

5 Conclusion

The paper presents a framework for recognizing faces in the cartoon domain. The proposed framework relies on DCNN for extracting representations and was able to achieve a recognition accuracy greater than other state-of-the-art descriptors and classifiers. The experiments performed in the dataset proved the vitality of using representations from both the real and the cartoon domain for building efficient cartoon face recognition systems. It was seen that the performance of the model improved on the incorporation of local features.

The below par performance of several handcrafted and deep descriptors clearly shows that recognizing cartoon images is a difficult task. According to the authors, the focus of the future cartoon recognition systems should be on building a bag of descriptors that focus on what set of facial representations are important for a given class. For example, several classes may have larger eyes while others might have a bigger nose. Therefore, cartoon face recognition systems should also consider what facial features should be given more prominence for a given class of data.

References

1. Mishra, A., Rai, S.N., Mishra, A., Jawahar, C.V.: IIIT-CFW: a benchmark database of cartoon faces in the wild. In: European Conference on Computer Vision, pp. 35–47. Springer (2016)
2. Glasberg, R., Samour, A., Elazouzi, K., Sikora, T.: Cartoon-recognition using video & audio descriptors. In: 2005 13th European Signal Processing Conference, pp. 1–4. IEEE (2005)
3. Takayama, K., Johan, H., Nishita, T.: Face detection and face recognition of cartoon characters using feature extraction. In: Image, Electronics and Visual Computing Workshop, p. 48 (2012)
4. Humphrey, E.: Cartoon Recognition and Classification. University of Miami (2009)
5. Taigman, Y., Polyak, A., Wolf, L.: Unsupervised cross-domain image generation (2016). arXiv:1611.02200
6. Liu, Y., Qin, Z., Luo, Z., Wang, H.: Auto-painter: cartoon image generation from sketch by using conditional generative adversarial networks (2017). arXiv:1705.01908
7. Mirza, M., Osindero, S.: Conditional generative adversarial nets (2014). arXiv:1411.1784
8. Sermanet, P., Eigen, D., Zhang, X., Mathieu, M., Fergus, R., LeCun, Y.: Overfeat: integrated recognition, localization and detection using convolutional networks (2013). arXiv:1312.6229
9. Fu, H., Cao, X., Tu, Z.: Cluster-based co-saliency detection. IEEE Trans. Image Process. 22(10), 3766–3778 (2013)
10. Shukla, P., Dua, I., Raman, B., Mittal, A: A computer vision framework for detecting and preventing human-elephant collisions. In: Proceedings of the IEEE Conference on Computer Vision and Pattern Recognition, pp. 2883–2890 (2017)
11. Shukla, P., Gupta, T., Saini, A., Singh, P., Balasubramanian, R.: A deep learning frame-work for recognizing developmental disorders. In: 2017 IEEE Winter Conference on Applications of Computer Vision (WACV), pp. 705–714. IEEE (2017)
12. Shukla, P., Sadana, H., Bansal, A., Verma, D., Elmadjian, C., Raman, B., Turk, M.: Automatic cricket highlight generation using event-driven and excitement-based features. In: Proceedings of the IEEE Conference on Computer Vision and Pattern Recognition Workshops, pp. 1800–1808 (2018)
13. Parkhi, O.M., Vedaldi, A., Zisserman, A., et al.: Deep face recognition. In: BMVC, vol. 1, p. 6 (2015)
14. Dalal, N., Triggs, B.: Histograms of oriented gradients for human detection. In: IEEE Computer Society Conference on Computer Vision and Pattern Recognition, 2005. CVPR 2005, vol. 1, pp. 886–893. IEEE (2005)
15. Ojala, T., Pietikainen, M., Maenpaa, T.: Multiresolution gray-scale and rotation invariant texture classification with local binary patterns. IEEE Trans. Pattern Anal. Mach. Intell. 24(7), 971–987 (2002)
16. Krizhevsky, A., Sutskever, I., Hinton, G.E.: ImageNet classification with deep convolutional neural networks. In: Advances in Neural Information Processing Systems, pp. 1097–1105 (2012)

Deep Learning Framework for Detection of an Illicit Drug Abuser Using Facial Image

Tanu Gupta, Meghna Goyal, Goutam Kumar
and Balasubramanian Raman

Abstract The detection of an illicit drug abuser by analyzing the subject's facial image has been an active topic in the field of machine learning research. The big question here is up to what extent and with what accuracy can a computer model help us to identify if a person is an illicit drug abuser only by analyzing the subject's facial image. The main objective of this paper is to propose a framework which can identify an illicit drug abuser just by giving an image of the subject as an input. The paper proposes a framework which relies on Deep Convolutional Neural Network (**DCNN**) in combination with Support Vector Machine (SVM) classifier for detecting an illicit drug abuser's face. We have created dataset consisting of 221 illicit drug abusers' facial images which present various expressions, aging effects, and orientations. We have taken random 221 non-abusers' facial images from available dataset named as, Labeled Faces in the Wild (LFW). The experiments are performed using both datasets to attain the objective. The proposed model can predict if the person in an image is an illicit drug abuser or not with an accuracy of 98.5%. The final results show the importance of the proposed model by comparing the accuracies obtained in the experiments performed.

Keywords Deep learning · Convolutional neural network · Face recognition

1 Introduction

The number of illicit drug abusers has been increasing rapidly on a global scale. As per the World Drug Report 2017 released by United Nations Office of Drugs and Crime [1], 255 million people were using illicit drugs in 2015 whereas there were 247 million people using the illicit drugs in 2014. Considering such a huge number and

T. Gupta (✉) · M. Goyal · G. Kumar · B. Raman
Indian Institute of Technology Roorkee, Roorkee, India
e-mail: tgupta@dm.iitr.ac.in

B. Raman
e-mail: balarfma@iitr.ac.in

© Springer Nature Singapore Pte Ltd. 2020   11
B. B. Chaudhuri et al. (eds.), *Proceedings of 3rd International Conference on Computer Vision and Image Processing*, Advances in Intelligent Systems and Computing 1022,
https://doi.org/10.1007/978-981-32-9088-4_2

Fig. 1 Sample images from our dataset representing effects of illicit drugs on human face

the growth at which the problem of illicit drug abuse is growing, it is becoming more apparent and challenging for humanity. Globally the most commonly used illicit drugs are cocaine, methamphetamine (meth), heroin, marijuana, and crack cocaine. This list is not exhaustive but will be used for the scope of this research.

As per the extensive research done by an organization named Rehab [2], the repeated and prolonged use of illicit drugs causes various side effects. A noticeable change in the face is the most prominent side effect of almost all illicit drugs as shown in Fig. 1. For instance, the physiological side effects of meth like acne, dry mouth, teeth clenching, dull skin, and self-inflicted wounds on the face and body are documented by the project Faces of Meth [3]. The regular consumption of certain illicit drugs causes physiological changes in the skin which make it possible to detect illicit drug abusers only by observing the facial changes. The detection of illicit drug abusers has its application in the areas like a job or army recruitment where the background verification of a person is required.

The detection of illicit drug abusers based on facial analysis is an interesting and useful subject, but has been little explored. The research in [4] has proposed a dictionary learning framework which will help one to recognize the face of a subject even after he has been abusing illicit drugs for a long time. In order to add value to that research, we have built a framework which can tell if a person is an illicit drug abuser or not just by analyzing his current image. In this research work, various experiments have done in order to detect illicit drug abusers from facial images. Initially, features have been extracted using handcrafted representations like HOG, GIST, BISF, LBP, BagOfWords (BoW), and **DCNN** and after that classifier algorithms like support vector machine (SVM), Random Forests (RF), and K-Nearest Neighbor (KNN) have been used to classify the images. In proposed method, SVM classifier is used with **DCNN**. To showcase the final results, comparison between accuracies obtained from the various handcrafted descriptors used with various classifiers and the proposed method has been done.

Our contribution through this paper is summarized below:

1. We have created an **Illicit Drug-Abused Face dataset** containing images of 221 illicit drug abusers which are collected from various sources from Internet.
2. We have proposed a **DCNN** framework with SVM classifier to detect an illicit drug abuser's face and a non-abuser's face in a combined dataset of images.

The rest of the paper has been organized as follows. A description of related work is presented in Sect. 2. The proposed method is described in Sect. 3. The dataset and scenarios that have been used for experimentation have been described in Sect. 4. The experimental results have been presented in Sect. 5. Section 6 concludes the paper.

2 Related Work

Since drug abuse is a serious concern in today's society, several types of research have been conducted on various impacts of illicit drugs on human health [5]. In order to alert people about the adverse effects of meth, the images of haggard and sunken-cheeked faces of meth addicts were released [3]. Further, images of cocaine, crack, and heroin addicts representing weight loss and fast aging effect were collected and disseminated, so that people could be made aware of the effects of these drugs.

An ample amount of research has been completed in order to use face recognition for various applications. In [6], face recognition is done using the LBP descriptor. The images in which the faces could be recognized may have different facial expressions, illumination, aging effects, and other external variations. These faces could be recognized with good accuracy. A deep learning framework was proposed in [7] for recognizing the developmental disorders. A fine-tuned DCCN was used for the feature extraction along with the SVM classifier. The model is tested on different experimental scenarios which include differentiation of the normal faces from those with cognitive disabilities. The dataset consisted of 1126 images of normal subjects and disabled subjects each. The accuracy obtained is 98.80%.

In spite of all the researches regarding illicit drug abuse, there is a little work done on the detection of illicit drug abusers. In 2017, [4] research was conducted which objectified the facial changes of a subject due to the consistent consumption of illicit drugs and the deterioration in the performance of two commercial face recognition systems. Yadav et al. [4] proposed a dictionary learning framework to recognize the effect of illicit drug abuse on face recognition. The accuracy of the proposed model was 88.88%.

3   Proposed Method

The objective of the paper is to efficiently detect an illicit drug abuser's face and a non-abuser's face. Machine learning based classification techniques can give good accuracy only if the features are extracted properly from the images. We have used basic handcrafted representations such as HOG [8], GIST [9–11], LBP [12–14], BISF [15], and BOWs [16] to get the best possible facial features. The classifiers RF, SVM, and KNN are trained with all the representations, respectively. Further improvement in the accuracy is done by proposing a deep learning framework CNN for representations and again train SVM classifier to classify the images. As shown in the block diagram in Fig. 2, two CNN architectures have been used in the framework. The facial image is given as an input to the first CNN and the salient features extracted from the facial images are given as an input to the second CNN. The features obtained from the above two branches are merged and then fed to the classifier.

3.1   Saliency-Based Feature Extraction

The cluster based co-saliency method has used to get the salient features of the face. Figure 3 shows the salient images corresponding to the input drug abused facial images. The highly salient parts and the lower salient parts represent the unique parts of the images and the background of the images, respectively.

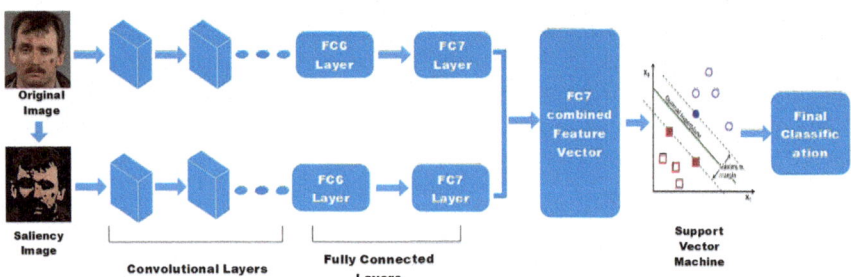

Fig. 2   A block diagram representing proposed architecture

Fig. 3 Salient images corresponding to the drug abused images

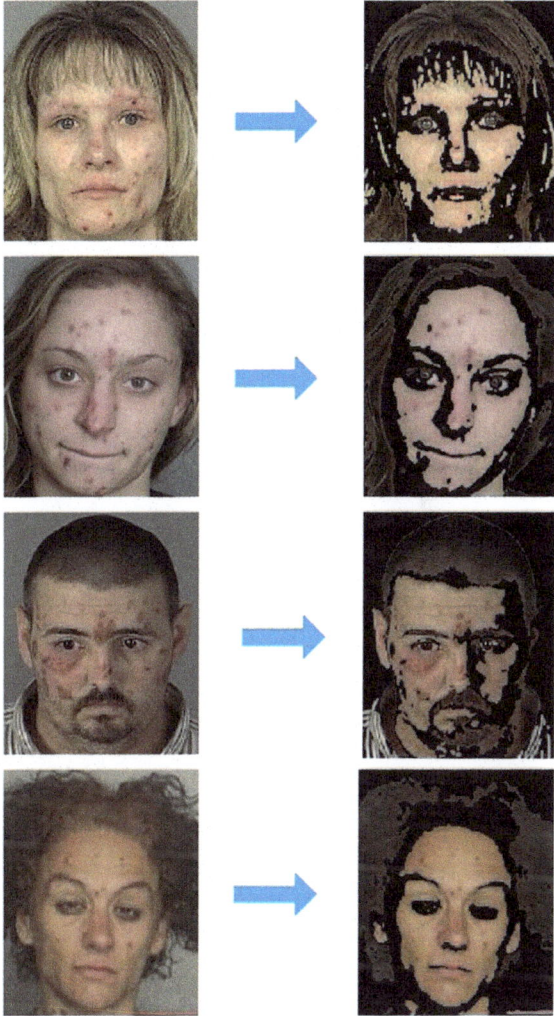

3.2 Using Deep CNNs for Extracting Features

We have used deep representations over other handcrafted representations mainly because of the superiority of **DCNN** over traditional descriptors for complex image recognition tasks. Also, it requires minimal processing [17]. Face recognition of drug abusers is done using deep CNNs fc6 and fc7 layers. The Alex-Net architecture [18] is used for the fine-tuning of all different CNNs. The input layer is of $227 \times 227 \times 3$, conv of $96 \times 11 \times 11$, max pooling of 3×3, conv of $128 \times 5 \times 5$, max pooling of 3×3, conv of $256 \times 3 \times 3$, conv of $192 \times 3 \times 3$, conv of $192 \times 3 \times 3$, 3×3 pooling 4096×1, fc $- 4096 \times 1$ fc 1000×1 fc layers. Before being fed into the respective

CNNs, facial images are resized to 227×227. The initial learning rate is set to 0.001 for the final fc layer and 0.0001 for the remaining layers while fine-tuning. The dropout is 0.5 and the momentum is 0.8 for a total 10,000 iterations. For capturing the different aspects of the facial images, two CNNs are used and 4096 representations are extracted overall from them. These representations are concatenated together to form a feature vector comprising of 67,840 elements.

3.3 Classifier

The representations extracted from the above methods are trained and classified with the help of SVM classifier. The SVM is trained on the linear kernel and the value of kernel offset is set to 0.

4 Experimental Setup

A description of the experimental scenarios on which the model is tested along with the details of the dataset has been elaborated below. The experiments are conducted using the MATLAB 2017a.

4.1 Description of Dataset

Illicit drug abuse is a shameful and confidential issue of an addict. As a result, dataset of limited size is available for the images of illicit drug abusers. However, few websites [2] have released images of drug abusers in order to put light on the devastating physiological side effects of illicit drugs. Moreover, few images and stories of drug abusers who have quit the drugs are available to motivate other drug abusers. We have collected the number of images in the database using facial images of Internet subjects. Illicit Drug abuse Database consists of 221 facial images of drug abusers. The database includes the addicts of drugs such as cocaine, heroin, meth, alcohol, crack, and coke. The images available on the web are of different facial orientations and few of them are blurred. So we have cropped them in order to get the better accuracy.

We have divided the dataset into the two categories, facial images of illicit drug abusers and images of the non-drug abusers (LFW). LFW dataset comprises facial images of various orientations. The images are different in many aspects such as face orientation and clarity of images. The dataset is divided in the ratio of 70:30 for training and testing purpose.

4.2 Experimental Scenario

In order to check the performance of the proposed model, we have performed experiments using handcrafted representations on the given dataset. The purpose of the experiments is to differentiate between an illicit drug abuser's face and non-abuser's face using the proposed model for the improved accuracy.

For the comparison purpose, we have performed the experiments in which the features extracted using HOG, BISF, GIST, BoW, and LBP are fed to the classifiers (KNN, SVM, RF). Then we have taken a record of accuracy obtained on the given dataset. Afterward, the same experiment is performed using the proposed model.

5 Experimental Results

The experimental results are recorded in the order described below. To begin with the experiments, we have extracted features using the HOG descriptor. The images in the database are classified using the extracted features with SVM, RF, and KNN classifier algorithms. The accuracies as recorded in Table 1 are 89.57%, 91.30%, and 84.35%, respectively. Next we have used GIST for the purpose of feature extraction. Again the images are classified using the extracted features with all the three earlier mentioned algorithms. This time the accuracy as recorded are 94.78, 93.91, and

Table 1 Comparison of the proposed model and handcrafted representation accuracies

Technique	Accuracy (%)
HOG+SVM	89.57
GIST+SVM	94.78
BSIF+SVM	72.17
LBP+SVM	91.30
BagOfWords+SVM	87.96
HOG+Random Forests	91.30
GIST+Random Forests	93.91
BSIF+Random Forests	89.57
LBP+Random Forests	93.91
BagOfWords+Random Forests	92.59
HOG+KNN	84.35
GIST+KNN	94.78
BSIF+KNN	74.78
LBP+KNN	93.91
BagOfWords+KNN	83.33
Proposed approach	98.5

Table 2 Confusion matrix for drug-abused and non-abused dataset

	Drug abused	Non-abused
Drug abused	0.925	0.075
Non-abused	0	1

Fig. 4 Samples of correctly classified images from the used dataset

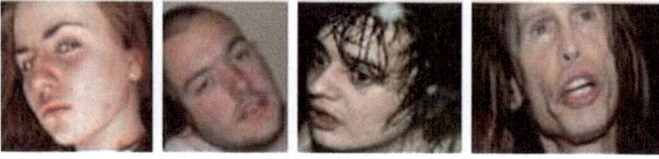

Fig. 5 Samples of wrongly classified images from the used dataset

94.78%. In the next experiment, LBP is used to extract the features which are then used with SVM, RF, and KNN classifiers. The accuracies obtained this time are 91.30%, 93.91%, and 93.91%, respectively. The performance of BSIF descriptor for feature extraction with each of the abovementioned classifier is measured as 72.17, 89.57, and 74.78%. Moreover, the accuracies obtained for image classification using BagOfWords for feature extraction with all the three classifiers are 87.96, 92.59, and 83.33%. In order to compare the accuracy of different traditional methods with that of the proposed method, the classification of abuser's and non-abuser's faces is performed using DCNN method with SVM classifier. The accuracy obtained in this case is 98.5%.

There are 60 images of illicit drug abusers which have been used for the testing the framework. The confusion matrix represented in Table 2 gives the accuracy up to which the model is able to do the recognition. Figure 4 represents the few examples of correctly detected images whereas Fig. 5 consists of some wrongly detected images.

6 Conclusions

Nowadays, the impact of illicit drug abuse is becoming a serious concern in the world. The ability to differentiate between an illicit drug abuser's face and a non-abuser's face with better accuracy is one of the challenges in the field of machine learning. Our work would facilitate in solving many real-life practical problems such as: it can be used by law enforcement to identify if any suspect is an addict, it can be used at airport terminals to be able to spot drug traffickers. As the facial changes are drastic when images from before and after the use of drugs are compared, a combination of several experiments is performed to differentiate between an illicit drug abuser's face and a non-abuser's face in a combined dataset. This paper presents a deep learning framework for face recognition, which will identify if the subject in a given image is a drug addict or not. The proposed DCNN framework is able to achieve an overall accuracy of 98.5% with SVM. Also, in this paper, we have presented a database consisting of 221 images of illicit drug abusers. These images cover variations in the features such as expressions, aging effects, and orientations. This database can further be used in future works for validating similar approaches.

References

1. United Nations Office on Drugs and Crime: World drug report 2017, Vienna. https://www.unodc.org/wdr2017/index.html (2017). Accessed 20 Jan 2018
2. Rehabs: Faces of addiction: image collection of drug abusers. https://www.rehabs.com/explore/faces-of-addiction. Accessed 20 Jan 2018
3. Multnomah County Sheriff Office: Faces of meth. http://www.mcso.us/facesofmeth/. Accessed 25 Jan 2018
4. Yadav, D., Kohli, N., Pandey, P., Singh, R., Vatsa, M., Noore, A.: Effect of illicit drug abuse on face recognition. In: 2016 IEEE Winter Conference on Applications of Computer Vision (WACV), pp. 1–7. IEEE (2016)
5. Devlin, R.J., Henry, J.A.: Clinical review: major consequences of illicit drug consumption. Crit. Care 12(1), 202 (2008)
6. Ahonen, T., Hadid, A., Pietikainen, M.: Face recognition with local binary patterns. In: Computer Vision—ECCV 2004. ECCV. Lecture Notes in Computer Science, vol. 3021. Springer, Berlin, Heidelberg (2004)
7. Saini, A., Singh, P., Shukla, P., Gupta, T., Raman, B.: A deep learning framework for recognizing developmental disorder. In: IEEE Winter Conference on Applications of Computer Vision (WACV), Santa Rosa, CA, pp. 1–10. IEEE (2017)
8. Dalal, N., Triggs, B.: Histograms of oriented gradients for human detection. In: IEEE Computer Society Conference on Computer Vision and Pattern Recognition, 2005. CVPR 2005, vol. 1, pp. 886–893. IEEE (2005)
9. Hays, J., Efros, A.A.: Scene completion using millions of photographs. Commun. ACM 51(10), 87–94 (2008)
10. Douze, M., Jégou, H., Sandhawalia, H., Amsaleg, L., Schmid, C.: Evaluation of gist descriptors for web-scale image search. In: Proceedings of the ACM International Conference on Image and Video Retrieval, p. 19. ACM (2009)
11. Oliva, A., Torralba, A.: Modeling the shape of the scene: a holistic representation of the spatial envelope. Int. J. Comput. Vis. 42(3), 145–175 (2001)
12. He, D.-C., Wang, L.: Texture unit, texture spectrum, and texture analysis. IEEE Trans. Geosci. Remote Sens. 28(4), 509–512 (1990)
13. Wang, L., He, D.-C.: Texture classification using texture spectrum. Pattern Recognit. 23(8), 905–910 (1990)
14. Ahonen, T., Hadid, A., Pietikainen, M.: Face description with local binary patterns: application to face recognition. IEEE Trans. Pattern Anal. Mach. Intell. 28(12), 2037–2041 (2006)
15. Kannala, J., Rahtu, E.: BSIF: binarized statistical image features. In: 2012 21st International Conference on Pattern Recognition (ICPR), pp. 1363–1366. IEEE (2012)

16. Zhang, Y., Jin, R., Zhou, Z.-H.: Understanding bag-of-words model: a statistical framework. Int. J. Mach. Learn. Cybern. **1**(1–4), 43–52 (2010)
17. Shamov, I.A., Shelest, P.S.: Application of the convolutional neural network to design an algorithm for recognition of tower lighthouses. In: 2017 24th Saint Petersburg International Conference on Integrated Navigation Systems (ICINS), pp. 1–2. IEEE (2017)
18. Krizhevsky, A., Sutskever, I., Hinton, G.E.: ImageNet classification with deep convolutional neural networks. In: Advances in Neural Information Processing Systems, pp. 1097–1105 (2012)

Pneumonia Detection on Chest X-Ray Using Machine Learning Paradigm

Tej Bahadur Chandra and Kesari Verma

Abstract The chest radiograph is the globally accepted standard used for analysis of pulmonary diseases. This paper presents a method for automatic detection of pneumonia on segmented lungs using machine learning paradigm. The paper focuses on pixels in lungs segmented ROI (Region of Interest) that are more contributing toward pneumonia detection than the surrounding regions, thus the features of lungs segmented ROI confined area is extracted. The proposed method has been examined using five benchmarked classifiers named Multilayer Perceptron, Random forest, Sequential Minimal Optimization (SMO), Logistic Regression, and Classification via Regression. A dataset of a total of 412 chest X-ray images containing 206 normal and 206 pneumonic cases from the ChestX-ray14 dataset are used in experiments. The performance of the proposed method is compared with the traditional method using benchmarked classifiers. Experimental results demonstrate that the proposed method outperformed the existing method attaining a significantly higher accuracy of 95.63% with the Logistic Regression classifier and 95.39% with Multilayer Perceptron.

Keywords Chest X-Ray · Consolidation · Pneumonia · Radiography · Thoracic disease · Pulmonary disease · Segmentation

1 Introduction

Medical imaging has a significant role in the categorization/classification of diseases. Chest X-ray is a medical imaging technology that is economical and easy to use. It produces an image of the chest, lung, heart, and airways (trachea) [1]. White consolidation on chest X-ray is the infection caused by bacteria, viruses, fungi, parasites, etc., within the small air spaces of the lungs results in an inflammatory response which is caused due of various abnormalities like pneumonia, tuberculosis, pneumothorax,

T. B. Chandra (✉) · K. Verma
Department of Computer Applications, National Institute of Technology, Raipur, India
e-mail: tejbahadur1990@gmail.com

K. Verma
e-mail: kverma.mca@nitrr.ac.in

© Springer Nature Singapore Pte Ltd. 2020
B. B. Chaudhuri et al. (eds.), *Proceedings of 3rd International Conference on Computer Vision and Image Processing*, Advances in Intelligent Systems and Computing 1022, https://doi.org/10.1007/978-981-32-9088-4_3

pleural effusion, etc. These inflammations can be easily seen as white patches on the chest X-ray (CXR). The vague appearance and resemblance with many other pulmonary abnormalities make it challenging to diagnose pneumonia for both radiologists as well as for automated computer-aided diagnostic (CAD) systems [2]. The invention of machine learning algorithms has motivated the researchers to develop a fully automated system that can assist radiologist as well as serve the purpose in resource-constrained remote areas. This paper presents a method for automatic detection of pneumonia by using lungs segmented ROI confined feature extraction method on CXR images. Instead of using the whole CXR image for analysis, which may lead to false diagnosis, the method considered only the lungs air space cavity by lungs segmented ROI selection. This type of automated system can be of great use in remote rural areas where the number of patients is very large as compared to available experienced radiologists.

1.1 Related Work

Automated computer-aided diagnosis (CAD) is becoming popular day by day. In the past few years, various methods have been proposed to improve the accuracy of pulmonary disease detection on CXRs. A significant work on abnormality detection (especially tuberculosis) on CXR can be found in [3–6]. Antani [7] has presented valuable work in the field of automated screening of pulmonary diseases including tuberculosis on digital CXR images. Ahmad et al. [8] compared the performance of different classifiers for abnormality analysis in CXRs. Karargyris et al. [9] proposed a new approach that combines both texture and shape feature for more accurate detection of tuberculosis and pneumonia. In a step toward building a mass screening system with high precision diagnosis, Wang et al. [10] proposed a new dataset namely "ChestX-ray8" and demonstrated thoracic disease localization using the deep convolutional neural network. This work has been further improved by Yao et al. [11] by considering the interdependency among the thoracic diseases and by Rajpurkar et al. [12] using deep learning approach with the 121-layer convolutional neural network.

Automatic lung field segmentation has drawn considerable attention of the researchers. The noteworthy contribution by Ginneken et al. [13] suggests that lung segmentation shows a significant improvement in the accuracy of the abnormality detection. Further, the author has compared the active shape, active appearance, and multi-resolution pixel classification model and observed nearly equal performance in each. A fully automatic technique for suppressing the ribs and clavicle in the chest X-ray is described in [14]. Shi et al. [15] have proposed patient-specific shape statistics and population-based deformable model to segment lung fields. Annangi et al. [16] used prior shape and low-level features based on the active contour method for lung segmentation. The groundbreaking work has been proposed by Candemir et al. [13] for automatic lung segmentation using anatomical atlases with nonrigid deformable registration.

Various features proven to be significant for the analysis of medical radiographs include shape feature [9, 16], texture feature: Histogram of Oriented Gradients (HOG) [5, 7, 17], Local Binary Patterns (LBP) [5, 7, 17, 18], Gray-Level Co-Occurrence Matrix (GLCM) based features [6], and Gabor feature [7, 8, 19].

1.2 Organization

The rest of the paper is organized as follows. Section 2 presents the detailed description of the dataset used to train and test the model, lungs segmentation, feature extraction methods, and different classifiers used to evaluate the classification accuracy. Section 3 elaborates the experimental setup followed by results and discussion in Sect. 4, and future scope in Sect. 5.

2 Materials and Methods

2.1 Dataset

In this paper, we used 412 CXR images containing 206 normal and 206 pneumonic cases from NIH ChestX-ray14 dataset [10] to train and test the classifier. The dataset contains a total of 112,120 frontal-view chest X-rays images of 30,805 unique patients with the text-mined fourteen thoracic disease labels. The disease labels are expected to have accuracy greater than 90%. All the CXR images used in this study are in PNG (Portable Network Graphics) grayscale format and having a resolution of 1024 dpi. Further, the lung atlas images from the set of reference images provided with the segmentation algorithm [13] are used to train the lung segmentation model.

2.2 Segmentation of Lungs Region and Preprocessing

Extracting features from the segmented portion of the lungs reduces the probable chances of the false-positive results. The organs like mediastinum, aortic knuckle, the apex of the heart, and the left and right diaphragm in CXR significantly affect the classification accuracy. This paper uses nonrigid registration-driven robust lung segmentation method proposed by Candemir et al. [13] for lungs segmented ROI selection. The method is organized in three stages as shown in Fig. 1.
Stage 1. In the first stage, the top five X-ray images that are most alike to the patient's X-ray are retrieved from a set of reference X-ray images using content-based image retrieval (CBIR) technique. The CBIR method employs partial radon transform and Bhattacharrya shape similarity measure to handle little affine distortion

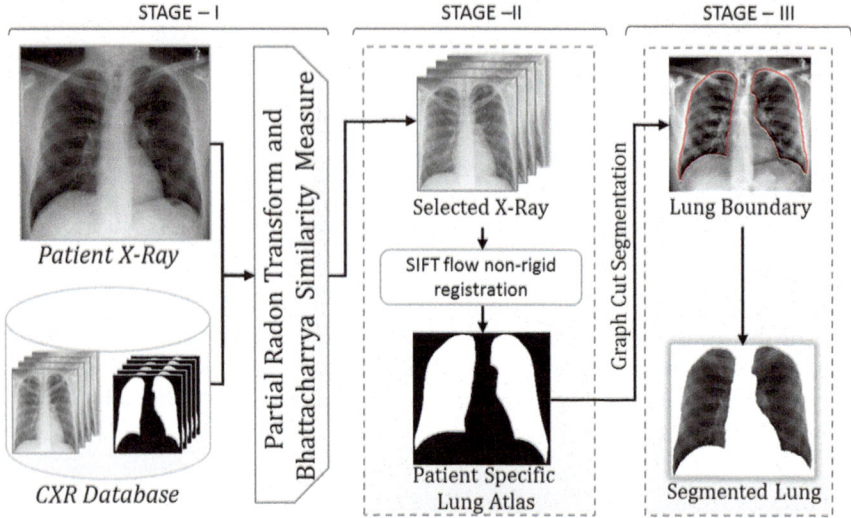

Fig. 1 Three stages of nonrigid registration-driven robust lung segmentation method (Candemir et al. 2014)

and to compute the degree of similarity between input and reference CXR images respectively.

Stage 2. In the second stage, the method employs SIFT flow deformable registration of the training mask to get the patient-specific modal of lung shape. The warped lung mask of top-ranked X-ray images is averaged to get the approximate lung modal.

Stage 3. Finally, in the third stage, the graph-cut method is used to extract the lungs boundary from patient-specific lung atlas using discrete optimization approach. The graph-cut method efficiently handles the problem of local minima using the energy minimization framework.

Subsequently, the binary mask extracted from the lung segmentation method is further used to cut the lungs region from the original CXR images.

ROI Selection: Image preprocessing is an important step in automated medical images analysis. The obtained segmented CXR images contain large white regions around the lungs segmented ROI which considerably affect the discriminating criteria of the features. Thus, in order to exclude it from feature extraction, the paper uses the lungs segmented ROI confined feature extraction technique which replaces all the pixels around the lungs segmented ROI with NaN (Not a Number) (as shown in Fig. 2).

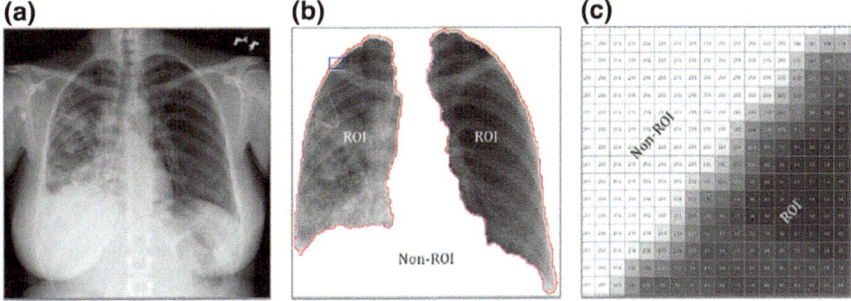

Fig. 2 **a** Full chest X-Ray image. **b** Segmented lungs showing bounded ROI Area. **c** Pixel values of blue patch showing ROI and non-ROI area

2.3 Feature Extraction

The presence of opaque consolidations on CXR due to pneumonia can be analyzed using statistical features. In this study, eight statistical features based on the first-order histogram of the original image [20] are used. The features extracted using this method does not take into account the neighboring pixel relationships. The features are summarized in Table 1. The $P(I)$ represents the first-order histogram at the gray level I of an image having total L gray levels.

$$P(I) = \frac{N(I)}{N} = \frac{\text{number of pixels with gray level I}}{\text{total number of pixels in the region}}$$

Table 1 Eight first-order statistic features (Srinivasan et al. 2008)

Code	Feature	Equation
F1	Mean (m)	$\sum_{I=0}^{L-1} IP(I)$
F2	Variance ($\mu 2$)	$\sum_{I=0}^{L-1} (I - m)^2 P(I)$
F3	Standard deviation (σ)	$\sqrt{\mu_2}$
F4	Skewness ($\mu 3$)	$\sum_{I=0}^{L-1} (I - m)^3 P(I)$
F5	Kurtosis ($\mu 4$)	$\sum_{I=0}^{L-1} (I - m)^4 P(I)$
F6	Smoothness (R)	$1 - \frac{1}{1+\sigma^2}$
F7	Uniformity (U)	$\sum_{I=0}^{L-1} P^2(I)$
F8	Entropy (e)	$\sum_{I=0}^{L-1} P(I) \log_2 P(I)$

2.4 Classification

The pneumonia detection performance of the proposed method is evaluated using five benchmarked classifiers namely Multilayer Perceptron (MLP), Random forest, Sequential Minimal Optimization (SMO), Classification via Regression, and Logistic Regression classifier available in Weka (Waikato Environment For Knowledge Analysis) tool, version 3.8.2 [21]. The classifiers' performance are evaluated with 10-fold cross-validation. Following are the short descriptions of each classifier:

a. *Multilayer Perceptron (MLP)*: It mimics the working of the human brain by utilizing backpropagation with adjustable weight w_{ij} for training its neurons. The trained model can solve very complex classification problems stochastically and can also deal with nonlinearly separable data [22, 23].
b. *Random Forest*: This supervised classification and regression method works by generating a forest (ensemble) of decision trees. It uses majority voting to decide the final class labels of the object. This classifier works efficiently on large databases and even handles missing data [24].
c. *Sequential Minimal Optimization (SMO)*: It is used to train the support vector machine (SVM) classifier by decomposing large quadratic programming problems into sub-problems. The multiclass problems are solved using pairwise classification [25].
d. *Classification via Regression*: It builds a polynomial classification model that uses regression method for classification through the given regression learner. For each binarized class labels, regression learner generates one regression model which predicts a value from a continuous set, whereas classification predicts the belongingness to a class [26].
e. *Logistic Regression*: It is the statistical method that uses a logistic sigmoidal function to compute probability value which can be mapped to two or more discrete classes. It is most widely used in medical fields for risk prediction based on observed characteristics of the patient [27].

3 Experimentation and Performance Evaluation

The proposed structural modal for machine learning based diagnosis of pneumonia is shown in Fig. 3. The modal works in two phases: training and testing. Both the phases include image preprocessing followed by feature extraction and classification. In the training phase, eight first-order statistical features are extracted from a selected set of 412 CXR images. The extracted features are now utilized to build training models using benchmarked classifiers whereas in the testing phase, the extracted features from the patient X-ray are fed to the classifier and the predicted class label is compared with the available ground truth data.

The performance of different benchmarked classifiers applied on the preprocessed dataset is evaluated and compared using six performance metrics, namely Accuracy,

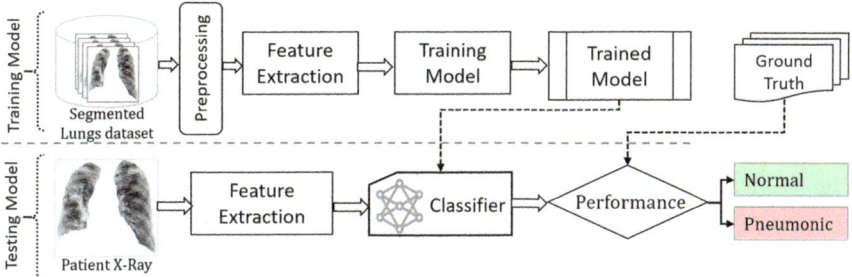

Fig. 3 Structural design of the experimental setup

Sensitivity or Recall, Specificity, Precision (PPV), area under the Receiver Operating Characteristic curve (AUROC), and F1 Score [22]. Short descriptions of the performance metrics are shown in Table 2 where TP = True Positive, TN = True Negative, FP = False Positive, FN = False Negative, P = TP + FN, and N = TN + FP.

Table 2 Short description of performance metrics [22]

Measure (in %)	Equation	Description
Accuracy	$\frac{TP+TN}{P+N} \times 100$	Percentage of correctly classified test cases by the classifier
Sensitivity/recall	$\frac{TP}{P} \times 100$	The proportion of pneumonia cases that are correctly identified to the total actual pneumonic cases. Also known as true-positive or recognition rate
Specificity	$\frac{TN}{N} \times 100$	The proportion of normal cases that are correctly identified to total actual normal cases. Also known as true-negative rate
Precision (PPV)	$\frac{TP}{TP+FP} \times 100$	The proportion of pneumonia cases that are correctly identified to total predicted pneumonic cases. Also known as positive predicted value
Area under curve (AUC)	$\frac{1}{2}\left(\frac{TP}{P} + \frac{TN}{N}\right)$	It is the common measure of the performance of the classifier using sensitivity and specificity
F1 score	$\frac{2 \times precision \times recall}{precision + recall}$	It is the harmonic mean of precision and recall and gives equal weight to both false-positive and false-negative cases. It is used to measure the test's accuracy

4   Results and Discussions

The automated computer-aided diagnosis of pneumonia using the whole chest
X-ray image suffers from false-positive results due to the presence of various organs
like heart, thymus gland, aortic knuckle, sternum, diaphragm, spine, etc., in thorax
regions. The proposed method overcomes this limitation by using segmented lungs
regions and ROI confined feature extraction. The experiments were conducted using
a full chest X-Ray image and segmented lungs region are shown in Tables 3 and
4, respectively. From the experimental results, it can be observed that the pneumo-

Table 3 Classifiers' performance prior to lung segmentation (using full chest X-ray image)

Classifier	Performance measures					
	Accuracy (%)	Sensitivity (%)	Specificity (%)	Precision (%)	AUC	F1 score (%)
Multilayer Perceptron	**92.233**	**86.408**	**98.058**	**97.802**	**0.922**	**91.753**
Random forest	90.534	86.408	94.660	94.180	0.905	90.127
Sequential Minimal Optimization	89.806	80.097	99.515	99.398	0.898	88.710
Classification via Regression	91.990	86.408	97.573	97.268	0.920	91.517
Logistic Regression	91.505	86.408	96.602	96.216	0.915	91.049

Table 4 Classifiers performance after lung segmentation (using ROI confined feature extraction)

Classifier	Performance measures					
	Accuracy (%)	Sensitivity (%)	Specificity (%)	Precision (%)	AUC	F1 score (%)
Multilayer Perceptron	**95.388**	**93.204**	**97.573**	**97.462**	**0.954**	**95.285**
Random forest	94.417	93.689	95.146	95.074	0.944	94.377
Sequential Minimal Optimization	93.689	89.320	98.058	97.872	0.937	93.401
Classification via Regression	94.660	91.262	98.058	97.917	0.947	94.472
Logistic Regression	**95.631**	**93.689**	**97.573**	**97.475**	**0.956**	**95.545**

nia detection performance using segmented lungs regions surpasses the traditional method that uses full chest X-ray image. This is because, in traditional methods, the extracted features are significantly affected by Non-ROI portions leading to misclassification while the proposed method extracts feature from the segmented lung ROI that are contributing more toward disease classification. To assess and compare the performance of pneumonia detection, five benchmarked classifiers are used. From Table 3, it can be observed that Multilayer Perceptron outperforms the other benchmarked classifiers achieving higher accuracy of 92.23% when used with features extracted from full chest X-ray image (without lung segmentation).

However, the overall disease detection performance is improved when the segmented lungs region is used (as shown in Table 4) as the pixels in ROI are contributing more toward pneumonia detection than the surrounding regions.

The classification performance of the proposed method using segmented lungs region shown in Table 4 suggests that the Logistic Regression classifier and Multilayer Perceptron outperform the others benchmarked classifiers. The highest performance with an accuracy of 95.63% is obtained using Logistic Regression classifier whereas the Multilayer Perceptron also gives nearly equal performance with an accuracy of 95.39% under the same configuration. The average percentage increase in performance metrics of the benchmarked classifiers using the segmented regions is shown in Table 5. From the table, it is found that the lung segmentation based ROI feature extraction approach shows an average increase in the accuracy of each classifier by 3.54%. The significant upswing in performance is due to ROI confined feature extraction which considers only the pixels in ROI on segmented lungs as shown in Fig. 2b. The Logistic Regression classifier shows comparably noteworthy

Table 5 Percentage increase in classifier performance after lung segmentation (using ROI confined feature extraction)

Classifier	Performance measures					
	Accuracy (%)	Sensitivity (%)	Specificity (%)	Precision (%)	AUC	F1 score (%)
Multilayer Perceptron	3.155	6.796	−0.485	−0.340	0.032	3.533
Random forest	3.883	7.282	0.485	0.894	0.039	4.250
Sequential Minimal Optimization	3.883	9.223	−1.456	−1.525	0.039	4.691
Classification via Regression	2.670	4.854	0.485	0.649	0.027	2.956
Logistic Regression	4.126	7.282	0.971	1.259	0.041	4.496
Average increase	**3.544**	**7.087**	**0.000**	**0.187**	**0.035**	**3.985**

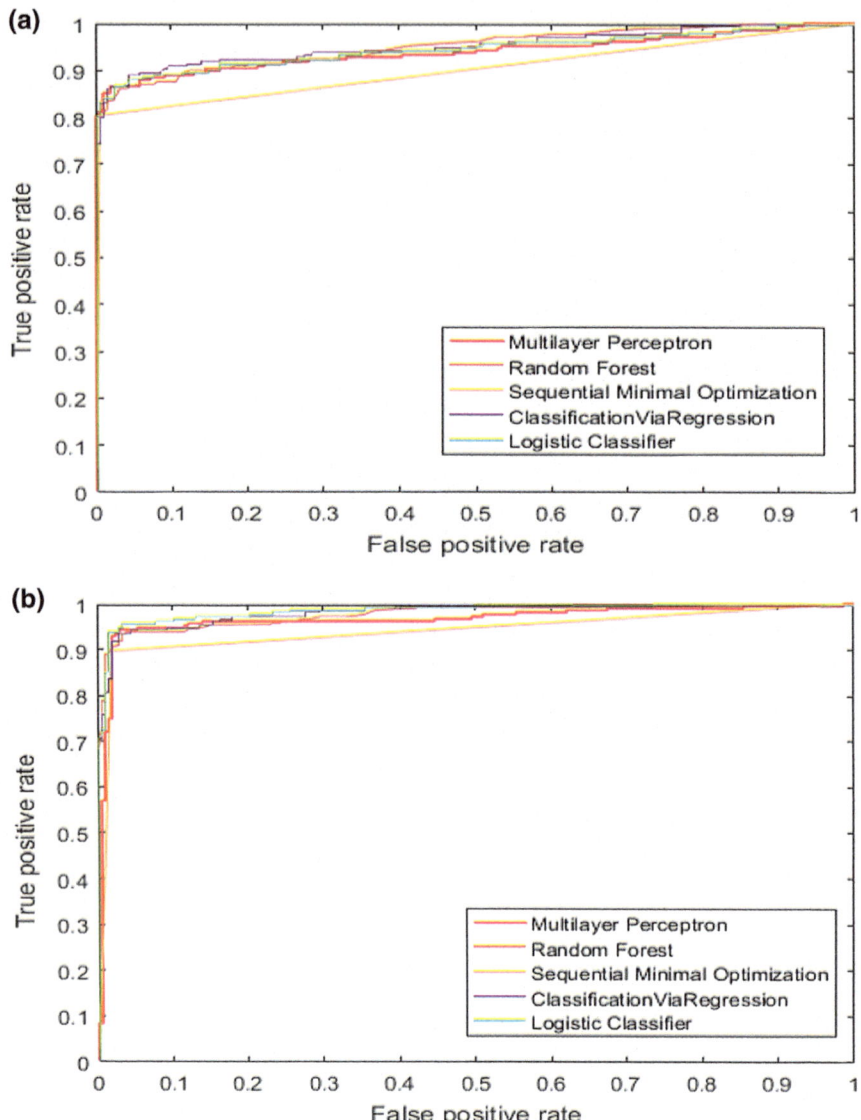

Fig. 4 **a** Area under ROC curve of five classifiers prior to lung segmentation (using full chest X-ray image). **b** The area under ROC curve of five classifiers after lung segmentation

improvement in performance than others with an increase in accuracy of 4.13%; furthermore, the sensitivity, specificity, precision, AUC, and F1 score are increased by 7.28%, 0.97%, 1.26%, 0.041, and 4.50%, respectively.

Performance metrics shown in Tables 3 and 4 provide us with measures to assess the classification performance with a fixed threshold (cutoff) value on class probability. However, different thresholds result in different accuracy, sensitivity, specificity, precision, and F1 score values. Thus, in order to measure the overall performance of the classifier model over its entire operating range, area under ROC (receiver operating characteristic) curve (AUROC) is used. The ROC analysis offers tools to make an optimal selection of classifier models discarding suboptimal ones independently from the cost context or the class distribution. The area under the curve (AUC) for the perfect classifier is 1.0 while the classifier with no power has the AUC of 0.5. The AUCs of most classifier fall between these two values. Figure 4a and b show the plots of the area under the ROC curve of five benchmarked classifiers using full chest X-ray image prior to lung segmentation and after lung segmentation, respectively. From the analysis of ROC plots, it is observed that AUC using the segmented lungs region is significantly higher than using full chest X-ray image. The AUC of the Logistic Regression classifier and Multilayer Perceptron with value 0.956 and 0.954, respectively, using segmented lungs regions fall very close to 1.0 achieving a higher rank among benchmarked classifier models.

5 Conclusion and Future Work

Automatic detection of pneumonia using chest radiograph is still challenging using CAD systems. This paper presents a method for automatic detection of pneumonia using the statistical feature of the lungs airspace. The method is based on the analysis of nonrigid deformable registration driven automatically segmented lungs regions and lungs segmented ROI confined feature extraction. Experiments performed on 412 chest X-ray images containing 206 normal and 206 pneumonic cases from ChestX-ray14 dataset suggest that the performance of the proposed method for automatic detection of pneumonia using segmented lungs is significantly better than traditional method using full chest X-ray images. The average accuracy of the proposed method is 3.54% higher than the traditional method. The Logistic Regression classifier with disease detection accuracy of 95.63% outperforms the other benchmarked classifiers using segmented lungs regions. However, despite high accuracy, the method needs more reliable feature analysis techniques and rigorous testing on thousands of real-time test cases.

The lung segmentation approach used in this paper needs further enhancement for pediatric cases and pleural effusion. The method completely ignores the clinical background of the patients; the integration of this method is a further aspect of research.

Acknowledgements The authors would like to thank Dr. Javahar Agrawal, Diabetologist and Senior Consulting Physician, Lifeworth Super Speciality Hospital, Raipur and Dr. A. D. Raje, Consulting Radiologist, MRI Diagnostic Institute, Choubey Colony, Raipur for their valuable guidance.

References

1. Dong, Y., Pan, Y., Zhang, J., Xu, W.: Learning to read chest X-Ray images from 16000+ examples using CNN. In: Proceedings of the Second IEEE/ACM International Conference on Connected Health: Applications, Systems and Engineering Technologies
2. Van Ginneken, B., Ter Haar Romeny, B.M., Viergever, M.A.: Computer-aided diagnosis in chest radiography: a survey. IEEE Trans. Med. Imaging **20**, 1228–1241 (2001)
3. Mohd Rijal, O., Ebrahimian, H., Noor, N.M.: Determining features for discriminating PTB and normal lungs using phase congruency model. In: Proceedings—IEEE-EMBS International Conference on Biomedical and Health Informatics: Global Grand Challenge of Health Informatics, BHI 2012, vol. 25, pp. 341–344 (2012)
4. Van Ginneken, B., Philipsen, R.H.H.M., Hogeweg, L., Maduskar, P., Melendez, J.C., Sánchez, C.I., Maane, R., dei Alorse, B., D'Alessandro, U., Adetifa, I.M.O.: Automated scoring of chest radiographs for tuberculosis prevalence surveys: a combined approach. In: Fifth International Workshop on Pulmonary Image Analysis, pp. 9–19 (2013)
5. Jaeger, S., Karargyris, A., Candemir, S., Folio, L., Siegelman, J., Callaghan, F., Xue, Z., Palaniappan, K., Singh, R.K., Antani, S., Thoma, G., Wang, Y., Lu, P., Mcdonald, C.J.: Automatic tuberculosis screening using chest radiographs. Stefan **33**, 233–245 (2014)
6. V, R.D.: Efficient automatic oriented lung boundary detection and screening of tuberculosis using chest radiographs. J. Netw. Commun. Emerg. Technol. **2**, 1–5 (2015)
7. Antani, S.: Automated detection of lung diseases in chest X-Rays. US Natl. Libr. Med. (2015)
8. Ahmad, W.S.H.M.W., Logeswaran, R., Fauzi, M.F.A., Zaki, W.M.D.W.: Effects of different classifiers in detecting infectious regions in chest radiographs. In: IEEE International Conference on Industrial Engineering and Engineering Management 2015–January, pp. 541–545 (2014)
9. Karargyris, A., Siegelman, J., Tzortzis, D., Jaeger, S., Candemir, S., Xue, Z., Santosh, K.C., Vajda, S., Antani, S., Folio, L., Thoma, G.R.: Combination of texture and shape features to detect pulmonary abnormalities in digital chest X-rays. Int. J. Comput. Assist. Radiol. Surg. **11**, 99–106 (2016)
10. Wang, X., Peng, Y., Lu, L., Lu, Z., Bagheri, M., Summers, R.M.: ChestX-ray8: hospital-scale chest X-ray database and benchmarks on weakly-supervised classification and localization of common thorax diseases, pp. 2097–2106 (2017)
11. Yao, L., Poblenz, E., Dagunts, D., Covington, B., Bernard, D., Lyman, K.: Learning to diagnose from scratch by exploiting dependencies among labels, pp. 1–12 (2017). arXiv preprint arXiv: 1710.10501
12. Rajpurkar, P., Irvin, J., Zhu, K., Yang, B., Mehta, H., Duan, T., Ding, D., Bagul, A., Langlotz, C., Shpanskaya, K., Lungren, M.P., Ng, A.Y.: CheXNet: radiologist-level pneumonia detection on chest X-rays with deep learning, pp. 3–9 (2017). arXiv preprint arXiv:1711.05225
13. Candemir, S., Jaeger, S., Palaniappan, K., Musco, J.P., Singh, R.K., Xue, Z., Karargyris, A., Antani, S., Thoma, G., McDonald, C.J.: Lung segmentation in chest radiographs using anatomical atlases with nonrigid registration. IEEE Trans. Med. Imaging **33**, 577–590 (2014)
14. Suzuki, K., Abe, H., MacMahon, H., Doi, K.: Image-processing technique for suppressing ribs in chest radiographs by means of massive training artificial neural network (MTANN). IEEE Trans. Med. Imaging **25**, 406–416 (2006)
15. Shi, Y., Qi, F., Xue, Z., Chen, L., Ito, K., Matsuo, H., Shen, D.: Segmenting lung fields in serial chest radiographs using both population-based and patient-specific shape statistics. IEEE Trans. Med. Imaging **27**, 481–494 (2008)
16. Annangi, P., Thiruvenkadam, S., Raja, A., Xu, H., Sun, X.S.X., Mao, L.M.L.: A region based active contour method for x-ray lung segmentation using prior shape and low level features. 2010 IEEE International Symposium on Biomedical Imaging From Nano to Macro, pp. 892–895 (2010)
17. Surya, S.J., Lakshmanan, S., Stalin, J.L.A.: Automatic tuberculosis detection using chest radiographs using its features abnormality analysis. J. Recent Res. Eng. Technol. **4** (2017)

18. Fatima, S., Irtiza, S., Shah, A.: A review of automated screening for tuberculosis of chest Xray and microscopy images. Int. J. Sci. Eng. Res. **8**, 405–418 (2017)
19. Scholar, P.G.: A robust automated lung segmentation system for chest X-ray (CXR) images. Int. J. Eng. Res. Technol. **6**, 1021–1025 (2017)
20. Srinivasan, G., Shobha, G.: Statistical texture analysis. In: Proceedings of World Academy of Science, Engineering and Technology, vol. 36, pp. 1264–1269 (2008)
21. Frank, E., Hall, M.A., Witten, I.H.: The WEKA workbench, 4th edn, pp. 553–571. Morgan Kaufmann (2016)
22. Han, J., Kamber, M., Pei, J.: Data mining: concepts and techniques (2012)
23. Haykin, S.: Neural networks: a comprehensive foundation. Prentice Hall (1998)
24. Breiman, L.: Random forests. Mach. Learn. **45**, 5–32 (2001)
25. Platt, J.C.: Sequential minimal optimization: a fast algorithm for training support vector machines. Adv. Kernel Methods, 185–208 (1998)
26. Frank, E., Wang, Y., Inglis, S., Holmes, G., Witten, I.H.: Using model trees for classification. Mach. Learn. **32**, 63–76 (1998)
27. Sperandei, S.: Understanding logistic regression analysis. Biochem. Medica. **24**, 12–18 (2014)

A Quantitative Comparison of the Role of Parameter Selection for Regularization in GRAPPA-Based Autocalibrating Parallel MRI

Raji Susan Mathew and Joseph Suresh Paul

Abstract The suitability of regularized reconstruction in autocalibrating parallel magnetic resonance imaging (MRI) is quantitatively analyzed based on the choice of the regularization parameter. In this study, L-curve and generalized cross-validation (GCV) are adopted for parameter selection. The results show that: (1) Presence of well-defined L-corner does not guarantee an artifact-free reconstruction, (2) Sharp L-corners are not always observed in GRAPPA calibration, (3) Parameter values based on L-curves always exceed those based on GCV, and (4) Use of a predetermined number of filters based on the local signal power can result in a compromise between noise and artifacts as well as better visual perception. It is concluded that appropriate use of regularized solutions facilitates minimization of noise build-up in the reconstruction process, without enhancing the effects of aliasing artifacts.

Keywords L-curve · GCV · g-factor · GRAPPA · Regularization

1 Introduction

Parallel imaging is a robust method for accelerating the acquisition of magnetic resonance imaging (MRI) data and has made possible many new applications of MR imaging. Parallel imaging (PI) works by acquiring a reduced amount of k-space data with an array of receiver coils. These undersampled data can be acquired more quickly, resulting in aliased images. Aliasing introduced due to skipping of phase-encode lines is corrected using PI reconstruction algorithms. While image domain-based algorithms require explicit knowledge of the coil sensitivity function, the class of k-space-based PI reconstruction methods use a data fitting approach to calculate the linear combination weights that reconstruct output or "target" data

R. S. Mathew (✉) · J. S. Paul
Medical Image Computing and Signal Processing Laboratory, Indian Institute of Information Technology and Management-Kerala (IIITM-K), Trivandrum, Kerala, India
e-mail: rajisusan.res15@iiitmk.ac.in

J. S. Paul
e-mail: j.paul@iiitmk.ac.in

© Springer Nature Singapore Pte Ltd. 2020
B. B. Chaudhuri et al. (eds.), *Proceedings of 3rd International Conference on Computer Vision and Image Processing*, Advances in Intelligent Systems and Computing 1022, https://doi.org/10.1007/978-981-32-9088-4_4

from neighboring input or "source" data. We refer to this second class of methods as "data-driven" reconstructions because they are based on limited knowledge of the underlying physical process and rely on training data to calibrate the relationship between input and output data (e.g., GRAPPA [1], SPIRiT [2], and PRUNO [3]). Both types of reconstruction techniques need calibration for either coil sensitivity estimation or interpolation coefficients for k-space reconstruction.

In general, calibration is performed by finding the solution of an overdetermined system of equations connecting the calibration matrix (Π) and the corresponding set of observations ($\mathbf{k_u}$) from each coil. Since the calibration problem is of discrete ill-posed nature, with the numerical rank of the matrix not well defined, the interpolation coefficients can be hopelessly contaminated by the noise in directions corresponding to small singular values of Π. Because of this, it is necessary to compute a regularized solution in which the effect of such noise is filtered out.

The most popular regularization method for GRAPPA is Tikhonov regularization (also called ridge regression or Wiener filtering in certain contexts) [4, 5]. In Tikhonov regularization, damping is added to each singular value decomposition (SVD) component of the solution, thus effectively filtering out the components corresponding to the small singular values. The choice of the regularization parameter is crucial to yield a good solution for all regularization methods.

As the performance of regularization depends on the appropriate choice of a regularization parameter (λ) in Tikhonov regularization, a quantitative analysis of reconstruction performance is required to analyze the appropriateness of the standard selection strategies as applicable to GRAPPA reconstruction. Over the last four decades, many quite different methods for choosing the optimum parameter have been proposed. Most of these methods have been developed with some analytical justification, but this is usually confined to a particular framework. In this article, we analyze: (1) Optimum regularization parameter for Tikhonov using L-curve method and GCV and (2) Quantitatively compare the regularized reconstruction in each case using root mean square error (RMSE) and g-factor maps [6]. To overcome the problems with L-curve or GCV-based realizations, multiple filter approach is used here instead of using a single filter. A set of filters (filter bank) computed using parameter values between L-curve parameter and GCV parameter are used for the reconstruction of missing points based on the local signal power.

2   Generalized Autocalibrating Partially Parallel
Acquisition (GRAPPA)

GRAPPA is a more generalized and improved version derived from Variable Density AUTOSMASH [7]. A linear combination of neighboring points is used to estimate the missing points in GRAPPA. Unlike SMASH [8] which reconstructs a composite image, GRAPPA reconstructs the coil images separately. GRAPPA uses autocalibrating signal (ACS) lines obtained from the fully sampled central portion of each channel k-space. The unacquired k-space points are then estimated using a linear combination of the acquired phase-encode (PE) lines using one or more kernels computed using the ACS data.

In general, the reconstruction procedure consists of a calibration step in which the coil coefficients (GRAPPA weights) are computed, followed by the estimation of the unacquired k-space points outside the ACS. The calibration step is performed using the samples in the ACS, forming the training data. With the number of coils $= n_C$, the weights are dependent on an index $\eta = 1, 2,..., R-1$ that indicates the distance of an unacquired line from its nearest acquired line. For example, if $R = 4$, with η taking values 1, 2, and 3 for the first, second, and third unacquired lines. Let k_y denotes the PE index of an acquired line, then the conventional GRAPPA model can be expressed as fitting the target data $K^l(k_y + \eta \Delta k_y, k_x)$ to the nearest source data $k(k_y, k_x)$ at the location (k_y, k_x).

$$K^l(k_y + \eta \Delta k_y, k_x) = z_\eta^l k(k_y, k_x) \tag{1}$$

where $\mathbf{k}(k_y, k_x) = [\mathbf{k}^1(k_y, k_x), \mathbf{k}^2(k_y, k_x), \ldots \ldots, \mathbf{k}^{nc}(k_y, k_x)]$ is a vector formed using vectors of measurement values in the training data from each coil. Each point in the ACS line contributes to a row $\mathbf{k}(k_y, k_x)$ in the calibration matrix denoted as Π. The observation vector $\mathbf{k}_u$ for calibration is obtained by the inclusion of the corresponding element in each coil $K^l(k_y + \eta \Delta k_y, k_x)$ from each training pair. The calibration process is determining the least squares (LS) solution.

$$z_\eta^l = (\Pi^H \Pi)^{-1} \Pi^H \mathbf{k}_u \tag{2}$$

Following the calibration, estimation of a missing k-space value $K^l(k_y + \eta \Delta k_y, k_x)$ is achieved by applying the filter weights z_η^l to the nearest acquired data vector denoted as $\mathbf{k}(k_y, k_x)$. In the further description, the terms *training dataset* and *estimation dataset* are used to refer to $\mathbf{k}(k_y, k_x)$ within and outside the ACS, respectively. The reconstruction of missing k-space lines in the j-th coil using GRAPPA can be expressed as

$$K^j(k_y + \eta \Delta k_y, k_x) = \sum_{l=1}^{n_c} \sum_{b=-P_l}^{P_h} \sum_{h=-F_l}^{F_h} K^l\left(k_y + bR\Delta k_y, k_x + h\Delta k_x\right) z_\eta^l \tag{3}$$

where b and h denote sampling indices of the neighboring points along phase- and frequency-encoding directions, respectively. The reconstruction processes are shown schematically in Fig. 1. The data acquired in each coil (black circles) are weighted by the respective GRAPPA coefficients fit to estimate the missing target data (gray circle). By increasing the size of the kernel, GRAPPA incorporates more information into the estimation, resulting in an improved fit.

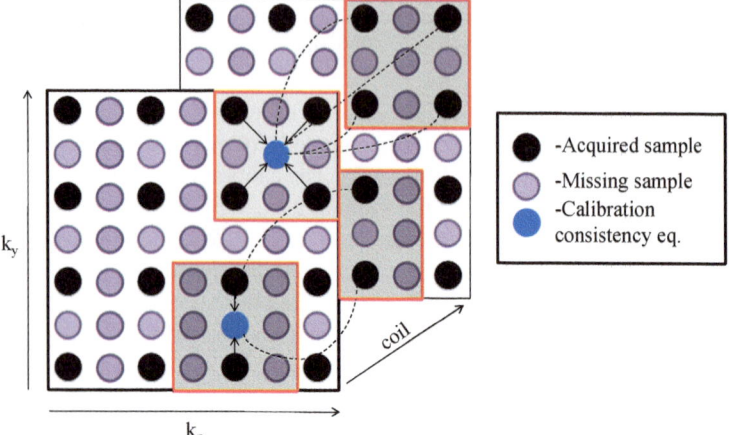

Fig. 1 Traditional 2D GRAPPA: missing k-space data are synthesized from neighboring acquired data

2.1 Optimum Parameter Selection Strategies for Regularization

A regularized parameter estimate **z** is affected by errors due to regularization (bias) as well as due to the error in the data called the perturbation error (variance). Increasing the strength of regularization results in increased regularization error and decreased perturbation error and vice versa. An optimum parameter is aimed at achieving an estimate that minimizes the sum of bias and variance. The existing parameter selection strategies include the discrepancy principle, the generalized cross-validation, and the L-curve method. In this work, the regularized versions of GRAPPA reconstructions obtained using different parameter selection strategies for Tikhonov regularization are compared. Here, parameter selection is carried out using any one of the approaches: (1) Discrepancy principle [9], (2) L-curve [10], and (3) GCV [11]. All parameter selection methods are implemented using Hansen's regularization toolbox [12].

According to the discrepancy principle, for a consistent ill-posed with perturbation only in the observation vector, the selection of regularization parameter can be achieved when the residual norm is equal to the upper bound δ_e of the noise levels present in data. Accordingly, the regularized solution is obtained such that [9]

$$\|\Pi z - \mathbf{k}_u\| = \delta_e \quad \text{where } \|\mathbf{e}\| \leq \delta_e \tag{4}$$

where **e** is the error in the measurement vector $\mathbf{k}_u$. However, as the noise levels are not well defined, it is not straightforward to determine the upper bound. In GRAPPA-based calibration, the determination of parameter based on these bounds determine an estimate of the highest amount of regularization that can be applied without potentially amplifying the artifacts in the reconstruction.

The GCV function approximately minimizes the expected mean-squared error of predictions of the transformed data with an estimated linear model [11]. Generally, the GCV function provides a minimum near zero regularization (i.e., at λ close to 0). Therefore, the generalized cross-validation leads to undersmoothed estimates [13]. Undersmoothed estimates can be avoided by fixing bounds for the regularization parameters, for example, using a priori guesses of the magnitude of the residuals [14]. Using the GCV method, the parameter is obtained as that minimizes the function given by

$$G(\lambda) = \frac{\|\Pi z_\lambda - \mathbf{k}_u\|_2^2}{T^2} \qquad (5)$$

where the numerator is the squared residual norm and the denominator is a squared effective number of degrees of freedom. There is no restriction that the effective number of degrees of freedom is an integer and can be written in terms of the filter factors as

$$T = n - \sum_{i=1}^{rank(\Pi)} \frac{\sigma_i^2}{\sigma_i^2 + \lambda} \qquad (6)$$

L-curve method chooses the optimum parameter based on as L-curve plot, which is a log–log plot of the solution norm versus residual norm, with λ as the parameter. For increasing λ values, $\|L z_\lambda\|$ decreases and residual norm $\|\Pi z_\lambda - \mathbf{k}_u\|$ increases. The flat and the steep parts of the L-curve denotes the solutions dominated by regularization errors and perturbation errors. The vertical part of the L-curve corresponds to solutions and this region is very sensitive to changes in the regularization parameter. The horizontal part of the L-curve corresponds to solutions where the residual norm is more sensitive to the regularization parameter. The reason is that z_λ is dominated by the regularization error as long as $\mathbf{k}_u$ satisfies the discrete Picard criterion. Therefore, $\|L z_\lambda\|$ is monotonically decreasing function of $\|\Pi z_\lambda - \mathbf{k}_u\|$.

In the multiple filter approach, a number of filters are computed using Tikhonov regularization with different parameter values. As the L-curve parameter in GRAPPA is found to be larger than that of GCV, the parameter values are chosen between that of L-curve parameter and GCV. A number of parameter values in between the two limits as set by GCV and L-curve is used in the calibration. A filter bank using the predetermined set of parameters is generated. During reconstruction, each point chooses one of the predetermined filters for obtaining the missing point. For mapping each point to the filter, the signal power of each point is computed as the squared sum of the neighboring acquired points at each missing location. Then the filter with a larger parameter is mapped to the location with the smallest signal power and vice versa. This method for reconstruction yields better noise reduction and lesser artifacts as compared to any of the single filter-based methods.

In each dataset, the final reconstruction to be considered for comparison is performed using an optimal kernel size. For each dataset and regularization method, a priori is determined by observing the root mean square error (RMSE) values for a prefixed set of kernel sizes. The g-factors and reconstruction errors used for comparison are measured using optimal kernel sizes.

In pMRI, there is a nonuniform loss in signal-to-noise ratio (SNR) as compared to fully sampled images. SNR is directly dependent on the acceleration factor and coil geometry dependent factor (g-factor). The g-factor is strongly related to the encoding capability of the receiver coil array. The g-factor is used as a quantitative measure to evaluate the nonuniform noise enhancement in the reconstructed image [6]. Therefore, an increase in the g-factor value indicates an increase in noise in the reconstructed image. Here, a mean g-factor value is used as a metric of comparison.

The reconstruction error is computed using the relative l_2-norm error (RLNE) [15]. The root sum-of-squares (rSoS) of the fully sampled reference image (I_{ref}) and reconstructed image (I_{acc}) are used for RLNE calculation.

$$\text{RLNE} = \left\| I_{ref_1} - I_{acc} \right\|_2 / \left\| I_{ref_1} \right\|_2. \tag{7}$$

3 Results

Raw k-space data were acquired from Siemens 1.5T Magnetom-Avanto clinical MR scanner at Sree Chitra Tirunal Institute of Medical Sciences and Technology, Trivandrum, India. All subjects were volunteers with prior written informed consent collected before scanning. The data consist of fully sampled brain images acquired using array head coils and spine images acquired using surface coils. All reconstructions are performed on retrospectively sampled data with an acceleration factor of 2 and 32 ACS lines. Figure 2 shows the L-curves as in Tikhonov regularization. One can explicitly locate the L-curve corner only for *datasets*-I and III.

Figure 3 shows the GCV values plotted as a function of λ. The optimum regularization parameter is obtained at the point where $G(\lambda)$ is minimum. The optimum parameter is indicated using the asterisk symbol. It is observed that the optimum parameter in some cases corresponds to the very first point in the curve. This is due to the presence of correlated errors in the measurement vector. The images reconstructed using the L-curve, GCV, and filter bank-based approaches for *datasets*—I and II are shown in Fig. 4. Here, a filter bank comprising of ten filters is used. There is significant aliasing in L-curve-based reconstructed images. Aliasing is introduced due to over-regularization introduced by the wrong choice of the L-curve corner.

The RLNEs of GRAPPA reconstructed images obtained using each type of regularization are summarized in Table 1. The mean and median values are also indicated to compare across methods. The mean g-factors for different selection strategies for GRAPPA reconstruction are summarized in Table 2.

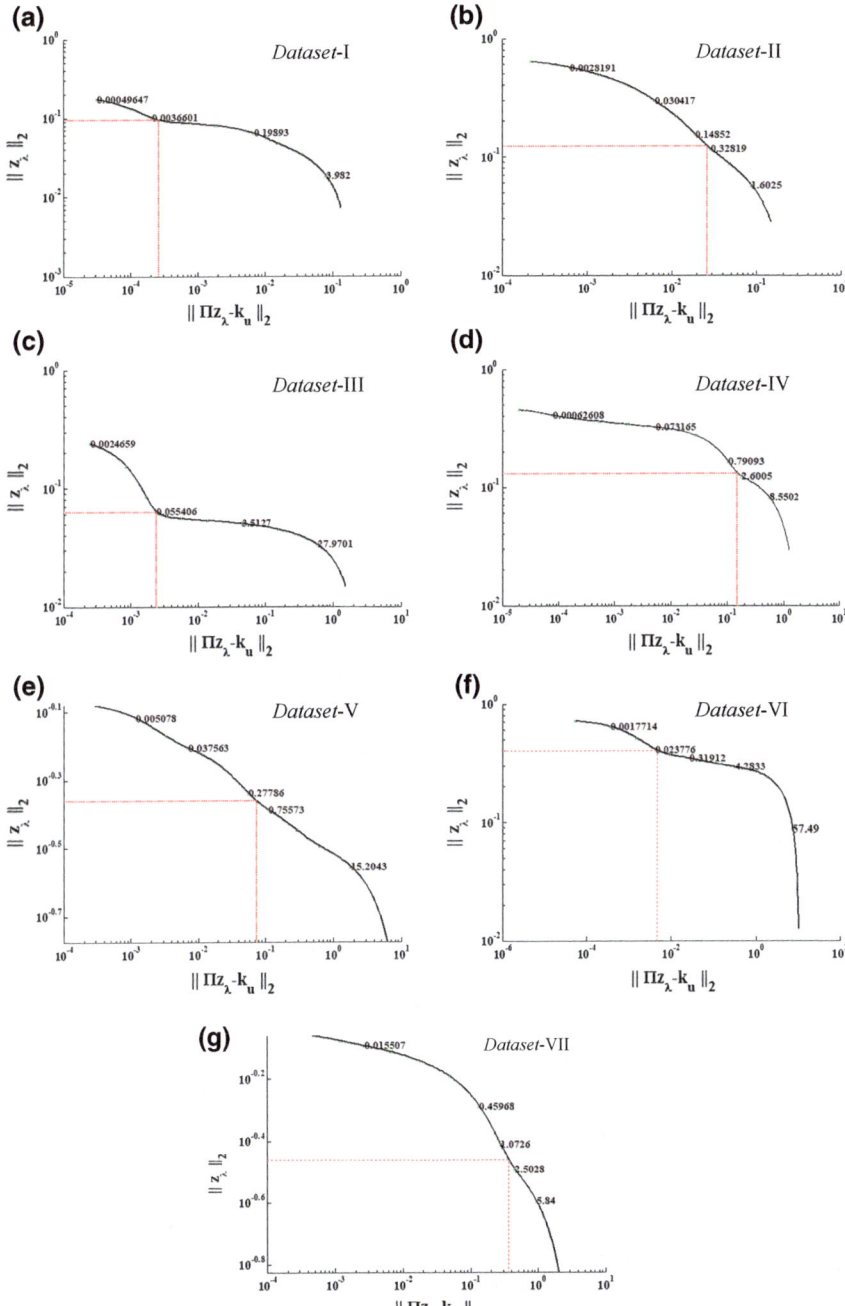

Fig. 2 L-curves for Tikhonov regularization in GRAPPA. The red indicates regularization parameter computed using regularization toolbox in MATLAB

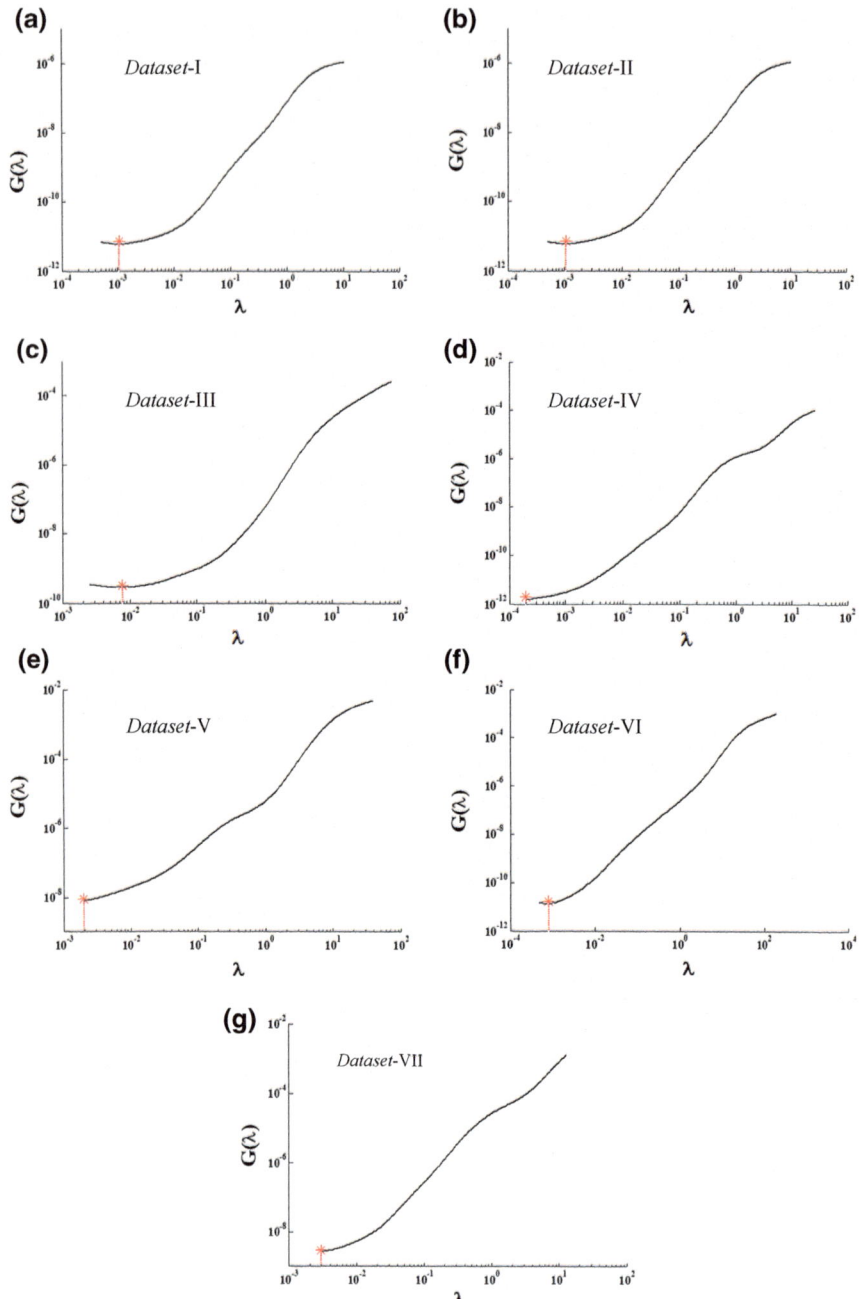

Fig. 3 GCV for Tikhonov regularization in GRAPPA. The red indicates regularization parameter chosen at the GCV minimum

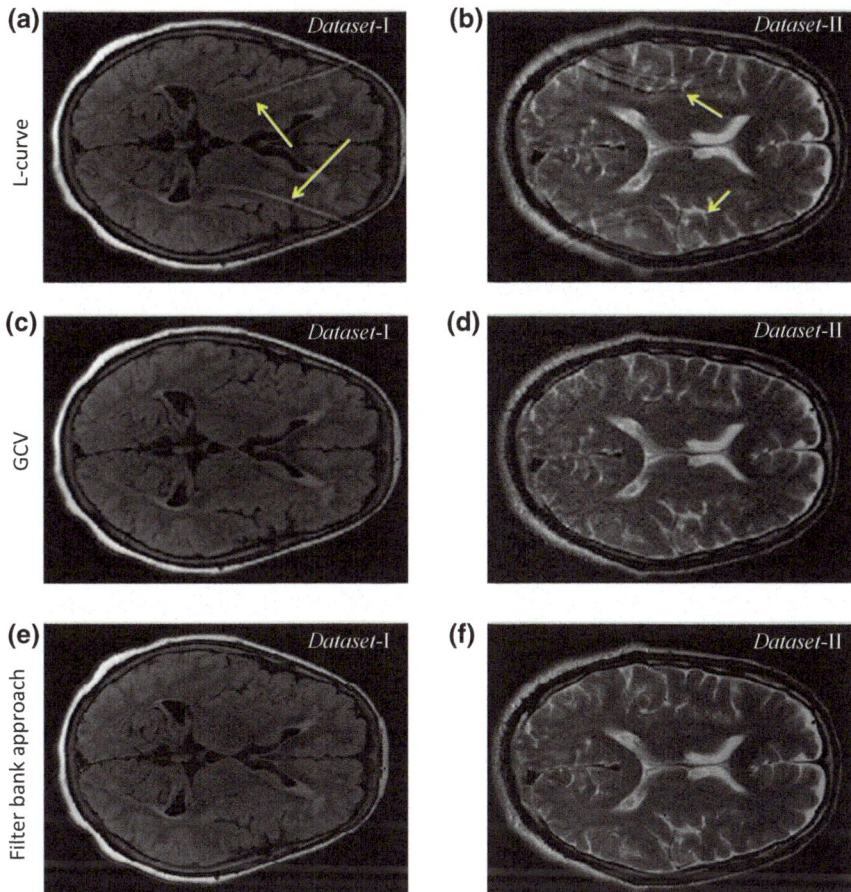

Fig. 4 GRAPPA reconstructed images using Tikhonov regularization

Table 1 RLNE of GRAPPA reconstructed images

Dataset/method	L-curve	GCV	Filter bank approach
I	0.1902	0.1711	0.1200
II	0.0641	0.0618	0.0410

Table 2 Mean g-factor of GRAPPA reconstructed images

Dataset/method	L-curve	GCV	Filter bank approach
I	0.4399	0.3955	0.2513
II	0.3435	0.3505	0.3010

4   Discussion

The accuracy of regularized output depends on the selection of appropriate regularization parameter. In the context of parallel MRI, the optimum parameter controls the degree of regularization and thus determines the compromise between SNR and artifacts. While under-regularization leads to residual noise or streaking artifacts in the image, over-regularization removes the image features and enhances aliasing. In this study, two approaches based on L-curve and GCV are used to locate the optimum parameter in Tikhonov regularization.

L-curve is used to locate the optimum parameter for both Tikhonov and spectral cut-off regularization. The L-curves exhibit typical L-corners only in a subset of the datasets. A mere presence of sharp L-corner does not guarantee an artifact-free reconstruction. As the matrix size involved in the problem increases, the parameter computed by L-curve may not behave consistently (asymptotic behavior) leading to some over-regularization. The extent of over-regularization is dependent on the decay rate of singular values, i.e., faster the decay, lesser will be the over smoothing. From our observations, it is also inferred that artifacts always appear in cases where a corner is difficult to locate. For Tikhonov regularization in GRAPPA, the L-curve corners were well defined in *datasets*—I, III, IV, and VI. This is possible because the performance of L-curve-based detection is better where the solution smoothness is small and input noise levels are high. L-curve breaks down if the exact solution in the underlying problem is very smooth; i.e., the SVD coefficients decay fast to zero such that the solution is dominated by the first few SVD components. Thus for smooth solutions, the L-curve computes worse parameter.

GCV is also used as a merit function to locate the optimum parameter for Tikhonov regularization. It is observed that GCV exhibits a minimum only in a subset of datasets reconstructed with spectral cut-off regularization. In the rest of the datasets, GCV is very flat with multiple local minima and the global minimum can be at the extreme endpoint for under-smoothing. This is because the GCV method is rather unstable when the numbers of singular values are less or noise is correlated. The GCV mostly performs well for Tikhonov for reasonably large data sets (number of singular values are large) with uncorrelated errors (white noise). It does not perform so well for Tikhonov regularization in cases affected by the rate of change of smoothness of solution being greater than the decay rate of singular values of the calibration matrix.

5   Conclusion

GRAPPA calibration involves solution of discrete ill-posed least squares, resulting in noisy reconstruction and aliasing artifacts. Proper use of regularized solutions facilitates minimization of noise build-up in the reconstruction process, without enhancing the effects of artifacts. However, the choice of proper regularization parameter is feasible only with precise knowledge of the error bounds. Any technique for choosing regularization parameters in the absence of information about error level and its statistics can pose difficulties in arriving at an optimal reconstruction. As a consequence, inferences obtained using different selection strategies can lead to suboptimal reconstructions. Usage of multiple filters based on the knowledge of the local signal power can yield better reconstructions compared to that of single filter-based reconstruction.

Acknowledgements The authors are thankful to Maulana Azad national fellowship by UGC and planning board of Govt. of Kerala (GO(Rt)No. 101/2017/ITD.GOK(02/05/2017)) for the financial assistance.

References

1. Griswold, M.A., Jakob, P.M., Heidemann, R.M., Nittka, M., Jellus, V., Wang, J., Kiefer, B., Haase, A.: Generalized autocalibrating partially parallel acquisitions (GRAPPA). Magn. Reson. Med. **47**, 1202–1210 (2002)
2. Lustig, M., Pauly, J.: SPIRiT: iterative self-consistent parallel imaging reconstruction from arbitrary k-space. Magn. Reson. Med. **64**, 457–471 (2010)
3. Zhang, J., Liu, C., Moseley, M.E.: Parallel reconstruction using null operations. Magn. Reson. Med. **66**, 1241–1253 (2011)
4. Tikhonov, A.N.: Solution of incorrectly formulated problems and the regularization method. Soviet Math. Dokl. **4**, 1035–1038 (1963)
5. Tikhonov, A.N., Arsenin, V.: Solutions of Ill-posed problems. Winston & Sons, Washington, DC, USA (1977)
6. Breuer, F.A., Kannengiesser, S.A., Blaimer, M., Seiberlich, N., Jakob, P.M., Griswold, M.A.: General formulation for quantitative g-factor calculation in GRAPPA reconstructions. Magn. Reson. Med. **62**, 739–746 (2009)
7. Jakob, P.M., Griswold, M.A., Edelman, R.R., Sodickson, D.K.: AUTO-SMASH: a self calibrating technique for SMASH imaging: SiMultaneous acquisition of spatial harmonics. Magma **7**, 42–54 (1998)
8. Sodickson, D.K., Manning, W.J.: Simultaneous acquisition of spatial harmonics (SMASH): fast imaging with radiofrequency coil arrays. Magn. Reson. Med. **38**, 591–603 (1997)
9. Morozov, V.A.: Methods for solving incorrectly posed problems. Springer, New York (1984)
10. Hansen, P.C.: Analysis of discrete ill-posed problems by means of the L-curve. SIAM Rev. **34**, 561–580 (1992)
11. Golub, G.H., Heath, M., Wahba, G.: Generalized cross-validation as a method for choosing a good ridge parameter. Technometrics **21**, 215–223 (1979)
12. Hansen, P.C.: Regularization tools, a Matlab package for analysis of discrete regularization problems. Numer. Algorithms **6**, 1–35 (1994)
13. Wahba, G., Wang, Y.: Behavior near zero of the distribution of GCV smoothing parameter estimates. Stat. Probabil. Lett. **25**, 105–111 (1995)
14. Hansen, P.C.: Rank-deficient and discrete Ill-posed problems: numerical aspects of linear inversion, p. 247. SIAM, Philadelphia (1998)
15. Qu, X., et al.: Undersampled MRI reconstruction with patch-based directional wavelets. Magn. Reson. Imag. **30**, 964–977 (2012)

Improving Image Quality and Convergence Rate of Perona–Malik Diffusion Based Compressed Sensing MR Image Reconstruction by Gradient Correction

Ajin Joy⊙ and **Joseph Suresh Paul**⊙

Abstract A memory-based reconstruction algorithm is developed for optimizing image quality and convergence rate of Compressed Sensing-Magnetic Resonance Image (CS-MRI) reconstruction using Perona–Malik (PM) diffusion. The PM diffusion works by estimating the underlying structure of an image and diffusing the image in a nonlinear fashion to preserve the sharpness of edge information. The edges due to undersampling artifacts are generally characterized as weak edges (false edges) identified by the comparatively smaller gradient magnitude associated with it. The convergence rate for CS-MRI reconstruction based on PM diffusion, therefore, depends on the extent by which the gradients attributed to false edges are diffused off per iteration. However, if the undersampling interference in high, gradient magnitudes of false edges can become comparable to that of true edges (boundaries of actual anatomical features). This might either lead to preservation of false edges or diffusion of true edges. This reduces the quality of reconstructed images. In such scenarios, we assume that the gradient information in past iterations contains useful structural information of the image which is lost while diffusing the image. Hence, we propose to overcome this problem by correcting the estimate of the underlying structure of the image using a combination of gradient information from a number of past iterations. This reduces the diffusion of weak edges by restoring the otherwise lost weak structural information at every iteration.

Keywords Perona–Malik diffusion · Compressed sensing · Gradient correction · Memory · Nonlinear diffusion

A. Joy (✉) · J. S. Paul
Medical Image Computing and Signal Processing Laboratory, Indian Institute of Information Technology and Management-Kerala (IIITM-K), Trivandrum, Kerala, India
e-mail: ajin.joy@iiitmk.ac.in

J. S. Paul
e-mail: j.paul@iiitmk.ac.in

1 Introduction

Magnetic resonance imaging (MRI) data are acquired in the frequency domain (k-space). It is a lengthy process leading to subject discomfort and motion artifacts. To accelerate the acquisition, generally, only a subset of the data is collected and the missing k-space information is estimated using various reconstruction algorithms. Recently, the application of compressed sensing (CS) theory showed that accurate reconstruction of MR images is possible by making use of its sparse representation [1–3]. This allows accelerating the data acquisition by collecting the frequency information only at random locations of the k-space. Since MR images are naturally compressible, CS-MRI uses transform domains like wavelet and discrete cosine to represent the images in their sparse form [4–6]. Basic assumption involved is that if the k-space is fully sampled, the data will be highly sparse in the transform domain. Due to the undersampling of the k-space, the data become less sparse [7–9]. Hence the reconstruction works by enforcing sparsity in the sparse domain (generally by minimizing its l_1-norm) and the correction of its k-space representation using known k-space values in an iterative fashion [10–14]. Improving the convergence rate of these algorithms without compromising the reconstruction quality is one of its main challenges. This is particularly important in the learning-based CS frameworks where redundant systems for sparsifying MR images are trained from an initial approximation of the image [14–17].

The initial noisy image in the MR image reconstruction has noise in the form of edge-like artifacts resultant from the undersampling of the k-space. Strength of these artifacts increases with a reduction in the degree of randomness in the sampling pattern. As a result, the gradient magnitudes of the image generated by artifacts tend to be similar to that of the true edges. Edge preserving reconstructions generally use the image gradient to preserve significant edges while removing artifacts. Since a faithful representation of true edges is relatively difficult to achieve in such cases, the reconstructed image can be noisy/blurry. To overcome this, we propose a memory-based gradient estimation technique where the combined information from the outputs of a number of iterations are used to identify the underlying structure of the image.

2 Related Work

The widely used sparsifying transforms in MRI include wavelet and finite difference. The Total Variation (TV) based approaches are generally used to for sparse reconstruction in finite difference domains. Recently, use of nonlinear (NL) approach based on the Perona–Malik (PM) diffusion is known to improve the reconstruction [8, 18–20] and has the advantage that the diffusion process will not affect the edges where gradient values are significantly high. While the regularization parameter plays a critical role in achieving the right amount of denoising in TV, NL diffusion gives more control over denoising by adjusting the contrast parameter. The statistical estimation of the contrast parameter in the biased PM diffusion reconstruction removes the need for searching the critical parameters compared to TV.

In PM diffusion, the identification of the underlying structure of the image plays an important role in preserving edges from diffusion. The edges are identified from the strength of the gradient magnitudes. Stronger gradients generally correspond to true edges that represent the boundaries of distinct anatomical features of the image while weaker ones (false edges) correspond to noise due to undersampling interference.

3 Theory

PM diffusion is a well-studied edge-preserving denoising technique [21, 22] which is also known to be an effective denoising tool in MRI [23, 24]. It works by denoising the uniform intensity regions while enhancing the edges in the image. In a CS framework, PM diffusion process reconstructs the image by a series of diffusion iterations given by

$$U_{diffused} = U + \gamma \, div(g(|\nabla U|, \alpha)\nabla U) + cF'_u(K - F_u U), \qquad (1)$$

where c is the bias relaxation factor, γ is the diffusion strength parameter, div is the divergence operator, and K is the set of values acquired in the undersampled k-space with locations of unacquired values filled with zeros. F_u operating on the image (U) computes the Fourier Transform of U followed by setting the unacquired frequency points to zero and F'_u is the inverse Fourier Transform. g represents the diffusivity function which is a function of image gradients (∇U) and a contrast parameter (α). The diffusion in Eq. (1) varies with the choice of g. For PM diffusion, it is chosen as

$$g(|\nabla U|) = \frac{1}{1 + (|\nabla U|/\alpha)^2}, \qquad (2)$$

where (α) is the contrast parameter that separates the gradient magnitudes of noise and edges in an image.

While the first two terms in Eq. (1) performs the PM diffusion, the third term corrects the diffused image using known partial data, thereby biasing the diffusion process. Here, the initial noisy image is not diffused continuously as in the case of a general denoising application. It is but diffused as a sequence of single-step diffusion processes where at every iteration, structural information is reinforced by biasing before diffusing it further. Biasing ensures that the frequency components are most consistent with the acquired k-space data. For a detailed description of PM diffusion-based reconstruction technique, we refer the reader to Ref. [18].

Strength of gradient magnitudes representing the underlying structure of the image plays an important role in the quality of the reconstructed image and speed of reconstruction. This is because better identification of the true edges would drive the iterations toward the biased solution thereby accelerating the process. Also, this ensures that the true edges are not diffused off, hence the improved image quality.

Contrast parameter defines the threshold for separating the gradient magnitudes attributed to signal and noise. When the undersampling artifacts are more edge-like, an estimate of the contrast parameter to diffuse these artifacts are also expected to remove some of the gradient magnitudes attributed to true edges as well. This would often result in over diffused (blurred) images.

We propose to overcome this problem by reintroducing the gradient information from past iterations into the current one so that the chances to preserve true edges is improved. Since the bias correction is more likely to reintroduce true edges than artifacts, combined gradient magnitudes from a number of iterations would strengthen the true edges compared to artifacts. For a prediction order n, gradient magnitude used in an iteration is predicted from previous n number of iterations as

$$\widehat{\nabla U}^{(q)} = \nabla U^{(q-1)} + \nabla U^{(q-2)} + \cdots + \nabla U^{(q-n)}$$
$$= \nabla \left(U^{(q-1)} + U^{(q-2)} + \cdots + U^{(q-n)} \right), \tag{3}$$

where q is the iteration number. Therefore, the biased NL diffusion reconstruction is now modified as

$$U_{diffused} = U + \gamma \, div \left(g \left(\left| \widehat{\nabla U} \right|, \alpha \right) \widehat{\nabla U} \right) + c F_u'(K - F_u U). \tag{4}$$

As the iterations progress, more gradient information is getting diffused off. If a pixel corresponds to true edge, it is less likely to be subjected to diffusion. Hence, the gradient strength will not decay fast at those locations. On the other hand, noise is expected to be diffused off continuously as the iterations progress. This will cause the gradient magnitudes in those locations to weaken rapidly. Hence, a spatially varying set of weights for the gradients give a better representation of the underlying structure in the image. The gradient calculation from a weighted combination of outputs from past iterations is calculated as

$$\widehat{\nabla U}^{(q)} = \nabla \left(\beta_1 U^{(q-1)} + \beta_2 U^{(q-2)} + \cdots + \beta_n U^{(q-n)} \right), \tag{5}$$

where β is a spatially varying weight having the same dimensions as that of U.

Furthermore, it is known that the quality of the reconstructed image in CS can be improved by sparsifying the image using multiple transform domains [25]. Incorporating the sparse representation of wavelet basis, the reconstruction in Eq. (4) can be improved by soft thresholding the wavelet representation of the output of biased PM diffusion at every iteration. Soft thresholding operation is defined as

$$soft(W, \tau) = \begin{cases} W + \tau, & \text{if } W \leq -\tau \\ 0, & \text{if } |W| < \tau \\ W - \tau, & \text{if } W \geq \tau \end{cases} \tag{6}$$

where W is the wavelet transform U and τ is a magnitude threshold.

4 Method

An initial estimate of the image ($U^{(0)}$) is obtained from the undersampled k-space data by setting the unacquired locations as zero and performing inverse Fourier Transform. It is then subjected to the diffusion as in Eq. (4). For the reconstruction of a prediction order "n", memory used consists of n number of cells. Output of every iteration is saved in a cell such that outputs from n number of previous iterations are always available in the memory. The memory is updated in a first-in-first-out manner.

A set of weights are derived proportional to the gradient magnitude of the images in the cells. At the locations where gradient magnitude is rapidly decreasing with iterations, the weight will proportionally decrease. Since the gradient magnitudes attributed to edges will not decay as fast as that of noise, the weight is high at these indices for all cells in the memory. The gradient values are then updated using the weights, as in Eq. (5). A work flow of the proposed method is shown in Fig. 1.

As iterations progress, the difference in output images from successive iterations becomes small. Consequently, the gradients from older iterations become less significant. Therefore, the memory size is reduced when the difference is small. When the

Fig. 1 Work flow of the proposed method

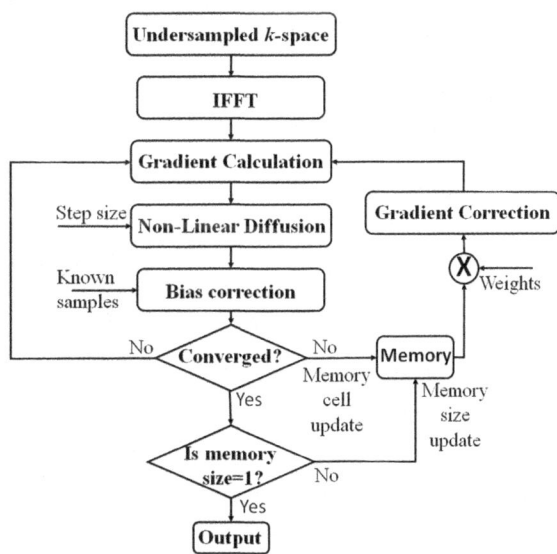

sparsity in wavelet basis is used along with PM diffusion, output of every iteration is transformed to wavelet domain and subjected to soft thresholding as in Eq. (6). Then it is transformed back to the image domain and proceeds to the next iteration.

4.1 Reconstruction Parameters

Different parameters controlling the reconstruction include relaxation parameter c, diffusion strength γ, prediction order n, and the contrast parameter α. α is estimated using mean/median absolute deviation (MAD) of ∇U [22]. When the gradients of an image are calculated along the vertical and horizontal directions, the numerical scheme of PM diffusion becomes stable when $0 < \gamma < 1/4$ [23, 26]. $c \in (0, 1]$ is set to 1 for maximum consistency with the acquired data and a prediction order of 8 is used for all reconstructions discussed in this paper.

4.2 Error Measure

The error in the reconstruction is computed using the relative l_2-norm error (RLNE) of the image.

$$RLNE^{(q)} = \frac{\left\| U^{(q)} - U_{ref} \right\|_2}{\left\| U_{ref} \right\|_2} \tag{7}$$

where U_{ref} is the ground truth image.

4.3 Datasets Used

Different datasets used in this work are shown in Fig. 2b and c. The dataset I (Fig. 2b) and II (Fig. 2c) are T2-weighted brain images of size 256×256 acquired from a healthy volunteer at a 3T Siemens Trio Tim MRI scanner (Siemens Healthineers) with 32 coils using the T2-weighted turbo spin echo sequence (TR/TE = 6100/99 ms, FOV = 220×220 mm^2, slice thickness = 3 mm), shared freely at [27]. An example of a Poisson disk sampling mask used to retrospectively undersample the k-space data is shown in Fig. 2a.

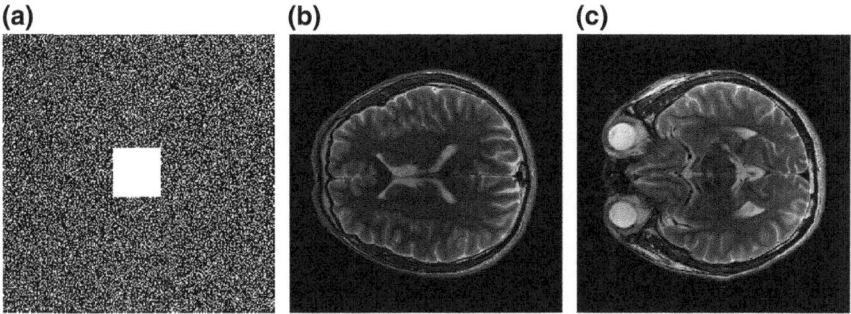

Fig. 2 **a** Poisson disk sampling pattern to collect 30% samples. **b, c** Magnitude images of brain data

5 Results and Discussion

Improvement in the computation of diffusion update resultant from the nth order gradient prediction is shown in Fig. 3. The panels in the first and second columns depict the diffusion updates calculated using the state-of-the-art PM diffusion process (2nd term in Eq. (1)) and the proposed method (2nd term in Eq. (4)), respectively. Row-wise panels correspond to diffusion updates at different iteration numbers. The corresponding iteration number is shown in the inset.

It indicates that the edges in diffusion update becomes better resolved with the use of the proposed reconstruction model. The resultant RLNEs are plotted as a function of iterations in Fig. 4 for three different reconstruction approaches: (A) PM diffusion without memory, (B) PM diffusion with memory, and (C) sparsity of both finite difference and wavelets with memory.

Plots in Fig. 4a, b correspond to dataset I and II, respectively. In both cases, minimum RLNE is achieved using method C. This is expected since the sparsity in both, the finite difference representation and the wavelet domain is utilized. Comparing B and A, it is clear that the use of memory in B resulted in faster convergence and smaller RLNE.

Reconstructed images of dataset I and II are shown in Fig. 5. Subpanels a–d depict the results of zero-filled reconstruction, method A, method B, and method C, respectively. The Region-Of-Interest (ROI) shown in white bounding box is enlarged for better visual comparison.

Compared to method C, edges appear sharper in methods A and B. However, the tendency of blocky effect is most in A and minimal in C.

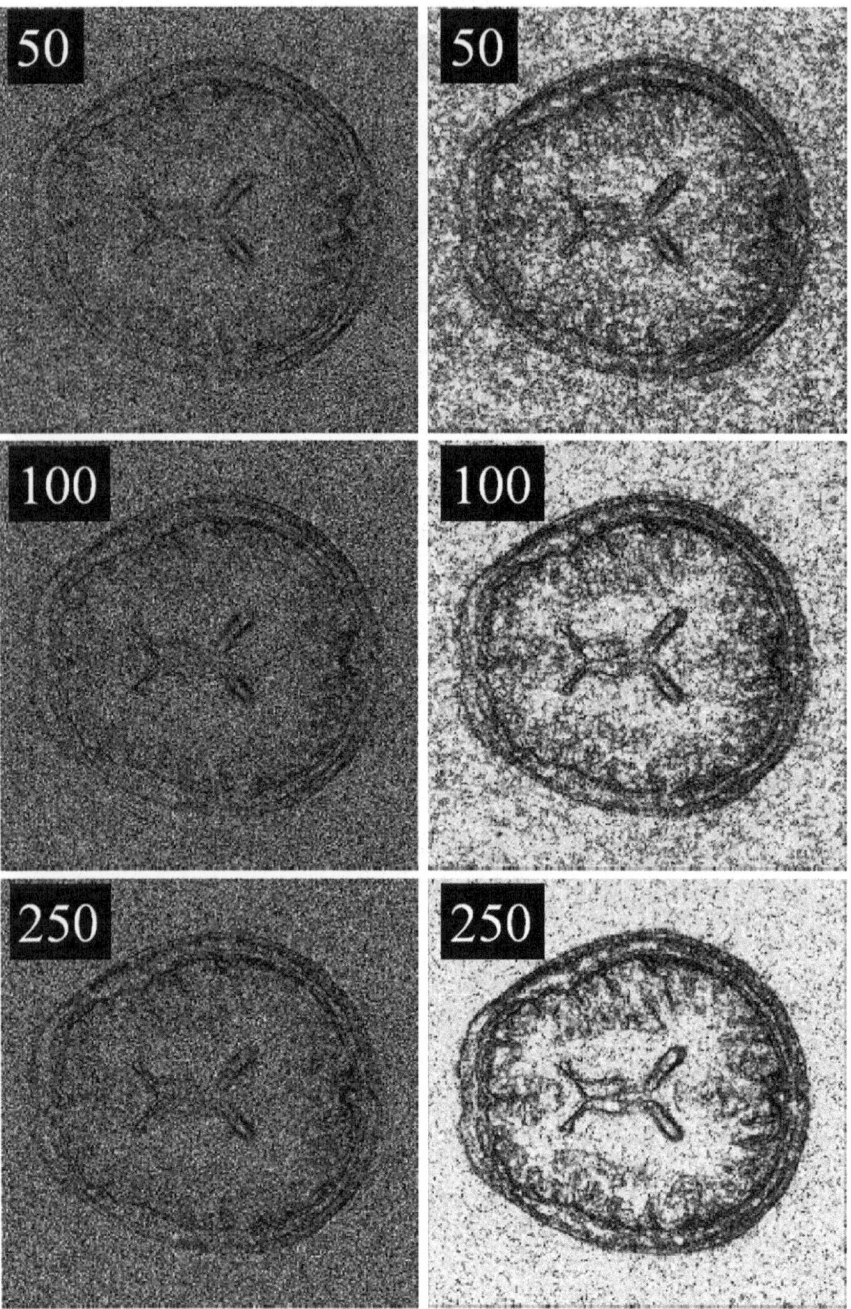

Fig. 3 Diffusion update for PM diffusion. Panels in the left and right columns correspond to PM diffusion without and with memory. Insets show the corresponding iteration number

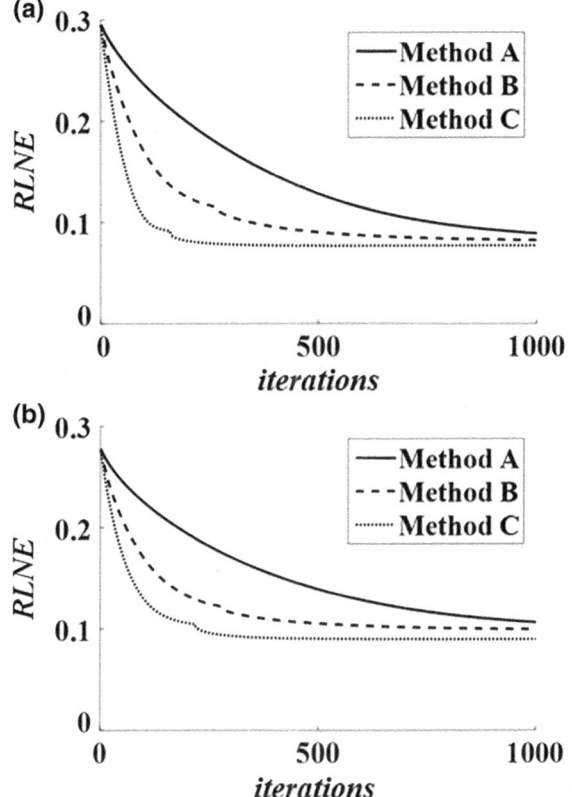

Fig. 4 RLNEs plotted as a function of iterations for Method A, B, and C. **a** RLNE for dataset I, **b** RLNE for dataset II

5.1 Computational and Time Complexity

From Fig. 4, it is obvious that method C converges faster than the other two methods. In fact, C converges in about 250 iterations while methods B and A take around twice and four times as many number of iterations. However, the computational complexity also increases from A to C, with A being the simplest and C being most complex among the three. It is observed that method B completes one iteration twice as fast as C and method A is around 3 times faster than C. However, since B and C take fewer iterations to complete the reconstruction, they converge faster than A.

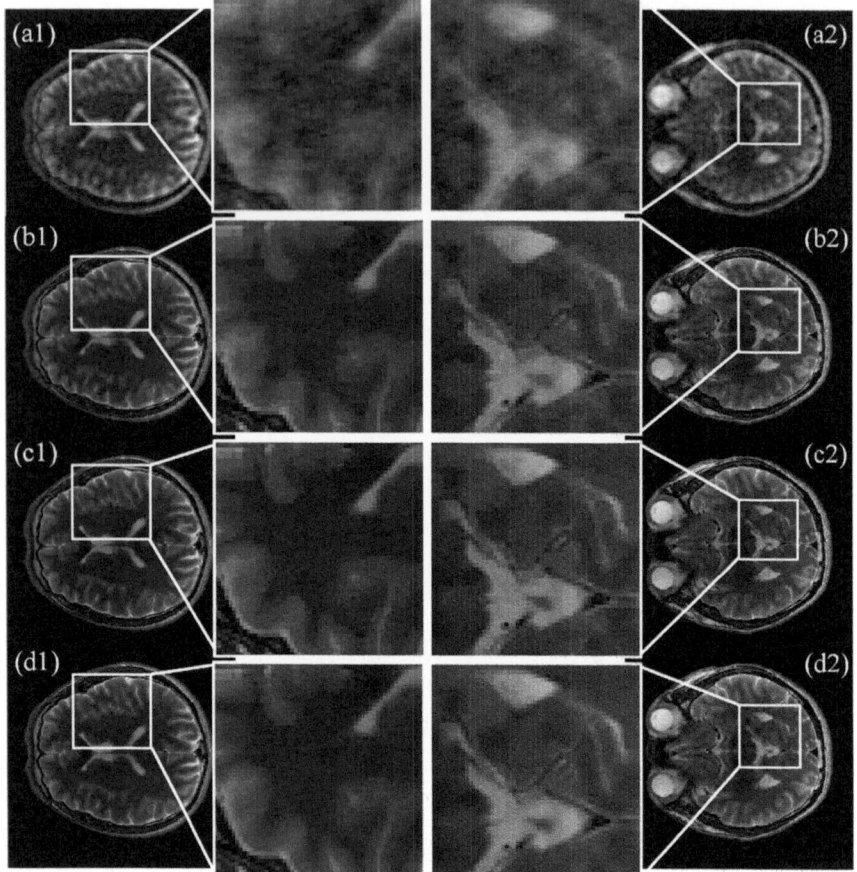

Fig. 5 Comparison of different reconstruction approaches. **a1–a2** Zero-filled reconstruction, **b1–b2** method A, **c1–c2** method B, and **d1–d2** method C

6 Conclusion

A memory-based reconstruction algorithm for better prediction of gradient magnitudes signifying true edges in the image is presented. It is shown that the proposed technique optimizes both image quality and convergence rate of CS reconstruction using PM diffusion. Reconstruction using the memory-based approach generated images with improved RLNE measure in less than half the number of iterations as that of the conventional approach.

Acknowledgements Authors are thankful to the Council of Scientific and Industrial Research-Senior Research Fellowship (CSIR-SRF, File No: 09/1208(0002)/2018.EMR-I) and planning board of Govt. of Kerala (GO(Rt)No.101/2017/ITD.GOK(02/05/2017)), for financial assistance. Authors also thank the researchers who publicly shared the datasets used in this work.

References

1. Candes, E.J., Romberg, J., Tao, T.: Robust uncertainty principles: exact signal reconstruction from highly incomplete frequency information. IEEE Trans. Inf. Theory **52**(2), 489–509 (2006)
2. Donoho, D.L.: Compressed sensing. IEEE Trans. Inf. Theory **52**(4), 1289–1306 (2006)
3. Lustig, M., Donoho, D., Pauly, J.M.: Sparse MRI: the application of compressed sensing for rapid MR imaging. Magn. Reson. Med. **58**(6), 1182–1195 (2007)
4. Guerquin-Kern, M., Haberlin, M., Pruessmann, K.P., Unser, M.: A fast wavelet-based reconstruction method for magnetic resonance imaging. IEEE Trans. Med. Imaging **30**(9), 1649–1660 (2011)
5. Liu, Y., Cai, J.F., Zhan, Z., Guo, D., Ye, J., Chen, Z., Qu, X.: Balanced sparse model for tight frames in compressed sensing magnetic resonance imaging. PLoS ONE **10**(4), E0119584 (2015)
6. Montefusco, L.B., Lazzaro, D., Papi, S.: Nonlinear filtering for sparse signal recovery from incomplete measurements. IEEE Trans. Signal Proc. **57**(7), 2494–2502 (2009)
7. Montefusco, L.B., Lazzaro, D., Papi, S.: Fast sparse image reconstruction using adaptive nonlinear filtering. IEEE Trans. Image Proc. **20**(2), 534–544 (2011)
8. Li, X.: The magic of nonlocal Perona-Malik diffusion. IEEE Signal Process. Lett. **18**(9), 533–534 (2011)
9. Liu, P., Xiao, L., Zhang, J.: Fast second degree total variation method for image compressive sensing. PLoS ONE **10**(9), E0137115 (2015)
10. Liu, Y., Zhan, Z., Cai, J.F., Guo, D., Chen, Z., Qu, X.: Projected iterative soft-thresholding algorithm for tight frames in compressed sensing magnetic resonance imaging. IEEE Trans. Med. Imaging **35**(9), 2130–2140 (2016)
11. Ravishankar, S., Bresler, Y.: MR image reconstruction from highly undersampled k-space data by dictionary learning. IEEE Trans. Med. Imaging **30**(5), 1028–1041 (2011)
12. Qu, X., Guo, D., Ning, B., Hou, Y., Lin, Y., Cai, S., Chen, Z.: Undersampled MRI reconstruction with patch-based directional wavelets. Magn. Reson. Imaging **30**(7), 964–977 (2012)
13. Ning, B., Qu, X., Guo, D., Hu, C., Chen, Z.: Magnetic resonance image reconstruction using trained geometric directions in 2D redundant wavelets domain and non-convex optimization. Magn. Reson. Imaging **31**(9), 1611–1622 (2013)
14. Lai, Z., Qu, X., Liu, Y., Guo, D., Ye, J., Zhan, Z., Chen, Z.: Image reconstruction of compressed sensing MRI using graph-based redundant wavelet transform. Med. Image Anal. **27**, 93–104 (2016)
15. Zhan, Z., Cai, J.F., Guo, D., Liu, Y., Chen, Z., Qu, X.: Fast multi-class dictionaries learning with geometrical directions in MRI. IEEE Trans. Biomed. Eng. **63**(9), 1850–1861 (2016)
16. Qu, X., Hou, Y., Lam, F., Guo, D., Zhong, J., Chen, Z.: Magnetic resonance image reconstruction from undersampled measurements using a patch-based nonlocal operator. Med. Image Anal. **18**(6), 843–856 (2014)
17. Baker, C.A., King, K., Liang, D., Ying, L.: Translational-invariant dictionaries for compressed sensing in magnetic resonance imaging. In: Proceedings of IEEE International Symposium on Biomedical Imaging, 2011, pp. 1602–1605
18. Joy, A., Paul, J.S.: Multichannel compressed sensing MR Image reconstruction using statistically optimized nonlinear diffusion. Magn. Reson. Med. **78**(2), 754–762 (2017)
19. Joy, A., Paul, J.S.: A mixed order nonlinear diffusion compressed sensing MR image reconstruction. Magn. Reson. Med. **00**, 1–8 (2018). https://doi.org/10.1002/mrm.27162
20. Saucedo, A., Joy, A., Daar, E.S., Guerrero, M., Paul, J.S., Sarma, M.K., Thomas, M.A.: Comparison of compressed sensing reconstruction for 3D echo planar spectroscopic imaging data using total variation and statistically optimized Perona-Malik non-linear diffusion. In: Proceedings of the Joint Annual Meeting ISMRM-ESMRMB 2018 June. International Society for Magnetic Resonance in Medicine (ISMRM), Paris, France (in press)
21. Weickert, J.: Anisotropic Diffusion in Image Processing. Teubner, Stuttgart (1998)
22. Tsiostsios, C., Petrou, M.: On the choice of the parameters for anisotropic diffusion in image processing. Pattern Recognit. **46**(5), 1369–1381 (2013)

23. Gerig, G., Kubler, O., Kikinis, R., Jolesz, F.A.: Nonlinear anisotropic filtering of MRI data. IEEE Trans. Med. Imaging **11**(2), 221–32 (1992)
24. Krissian, K., Aja-Fernández, S.: Noise-driven anisotropic diffusion filtering of MRI. IEEE Trans. Image Process. **18**(10), 2265–74 (2009)
25. Chen, G., Zhang, J., Li, D.: Fractional-order total variation combined with sparsifying transforms for compressive sensing sparse image reconstruction. J. Vis. Commun. Image Represent. **38**, 407–22 (2016)
26. Perona, P., Malik, J.: Scale-space and edge detection using anisotropic diffusion. IEEE Trans. Pattern Anal. Mach. Intell. **12**(7), 629–639 (1990)
27. Qu, X.: T2 weighted brain. https://sites.google.com/site/xiaoboxmu/publication. Accessed 04 Aug 2018

Blind Quality Assessment of PFA-Affected Images Based on Chromatic Eigenvalue Ratio

Kannan Karthik and Parveen Malik

Abstract Quality assessment of Purple Fringing Aberrated (PFA) images remains an unsolved problem because the original untainted natural image of the scene is not available at the point of analysis. As a result, the problem assumes the form of a blind assessment. This PFA is a false coloration localized around the edge regions where the contrast differential is high. One can, therefore, surmise that if this coloration is largely homogeneous in the chrominance space, the edge is expected to be crisp and the image sharp and clear. However, if this coloration pattern is diverse in the chrominance space, the edges will be fuzzy and this, in turn, will have an impact on the visual clarity of the image. The fringe diversity, therefore, becomes a measure of PFA image quality, provided the fringes are distributed in several parts of the image. This diversity has been captured by first characterizing the chrominance space spanned by the PFA pixels and then using the eigenvalue ratio as a measure of color diversity.

Keywords Blind · Quality assessment · Purple fringing aberration (PFA) · Chrominance

1 Introduction and Formulation

Defects in imaging devices such as regular digital cameras can lead to annoying visual artifacts in the image frame. These visual artifacts are either distributed locally or are diffused over the entire the image frame (i.e., has a global impact). In cases where the visual artifacts are perceptible and jarring to the human eye, it is important to declare these images to be unfit for further use and processing particularly from an esthetic viewpoint. One such defect is Purple Fringing Aberration (PFA), which occurs predominantly in the sensor grid [6, 8] because of a phenomenon called

K. Karthik (✉) · P. Malik
EEE Department, Indian Institute of Technology Guwahati, Guwahati 781039, Assam, India
e-mail: k.karthik@iitg.ernet.in

P. Malik
e-mail: parveen@iitg.ernet.in

© Springer Nature Singapore Pte Ltd. 2020 59
B. B. Chaudhuri et al. (eds.), *Proceedings of 3rd International Conference on Computer Vision and Image Processing*, Advances in Intelligent Systems and Computing 1022,
https://doi.org/10.1007/978-981-32-9088-4_6

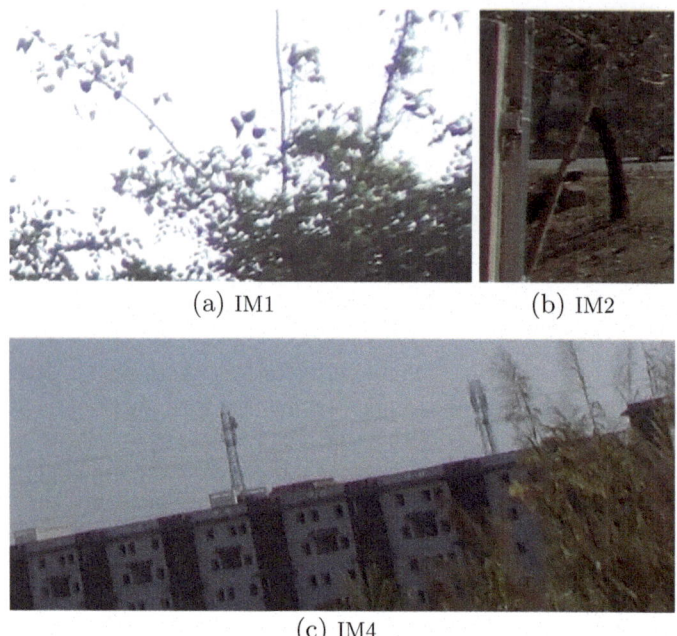

(a) IM1   (b) IM2

(c) IM4

Fig. 1 **a** PFA-affected image of leaves and branches with the sky in the backdrop (IM1); **b** PFA-affected image with trees and foliage (IM2); **c** Building shot from a distance (IM4)

"sensor blooming". Figure 1 shows some examples of these fringe-affected images to different degrees taken using the same NIKON Coolpix L23 camera. The extent to which the images (IM1–IM6) appear distorted, not just because of the presence of the purple fringes but also due to the presence of other forms of mild color distortions, varies from scene to scene. We may raise the following questions:

- Is it possible to rank these images (IM1, IM2, and IM4) in the order of decreasing visual quality.
- Is it possible to use the information available with the purple fringes alone to establish a global quality measure?

While PFA is a camera defect, it has not found its way as a viable forensic tool [14] for a variety of reasons discussed in part in the counterview to Ido et al. [9]:

– Different scenes captured with the same midrange camera are expected to exhibit diverse fringe coloration patterns, depending on relative locations of light sources, their compositions, and local texture/colors reflected from the scenes with respect to the camera angle or perspective. This means that the nature of the fringe cannot be traced back to camera model or brand or the original parameter camera setting [8].

Much of existing research has, therefore, focussed on the detection of these purple fringes [4, 5, 7]. However, of late, the focus has expanded to include fringe correction with detection [1, 3, 12]. In all these correction methods, the images taken up for correction are of a simplistic type wherein fringe region is formed in the shadows of objects. In other words, the implicit assumption is that the fringes do not eat into the objects or do not overlap with the objects. This makes it easier to detect and replace fringe regions with a plain color substitute. Images shown in Fig. 1 are of a much more complex type, wherein in some cases, the fringes are formed in the backdrop of complex texture and a different color. This causes a greater degradation in the visual quality of the image and it is quite possible that this complex mixing phenomenon is helped by the combination of an optical defect with a sensor one. It is precisely this form of distortion, we attempt to evaluate and quantify in this paper subject to perceptual constraints. The biggest problem with PFA-affected images, is that there is no references on which one can pivot and evaluate the extent of the fringe distortion and its eventual impact on perceptual quality.

One way by which this problem can be circumvented, is by attempting a complete removal of these purple fringes and then using the corrected image as a reference for computing quality metrics such as Signal-to-Noise Power (SNR) ratio [2]. This approach works provided the fringes can be removed fully without carrying with it portions of the regular image. The images chosen by Ju et al. [2], were of a simplistic type, wherein local neighborhoods around the fringes were largely plain in nature, restricting the color correction procedure to a color mapping process, through some form of a color proximity measurement. However, such correction approaches are expected to fail, when applied to images of the type presented in Fig. 1, which are not only diverse with respect to the fringe coloration, but some of them have complex texture in the backdrop (such as the random orientations of leaves in image IM1).

The rest of the paper is organized as follows: In Sect. 2, we present the motivation for our work connected with quality assessment of PFA-affected images. In Sect. 3, we discuss the details of our proposed algorithm which examines the diversity profile of the fringe pattern to form an assessment regarding the PFA-affected color image quality. Finally, we apply this algorithm to a variety of test images to obtain the corresponding quality scores in Sect. 4. These scores are corroborated with a subjective assessment concerning select groups of images.

2 Blind Quality Assessment of PFA-Affected Images

There are two elements which constitute a fringe in a PFA-affected image:

- Locality of the phenomenon, tied to the near saturation or high contrast zones in the image because of a charge leakage in the sensor grid.
- The fringes have a purple base (narrowband or broadband). The reason for this is not exactly known, while one can possibly attribute this to the nature of the color filter array and the relative positioning of the color filter pellets.

Figure 2d shows the location of the purple fringe in relation to the high contrast region and with reference to the gradient profile. Fringes are general produced toward

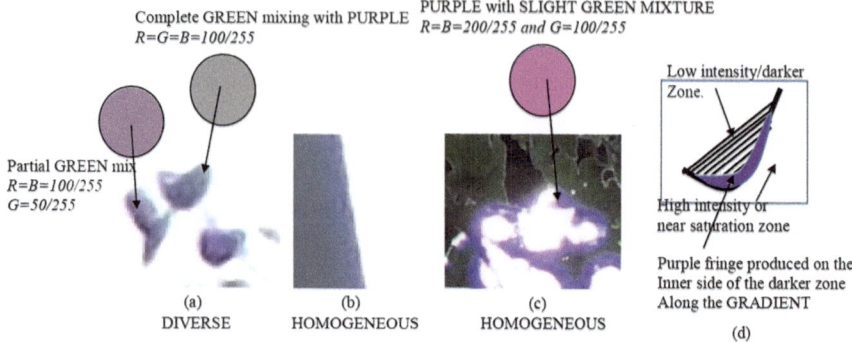

Fig. 2 **a** Two distinct fringe coloration patterns on the same leaf structure (cropped from image IM1 Fig. 1a); Fringes produced in this case are very diverse; **b** Fringes produced near the roof edge because of the backlit sky (fringes largely homogeneous in nature); **c** Purple fringes produced near the leaf endings with a slight green mix (fringes are of a homogeneous type); **d** Location, color, and positioning of the purple fringes in general

the darker side along the gradient and tend to fade away as one moves deeper into the darker zone. The girth or width of the position is a function of several factors such as (a) Magnitude of differential intensities across the gradient; (b) Intensity of the backlight (in the case of images involving leaves with the sky and sun in the backdrop—Fig. 2a, c). The greater the intensity of the backlight, greater is the expected spillover, leading to a deeper and more expansive fringe pattern. Note that the fringe patterns in the two leaf images are quite different, as the leaf arrangement in the case of Fig. 2a is more randomized (smaller leaves with different curvatures and different orientations) in comparison with the leaf pattern in Fig. 2c, wherein the leaf structure is largely planar.

In any natural scene, the sharpness of the edge is a good indication of image quality. When there is a color fringe aberration of the PFA type, the edges corresponding to the high contrast zones, tend to become fuzzy because of the false purple coloration toward the darker side of these edges. The sharpness in the image therefore drops considerably. The greater the color diversity of the fringe, the more fuzzy the corresponding edge is likely to appear, visually. This aspect can be witnessed by comparing Fig. 2a with Fig. 2c. Hence, the color diversity of the fringes are a good indicator of the color precision and subsequently the overall quality of the image.

To facilitate this quality assessment, we first extract the fringes using the content-adaptive purple fringe detection algorithm in Karthik and Malik [5] and Malik and Karthik[8]. Once these fringe regions have been identified and the fringes extracted, we devise an algorithm to quantify the extent of color diversity in these fringes. This score is expected to mimic the subjective score for a variety of images.

3 Proposed Chromatic Eigenvalue Ratio-Based Quality Measure

Let I_R, I_G, and I_B be the red, green, and blue channels in a specific color image I_{TEST} undergoing a test and analysis. Let $(x, y) \in \{1, 2, \ldots, N_1\} \times \{1, 2, \ldots, N_2\}$ be a specific spatial location within the image frame. The corresponding normalized red, green, and blue intensity levels at location (x, y) are $I_R(x, y)$, $I_G(x, y)$, and $I_B(x, y)$, respectively. From these intensity levels, the luminance and chrominance components (Y-Cb-Cr) are computed as [13]

$$\begin{pmatrix} Y(x, y) \\ C_B(x, y) \\ C_R(x, y) \end{pmatrix} = \begin{pmatrix} 0.3 & 0.6 & 0.1 \\ -0.2 & -0.3 & 0.5 \\ 0.5 & -0.4 & -0.1 \end{pmatrix} \begin{pmatrix} I_R(x, y) \\ I_G(x, y) \\ I_B(x, y) \end{pmatrix} \tag{1}$$

A $M \times M$, Gaussian derivative kernel, $\mathbf{D}$, is created with a standard deviation σ_{KER} with value of σ_{KER} chosen in such a way that only thick edges or edges which culminate in plateaus are picked up. Too small a σ_{KER} will pick up the finer edges which are of little use since the fringes are expected to positioned near the thicker edges. Too large a σ_{KER} might eventually lead to excessive smoothing and the loss of detail, which may result in a few fringe regions not being picked up. Hence, a moderate σ_{KER} is advised. In our calibration process as per the work related to content adaptive fringe detection, we have chosen $\sigma_{KER} = 3$ [8]. If $\mathbf{D}_x$ is the X-gradient matrix and $\mathbf{D}_y$ the Y-gradient matrix inherited from the derivative of the Gaussian parameterized by σ_{KER}, the X- and Y-derivatives of the zero padded version of the luminance image $\mathbf{Y}$ are

$$\mathbf{Y}_x = \mathbf{Y} \star \mathbf{D}_x$$
$$\mathbf{Y}_y = \mathbf{Y} \star \mathbf{D}_y \tag{2}$$

where "$\star$" is the convolution operator. The gradient magnitude is computed as

$$\mathbf{M}_{xy}(x, y) = \sqrt{\mathbf{Y}_x(x, y)^2 + \mathbf{Y}_y(x, y)^2} \tag{3}$$

The region of significant interest, where the fringes are expected to lie is then obtained by thresholding this gradient magnitude:

$$\mathbf{R}_{SIG} = \left\{ (x, y), s.t. \frac{\mathbf{M}_{xy}(x, y)}{\mu_G} \geq T_0 \right\} \tag{4}$$

where T_0 is the binarization threshold set to "0.25" to enhance the interest zone (generally, this is set as "1" [8]). The normalizing factors is the mean over all the gradient magnitudes μ_G. The fringe pixels contained within this significant region $\mathbf{R}_{SIG}$ are then picked up based on a simple ratio test executed in the chrominance space as

$$\mathbf{R}_{PFA} = \left\{ (x, y) \in \mathbf{R}_{SIG}, s.t. \frac{\sqrt{(C_B(x, y) - 0.3)^2 + (C_R(x, y) - 0.4)^2}}{\sqrt{(C_B(x, y) + 0.3)^2 + (C_R(x, y) + 0.4)^2}} \leq T_1 \right\} \tag{5}$$

where T_1 is another relativistic threshold set to 0.95 [8]. Once the fringe region is extracted, the chrominance vectors corresponding to the estimated fringe pixels are accumulated as follows:

$$\mathbf{P}_{x,y} = \begin{bmatrix} C_B(x, y) \\ C_R(x, y) \end{bmatrix} \text{ with } (x, y) \in \mathbf{R}_{PFA} \tag{6}$$

The eigenspace spanned by the chrominance vectors $\mathbf{P}_{x,y}$ within the PFA region along is characterized by determining the mean vector and the chrominance covariance matrix:

$$\bar{\mu}_P = \frac{1}{Card(\mathbf{R}_{PFA})} \sum_{(x,y)\in\mathbf{R}_{PFA}} \mathbf{P}_{x,y} \tag{7}$$

$$\bar{\Sigma}_P = \frac{1}{Card(\mathbf{R}_{PFA})} \sum_{(x,y)\in\mathbf{R}_{PFA}} \left[(\mathbf{P}_{x,y} - \bar{\mu}_P)(\mathbf{P}_{x,y} - \bar{\mu}_P)^T \right] \tag{8}$$

where $Card(\mathbf{R}_{PFA})$ is the cardinality of the set $\mathbf{R}_{PFA}$ (or the number of points included the set). The eigenvectors and eigenvalues representing the chrominance space contained within the PFA region can be determined by decomposing the covariance matrix as

$$\bar{\Sigma}_P = \mathbf{V}\mathbf{D}_v\mathbf{V}^T \tag{9}$$

where, the matrix $\mathbf{V}$ comprises of the two eigenvectors spanning the CB-CR space over the PFA region and $\mathbf{D}_v$ represents the eigenvalue matrix corresponding to the two eigen-directions specified in the eigenvector matrix $\mathbf{V}$. Since the covariance matrix is symmetric and real valued, the eigenvalues λ_1, λ_2 are strictly positive and indicate the variances along those directions. If we define,

$$\rho_{EV} = \frac{\max(\lambda_1, \lambda_2)}{\min(\lambda_1, \lambda_2)} \tag{10}$$

this ratio gives an indication regarding the spread of the fringe cluster within the chrominance space. The closer the ratio is to UNITY, the more diverse the cluster in terms of color. The more diverse the cluster, the more fuzzy the edges around the fringes are expected to appear, leading to a lower quality image. Hence, this diversity parameter translates to quality measure. We thus define the quality metric as a natural logarithm of this EIGENRATIO,

$$Q(I_{TEST}) = ln(\rho_{EV}) \tag{11}$$

Larger the skew in the chrominance space, greater is the purity of the color in the fringe region and subsequently, the image clarity is expected to be higher.

4 Results and Discussions

Three PFA images have been compared in Fig. 3 and analyzed both subjectively and quantitatively as follows:

– $IM_{TEST}(1)$, Fig. 3a, Leaves with the sky in the backdrop, wherein the purple fringes are produced because of the backlight. Because of the random orientations of the leaves and their sharp curvatures, the colorations are nonuniform and arbitrary. There is a varied mixing of GREEN and PURPLE across the entire image (which was discussed earlier in Fig. 2). This is the worst one among the three images in terms of clarity as there is a fuzzy color induced blurring and loss of detail, especially, around the edge regions of the leaves. This is captured in the lowest quality score among the three, viz., $Q(IM_{TEST}(1)) = 0.9763$.
– $IM_{TEST}(2)$, Fig. 3b, large-sized leaves with backlight. The difference from the previous case is that the orientation of the leaves with respect to the backlight is almost fixed. The arrangement is largely planar. Hence, the fringe patterns are predictable and the cluster in the CB-CR space is much more skewed Fig. 3k as compared to Fig. 3j, leading to a mid-level score of $Q(IM_{TEST}(2)) = 1.8898$.
– In $IM_{TEST}(3)$, Fig. 3c, although the PFA sample extracted is not quite representative of the entire scene, the purity of the purple fringe is far greater when compared with the other two images. This is clearly visible in the skew associated with the CB-CR cluster shown in Fig. 3l, leading to a high-quality score of $Q(IM_{TEST}(3)) = 2.1116$. Perceptually, this image has the highest clarity among the three because of the amount of detail. The sharpness measure, if evaluated using some other algorithm such as BRISQUE [10] is expected to be the highest for this image. Hence, subjectively the third image is expected to score higher than the other two purely based on clarity and detail.

Since, it is extremely difficult to do a subjective comparison and blind quality assessment algorithms such as BRISQUE [10] and NIQE [11] are tailored toward assessment of gray scale images based on natural scene deviation statistics, it would be inappropriate to compare the proposed algorithm which focusses on color distortions with ones which are geared to detect sharpness, clarity, and structure in the *luminance domain alone*. To circumvent this problem, we compare PFA images in a pairwise fashion using the proposed algorithm and pad this with a subjective assessment. We compare images of a similar type (leaves images with leaves images; buildings with buildings; lampposts and other objects with similar objects, etc.). A subjective ranking was generated purely based on visual inspection for triplets of images of a certain type: (a) TYPE-1: Leaves with sky in the backdrop; (b) TYPE-2: Buildings; (c) TYPE-3: Objects with a planar background; The subjective ranking is enlisted in Table 1.

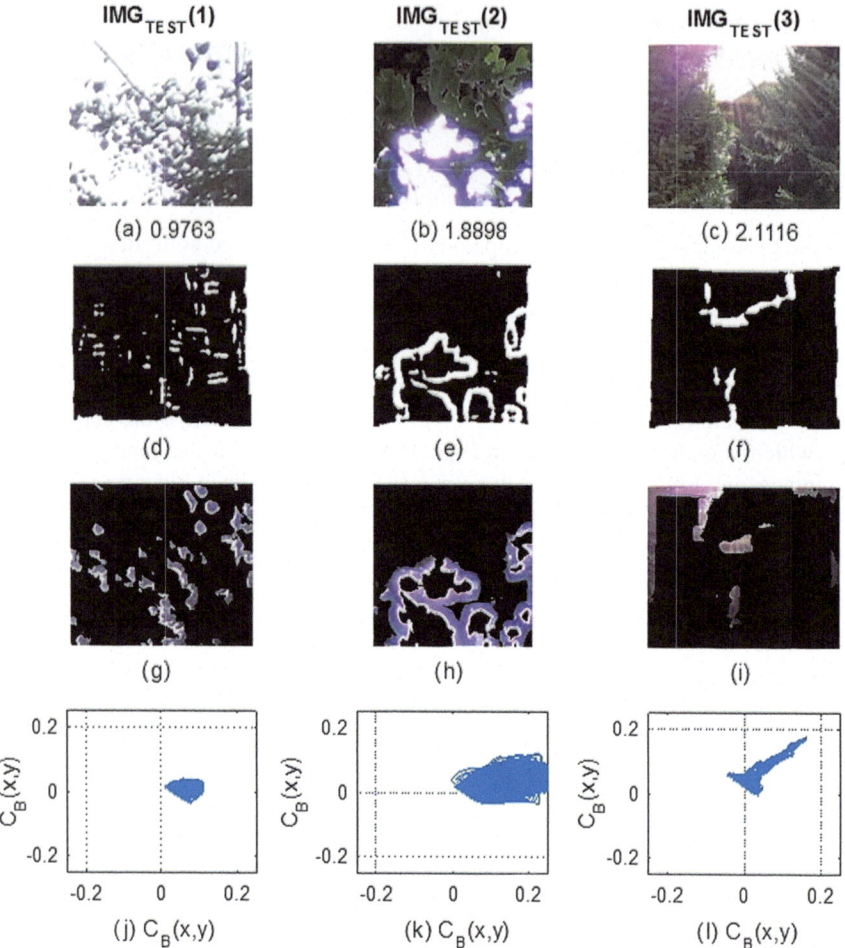

Fig. 3 **a–c** Test images being compared; **d–f** Thresholded gradient profiles for the corresponding test images ($\sigma_{KER} = 3$ and $T_0 = 0.25$); **g–i** Detected PFA patterns (Ratio threshold $T_1 = 0.95$); **j–l** Clusters of points in the chromatic space corresponding to the PFA region; Note the images have scored using the proposed algorithm in the order $Q(IM_{TEST}(1)) < Q(IM_{TEST}(2)) < Q(IM_{TEST}(3))$, which is also true subjectively when checked in terms of clarity

Only one individual was involved in the ranking process of these triplets to ensure that there was a consistency in the choice of subjective parameters on the basis of which these images were ranked on a relativistic scale. If more than one individual were included, there would have been a divergence in the decision-making process and an inconsistency in the choice of subjective parameter sets.

The actual scores obtained from the proposed algorithm have been demonstrated through the sets of figures, Fig. 5 involving images of buildings, Fig. 4 involving leaf images and Fig. 6 involving objects with a planar background.

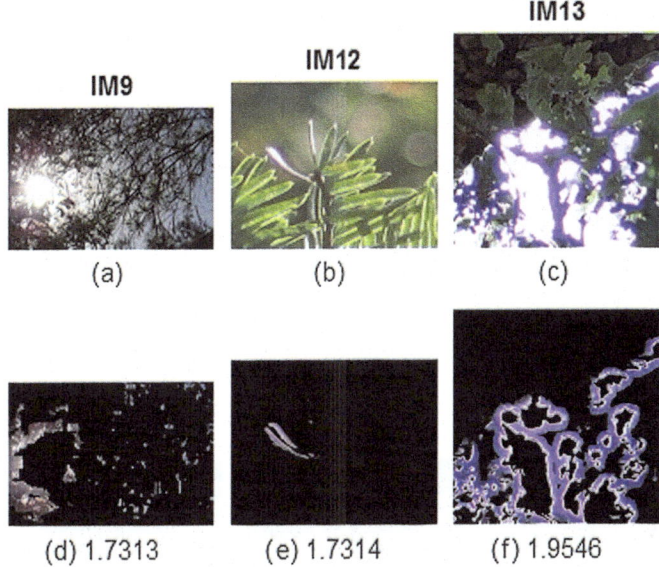

Fig. 4 As per these algorithmic scores the relative ranks within the group are 3, 2, and 1 respectively. Corroboration with corresponding subjective assessment is 33% (Table 1)

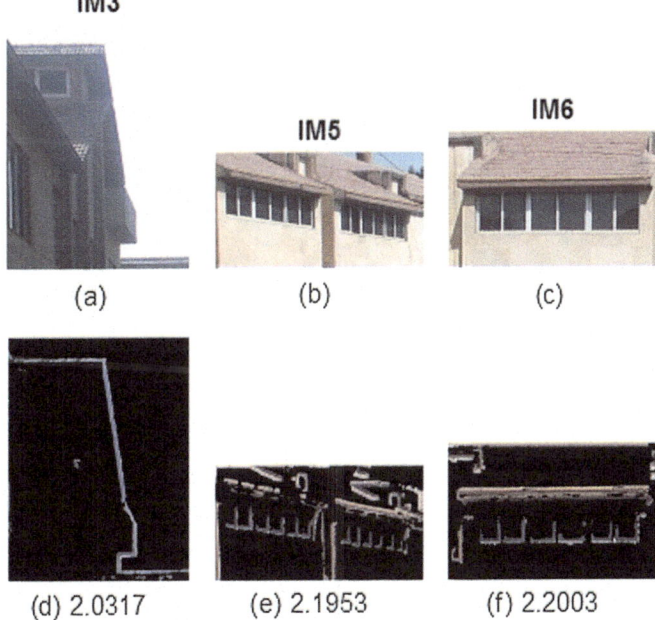

Fig. 5 As per these algorithmic scores the relative ranks within the group are 3, 2, and 1 respectively. Corroboration with corresponding subjective assessment is 100% (Table 1)

IM16

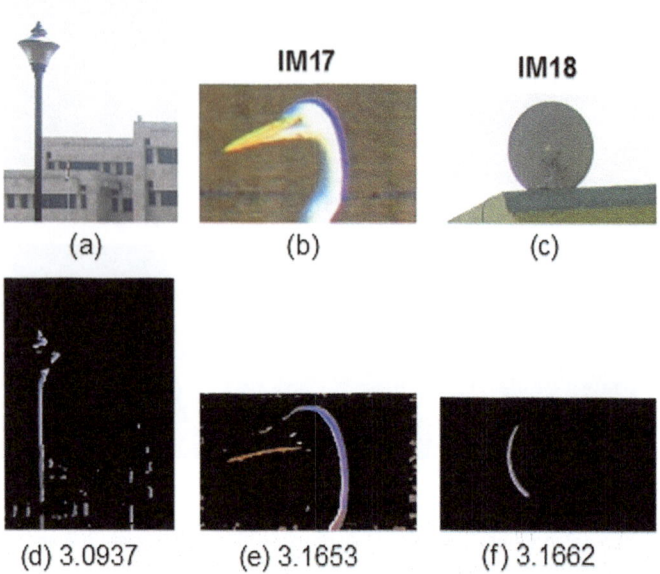

(a)   (b)   (c)

(d) 3.0937   (e) 3.1653   (f) 3.1662

Fig. 6 As per these algorithmic scores the relative ranks within the group are 3, 2, and 1 respectively. Corroboration with corresponding subjective assessment is 0% (Table 1), possibly because the fringes are NOT distributed throughout the image. Thus they do not form a representative element for overall scene

These scores have been transformed into algorithm-specific rankings (on a relativistic platter) and finally corroborated with the corresponding subjective ranking, using the following association formula:

$$CS(Imagetype) = Corroboration\,[SR(ImageType), PR(ImageType)] \tag{12}$$

If $\{IM1, IM2, \ldots, IM(t)\}$ are t images from a certain type having subjective ranks $\{SR(1), SR(2), \ldots, SR(t)\}$ and rankings using the proposed algorithm as $\{PR(1), \ldots, PR(t)\}$, with $SR(i) \in 1, 2, 3, \ldots, t$ and $PR(i) \in 1, 2, 3, \ldots, t$, the similarity between the two scores can be quantified as

$$SIMI(\bar{SR}, \bar{PR}) = \begin{bmatrix} \tilde{\delta}(SR(1) - PR(1)) \\ \tilde{\delta}(SR(2) - PR(2)) \\ \ldots \\ \tilde{\delta}(SR(t) - PR(t)) \end{bmatrix} \tag{13}$$

where $\tilde{\delta}(m) = 1$ if $m = 0$ and $\tilde{\delta}(m) = 0$ if $m \neq 0$, viz., it is the discrete delta function. The final corroboration score linking the two sets of rankings is obtained

Table 1 Subjective and Algorithmic rankings for three different image groups (TYPE-1: Leaves; TYPE-2: Buildings, and TYPE-3: Objects with a planar background). Corresponding CORROBO-RATION SCORES are also given

TYPE	Image number	Subjective rank	Proposed rank	Corroboration score (%)
TYPE-1	IM9	3	3	**33**
	IM12	1	2	
	IM13	2	1	
TYPE-2	IM3	3	3	**100**
	IM5	2	2	
	IM6	1	1	
TYPE-3	IM16	1	3	**0**
	IM17	3	2	
	IM18	2	1	

by simply computing the normalized Hamming weight of the *similarity* vector, $SIMI(\bar{SR}, \bar{PR})$. Note that in the case of a point to point match the final score will be 100% while in the case of a perfect shuffle (complete mismatch), the score will be zero. This comparison is shown in Table 1. Note that in the case of TYPES-1 and 2, in almost all images, the fringes are positioned in different parts of the frame and so form a representative sample of the overall content (leading to subjective corroboration scores of 33% and 100% respectively). In the case of TYPE-3 involving objects in a plain background, the effect of the fringe is largely localized and hence does not represent the whole frame (here the score is 0%). The rankings for TYPE-3 are therefore arbitrary. In summary, if the fringe regions are distributed throughout the natural scene, they can be used very effectively to gauge the extent of color distortion and subsequently image quality.

5 Conclusions

In this paper, we have exploited the diversity profile of the purple fringes to perform a blind quality assessment of PFA-affected images. We have observed that results corroborate well with subjective assessment, provided the fringes are distributed uniformly throughout the image.

Acknowledgements We thank Google Inc., particularly its image search section, for providing us with the links to several web-based discussions and forums involving PFA and its extreme effects, from which we obtained several exemplar PFA-corrupted images generated by several midrange cell phones for further quality analysis.

References

1. Jang, D.W., Park, R.H.: Color fringe correction by the color difference prediction using the logistic function. IEEE Trans. Image Process. **26**(5), 2561–2570 (2017)
2. Ju, H.J., Park, R.H.: Colour fringe detection and correction in YCbCr colour space. IET Image Process. **7**(4), 300–309 (2013). https://doi.org/10.1049/iet-ipr.2012.0524
3. Jung, C.D., Jang, D.W., Kim, H.S., Park, R.H.: Color fringe correction using guided image filtering. In: The 18th IEEE International Symposium on Consumer Electronics (ISCE 2014), pp. 1–2. IEEE (2014)
4. Kang, S.: Automatic removal of purple fringing from images. US Patent App. 11/322,736 (2007)
5. Karthik, K., Malik, P.: Purple fringing aberration detection based on content adaptable thresholds. In: International Conference on Smart Systems, Innovations and Computing (SSIC 2017) (2017)
6. Kim, B., Park, R.: Automatic detection and correction of purple fringing using the gradient information and desaturation. In: Proceedings of the 16th European Signal Processing Conference, vol. 4, pp. 1–5 (2008)
7. Kim, B.K., Park, R.H.: Detection and correction of purple fringing using color desaturation in the xy chromaticity diagram and the gradient information. Image Vis. Comput. **28**(6), 952–964 (2010). https://doi.org/10.1016/j.imavis.2009.11.009
8. Malik, P., Karthik, K.: Iterative content adaptable purple fringe detection. Signal Image Video Process. (2017). https://doi.org/10.1007/s11760-017-1144-1
9. Malik, P., Karthik, K.: Limitation of PFA-events as a forensic tool. In: 4th International Conference on Signal Processing and Integrated Networks (SPIN), pp. 3905–3908 (2017). https://doi.org/10.1109/SPIN.2017.8049940
10. Mittal, A., Moorthy, A.K., Bovik, A.C.: No-reference image quality assessment in the spatial domain. IEEE Trans. Image Process. **21**(12), 4695–4708 (2012). https://doi.org/10.1109/TIP.2012.2214050
11. Mittal, A., Soundararajan, R., Bovik, A.C.: Making a completely blind image quality analyzer. IEEE Signal Process. Lett. **20**(3), 209–212 (2013). https://doi.org/10.1109/LSP.2012.2227726
12. Tomaselli, V., Guarnera, M., Bruna, A.R., Curti, S.: Automatic detection and correction of purple fringing artifacts through a window based approach. In: 2011 IEEE International Conference on Consumer Electronics-Berlin (ICCE-Berlin), pp. 186–188. IEEE (2011)
13. WIKIPEDIA (Aug 2019) YCbCr. https://en.wikipedia.org/wiki/YCbCr
14. Yerushalmy, I., Hel-Or, H.: Digital Image Forgery Detection Based on Lens and Sensor Aberration, vol. 92, pp. 71–91. Springer (2011)

Investigation on the Muzzle of a Pig as a Biometric for Breed Identification

Shoubhik Chakraborty, Kannan Karthik and Santanu Banik

Abstract Visual features associated with cattle and pigs cannot just be used for individual biometric identification, they can also be used to organize the domain into distinct breeds. In this paper, we use cropped muzzle images of pigs for breed identification. Gradient patch density maps are first created, and then the patch density profile distribution tailored to a specific breed is learnt to characterize the feature space for each of the four pig breeds: Duroc, Ghungroo, Hampshire, and Yorkshire. A Maximal Likelihood (ML) inferencing at the patch level followed by a second-tier decision fusion based on majority vote has been used to detect the breed from any query feature computed from an arbitrary muzzle image. Duroc, Ghungroo, and Yorkshire show good classification accuracies of 75.86%, 70.59%, and 100%, respectively, while the accuracy drops for Hampshire to 58.78% on account of the intrinsic white patch diversity in the breed and its similarity to Duroc and Ghungroo on some counts.

Keywords Gradient significance map · Pig · Breed identification · Biometric · Maximal likelihood · Conditional density function · Decision fusion · Majority vote

S. Chakraborty (✉) · K. Karthik
Department of Electronics and Electrical Engineering, Indian Institute of Technology Guwahati, Guwahati 781039, Assam, India
e-mail: shoubhik@iitg.ac.in

K. Karthik
e-mail: k.karthik@iitg.ac.in

S. Banik
Animal Genetics and Breeding ICAR National Research Centre on Pig, Guwahati 781015, Assam, India
e-mail: sbanik2000@gmail.com

© Springer Nature Singapore Pte Ltd. 2020
B. B. Chaudhuri et al. (eds.), *Proceedings of 3rd International Conference on Computer Vision and Image Processing*, Advances in Intelligent Systems and Computing 1022, https://doi.org/10.1007/978-981-32-9088-4_7

1   Introduction

Analysis of visual features of domestic and farm animals from the point of view of both biometric as well as breed identification is a relatively new field. In this work, we focus on the analysis of visual images associated with pigs not from a biometric standpoint, but from the point of view of breed identification. Every pig breed has distinct characteristics concerning its morphology, size, immunity to common epidemics as a function of the environment in which they are bred and finally in terms of their appeal to the social palate as food. Hence, breed selectivity, identification, and tracking, plays a role in ensuring commercial viability and stability across the supply chain system.

A majority of the available literature has focussed on biometric analysis and more specifically driven toward cattle (mainly cows). Initially, the main challenge was in the identification and selection of an appropriate portion of the animal in the imaging domain, which should contain a biometric identifier. Once this region of interest in the visual domain and the mode of acquisition is established, further processing can be done to extract interest points for matching these derived features from different animals.

For cattle biometrics, Localized Binary Patterns (LBP) descriptors from facial images were used by Cai and Li [1]. Semiautomatic or partially manual alignment and cropping were required to ensure that the images were perfectly registered before being subjected to LBP analysis. Problems in spatial registration of the patch LBP descriptors affected the classification results.

An attempt was made to use the MUZZLE palette of cows for biometric analysis by Awad et al. [2]. SIFT [3], points were extracted from the muzzle and were matched using the RANSAC algorithm [4]. SIFT as a feature is effective as long as the neighborhood profile is characterized distinctively around each interest point of significance. The matching between two interest points from two different muzzle images is done by comparing their abstracted neighborhood profiles. This registration process is extremely difficult because of the diversity with which pores and ciliary patterns within the same muzzle are expected to be captured owing to pose/camera variations, illumination changes and self-shadowing effects. The problems exist even for SURF features as these are also interest point based [5].

In comparison with Cai and Li [1], LBP codes were generated for every pixel and eventually concatenated to form a histogram over the entire muzzle. This histogram was selected as the base feature. The intra-covariance and cross/interclass feature covariance matrices were then determined and then the optimal linear transform which maximized separability across classes was computed. Results from this approach were promising as the training set was augmented by rotational variations of the muzzle images from the same subject.

While all these papers provide useful cues toward the selection of features for biometric identification, breed classification in pigs becomes a different ball game for the following reasons:

– While it is possible that visual descriptors or features selected for Biometric recognition offer good classification rates, it remains to be seen whether these features are stable over time. Only when these descriptors remain largely static for several months and years together do they qualify as viable biometric identifiers.

– It is hard to tell whether the biometric identifiers from the earlier papers happen to be a current (one shot) representation of the face or the muzzle in farm animals as their progression over time have not been examined.

– However, in the case of breed identification, there is potential to identify common traits across animal subjects within the breed which qualify as ancestral and genetically driven traits. Therefore, in some sense, breed identification should precede individual identification as this temporal element associated with the stability of the feature is now displaced by an ensemble statistic or characteristic.

Genetic confirmation of any visual descriptor (or statistic) is possible from a breed analysis and when this is augmented by an implicit assumption regarding ERGODICITY, the same visual descriptor can be used for biometric identification. Simply put:

Breed analysis is a prerequisite for Biometric analysis.

The rest of the paper is organized as follows: In Sect. 2, we provide the motivation for selecting the muzzle particularly from the point of view of breed analysis. The mechanism for first amplifying the key features in the muzzle, detecting them (or their positions) and computing the patch descriptors, is given in Sect. 3. Distinctiveness of the density maps across breeds are discussed in Sect. 4. Finally, the inferencing procedure and classification test results are presented in Sect. 5.

2   Problem Statement: MUZZLE for Breed Analysis of Pigs

The front portion of the nose of a pig is called the muzzle. The muzzle of a pig is a relatively smooth area and consists of two nostrils. These nostrils, coupled with the nasal linings have the following functionalities:

– Breathing: Inhalation and Exhalation.
– Regulation of body temperature.
– Sampling and discriminating between different types of odors.

Hair follicles and pores are distributed all over the muzzle surface. The density of these follicles tends to vary from breed to breed. For instance, the density of follicles found on the muzzle of the Ghungroo (which is a West Bengal breed [6]) is the highest, while the density found on the muzzle of Yorkshire (which is an exotic breed originating in Europe [7]) is the least. As far as Ghungroo is concerned, this increased hair density can be attributed to the relatively hot and humid areas in which they are reared. As the humidity increases, the rate of evaporation from the skin tends to drop. To counterbalance this and to ensure balanced release of internal heat, animals, and humans from high humidity zones tend to have more hair follicles on the exposed and active areas of their skin. However, in the case of breeds like Yorkshire, which are reared in colder areas, to prevent excessive body heat loss, the density of hair follicles are found to be considerably lower, as compared to those breeds which have originated closer to the equator: Ghungroo (West Bengal) and Duroc (originally African [8]).

Furthermore, the relatively narrow nostrils of Hampshire [9] and Yorkshire as compared to the Ghungroo and Duroc can be justified on account of the climatic conditions in which they have been reared. The Hampshire and Yorkshire breed of pigs have come from the colder regions of Europe and North America. It is common knowledge that the nose plays a critical role in preparing air for our lungs. Ideally, air should be warm and moist before it enters the lungs. This is because the microscopic hairs in the nasal passage, called cilia, that help to keep pathogens and dust from entering the lungs work better with warm moist air, as opposed to dry, cold air. For pigs from hot, wet climates, there is less of a need to prepare the air for the lungs, so they retained wider nostrils. Having longer, narrower nostrils increases contact between the air and the mucosal tissue in the nose. This friction plays a role in warming up and moistening the air before it enters the lungs [10]. Hence, the overall shape of the muzzle, relative sizes of the nostrils and the density of pores and cilia tend to be a function of the environment in which these pigs have been reared.

In the following sections, we propose our algorithm for amplifying and extracting relevant features from the muzzle image which contain information about the position and distribution of nostrils, hair follicles, and pores on the muzzle surface. These features are then refined to learn breed-specific characteristics which can be further deployed toward breed identification.

3   Gradient Profiling and Patch Statistics

As discussed in the previous section, this paper attempts breed identification of pigs based on the density profile of hair follicles (or cilia) and pores distributed over the surface of the muzzle and its periphery. As a preprocessing step, these pores and cilia are highlighted so that we can extract relevant features from them. The process starts with smoothing the image first to remove any noise present and then taking horizontal and vertical derivatives to highlight the pores and cilia. A suitable function for these two operations can be a derivative of a Gaussian which is shown below

$$\frac{\mathrm{d}}{\mathrm{d}x}\left[\frac{1}{\sqrt{2\pi\sigma^2}}\exp^{\frac{-x^2}{2\sigma^2}}\right] = -\frac{x}{\sqrt{2\pi}\sigma^{3/2}}\exp^{-\frac{x^2}{2\sigma^2}}$$

$$= -c_0 x \exp^{\frac{-x^2}{2\sigma^2}} \tag{1}$$

Two discrete kernels K_x and K_y (corresponding to horizontal and vertical gradients) obtained by sampling the above function are convolved with the image to obtain the horizontal and vertical gradient maps G_x and G_y respectively. The gradient magnitude profile is then computed from G_x and G_y according to

$$G_{mag}(i, j) = \sqrt{G_x^2(i, j) + G_y^2(i, j)} \qquad (2)$$

The gradient magnitude profile thus computed is normalized with respect to its mean and then thresholded to obtain a binary image, B.

$$\mu_G = \frac{1}{N_1 N_2} \sum_{i=1}^{N_1} \sum_{j=1}^{N_2} G_{mag}(i, j) \qquad (3)$$

$$B(i, j) = \begin{cases} 1 \text{ IF } \frac{G_{mag}(i,j)}{\mu_G} > \delta_G \\ 0 \text{ IF } \frac{G_{mag}(i,j)}{\mu_G} < \delta_G \end{cases} \qquad (4)$$

where δ_G is a relativistic threshold generally set to "1" as a compromise between picking up noise versus amplifying relevant information pertaining to pores and cilia on the muzzle. The Gaussian standard deviation parameter σ for smoothing, on the other hand, is picked as "1" (7×7 window) to ensure that sufficient detail is captured while taking a derivative of the image. This binary matrix is what we call the Gradient Significant Map (GSM). Figure 1 shows the muzzle images of several pigs: The first two images belong to Duroc (a, b), the second two belong to Ghungroo (c, d), the next two belong to Hampshire (e, f), and the last two to Yorkshire (g, h). If the threshold δ_G is too small (say 0.25), there will be an uncontrolled classification of pixels as significant ones and all binary images irrespective of their breed will appear the same Fig. 3. One the other hand if δ_G is too large (say 1.5), very little detail will be captured and once again all the density images will begin to appear the same, irrespective of their breed Fig. 4. As a compromise to facilitate sufficient base feature separation and a representation of the actual concentration of pores and cilia on the muzzle, this threshold is set to an in-between value "1" (Fig. 2).

The GSMs are then divided into equal size patches and a suitable statistic is calculated for each patch as shown in Fig. 5. The statistic calculated for each patch is the percentage of significant pixels in that patch. Thus, if the patch size is $m \times n$ and n_s is the no. of significant pixels in that patch as obtained from the GSM, then the corresponding patch statistic is obtained as

$$S(patch) = \frac{n_s}{m \times n} \times 100\% \qquad (5)$$

This patch statistic computed from each of the patches in Fig. 5, is eventually concatenated to form a feature vector for that muzzle image.

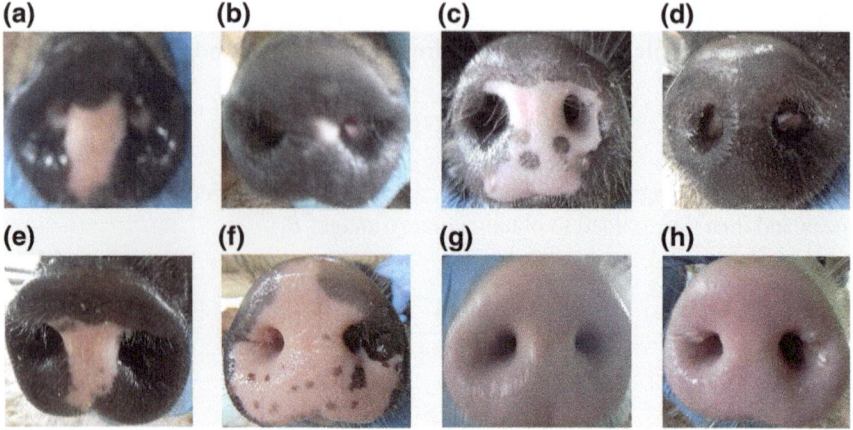

Fig. 1 Muzzle images of several pigs corresponding to different breeds: **a**, **b** Duroc, **c**, **d** Ghungroo, **e**, **f** Hampshire, **g**, **h** Yorkshire

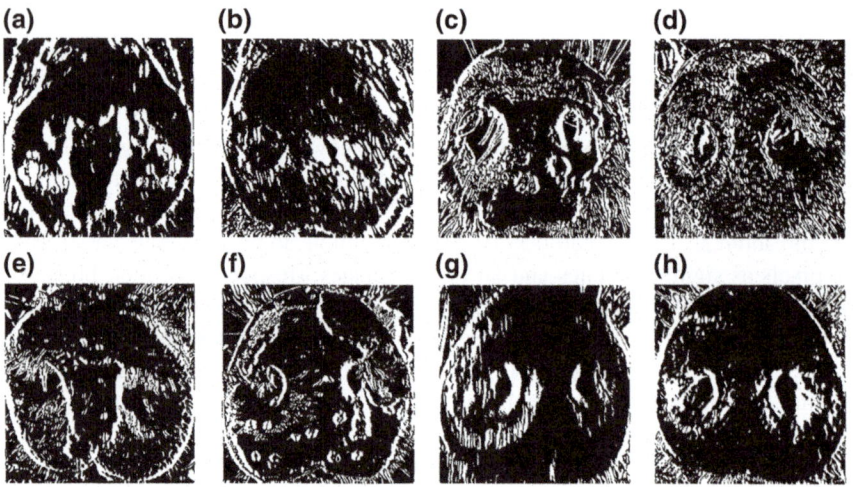

Fig. 2 Gradient Significance Map (GSM) for optimal choice of threshold $\delta_G = 1$ (Gradient smoothing parameter, $\sigma = 1$)

4 Diversity in Patch Statistics Across Breeds

At first, we formulate a patch diversity conjecture across breeds as follows:

Patch diversity conjecture: *The patch density profile is expected to be a function of the environment in which these animals are reared and is therefore expected to vary from breed to breed. More importantly, the density maps derived from different spatial locations are expected to be different. It is precisely this diversity in these*

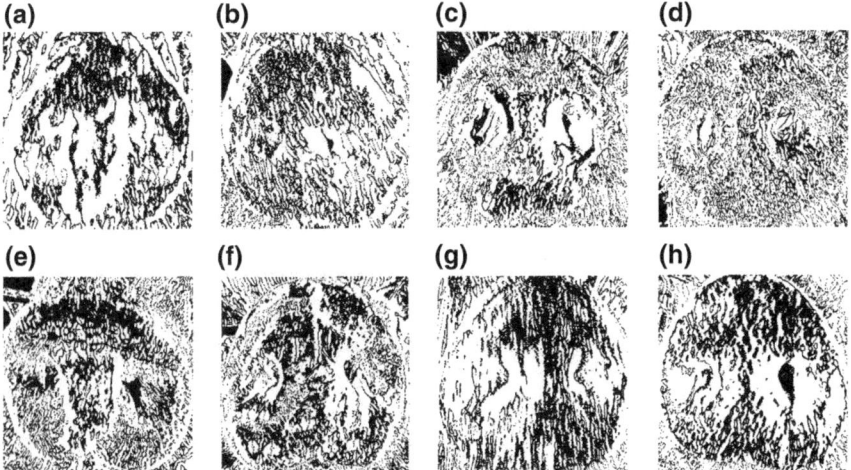

Fig. 3 Gradient Significance Map (GSM) for low threshold $\delta_G = 0.25$ (Gradient smoothing parameter, $\sigma = 1$). Almost every point is treated as a significant point and hence all binary images turn white to appear identical

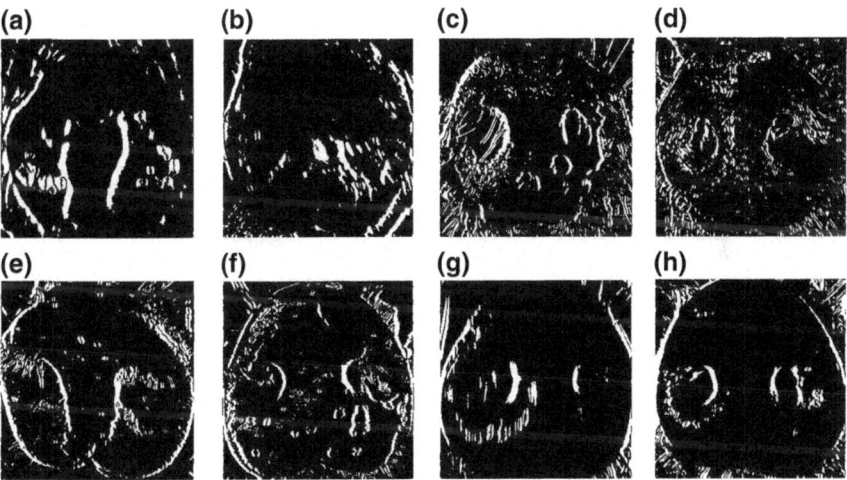

Fig. 4 Gradient Significance Map (GSM) for optimal choice of threshold $\delta_G = 2$ (Gradient smoothing parameter, $\sigma = 1$). Very few significant points are picked up and the patch densities corresponding to the actual concentration of pores and cilia are not captured accurately

patch distributions that we wish to use for our final inferencing and decision-making procedure related to breed identification.

The database contained a total of 311 muzzle prints corresponding to 30 animals across four breeds (Duroc, Ghungroo, Hampshire, and Yorkshire). There were around

Fig. 5 Division of the GSM
into patches and the
corresponding patch statistic

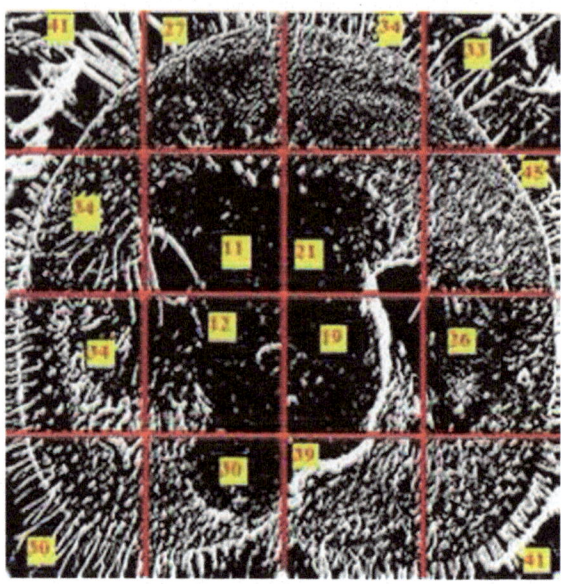

8–15 muzzle variations from each animal. Duroc and Ghungroo had six animals in
their set while Hampshire and Yorkshire had 9 animals each. The muzzle images
were taken with a high- resolution handheld camera with another person holding the
nose of the pig tightly to avoid excessive blurring of the image due to relative motion
between the camera and muzzle surface.

First, the RGB colored image is converted to gray scale, resized to 1000×1000
and the GSM is constructed. Regarding the patch size N_P, if the patch size is too
small, it will make the patch statistic too sensitive to camera panning and pig head
movement (pose variation) and illumination changes, during image acquisition. On
the other hand, large patch sizes are not desirable because spatial details are lost. As
a compromise, purpose we choose a patch size $N_P = 250$. With this patch size, and
with images resized to 1000×1000 as mentioned earlier, $4 \times 4 = 16$ patches are
generated, out of which the top-left and the top-right were left out of the analysis
owing to background interference.

This training phase was setup as follows:

– Using the proposed feature extraction algorithm the patch density statistics were
  computed for every muzzle image within the training set, which comprised of
  159 muzzle prints (out of a total of 311 muzzle images) coming from 30 different
  animals across four breeds. The patch statistics were computed for a patch size of
  25% (viz. the muzzle prints were split into 4×4 grids).
– Thus each print was mapped to a 4×4 matrix:

$$\mathbf{P}(Image - k) = \begin{pmatrix} S_{11}(k) & S_{12}(k) & S_{13}(k) & S_{14}(k) \\ S_{21}(k) & S_{22}(k) & S_{23}(k) & S_{24}(k) \\ S_{31}(k) & S_{32}(k) & S_{33}(k) & S_{34}(k) \\ S_{41}(k) & S_{42}(k) & S_{43}(k) & S_{44}(k) \end{pmatrix} \tag{6}$$

with $S_{ij}(k) \in [0, 1]$ indicating the fraction of significant points in patch located at position (i, j) and $i, j \in \{1, 2, 3, 4\}$ corresponding to Pig-k.

– From the $N_T = 159$, training muzzle prints across four breeds, the 4×4 patch matrices for each breed were concatenated.

$$\mathbf{DUROC}_T = \{\mathbf{D}_1, \mathbf{D}_2, \ldots, \mathbf{D}_{N_D}\}$$
$$\mathbf{GHUNG}_T = \{\mathbf{G}_1, \mathbf{G}_2, \ldots, \mathbf{G}_{N_G}\}$$
$$\mathbf{HAMP}_T = \{\mathbf{H}_1, \mathbf{H}_2, \ldots, \mathbf{H}_{N_H}\}$$
$$\mathbf{YORK}_T = \{\mathbf{Y}_1, \mathbf{Y}_2, \ldots, \mathbf{Y}_{N_Y}\} \tag{7}$$

with, $N_D + N_G + N_H + N_Y = N_T = 159$.

– If $\mathbf{BR}_k$ corresponds to a patch matrix from PIG-k in breed type $\mathbf{BR}$, this can be written down as

$$\mathbf{BR}_k = \begin{pmatrix} S_{11}(BR_k) & S_{12}(BR_k) & S_{13}(BR_k) & S_{14}(BR_k) \\ S_{21}(BR_k) & S_{22}(BR_k) & S_{23}(BR_k) & S_{24}(BR_k) \\ S_{31}(BR_k) & S_{32}(BR_k) & S_{33}(BR_k) & S_{34}(BR_k) \\ S_{41}(BR_k) & S_{42}(BR_k) & S_{43}(BR_k) & S_{44}(BR_k) \end{pmatrix} \tag{8}$$

where, $S_{ij}(BR_k) \in [0, 1]$. All the patch statistics from a breed corresponding to a spatial index (i, j) were concatenated to create a location specific conditional histogram. For instance, the conditional histograms for patch location $(i, j) \in \{1, 2, 3, 4\}$ for the four breeds can be created by first sorting the values from the patch statistics (from a specific breed), in ascending order and then binning the count of the values falling within a fixed range. If M is the number of histogram bins, this process is represented as follows:

$$\hat{\mathbf{f}}_{S_{ij}/DUROC}(x) = BIN_M \left[Sort \left(\{S_{ij}(D_1), S_{ij}(D_2), \ldots, S_{ij}(D_{N_D})\} \right) \right]$$
$$\hat{\mathbf{f}}_{S_{ij}/GHUNG}(x) = BIN_M \left[Sort \left(\{S_{ij}(G_1), S_{ij}(G_2), \ldots, S_{ij}(G_{N_G})\} \right) \right]$$
$$\hat{\mathbf{f}}_{S_{ij}/HAMP}(x) = BIN_M \left[Sort \left(\{S_{ij}(H_1), S_{ij}(H_2), \ldots, S_{ij}(H_{N_H})\} \right) \right]$$
$$\hat{\mathbf{f}}_{S_{ij}/YORK}(x) = BIN_M \left[Sort \left(\{S_{ij}(Y_1), S_{ij}(Y_2), \ldots, S_{ij}(Y_{N_Y})\} \right) \right]$$

where $Sort(.)$ sorts the array of scalars in the ASCENDING order and $BIN_M(.)$ generates the fractional count of values in M equi-spaced bins over the range $[0, 1]$.

– To ensure there is some form of a polynomial fit for the histograms, the Gaussian density function, which has two degrees of freedom, has been chosen as a reference:

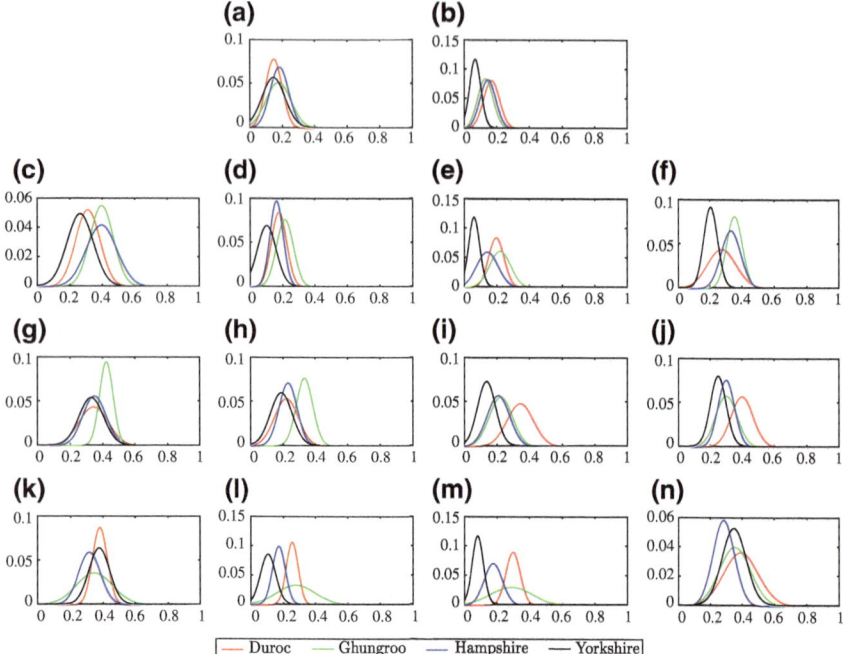

Fig. 6 Patchwise conditional densities for each of the four breeds

$$f_{S_{ij}/BR}(x) = \frac{1}{\sqrt{2\pi\sigma^2}}e^{-\frac{(x-\mu)^2}{2\sigma^2}} \qquad (9)$$

with, $BR \in$ {DUROC, GHUNG, HAMP, YORK}. In the training phase, we learn the parameters of this Gaussian fit, i.e., μ, σ. A plot of all the learnt Gaussian distributions are shown in Fig. 6. Note each sub-figure in the set (a–n) comprises of a parametric fit for each of the four histograms corresponding to a specific patch location (i, j) with patch locations $(1, 1)$ and $(1, 4)$ not considered on account of extreme background information and irrelevant details.

The following are some observations regarding the sub-figures:

– Sub-figures Fig. 6a, b still show considerable overlap in the conditional density functions on account of prevalent background information in all the patches extracted from these two spatial locations irrespective of the breed type.
– A clear discrimination between the conditional densities begins with sub-figure (c) and continues all the way to sub-figure (n).
– As predicted by the patch density conjecture, since Yorkshire has been reared in colder areas, the fraction of significant points corresponding to pores and cilia is much smaller as compared to the other breeds (corroborated by Fig. 2g, h). This is depicted by the "black" conditional density function in Fig. 6g, h, k, n), which has the smallest mean in almost all the patches.

– On the other hand, because of the high density of pores and cilia for Ghungroo, the conditional mean is much higher than the other breeds for most of the patches (Green Gaussian curve in Fig. 6d–l and corroborated by Fig. 2c, d).

5   Testing and Inferencing Procedure

When a query muzzle template is supplied to this spatial conditional patch distribution model, the patch density statistics are first computed using the procedure outlined in Sect. 3.

Thus, this query muzzle print becomes a 14-point vector:

$$\bar{Q} = [q_{1,2}, q_{1,3}, q_{2,1}, q_{2,2}, q_{2,3}, q_{2,4}, q_{3,1},$$
$$q_{3,2}, q_{3,3}, q_{3,4}, q_{4,1}, q_{4,2}, q_{4,3}, q_{4,4}] \tag{10}$$

The patchwise inferencing is done as follows: For each patch corresponding to the spatial location (i, j), the breed with which the corresponding query patch is most closely associated is extracted through a simple MAXIMAL LIKELIHOOD test.

$$\hat{BR}_Q(i, j) = ARG\ \underset{BR}{MAX}\ \{f_{S_{ij}/DUROC}(q_{ij}), f_{S_{ij}/GHUNG}(q_{ij}), f_{S_{ij}/HAMP}(q_{ij}), f_{S_{ij}/YORK}(q_{ij})\} \tag{11}$$

where $BR \in \{DUROC, GHUNG, HAMP, YORK\}$. The overall association of the query vector with one of the breeds is obtained by taking a MAJORITY VOTE across all patch decisions:

$$BR_Q(FINAL) = MAJORITY_{VOTE}\left[\hat{BR}_Q(1, 2), \hat{BR}_Q(1, 3), \hat{BR}_Q(2, 1), \dots, \hat{BR}_Q(4, 4)\right] \tag{12}$$

If $BR_Q(FINAL)$ is the same as the original breed, then the query has been IDENTIFIED correctly, otherwise there is a misclassification. For testing: 29 muzzle prints from Duroc, 34 from Ghungroo, 45 from Hampshire, and 44 from Yorkshire were deployed out of the total of 311 and the number of correct detections were noted in Table 1. Since Yorkshire demonstrated a low conditional mean and a small variance across most patches, as expected the classification percentage was high (100%).

Table 1 Results of the breed classification algorithm with patch size $N_P = 250$

Breed	Classification accuracy (%)
Duroc	75.86
Ghungroo	70.59
Hampshire	57.78
Yorkshire	100

Table 2 Confusion matrix associated with the breed classification algorithm: patch size set as $N_P = 250$. Ideally the diagonal elements must be as close to "1" as possible

	Duroc	Ghungroo	Hampshire	Yorkshire
Duroc	22/29	5/29	1/29	1/29
Ghungroo	4/34	24/34	6/34	0
Hampshire	4/45	7/45	26/45	8/45
Yorkshire	0	0	0	44/44

Duroc and Ghungroo showed moderate classification (slightly poorer) results of 75% and 70% respectively, as the variances in their conditional density functions were larger leading to a significant overlap in the functions. Since Duroc, Ghungroo, and Hampshire all have partial white patches in some pigs, Duroc tends to be confused for Ghungroo and Hampshire (Ghungroo higher, because of the similarity in the contour and overall muzzle structure) and Yorshire (least), while, Ghungroo tends to be confused for Duroc and Hampshire (both high) and Yorkshire (least). This can be witnessed in the confusion matrix Table 2.

Hampshire shows the worst classification result of 58% as the white patch diversity in terms of size and distribution across the muzzle is maximum within the class. This is confirmed by the fact that a significant fraction of Hampshire muzzle prints have been misclassified as Yorkshire (8/45) Table 2.

6 Conclusion

In this paper, we propose a location-specific feature learning algorithm, for breed classification of pigs, based on muzzle images. Gradient Significance Map was constructed from the muzzle images by thresholding normalized Gaussian gradients. This Gradient Significance Map was divided into patches and patch statistic values were computed for each patch. For each of the patches, four Gaussian distributions are learnt from the training data corresponding to the four classes. In the testing phase, Maximum Likelihood Estimation was used to assign each patch to a particular breed. A majority voting across all the patches was carried out to assign the final class label to a muzzle image. The classification rates for Duroc, Ghungroo, Hampshire, and Yorkshire are 75.86%, 70.59%, 58.78%, and 100% respectively.

Acknowledgements This work is a byproduct of an IMAGE IDGP subproject, under the purview of a broader initiative, related to Identification, Tracking, and Epidemiological analysis of pigs and goats. This comes under the ITRA frame, which is a national-level research initiative controlled and sponsored by the Ministry of Information Technology, Government of India. The authors thank Dr. Santanu Banik and his ICAR RANI team (Rani, Guwahati, and Assam) for collaborating with us [the signal processing partners from IIT Guwahati] and for capturing samples of Annotated Muzzle Images from the ICAR Rani pig farm, which eventually formed the repository for the breed analysis.

References

1. Cai, C., Li, J.: Cattle face recognition using local binary pattern descriptor. In: Signal and Information Processing Association Annual Summit and Conference (APSIPA). IEEE, Asia-Pacific, pp. 1–4 (2013)
2. Awad, A.I., Zawbaa, H.M., Mahmoud, H.A., Abdel Nabi, E.H.H., Fayed, R.H., Hassanien, A.E.: A robust cattle identification scheme using muzzle print images. In: 2013 Federated Conference on Computer Science and Information Systems (FedCSIS). IEEE, pp. 529–534 (2013)
3. Lowe, D.G.: Distinctive image features from scale-invariant keypoints. Int. J. Comput. Vision **60**(2), 91–110 (2004)
4. Shi, G., Xu, X., Dai, Y.: Sift feature point matching based on improved RANSAC algorithm. In: Proceedings of the 2013 5th International Conference on Intelligent Human-Machine Systems and Cybernetics—Volume 01, IHMSC'13. IEEE Computer Society, Washington, DC, USA, pp. 474–477 (2013)
5. Noviyanto, A., Arymurthy, A.M.: Automatic cattle identification based on muzzle photo using speed-up robust features approach. In: Proceedings of the 3rd European Conference of Computer Science, ECCS, vol. 110, p. 114 (2012)
6. Borah, S., Samajdar, T., Das, T.K., Marak, G.R.: Success story on income generation from pig: Ghungroo. http://www.kiran.nic.in/ss_income_generation_pig_ghungroo.html
7. National Swine Registry.: Yorkshire. http://nationalswine.com/about/about_breeds/yorkshire.php
8. WIKIPEDIA.: Duroc Pig. https://en.wikipedia.org/wiki/duroc_pig
9. National Swine Registry.: Hampshire. http://nationalswine.com/about/about_breeds/hampshire.php
10. Raymond, J., Scientific American.: The Shape of a Nose (2011). https://www.scientificamerican.com/article/the-shape-of-a-nose/

Video Key Frame Detection Using Block Sparse Coding

Hema Sundara Srinivasula Reddy GogiReddy and Neelam Sinha

Abstract In the advent of video data explosion, to understand the concept of the video, knowledge of representative data selection and summarization has become essential. In this regard, application of video key frame detection is becoming increasingly critical. Key frame selection of videos is the process of selecting one or more informative frames that depict the essence of the video. In state of the art, researchers have experimented with shot importance measure [1], epitome-based methods [2], and sparse coding techniques [3] to find informative frames of video. We propose block sparse coding formulation, which exploits the temporal correlation of video frames within the sparse coding framework for key frames selection. We solved the block sparse coding formulation using the Alternating Direction Method of Multipliers (ADMM) optimization. We show the comparison of results obtained with the proposed method, state-of-the-art algorithm [3] and ground truth on TRECVID 2002 [4] dataset. Comparison results show 8x run time and 6% F-score improvement compared to state of the art.

Keywords Key frame detection · Video summarization · Block sparse coding

1 Introduction

Key frame selection of videos is the process of selecting one or more informative k frames in the video. Key frame extraction is useful in applications like video summarization, video indexing, etc.

Vision-based key frame detection algorithms can be broadly divided into two approaches as given below:

H. S. S. Reddy GogiReddy (✉)
Samsung R&D Institute India Bangalore, Bangalore 560037, Karnataka, India
e-mail: srinivas.ghs@samsung.com

N. Sinha (✉)
IIIT-Bangalore Electronics, Electronics City, Bangalore 560100, Karnataka, India
e-mail: neelam.sinha@iiitb.ac.in

© Springer Nature Singapore Pte Ltd. 2020
B. B. Chaudhuri et al. (eds.), *Proceedings of 3rd International Conference on Computer Vision and Image Processing*, Advances in Intelligent Systems and Computing 1022,
https://doi.org/10.1007/978-981-32-9088-4_8

– Shot based: In this approach, first shot boundaries of the video are found, and then a key frame is extracted from selected important shots [1].
– Segment based: In this approach, a video is segmented into smaller video components, where each component can be a scene, one or more shots, etc. A key frame is selected from important components.

Another way for detecting key frame is by dimensionality reduction. In this approach, we reduce the *Available Frames* (object space) dimension of data by selecting only a subset of *key frames* (data points) that represents whole data.

This paper is organized as follows. Section 2 reviews key frame extraction methods. Section 3 reviews key frame detection method using sparse coding techniques. In Sect. 4, the proposed key frame extraction algorithm using block sparse coding is described, while Sect. 5 presents benchmarking results compared to ground truth data. Finally, concluding remarks are given in Sect. 6.

2 Review of Key Frame Extraction Methods

Video key frame detection is handled using two diverse approaches, namely, signal processing and video processing domains. In case of signal processing domain, feature vectors such as color histogram are extracted from video frames. Each feature vector is represented as a data point. Data points are processed without considering the intrinsic structure of signal (video data).

In video processing domain, video-specific features such as flow, edges, etc. are considered as feature vectors. These feature vectors capture the intrinsic structure of video. Key frames are detected by processing these feature vectors.

2.1 Prior Art

Diwakaran et al. [5] proposed a method for key frame detection based on the hypothesis that more key frames are required if there is more motion in the video. In this method, equal motion activity regions are detected in a video and then key frames are extracted from each activity region. Sikora et al. [6] approach treats video sequence key frame extraction is analogous to existing image search engine. In their approach, key frames are extracted using visual-attention-seeking features like lighting, camera motion, face, and appearance of text. Borth et al. [7] proposed key frame extraction method based on shot boundary detection. They also extended their approach to video summarization using key frames extracted from shots detected. Dang et al. [2] proposed a method that focuses on key frame extraction from consumer videos. Their method uses image epitome to extract key frames from videos.

Elhamifar et al. [3] used self-expressiveness property of data along with the sparse coding method to find representative data point selection. One of the use cases they showed was key frame extraction from videos. Dang et al. [8] used spectral clustering techniques to cluster ℓ_1-norm graph. Each cluster formed is considered as a point in Grassmann manifold, and geodesic distance between points is measured. Further, using the min-max algorithm and principal component centrality [9] key frames are detected.

3   Representative Selection Using Sparse Coding

In this section, we describe key data point detection using dimensionality reduction techniques using the sparse coding framework. When we consider features of frame as data points, this method can be extended to finding key frames of videos.

Denote $Y = [y_1, \ldots, y_n] \in \mathcal{R}^m$ is the set of data points, where m is the dimensionality of data points and n is number of data points. For key feature detection, one tries to learn dictionary D and coefficient matrix X that can efficiently represent data Y. We formulate an optimization problem to find the best representation of the data Y subject to appropriate constraints.

$$\sum_{i=1}^{N} \|y_i - Dx_i\|_2^2 = \|Y - DX\|_F^2 \tag{1}$$

By using the sparse dictionary learning framework, we solve the optimization problem to obtain sparse coefficient matrix X [10].

$$\min_{D,X} \|Y - DX\|_F^2 \quad \text{s.t.} \quad \|x_i\|_0 \leq s, \|d_j\|_2 \leq 1, \forall i, j \tag{2}$$

3.1   Finding Representative Data

Atoms of the dictionary D obtained by optimizing (2) are the linear combination of data points due to which they are not similar to original data [11], and hence they cannot be used as key frames. To find representative points that are from the original data points, we modify sparse dictionary learning framework as [3]

$$\sum_{i=0}^{N} \|y_i - Y c_i\|_2^2 = \|Y - Y C\|_F^2 \tag{3}$$

with respect to the coefficient matrix $C \triangleq [c_1 \ldots c_N] \in R^{N \times N}$, subject to additional constraints. The constraints are as follows:

- Minimize the reconstruction error of each data point.
- Enforce affine constraint $\mathbf{1}^\top C = \mathbf{1}^\top$ to make the selection invariant to global transformations.

This formulation has two distinct advantages. This formulation is convex as we are solving for one unknown matrix instead of two. Second, it ensures representatives are selected from the actual data points.

Finding ℓ_0/ℓ_q-norm is NP-hard problem, and hence we relax it to ℓ_1/ℓ_q as

$$\min \|Y - Y C\|_F^2 \quad \text{s.t.} \quad \|C\|_{1,q} \le \tau, \mathbf{1}^\top C = \mathbf{1}^\top \tag{4}$$

Using Lagrange multipliers, optimization problem (4) can be written as

$$\min \lambda \|C\|_{1,q} + \frac{1}{2} \|Y - Y C\|_F^2 \quad \text{s.t.} \quad \mathbf{1}^\top C = \mathbf{1}^\top \tag{5}$$

4 The Proposed Method for Shot Detection and Key Frame Extraction Using Block Sparse Coding

Finding key frames using Eq. (5) has two problems which are as follows:

– Run time of algorithm increases exponentially with the number of frames in the video.
– If a data point repeats after some time gap, it will be represented by an existing data point. This behavior is counterintuitive in video summarization, as we may have to consider the same frame as key frame in two different parts of video depending on context.

Our proposed method addresses problems with (5). Proposed approach for representative selection has two parts.

– Define an objective function which considers multi-block nature of data.
– Define post-processing of data to refine the representatives.

4.1 Sparsity-Inducing Norm for Multi-block-Natured Data

The optimization problem for representative selection in Eq. (5) does not consider the multi-block nature of data, i.e., each shot data/frame is closely related and they reside in the same subspace. From this intuition, if we assume that the sparse representations of signals will be in the same subspace. To take advantage of this behavior, we divide the data into nonoverlapping neighborhood blocks. We can write Y as $Y = [Y[1] \ldots Y[m]]$, where $Y[i]$ represents block of input data Y.

Selection of representative can be rewritten as

$$\min \quad \lambda \|C\|_{1,q} + \frac{1}{2} \|Y - [Y[1] \ldots Y[m]] C\|_F^2$$
$$\text{s.t.} \quad \mathbf{1}^\top C = \mathbf{1}^\top \tag{6}$$

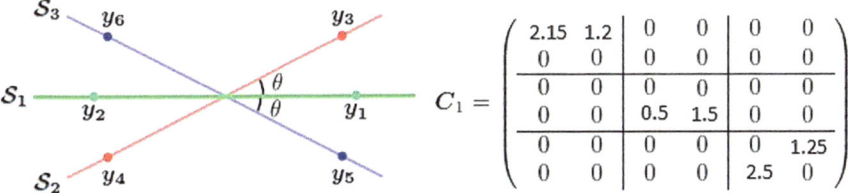

Fig. 1 Left: three one-dimensional subspaces in $\Re^2$ with normalized data points. Right: C_1 corresponds to the solution of Eq. (7). C_1 has three components corresponding to the three subspaces

Now we modify our objective functions to select representative data from its own block. The main difference with respect to classical representative selection [3] is that the desired solution of Eq. (6) corresponds to a few nonzero data points of own *block* rather than a few nonzero *elements* of Y, as depicted in Fig. 1. We say that a vector $C^\top = \left[C[1]^\top \cdots C[n]^\top \right]$ is *block sparse*, if every block $C[i] \in \Re^{m_i}$ chooses its representatives from own block. Note that, in general, a block sparse vector is not necessarily the sparse representation of data and vice versa. The problem of finding a representation of data y that uses the block data nature of Y can be cast as the following optimization program:

$$\min \quad \lambda \sum_{i=1}^{m} \|C[i]\|_{1,q} + \frac{1}{2} \|Y - [Y[1] \dots Y[m]]\, C\|_F^2 \tag{7}$$
$$\text{s.t.} \quad \mathbf{1}^\top C = \mathbf{1}^\top, diag(C) = 0$$

4.2 Selection of Representatives

After finding the probable candidates from (7), we need to refine representatives as the nature of block selection is heuristic. Two consecutive blocks in the data may represent the same subspace. To prune consecutive representatives from the same subspace, post-processing is done in two phases as given below:

- *Representative selection*: Optimization of Eq. (7) yields C which is of size $N \times N$, we take the norm of each column and select the top-most columns whose sum constitute 95% of total norm sum.
- *Refinement*: During the refinement process, we form a matrix with the distance between all pairs of probable candidates and difference of their feature vectors. We select the entries which have the highest difference.

5 Experiments and Results

In order to find the accuracy and superiority of the proposed framework, extensive experiments are conducted on various videos from TRECVID 2002 [4] dataset. In order to evaluate the performance of key frame extraction method presented in Sect. 4, the precision and recall methods are used. Let "GT" denote the ground truth key frames given along with the dataset and "Pred" is the key frame detected using our algorithm.

$$Recall = \frac{|Pred \cap GT|}{|GT|} \tag{8}$$

where $|GT|$ denotes number of key frames in ground truth.

The precision measure corresponds to the accuracy of the proposed method.

$$Precision = \frac{|Pred \cap GT|}{|Pred|} \tag{9}$$

where $|Pred|$ denotes number of key frames detected by proposed algorithm.

5.1 Experimental Setup

For a given test video, we take color histogram, with each color containing 16 bins. The feature vector of a frame is formed by stacking bins of its color histogram. Feature vectors of all video frames form columns of the data matrix. We use the block sparse doing formulation (6) with Alternating Direction Method of Multipliers (ADMM) optimization approach to find the representative frames. To avoid too-close representatives from neighboring blocks, we pruned the set of representatives from having too-close data points in post-processing. Block diagram of this flow is shown in Fig. 2.

5.2 Results

Table 1 shows the ground truth of video sequences used for validating the proposed algorithm. The "Consuming Women" video contains cuts along with transitions. "Looking Ahead Through Rohm Haas Plexiglas" video is an indoor video with

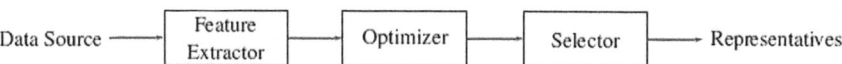

Fig. 2 System block diagram for representative selection

Table 1 Key frames ground truth data for TRECVID 2002 videos

Video name	Duration mins:secs	Number of key frames
Consuming women	3:25	14
Vision in the forest	5:11	42
To market, to market	4:35	28
Breakfast pals	2:0	17
Looking ahead through rohm	4:39	25

Table 2 Key frames detection results obtained using the proposed approach

Video name	Number of detected key frames	Precision	Recall
Consuming women	15	0.857	0.8
Vision in the forest	32	0.93	0.76
To Market, to market	21	0.857	0.71
Breakfast pals	14	0.85	0.75
Looking ahead through rohm	21	0.857	0.72

Fig. 3 Consuming women (women consumer's) detected sequence key frames, thumbnail 3 and 4 are wrong selections

panning and zooming. "Breakfast pals" is an animation video with cuts. "To Market, To Market" is an outdoor video of city tour. "Vision in the Forest" is an outdoor video with a lot of textures.

The precision and recall values of the respective video sequences compared with the ground truth are shown in Table 2 and the key frames of Consuming Women (Women consumer's) sequence detected by proposed algorithm are shown in Fig. 3.

5.3 Computational Time Comparison

Comparison of time taken to find key frames for block-based approach (6) and without block-based approach (5) is shown in Fig. 4. Time taken for key frame detection in block-based approach increases linearly with the number of frames, whereas time taken for solving (5) increases exponentially with the number of frames.

Fig. 4 Computational time comparison of proposed and non-block-based [3] approaches

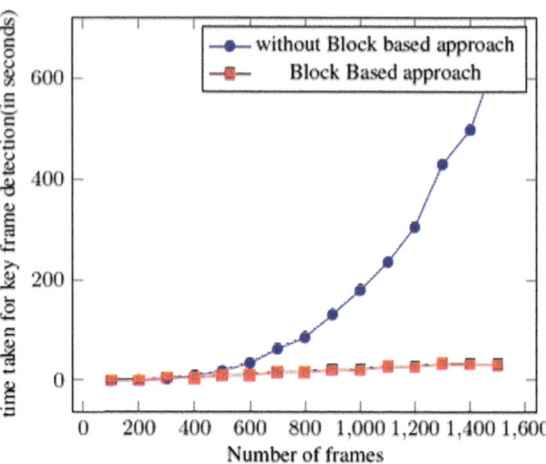

6   Conclusion

We proposed an algorithm for finding key frames in a video using sparse coding techniques. We assumed that each data point/frame can be expressed efficiently as a combination of the representatives that belong to the same block. We frame the problem as a block sparse vector learning problem where dictionary elements are the same as the input video frame features and the unknown sparse codes select the key frames. We showed the results for the different types of video, which show significant improvement in run time and visual results over the baseline algorithm.

References

1. Foote, J., Uchihashi, S.: Summarizing video using a shot importance measure and a frame-packing algorithm. In: ICASSP, vol. 6, pp. 30410–3044 (1999). https://doi.org/10.1109/ICASSP.1999.757482
2. Dang, C.T., Kumar, M., Radha, H.: Key frame extraction from consumer videos using epitome. In: ICIP, pp. 93–96 (2012). https://doi.org/10.1109/ICIP.2012.6466803
3. Elhamifar, E., Sapiro, S.G., Vidal, R.: See all by looking at a few: Sparse modeling for finding representative objects. In: CVPR, pp. 1600–1607 (2012). https://doi.org/10.1109/CVPR.2012.6247852
4. Yavlinsky, A., Magalhães, J., Jesus, R., Rüger, S.: Video Retrieval using Search and Browsing TREC Video Retrieval Evaluation Online Proceedings (2004)
5. Divakaran, A., Peker, K.A., Radhakrishnan, R.: Motion activity-based extraction of key-frames from video shots. In: ICIP, vol. 1, pp. I–I (2002). https://doi.org/10.1109/ICIP.2002.1038180
6. Kelm, P., Schmiedeke, S., Sikora, T.: Feature-based video key frame extraction for low quality video sequences. In: WIAMIS, pp. 25–28 (2009). https://doi.org/10.1109/WIAMIS.2009.5031423

7. Borth, D., Breuel, T., Schulze, C., Ulges, A.: Navidgator-similarity based browsing for image and video databases. In: Advances Artificial Intelligence, pp. 22–29 (2008). https://doi.org/10.1007/978-3-540-85845-4_3
8. Al-Qizwini, M., Dang, C., Radha, H.: Representative selection for big data via sparse graph and geodesic Grassmann manifold distance. In: ACSSC, pp. 938–942 (2014). https://doi.org/10.1109/ACSSC.2014.7094591
9. Ilyas, M.U., Radha, H.: A klt-inspired node centrality for identifying influential neighborhoods in graphs. In: CISS, pp. 1–7 (2010). https://doi.org/10.1109/CISS.2010.5464971
10. Aharon, M., Bruckstein, A., Elad, M.: k-svd: an algorithm for designing overcomplete dictionaries for sparse representation. IEEE Trans. Signal Process. 4311–4322 (2006)
11. Bach, F., Mairal, J., Ponce, J., Sapiro, G., Zisserman, A.: Discriminative learned dictionaries for local image analysis. In: CVPR, pp. 1–8 (2008). https://doi.org/10.1109/CVPR.2008.4587652

Robust Detection of Defective Parts Using Pattern Matching for Online Machine Vision System

Namita Singh, Abhishek Jaju and Sanjeev Sharma

Abstract Reliable detection of defective parts is an essential step for ensuring high-quality assurance standards. This requirement is of primary importance for online vision-based automated pellet stacking system. During nuclear fuel pin manufacturing, the image of a single row (consecutively placed components with no gap) is processed and analyzed to extract meaningful edges. Generally, these edges follow a regular pattern; however, the presence of surface cracks and chips can alter this pattern. In this paper, we formalize the detection of defective parts as a pattern matching problem. Three different patterns are proposed and evaluated for sensitivity, specificity, and accuracy. An experiment performed with the proposed pattern matching techniques show that multi-pattern matching is the most effective method for identifying defective parts.

Keywords Multi-pattern matching · Aho–Corasick algorithm · Finite state automata · Fuel pellet · Stack · Defects

1 Introduction

Pattern matching is required in many applications like information extraction, antivirus scanner, DNA matching, etc. Also, many algorithms in artificial intelligence utilize string matching algorithms for processing information. Depending on the desired objective, single or multiple matching may be needed. Further, patterns may be required to be matched exactly or approximately. In this case, pattern matching has come up as the key aid for providing information about the presence, number,

N. Singh · A. Jaju · S. Sharma (✉)
Division of Remote Handling and Robotics, Bhabha Atomic Research Centre, Mumbai 400085, India
e-mail: ssharma@barc.gov.in

N. Singh
e-mail: namita@barc.gov.in

A. Jaju
e-mail: ajaju@barc.gov.in

© Springer Nature Singapore Pte Ltd. 2020
B. B. Chaudhuri et al. (eds.), *Proceedings of 3rd International Conference on Computer Vision and Image Processing*, Advances in Intelligent Systems and Computing 1022, https://doi.org/10.1007/978-981-32-9088-4_9

and location of valid pellets within the images for robust automated pellet stacking job. An input string sequence is first generated through mapping of consecutive lengths after processing of images, and thereafter, pattern matching is used for the detection of anomalous sequences.

This paper is organized as follows. Related literature is briefly overviewed in Sect. 2. Generation of input strings for processing is described in Sect. 3. Thereafter, in Sect. 4, we present patterns that can be used for identifying pellets without any defect. Evaluation criterion for pattern matching results is explained in Sect. 5. Experimental results and discussion is presented in Sect. 6. The paper is concluded in Sect. 7.

2 Related Work

Single pattern matching algorithm can be classified into three types: prefix-based, suffix-based, and substring-based according to the part being searched for [1]. In complex or practical cases, a pattern can be a set of strings and not just a single word which comes under multi-pattern matching. Algorithms that are used for matching can be heuristics-based, hashing-based, bit-parallelism-based, or automaton-based [1]. Automaton-based algorithms build finite state automaton from the patterns during preprocessing and track the pattern in the text using state transitions.

Heuristics-based algorithms accelerate the pattern search by skipping some characters. Knuth Morris Pratt (KMP) [2] comes under this type. Boyer–Moore algorithm [3] believes in skipping based on bad character and good suffix heuristics. They perform well with a large-sized pattern which allows them to skip through a larger length.

Hashing-based algorithms compare the hash values of patterns against the part of the text under scanning. Famous algorithm under this type is Rabin–Karp algorithm [4]. Bit-Parallelism-based algorithms are based on a nondeterministic finite automaton that tracks the prefix of the pattern using parallel bit operations. SOG (Shift-OR with q-Grams) algorithm is bit-parallelism based where multi-patterns are considered as a single pattern of classes of characters.

Hashing-based algorithms like Rabin–Karp can be made to handle multiple patterns by maintaining a hash for each pattern but they greatly depend on the quality of hash function. Their scalability for multi-patterns cannot be exploited well if the multi-pattern sets have less distinct characters and have higher matches in the input text as in our case. Bit-Parallelism based algorithms are fast but they are prone to giving false positives.

Automaton based algorithms are the best choice when the pattern set is small [5]. A lot of work has been done in improving and optimizing the Aho–Corasick recently [6]. Applications of Aho–Corasick to pattern matching problems involving equivalences like palindrome structures have been explored [7]. Applications of the same have also been explored for network security [8].

In our case, as a pattern set is small, hashing- and heuristic-based algorithms are not suitable. Hence, automaton-based pattern matching is proposed for processing input string information.

3 Pattern Formulation

A nuclear fuel pin is a stack of linearly arranged cylindrical pellets [9]. Gray scale images of pellet stacks are captured from a calibrated line-scan camera based experimental system and are stored in a database [10]. Figure 1a shows an image captured by high-resolution camera that has a dark region between the consecutive pellets (P) and Fig. 2 shows the corresponding automata representation. The dark region here is referred to as gap (G). Generally, the sequence of pellet, gap, and pellet is repeated. The general pellet length distribution is known to us and is found to lie in a band. The distribution of gap lengths between any consecutive pellets is analyzed using the calibrated Camera system. Lengths obtained other than these are classified as erroneous.

$$Ideal\ input\ sequence = \{PGPGPG \ldots \ldots \ldots PGP\}$$

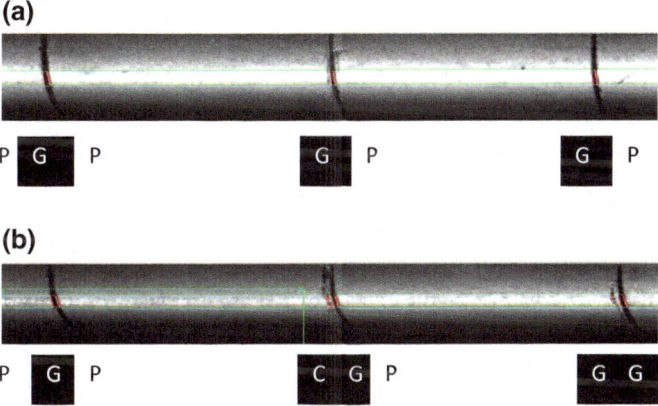

Fig. 1 Sequence of pellet (P) and gap (G) sequence. **a** Ideal sequence. **b** Altered sequence because of defects like cracks (C) and multiple gaps (G) in succession

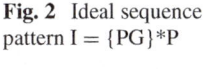

Fig. 2 Ideal sequence pattern I = {PG}*P

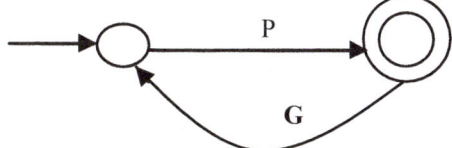

Table 1 Classification of consecutive lengths

S. no.	Length (mm)	Symbol	Description
1	<0.1	C	Crack
2	$0.1 \leq L \leq 0.5$	G	Gap
3	$0.5 < L \leq 13$	X	Invalid short length
4	$13 < L \leq 18$	P	Pellet length
5	>18	O	Invalid long length

However, defects like cracks and chips may alter this sequence as depicted in Fig. 1b. Different lengths that are likely because of defects are classified by assigning different symbols as shown in Table 1. Following is a typical case of the real input sequence.

$$Real\ input\ sequence = \{PGPPGC\ldots\ldots PGP\}$$

The application of pattern matching algorithm on real input sequence helps in the identification of altered sequence and therefore identification of defective pellets [11].

The input to the pattern matching algorithm is the sequence of characters over a finite alphabet $\sum = \{P, G, X, C, O\}$. Table 1 shows the description of letters in $\sum$. We can further define input sequence ($\sum^n$) as a sequence of all the symbols derived from n length measurements of a pellet stack.

4 Pattern Selection

Correct identification of the non-defective pellets marks them as candidates for exchange in the available pellets set as their lengths can be obtained with high accuracy. Defective pellets are those pellets for which the lengths cannot be obtained with accuracy due to deformation at the edges. This step is crucial for preparing a stack of defined length within a strict tolerance [10]. Robustness in our case means including only the valid pellets for use in stack preparation. In order to have robust and reliable identification, pattern matching algorithm shall satisfy the following objectives:

1. Minimum false positives: Defective pellets shall not be accepted after processing with pattern matching algorithms. It is an important criterion for maintaining high-quality assurance standards in production environments.
2. Minimum false negatives: Acceptable pellets with correct length shall not be rejected out after pattern matching. This will ensure higher throughput and reduce the load on the handling of rejected pellets.

Identification of non-defective pellets is performed in two stages. First, pattern matching is carried out for extracting pattern indices. Thereafter, the valid pellet index is retrieved from the matched patterns.

In this paper, pattern matching methodologies for the identification of non-defective pellets are proposed. These are analyzed with the input strings derived from the images captured by the experimental setup. These methodologies are described as follows.

4.1 Single Pattern Matching

Single Doublet Matching (SDM). This strategy uses a single pattern matching technique. It is the simplest and most intuitive wherein pattern $\{PG\}$ shown in Fig. 2 is matched from left to right until the first mismatch is found. This is done in order to avoid any ambiguity related to the classification of incoming lengths for validness. However, this method has the disadvantage that the rest of the sequence needs to be discarded because the scheme has no way to find out whether the length being mapped as P is valid or not if symbols like C, X, and O are present as either prefix or suffix.

Single Triplet Matching (STM). Here, for an input Sequence $\sum^n$ of finite length n, pattern $\{GPG\}$ is matched. This rule is based on the assumption that between any two valid gaps, lies a valid pellet with an exception to the first and the last pellet.

KMP algorithm is used for pattern matching which compares the pattern with the input sequence from left to right. It converts the pattern into a finite state automaton comprising of three parts:

1. Nodes: All the prefixes of the pattern are the states of the automaton. For a pattern with length p, there are $p + 1$ states in the automaton. Nodes are shown in Fig. 3.

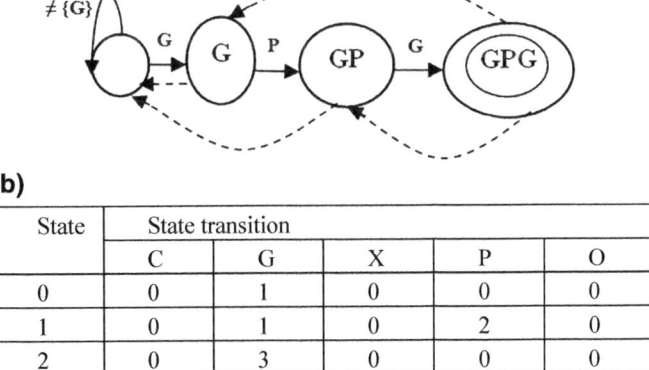

(a)

(b)

State	State transition				
	C	G	X	P	O
0	0	1	0	0	0
1	0	1	0	2	0
2	0	3	0	0	0
3	1	1	1	1	1

Fig. 3 **a** Finite deterministic automaton for the pattern "GPG". **b** Transition table showing success and failure transitions

2. Success transitions: Consider u is a state and L(u), a label on the state showing the alphabet sequence. For a state representing the final state, the label on the state shows the pattern and for all others, it shows the pattern prefix. The transition from the current state u to the next state v on the input of x is a success transition if label L(u)x $= $ L(v).
3. Failure transitions: Transition from the current state u (label L(u)) on the input of x to a previous state v (label L(v)), such that L(v) is the longest prefix of L(u).

4.2   Multi-pattern Matching

Single Pattern Matching can result in false negatives in case of SDM due to its conservative approach and false positives in case of STM due to its loose approach. Therefore, we search for a finite set of predefined patterns for robustly locating valid pellet length locations. We are proposing four patterns formed from *P, G,* and *O* (Table 2) for reducing false positives. Aho–Corasick is the most widely accepted algorithm for multi-pattern exact matching and is especially suitable for applications like ours where the pattern set K is known and fixed in advance, while the input LengthSequence $\sum^n$ is varying. Aho–Corasick proposed the use of automaton as the classic data structure to solve the problem. The running time of Aho–Corasick is found to be independent of the number of patterns. Automaton is constructed in $O\left(m\log|\sum|\right)$ preprocessing time and has $O\left(n\log|\sum|+k\right)$ search time where k represents the number of occurrences of patterns in the input sequence, m is the size of the input text, and n is the cumulative size of patterns [12]. Aho–Corasick extend the single pattern matching KMP algorithm to handle multi-patterns by combining it with automata. It creates a finite state machine to match multiple patterns in the sequence in one pass.

The finite and deterministic String Matching Automaton G (Q, $\sum$, g: Q X $\sum$ $\rightarrow$ Q, $\emptyset \in$ Q, F $\in$ Q) is built accepting all the words containing k $\in$ K as a suffix and the language [7].

$$L(G(k)) = \sum\nolimits^* k$$

where

K   Set of all patterns

Table 2 Patterns for identifying valid pellets

Pattern	Pattern sequence
I	PGPGP
II	PGPGO
III	OGPGP
IV	OGPGO

Q   Set of all prefixes of k in K
Ø   State representing the empty prefix
F = {k}   State representing the prefixes of all patterns in K
g   Mapping defining current State with the input character to state transition
           function.

   Three main stages: Goto function, Failure function, and Output function used for
pattern identification are shown in Fig. 4a and are described below.

(a)

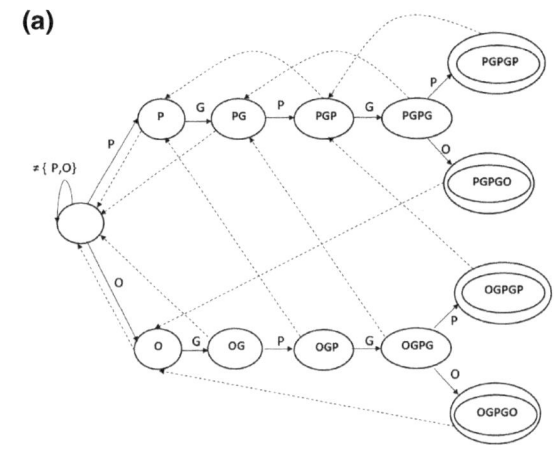

(b)

State	State transition				
	C	G	X	P	O
0	0	0	0	1	7
1	0	2	0	1	0
2	0	0	0	3	0
3	0	4	0	1	0
4	0	2	0	5	6
5	3	3	3	3	3
6	7	7	7	7	7
7	0	8	0	0	7
8	0	0	0	9	0
9	1	10	1	1	1
10	2	2	2	11	12
11	3	3	3	3	3
12	7	7	7	7	7

Fig. 4 **a** Aho–Corasick automaton G for set K. Failure links from any state to itself are not shown in the graph to keep it simple but are clear from the transition table. **b** Transition table

Goto Function. A trie is constructed using the given set of patterns. It maps a pair consisting of state and input symbols into a new state or reports fail in case of mismatch. States in the machine are the prefixes of the given set of patterns like for pattern $PGPGP$, Prefix $\{PGPGP\} = \{``P", ``PG", ``PGP", ``PGPG", ``PGPGP"\}$ are valid states in the automata.

These are followed by an empty state also called as start state, the state representing the empty prefix. For any input symbols s in $\sum$ and start state $\emptyset$, Goto function, $g(\emptyset, s) = \emptyset$ if $s = \{C, X, G\}$. Each node of the trie is a prefix of one or more patterns and the leaf of the trie represents a pattern.

Failure Function. For $q \in Q$ and $s \in \sum$, $g(q,s) = qs$ iff $qs \in Q$ else $g(q,s) = r$ such that r is the longest suffix of qs, which is also a prefix of x and thus r is a valid state in Q. In Fig. 4a, the failure transitions are marked with dashed arrows for the constructed automaton and the transition table is shown in Fig. 4b.

Output Function/Valid Pellet Index Retrieval. Automaton construction is the step for preprocessing of patterns. It is done once prior to the online measurement. Index of the valid pellet is retrieved once the index list for all the matched patterns is obtained from the previous stage. For each of the matched patterns matched at index j in the input sequence, j−2 is the index of the valid pellet. The valid status of the corresponding pellet is set to 1. For the first and last pellets, the case is handled using "PGP".

5   Evaluation Criterion

In order to evaluate the performance of different patterns, three statistical measures are applied. These are *sensitivity, specificity,* and *accuracy.*

Sensitivity, also called as a true positive rate measures the proportion of pellets that are correctly identified as valid among all valid pellets in the sequence. *Specificity*, also referred as true negative rate describes quantitatively how efficiently a method correctly excludes the pellets having incorrect lengths from a valid pellet list. *Accuracy* is defined as the probability that the output result is correct. For evaluating the mentioned parameters, it is essential to have the ground truth information about individual pellets in each of the test images. The software developed has a provision for visual inspection which facilitates the user to easily identify the defective pellet for each loaded test image.

If TP denotes the number of true positives, FP denotes the number of false positives; TN denotes the number of true negatives and FN denotes the number of false negatives, then sensitivity, specificity, and accuracy can be defined as explained in [13]

$$sensitivity = \frac{TP}{TP + FN}$$

$$\text{specificity} = \frac{TN}{FP + TN}$$

$$\text{Accuracy} = \frac{TN + TP}{TP + FP + TN + FN}$$

If a valid pellet gets detected and it also exists in the ground truth, then a value of 1 is added to TP. An undetected valid pellet which exists in the ground truth adds a value of 1 to FN. Similarly, a detected defective pellet which does not exist in the ground truth adds a value to FP and an undetected defective pellet which exists in the ground truth adds a value to TN.

6   Experiments, Results, and Discussion

Input strings for the pattern matching are generated from 100 stack images. In most of the cases, the sequence of the pellet length followed by a gap is observed (Fig. 5a). However, defects like chips at pellet boundaries and cracks on the surface of the pellets result in the following cases:

1. Both edges related to inter-pellet gap are missing. This gives rise to an increased length which combines multiple pellet lengths and is denoted as O in the sequence (Fig. 5b).
2. Only one edge corresponding to the inter-pellet gap is present which comes in the sequence as pellet length followed by another pellet length instead of gap (PP instead of PG) as shown in Fig. 5c.
3. More than two edges corresponding to the inter-pellet gap are present which gives rise to cases like PCGP, PGCP, PGGP, PCCP, etc. Image corresponding to one such case is shown in Fig. 5d.
4. Unwanted edge can come due to crack on the surface of the pellet. This can give rise to cases like GXXG, GXPG, or GPXG. Nevertheless, invalid length (X) can also occur because of chip defect. Figure 5d depicts invalid length because of a chip defect on adjacent pellets.

Strings generated from the images are used as input for testing the proposed patterns. Further, careful offline analysis of images is carried out for computing the ground truth information. Proposed pattern matching methodologies are then applied and evaluated for errors (false positives and false negatives) in detecting valid pellets. In our case, it is more important to exclude any erroneous pellet length considered to be valid than to miss a valid pellet length. Thus, false positives are costlier than false negatives.

Figure 6 shows the analysis of a typical input string generated from a test image. Valid pellet count along with indices for this string as shown in the ground truth is shown in Fig. 6d while results from the proposed methodologies are shown in Fig. 6a–c. Pellets classified as valid are marked in red for each strategy. It can be seen that *SDM* is very sensitive to the location of mismatch, i.e., the number of valid

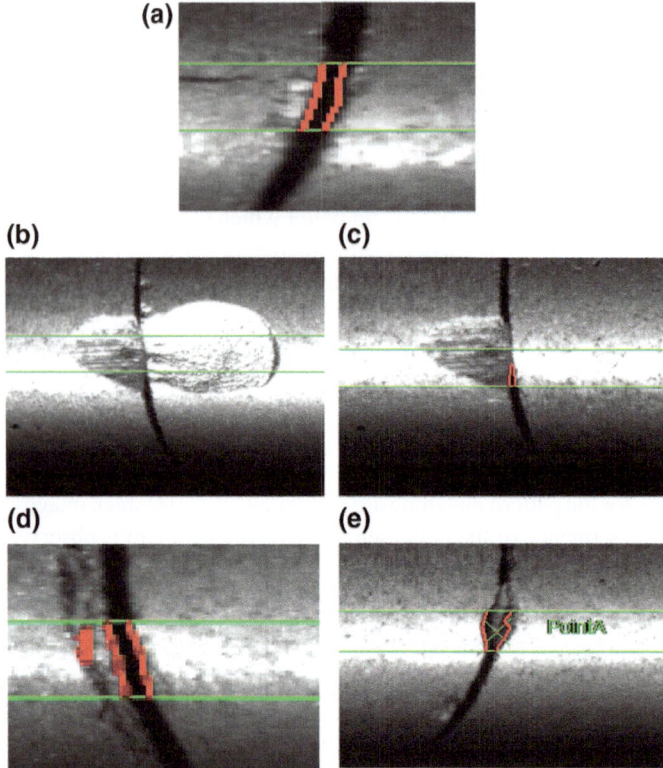

Fig. 5 a Perfect edges that result in acceptable pattern. **b** Both edges missing causing invalid long length referred to as *O*. **c** Only one edge detected because of the chip which causes defective sequence *PP*. **d** More than two edges that can give rise to cases like PGGP **e** Invalid length (*X*) because of chip defect

pellets retrieved is dependent on the mismatch position. In this case, only one pellet is found to be valid (Fig. 6a). Other two strategies are far more efficient in their approach to identifying valid pellets.

SDM performance degrades rapidly in practical scenarios when dealing with real production lot pellets. *SDM* ensures that no erroneous pellet is classified as valid but it becomes a less popular choice because of rejecting so many valid pellets. Its performance is position-dependent on the location of the defect. *STM* performs well in terms of its capability to classify valid pellets. In the sequence shown, 24 pellets were classified as valid pellets with one pellet classified wrongly as valid (pellet index 47). *MPM* also performs nicely with 22 pellets classified as valid and no pellet classified wrongly. We evaluate the performance of the three strategies based on their sensitivity, specificity, and accuracy based on the data from 100 stacks.

It can be seen that the specificity of *SDM* and MPM is reaching 100%. Accuracy-wise STM and MPM are the obvious choices.

Length Sequence	PGPPGPGPGPGPGPGPGPGPGPGPGPGPGPGPGPGPGPGPPGPGPG PGPGPGGPGXPGOGPGP
Valid Pellet count	1
Valid Pellet Indices	{1}
(a) SDM	
Length Sequence	PGPPGPGPGPGPGPGPGPGPGPGPGPGPGPGPGPGPGPGPPGPGPG PGPGPGGPGXPGOGPGP
Valid Pellet count	24
Valid Pellet Indices	{1,6,8,10,12,14,16,18,20,22,24,26,28,30,32,34,39,41,43,45,47,50,57,59}
(b) STM	
Length Sequence	PGPPGPGPGPGPGPGPGPGPGPGPGPGPGPGPGPGPGPGPPGPGPG PGPGPGGPGXPGOGPGP
Valid Pellet count	22
Valid Pellet Indices	{1,6,8,10,12,14,16,18,20,22,24,26,28,30,32,34,39,41, 43,45,57,59}
(c) MPM	
Length Sequence	PGPPGPGPGPGPGPGPGPGPGPGPGPGPGPGPGPGPGPGPPGPGPG PGPGPGGPGXPGOGPGP
Valid Pellet count	23
Valid Pellet Indices	{1,6,8,10,12,14,16,18,20,22,24,26,28,30,32,34,39,41,43,45,50,57,59}
(d) Reference correct outcome	

Fig. 6 Pattern comprising of lengths classified as per Table 1 generated from experimental setup (Red indices indicate detection). **a** Results from *SDM*. **b** Results from *STM* **c** Results from *MPM*. **d** Reference data

7 Conclusion

This paper shows the significance of different patterns for the identification of defect-less pellets from the images. Identified pellets are then used for making stacks of exact lengths.

Three methods for the identification of valid pellets suitable for performing robotic stacking are presented in this paper. *SDM* is good only in ideal cases which is rare. This has a very high specificity, i.e., we are assured that the pellets it provided for exchanging will be only valid pellets. Its downfall is that it has low sensitivity, i.e., its output in terms of valid pellets is low which may prove fatal for the optimality in the exchange algorithm being used for stacking job. *STM* does better but due to its less specificity, it takes a back seat to MPM which is best in all the cases and robustly handles all kinds of cases. Figure 7 shows that MPM has resulted in three times improvement as compared to SDM (from 24.6 to 98.4%) in the number of valid pellets available for exchange operations.

Valid pellets obtained from the proposed method have been employed to make stacks of definite lengths with a tolerance of $\pm 900 \mu$ [10].

Fig. 7 Obtained results for
the three discussed
approaches SDM, STM, and
MPM

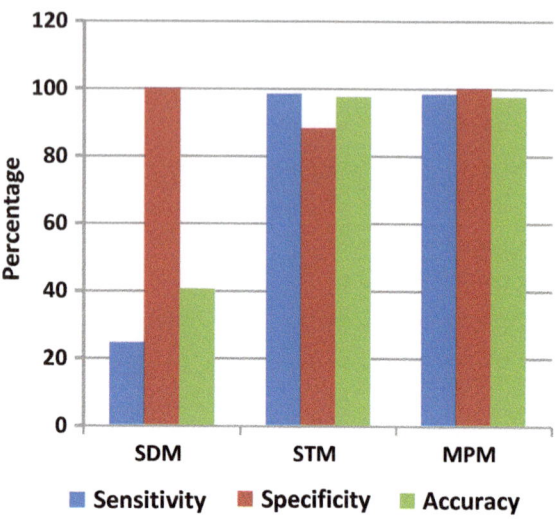

References

1. Navarro, G., Raffinot, M.: Flexible pattern matching in strings. Cambridge University Press (2002)
2. Knuth, D.E., Morris Jr, J.H., Pratt, V.R.: Fast pattern matching in strings. SIAM J. Comput. (1997)
3. Boyer, R., Moore, S.: A fast string searching algorithm. Commun. ACM **20**(10), 762–772 (1977)
4. Karp, R., Rabin, M.: Efficient randomized pattern-matching algorithms. IBM J. Res. Dev. **31**(2), 249–260 (1987)
5. Yao, A.C.: The complexity of pattern matching for a random string. SIAM J. Comput. 368–387 (1979)
6. Tran, N-P., Lee, M, Hong, S., Choi J.: High throughput parallel implementation of Aho-Corasick algorithm on a GPU. In: IEEE 27th International Symposium on Parallel and Distributed Processing Workshops and Ph.D. Forum (2013)
7. Kim, Hwee, Han, Yo-Sub: OMPPM: online multiple palindrome pattern matching. Bioinformatics **32**(8), 1151–1157 (2016)
8. Lin, P-C., Li, Z-X., Lin, Y-D.: Profiling and accelerating string matching algorithms in three network content security applications. In: IEEE Communications Surveys 2nd Quarter, vol. 8, no. 2 (2006)
9. Sharma, S., Jaju, A., Singh, N., Pal, P.K., Raju, Y.S., Rama Krishna Murthy, G.V.: Robotic system for stacking PHWR nuclear fuel pellets using machine vision. In: Characterization and Quality Control of Nuclear Fuels (CQCNF-2013) (2013)
10. Singh, N., Jaju, A., Sharma, S., Pal, P.K.: Online vision-based measurement of stacks of nuclear fuel pellets in a tray. In: Advances in Robotics Conference (2015)
11. Bulnes, F.G., Usamentiaga, R., Garcia, D.F., Molleda, J.: Detection of periodical patterns in the defects identified by computer vision systems. In: 11th International Conference on Intelligent Systems Design and Applications (2011)

12. Dori, S., Landau, G.M.: Construction of Aho Corasick automaton in linear time for integer alphabets. Elsevier (2006)
13. Altman, D.G., Bland, J.M.: Diagnostic tests. 1: Sensitivity and specificity. BMJ **308**(6943), 1552 (1994)

On the Choice of Coil Combination Weights for Phase-Sensitive GRAPPA Reconstruction in Multichannel SWI

Sreekanth Madhusoodhanan and Joseph Suresh Paul

Abstract The feasibility of applying Generalized Auto-calibrating Partially Parallel Acquisition (GRAPPA) techniques has been established, with a two-fold or more reduction in scan time without compromising vascular contrast in Susceptibility Weighted Imaging (SWI) by choosing an optimal sensitivity map for combining the coil images. The overall SNR performance in GRAPPA is also dependent on the weights used for combining the GRAPPA reconstructed coil images. In this article, different methods for estimating the optimal coil combination weights are qualitatively and quantitatively analysed for maximizing the structural information in the tissue phase. The performance of various methods is visually evaluated using minimum Intensity Projection (mIP), Among the three methods, sensitivity estimated using the dominant eigenvector mentioned as ESPIRiT-based sensitivity in this article shows superior performance over the other two methods including estimating the sensitivity from the centre k-space line and from reconstructed channel images. Combining channel images using ESPIRiT sensitivity shows its ability to preserve the local phase variation and reduction in noise amplification.

Keywords Coil combination · Parallel imaging · Sensitivity · SWI

1   Introduction

Susceptibility Weighted Imaging (SWI) is an MR imaging technique which utilizes the susceptibility variations between tissues in human body to image them [1–8]. These variations in the susceptibility of tissues are embedded in the phase of the MR image [1, 9]. The acquisition parameters are so chosen such that sufficient susceptibility contrast was obtained from vessels while simultaneously reducing

S. Madhusoodhanan (✉) · J. S. Paul
Medical Image Computing and Signal Processing Group, Indian Institute of Information Technology and Management-Kerala (IIITM-K), Trivandrum, Kerala, India
e-mail: sreekanth.m@iiitmk.ac.in

J. S. Paul
e-mail: j.paul@iiitmk.ac.in

© Springer Nature Singapore Pte Ltd. 2020
B. B. Chaudhuri et al. (eds.), *Proceedings of 3rd International Conference on Computer Vision and Image Processing*, Advances in Intelligent Systems and Computing 1022, https://doi.org/10.1007/978-981-32-9088-4_10

contrast among white matter, grey matter and ventricles [10]. As a result, the total acquisition time for SWI remains long. This results in motion-induced artefacts and patient discomfort. Thus, there is a need for faster acquisition time and efficient ways of combining multichannel coil data without losing the phase information.

Partially Parallel Imaging (PPI) acquisitions and reconstruction algorithms have been used commonly in clinical applications to speed up the MR acquisition without significantly reducing the Signal-Noise-Ratio (SNR) and contrast between vessels and other brain regions. SENSitivity Encoding (SENSE) [11] is an image domain based parallel imaging method which is highly subjected to B0 and B1 field inhomogeneity. Auto-calibrating techniques such as GeneRalized Auto-calibrating Partially Parallel Acquisitions (GRAPPA) can better handle the field inhomogeneities because GRAPPA reconstruction makes use of the fully sampled centre k-space data known as the auto-calibrating lines, whereas SENSE relies on coil sensitivity maps to estimate the unaliased image [12, 13]. Recent work has demonstrated the application of auto-calibrated data-driven parallel imaging methods to phase-sensitive imaging without any apparent impact on image phase. Lupo et al. [14] have demonstrated that GRAPPA reconstruction shows improved vessel contrast as compared to SENSE when applied to SWI Brain vasculature data acquired at 7T with reduction factor (R = 2).

The feasibility of accelerating SWI acquisitions by using parallel imaging techniques has shown that GRAPPA has an advantage over SENSE in terms of robustness in detecting small vessels [14]. In this work, GRAPPA reconstruction is applied on undersampled SWI images, which are then coil combined using three different types of coil combination weights suggested in the literature [15, 16]. The quality of phase-sensitive reconstruction is assessed by the application of venous enhancement filtering on the final combined image obtained using each coil combination weights.

2   Materials and Methods

2.1   Data Acquisition

Dataset-1: SWI is collected from volunteers using a Siemens 1.5T scanner with uniform excitation by a volume transmitter and reception by 32-channel head array coil. TE/TR values were 40/49 ms with a nominal flip angle (FA) of 20° and slice thickness 2.1 mm. The collected data was retrospectively undersampled by a factor of three prior to performing GRAPPA. The number of auto-calibrating (nACS) lines used is 24; Field of view (FOV) = 203×250 mm and the reconstructed matrix size is 260×320.

Dataset-2: SWI data is from a patient with multiple micro haemorrhages seen in the cerebellum and left temporal lobe. The acquisition was performed with a Siemens 1.5T scanner using 32 channel head array coils. The scan parameters were TE = 40 ms, TR = 49 ms, FA = 20°, slice thickness 2.1 mm, FOV = 203×250 and the reconstructed matrix size is 260×320.

Dataset-3: Multi-echo SWAN data was acquired on GE 3T MRI system equipped with a 12-channel head array coil using a three-dimensional gradient echo sequence. Scan parameters were as follows: repetition time (TR) = 42.6 ms, TE = 24.7 ms, flip angle (FA) = 15°, slice thickness of 2.4 mm acquisition matrix of 384 × 288, bandwidth (BW) = 325.52 Hz/pixel and FOV = 260 × 260 mm.

2.2 Reconstruction and Post-processing of Complex Multichannel Data

SWI data was acquired and retrospectively undersampled by an acceleration factor (R) while retaining central Auto-Calibrating Signal (ACS) lines. The GRAPPA-based parallel imaging technique employed a two-dimensional 2 × 7 interpolation kernel using two neighbouring points in the Phase Encode (PE) direction, placed symmetrically around the missing line and seven neighbouring data points in the Frequency Encode (FE) axis. After reconstruction, various post-processing techniques were employed as shown in Fig. 1 to obtain the final image.

2.3 GRAPPA Reconstruction

GRAPPA is a parallel MRI technique to estimate the missing k-space points by linearly combining more than one acquired line. GRAPPA uses the fully sampled centre k-space region (ACS lines) of each channel for training to estimate the unacquired data. Using a linear combination of the acquired PE lines and GRAPPA weights computed from ACS data, the unacquired points are estimated. The calibration step for GRAPPA weight calculation and estimation of unacquired points in the outer k-space are the two main steps in the reconstruction procedure. Calibration is performed using samples from ACS lines, forming the training data. Let k_y be the index of the acquired line in the phase encode direction, then the target data $K^l(k_y + \eta \Delta k_y, k_x)$ in the basic GRAPPA model with acceleration factor R and η varies from 1 to R−1, which can be defined from the source data $\mathbf{k}(k_y, k_x)$ at the location (k_y, k_x) as,

$$K^l\left(k_y + \eta \Delta k_y, k_x\right) = \mathbf{z}_\eta^l \mathbf{k}\left(k_y, k_x\right) \tag{1}$$

where $\mathbf{k}(k_y, k_x)$ represents a collection of acquired neighbouring samples for a particular location from all coils and $\mathbf{z}_\eta^l$ is the GRAPPA weights. For all training pairs in ACS $\{\mathbf{k}(k_y, k_x), K^l(k_y + \eta \Delta k_y, k_x)\}$ represented the calibration weights $\mathbf{z}_\eta^l$ are first estimated by inversion of (1). Though all locations within ACS are truly acquired, k_y refers to only those locations that strictly follow the sampling pattern in the outer k-space. The calibration matrix Π is now obtained by row-wise appending of the training set $\mathbf{k}(k_y, k_x)$ corresponding to each acquired ACS location following the

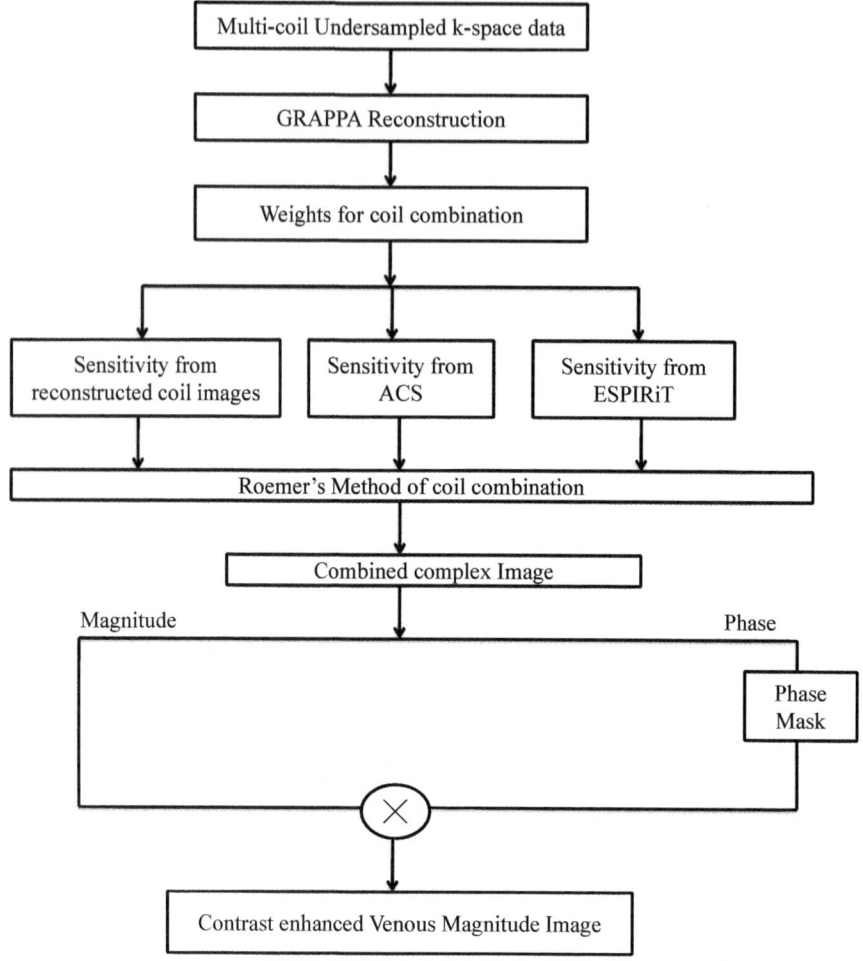

Fig. 1 Work flow showing venous enhancement

sampling pattern. The observation vector $\mathbf{k}_u$ for calibration is obtained from each training pair by considering the elements $K^l(k_y + \eta \Delta k_y, k_x)$. By determining the least squares (LS) solution, the calibration process is defined as

$$\mathbf{z}_\eta^l = \left(\Pi^H \Pi\right)^{-1} \Pi^H \mathbf{k}_u \tag{2}$$

Missing k-space value $K^l(k_y + \eta \Delta k_y, k_x)$ in the outer k-space is estimated by applying the filter obtained from calibration $\mathbf{z}_\eta^l$ to the acquired dataset $\mathbf{k}(k_y, k_x)$. In GRAPPA reconstruction, the unacquired points in the sample k-space for jth coil is estimated as

$$K^j\left(k_x, k_y + \eta \Delta k_y\right) = \sum_{l=1}^{n_c} \sum_{b=-P_l}^{P_k} \sum_{h=-F_l}^{F_k} K^l\left(k_y + bR\Delta k_y, k_x + h\Delta k_x\right)z_\eta^l \quad (3)$$

where h and b denote sampling indices along the frequency and phase-encoding directions, respectively.

2.4 Methods for Coil Combination

Roemer et al. demonstrated an optimal way to combine the data from multiple channels using an estimated or true sensitivity profile [15]. The combined complex image (I_R) is obtained as

$$I_R = \frac{\sum_{k=1}^{n_c} P_k^* S_k}{\sum_{k=1}^{n_c} |P_k|^2} \quad (4)$$

Sensitivity estimate from coil images. After estimation of unacquired points using GRAPPA reconstruction, the k-space data is inverse Fourier transformed to obtain the channel images. Absolute value of each coil image is divided by root-Sum-of-Squares (rSoS) image to estimate the coil sensitivity map from the reconstructed channel image.

Sensitivity from ACS. One of the most common ways of obtaining the sensitivity is from the Auto-Calibrating lines (ACS) of the undersampled k-space. The central ACS data (e.g. 24 × 24) is zero-padded to the size of the image and inverse Fourier transformed to obtain the low-resolution sensitivity map.

ESPIRiT-based sensitivity estimation. This procedure for estimating the coil sensitivity has evolved from efficient eigenvector-based implementation in the calibration steps of the original SPIRiT approach and its parallel imaging implementation is referred to as ESPIRiT [16]. In ESPIRiT, the calibration matrix is first decomposed using Singular Value Decomposition (SVD) into a left singular matrix U, a right singular matrix V and a diagonal matrix S of singular values [10]. The basis for the rows of the calibration matrix Π is obtained from the columns of the V matrix in the SVD of calibration data. The matrix V is then separated into V⊥ which spans the null space of Π and V. In ESPIRiT, the calibration matrix is first decomposed using Singular Value Decomposition (SVD) into a left singular matrix U, a right singular matrix V and a diagonal matrix S of singular values [17]. The basis for the rows of the calibration matrix Π is obtained from the columns of the V matrix in the SVD of calibration data. The matrix V is then separated into V⊥ which spans the null space of Π and V∥ which spans its row space. In ESPIRiT, the coil combination weights are derived from the dominant eigenvectors of the subspace spanned by the row space of

Π. Similarly, these weights from dominant eigenvectors can be used for combining the channel images after GRAPPA reconstruction, which spans its row space.

2.5 Homodyne Filtering

Homodyne filter is a type of high-pass filter which is widely used to remove the phase variations due to main field inhomogeneities and air-tissue interface. In Susceptibility imaging, the tissue susceptibility effects are local and exist primarily in high spatial frequencies, while external field effects exist primarily in the low spatial frequencies [18]. The high-pass filter is obtained by taking the original image $\rho(r)$, truncating it to the central complex image $\rho_n(r)$, creating an image by zero filling the elements outside the central elements and then complex dividing $\rho(r)$ by $\rho_n(r)$ to obtain a new image.

$$\rho'(r) = \rho(r)/\rho_n(r) \tag{5}$$

The phase masks scaled between 0 and 1 are designed to enhance the contrast in the original magnitude image by suppressing pixels having certain phase values. The venous enhanced image is obtained by element-by-element multiplication of the original magnitude image with the phase mask. Usually, the phase mask is multiplied 'r' times to increase the susceptibility-related contrast. In this article, results are shown with r = 4.

3 Results

Figure 2a–c shows mIP SWI obtained using from sensitivity estimated from GRAPPA reconstructed images, sensitivity from ACS data and ESPIRiT-based sensitivity,

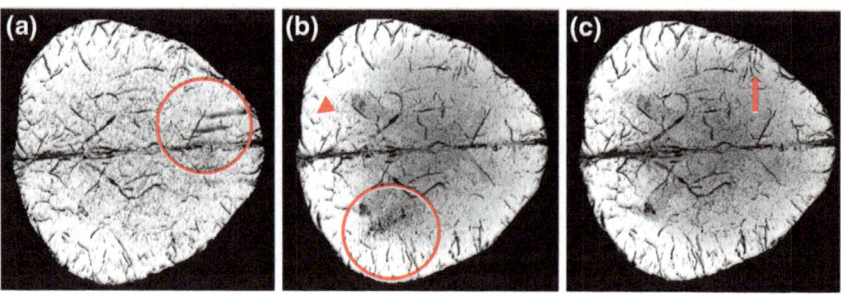

Fig. 2 mIP SWI image obtained after combining the channel images using coil sensitivity estimated from **a** reconstructed coil images. **b** ACS. **c** ESPIRiT sensitivity

respectively. GRAPPA reconstruction was performed on undersampled data with an acceleration of R = 3. From visual inspection, it is observed that combing the channel information using sensitivity estimated from GRAPPA reconstructed images in 2a is highly subjected to streaking artefacts and background noise. The streaks are highlighted using red circles. The reconstruction in 2b is comparatively less noisy. However, small vessels (see arrows) are not detected. ESPIRiT reconstruction in 2c detects more vessels (bold red arrow). Furthermore, background noise appearance is also minimal.

The effect of sensitivity in channel combination is shown using patient data with multiple microbleeds mainly in the left temporal region (Dataset-2). Figure 3 shows the mIP over 16.8 mm (8 slices) of the magnitude SWI obtained after GRAPPA reconstruction with R = 2. Left to right panel shows channel combined image using sensitivity estimated from a GRAPPA reconstructed images, b ACS and c ESPIRiT-based sensitivity, respectively. From the figure, it is observed that mIP SWI images obtained from combined channel images with ESPIRiT sensitivity show better SWI blooming and susceptibility-related contrast. Yellow arrows are used to highlight the better delineation of microbleed when the sensitivity is estimated from ESPIRiT. Red arrows are used to point out the improved resolution in venous structures when the channel combination weights are estimated from dominant eigenvectors of the calibration matrix.

Figure 4 shows the mIP over 9.6 mm (4 slices) of magnitude SWI images of a patient with bleeding in the left frontoparietal convexity (Dataset-3). a–c shows the channel combined image using sensitivity estimated from GRAPPA reconstructed images, centre k-space data and ESPIRiT-based sensitivity, respectively. SWI blooming is observed to be better in panel c compared to both a and b. Red arrows are used to highlight the improvement in venous contrast of the mIP SWI image using ESPIRiT-based sensitivity for channel combination. Even though the superficial veins are visible in panel a and b, better delineation and contrast are accomplished with chan-

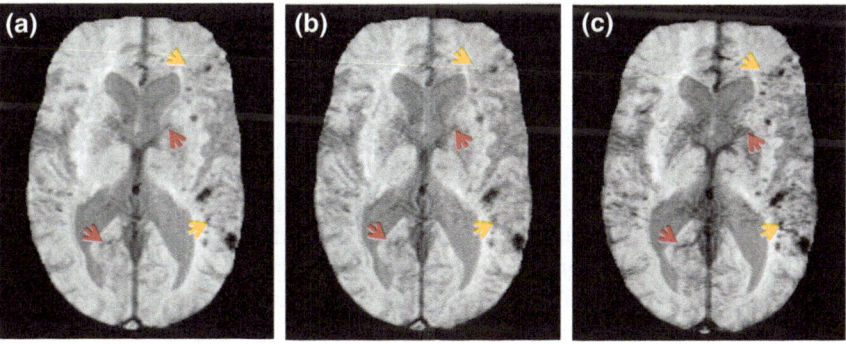

Fig. 3 Reconstructed mIP images from Dataset-2 using sensitivity estimated from **a** reconstructed coil images **b** ACS. **c** ESPIRiT sensitivity. Red arrows are used to indicate the improvement in resolution of faint venous structures and yellow arrows highlight the blooming due to microbleeds in the coil combined image with sensitivity estimated from ESPIRiT

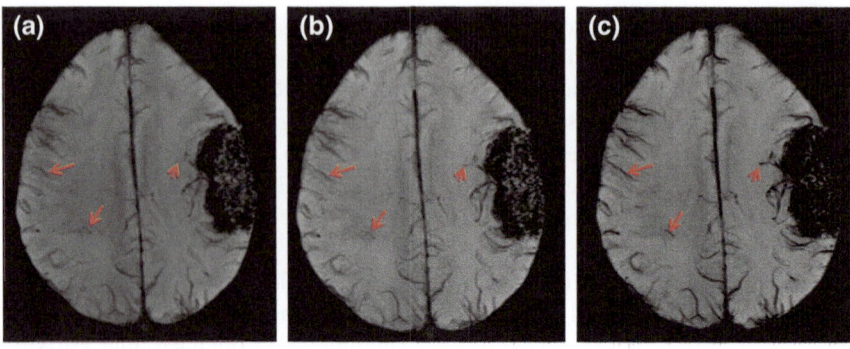

Fig. 4 Reconstructed mIP images from Dataset-3 using sensitivity estimated from **a** reconstructed coil images **b** ACS. **c** ESPIRiT sensitivity. Red arrows indicate the clear visualization of venous structures in panel (**c**)

nel combination using the high-quality sensitivity maps estimated from the SVD of the calibration matrix.

4 Discussion

In recent years, several methods for combining the channel images from array coils have been proposed based on computational requirement and the properties of the combined image. The most common and computationally less intense method for combining the multichannel phase data is the Roemer's method which requires an additional scan with a volume reference coil such as body coil to estimate the coil sensitivity map. The quality of the phase image is directly related to the quality of the sensitivity map used. Improper selection of the sensitivity map may cause phase artefacts and signal loss in the channel combined image. So the main aim of every channel combination method is to develop an SNR optimal coil combination method that retrieves the relative change in a local magnetic field which is related to the local phase change without any additional measurements or scanning.

Three methods have been presented for generating the sensitivity map for combining channel images with array coils. Neither of these methods requires an additional scan. The first method generates the sensitivity after reconstructing the channel images using GRAPPA. The second and third methods do not require the reconstructed channel images for generating the sensitivity map, but they use the auto-calibration lines. In the first method, since the sensitivity maps are generated using the reconstructed channel image, noise amplification from GRAPPA causes error in the channel combined image. This effect becomes more critical when a higher acceleration factor is used.

In the second method, the coil sensitivities are generated from centre k-space data which contains the low-frequency component of the channel image together with some noise components. These noise components may affect the quality of the channel combined image especially when more importance is given to the visualization of finer vascular structures as in SWI. So it is important to suppress the noise component in the sensitivity map. This points to the need for estimating the sensitivity map with high precision. So a well-established method like ESPIRiT can be used to estimate the sensitivity maps from the calibration data which is fully sampled. The sensitivity maps estimated from the dominant eigenvectors of the subspace spanned by the row space of calibration matrix is suitable for combining the coil information with high frequency phase variations as in SWI. Moreover, by using the principle of SVD and choosing the dominant eigenvector of calibration matrix, sensitivity estimation can be made noise-free when compared to that of estimating the sensitivity directly from centre k-space region.

Combination of channel images effectively by preserving the phase variation and reduce the phase artefacts will be beneficial for higher resolution gradient echo scan. Potential of this channel combination method is that it can be used for improving the relative phase change between different ROIs and creating susceptibility maps. With proper background suppression in the channel combined phase image of SWI/SWAN acquisition, quantification of iron deposition in patients with Alzheimer's and Parkinson's can be made robust and reliable.

5 Conclusion

We have evaluated the three methods for generating the sensitivity map for combining channel images from multichannel phased array coils. Neither of these methods requires additional measurements or a volume reference coil making the methods suitable for all field strength. Among the three methods, sensitivity estimated using the dominant eigenvector mentioned as ESPIRiT-based sensitivity in the article shows superior performance over the other two in terms of its ability to preserve the local phase variation and reduction in noise amplification. mIP SWI maps shown the superiority of ESPIRiT-based sensitivity to that achieved with sensitivity estimated from the reconstructed images and centre k-space data.

Acknowledgements The authors are thankful to the Council of Scientific and Industrial Research-Senior Research Fellowship (CSIR-SRF, File No: 09/1208(0001)/2018.EMR-I) and planning board of Govt. of Kerala (GO(Rt) No. 101/2017/ITD.GOK(02/05/2017)), for financial assistance.

References

1. Haacke, E.M., Xu, Y., Cheng, Y.C., Reichenbach, J.R.: Susceptibility weighted imaging (SWI). Magn. Reson. Med. **52**(3), 612–618 (2004)

2. Wycliffe, N.D., Choe, J., Holshouser, B., Oyoyo, U.E., Haacke, E.M., Kido, D.K.: Reliability in detection of hemorrhage in acute stroke by a new three-dimensional gradient recalled echo susceptibility-weighted imaging technique compared to computed tomography: a retrospective study. J. Magn. Reson. Imaging **20**(3), 372–377 (2004)
3. Sehgal, V., Delproposto, Z., Haacke, E.M., Tong, K.A., Wycliffe, N., Kido, D.K., Xu, Y., Neelavalli, J., Haddar, D., Reichenbach, J.R.: Clinical applications of neuroimaging with susceptibility-weighted imaging. J. Magn. Reson. Imaging **22**(4), 439–450 (2005)
4. Haacke, E.M., Cheng, N.Y., House, M.J., Liu, Q., Neelavalli, J., Ogg, R.J., Khan, A., Ayaz, M., Kirsch, W., Obenaus, A.: Imaging iron stores in the brain using magnetic resonance imaging. Magn. Reson. Imaging **23**(1), 1–25 (2005)
5. Sehgal, V., Delproposto, Z., Haddar, D., Haacke, E.M., Sloan, A.E., Zamorano, L.J., Barger, G., Hu, J., Xu, Y., Prabhakaran, K.P., Elangovan, I.R.: Susceptibility-weighted imaging to visualize blood products and improve tumor contrast in the study of brain masses. J. Magn. Reson. Imaging **24**(1), 41–51 (2006)
6. Haacke, E., Makki, M.I., Selvan, M., Latif, Z., Garbern, J., Hu, J., Law, M., Ge, Y.: Susceptibility weighted imaging reveals unique information in multiple-sclerosis lesions using high-field MRI. In: Proceedings of International Society for Magnetic Resonance in Medicine, vol. 15, p. 2302 (2007)
7. Tong, K.A., Ashwal, S., Obenaus, A., Nickerson, J.P., Kido, D., Haacke, E.M.: Susceptibility-weighted MR imaging: a review of clinical applications in children. Am. J. Neuroradiol. **29**(1), 9–17 (2008)
8. Roh, K., Kang, H., Kim, I.: Clinical applications of neuroimaging with susceptibility weighted imaging. J. Korean Soc. Magn. Reson. Med. **18**(4), 290–302 (2014)
9. Haacke, E.M., Mittal, S., Wu, Z., Neelavalli, J., Cheng, Y.C.: Susceptibility-weighted imaging: technical aspects and clinical applications, part 1. Am. J. Neuroradiol. **30**(1), 19–30 (2009)
10. Wang, Y., Yu, Y., Li, D., Bae, K.T., Brown, J.J., Lin, W., Haacke, E.M.: Artery and vein separation using susceptibility-dependent phase in contrast-enhanced MRA. J. Magn. Reson. Imaging **12**(5), 661–670 (2000)
11. Pruessmann, K.P., Weiger, M., Scheidegger, M.B., Boesiger, P.: SENSE: sensitivity encoding for fast MRI. Magn. Reson. Med. **42**(5), 952–962 (1999)
12. Griswold, M.A., Jakob, P.M., Heidemann, R.M., Nittka, M., Jellus, V., Wang, J., Kiefer, B., Haase, A.: Generalized autocalibrating partially parallel acquisitions (GRAPPA). Magn. Reson. Med. **47**(6), 1202–1210 (2002)
13. Wang, Z., Wang, J., Detre, J.A.: Improved data reconstruction method for GRAPPA. Magn. Reson. Med. **54**(3), 738–742 (2005)
14. Lupo, J.M., Banerjee, S., Kelley, D., Xu, D., Vigneron, D.B., Majumdar, S., Nelson, S.J.: Partially-parallel, susceptibility-weighted MR imaging of brain vasculature at 7 Tesla using sensitivity encoding and an autocalibrating parallel technique. In: 28th Annual International Conference of the IEEE Engineering in Medicine and Biology Society, 2006. EMBS'06, pp. 747–750 (2006)
15. Roemer, P.B., Edelstein, W.A., Hayes, C.E., Souza, S.P., Mueller, O.M.: The NMR phased array. Magn. Reson. Med. **16**(2), 192–225 (1990)
16. Uecker, M., Lai, P., Murphy, M.J., Virtue, P., Elad, M., Pauly, J.M., Vasanawala, S.S., Lustig, M.: ESPIRiT—an eigenvalue approach to autocalibrating parallel MRI: where SENSE meets GRAPPA. Magn. Reson. Med. **71**(3), 990–1001 (2014)
17. Horn, R. A., Johnson, C. R.: Matrix analysis. Cambridge University Press (1985)
18. Abduljalil, A.M., Schmalbrock, P., Novak, V., Chakeres, D.W.: Enhanced gray and white matter contrast of phase susceptibility-weighted images in ultra-high-field magnetic resonance imaging. J. Magn. Reson. Imaging **18**(3), 284–290 (2003)

Detection of Down Syndrome Using Deep Facial Recognition

Ankush Mittal, Hardik Gaur and Manish Mishra

Abstract Down syndrome is a genetic disorder that affects 1 in every 1000 babies born worldwide. The cases of Down syndrome have increased in the past decade. It has been observed that humans with Down syndrome generally tend to have distinct facial features. This paper proposes a model to identify people suffering from Down syndrome based on their facial features. Deep representation from different parts of the face is extracted and combined with the aid of Deep Convolutional Neural Networks. The combined representations are then classified using a Random Forest-based pipeline. The model was tested on a dataset of over 800 individuals suffering from Down syndrome and was able to achieve a recognition rate of 98.47%.

Keywords KNN · CNN · HOG · Machine learning · Computer vision · Random forest · Healthcare · Down syndrome

1 Introduction

Down Syndrome (DS) is a genetic disorder caused by an extra copy of chromosome 21 in humans. The problem being addressed here is to identify people affected with Down syndrome so that they can be provided with special care and help. The cases of Down syndrome have been steadily on the rise for the past few years. According to data available, 1 in every 691 babies in the USA is born with Down syndrome, which is approximately 6000 infants per year. Currently, there are about 40,000 people with Down syndrome living in the USA. Therefore, there must be some efficient way to recognize people affected with Down syndrome so that they may be given special attention. The proposed research focuses on building a model capable

A. Mittal · H. Gaur (✉) · M. Mishra
Graphic Era (Deemed to be University), Dehradun, India
e-mail: hardikgaur@geu.ac.in

A. Mittal
e-mail: dr.ankush.mittal@gmail.com

M. Mishra
e-mail: 21manishmishra@gmail.com

© Springer Nature Singapore Pte Ltd. 2020
B. B. Chaudhuri et al. (eds.), *Proceedings of 3rd International Conference on Computer Vision and Image Processing*, Advances in Intelligent Systems and Computing 1022, https://doi.org/10.1007/978-981-32-9088-4_11

of recognizing humans with Down syndrome based on their facial attributes. The proposed model can be used for diagnosis as well as for assistance in tracking and recognizing subjects with Down syndrome. The intuition behind the proposed model is that if we as humans can identify humans with Down syndrome based on their facial features why can't computer systems do the same?

Most of the earlier works on recognizing Down syndrome have been trained and tested on a relatively small dataset. The prime reason is that images of humans with Down syndromes are not that easily available. The proposed model takes into consideration a dataset of 853 subjects with Down syndrome. This model contributes to the development based on the detection of Down syndrome using deep learning algorithms [9]. For this purpose, the model makes use of the Convolution Neural Network as its base algorithm. A dataset consisting of the images of normal people as well as images of people suffering from Down Syndrome has been used for training as well as testing purposes.

Facial dysmorphism is a common feature in subjects with Down syndrome. In a study conducted by Cohen and Winer [3], it was observed that subjects with Down syndrome have significant distortion near the eye and the mouth region. Taking the analogy into consideration, the proposed model builds a model to extract those regions. Deep representations are then extracted from these regions and are used for building the model.

The detection of Down syndrome poses challenges due to the following reasons:

- There are no common facial attributes for the people suffering from Down syndrome. Each person's face is affected differently and thus model must be efficient enough to identify people correctly.
- People suffering from Down Syndrome are increasingly leading a normal life due to awareness about this condition. They are increasingly being incorporated in the workforce which requires them to travel to other places. To recognize such people from the crowd so that they can be administered correct treatment when required calls for an efficient model capable enough to identify the subjects accurately.

The remaining part of the paper has been organized as follows: Sect. 2 describes the various causes, symptoms, and attributes of humans with Down syndrome. A discussion of the existing work on recognizing humans with Down syndrome has been carried out in Sect. 3. The proposed architecture has been described in Sect. 4. The various experiments that were conducted have been presented and discussed in Sect. 5.

2 Down Syndrome

Down Syndrome is a chromosomal disorder. It is caused due to the presence of extra chromosome 21 in the cells of the affected person. The extra copy can be partial or full. The person with Down syndrome has forty-seven chromosomes instead of the normal forty-six chromosomes in a healthy individual. When a baby develops it receives twenty-three pairs of chromosomes (half from father and half from mother). When the cells divide these pairs of chromosomes are equally partitioned. But in some people, chromosome 21 does not get partitioned properly. This gives rise to abnormal chromosome numbers in the individual resulting in Down syndrome. Worldwide 1 in every 1000 babies is born with Down syndrome. People with Down syndrome develop various physical and mental characteristics which include the following. The physical characteristics are as follows:

- Flat facial features
- Small head and ears
- Short neck
- Abnormally large tongue
- Eyes that slant upward
- Poor muscle tone
- Short height.

The mental characteristics of humans with Down syndrome are as follows:

- Short attention span
- Impulsive behavior
- Slow learning
- Delayed language and speech development.

Down syndrome is peculiarized by cognitive disabilities, i.e., intellectual and developmental delays which include learning disabilities and speech impairment. This is because Down syndrome affects the hippocampus region of the brain that is responsible for memory and learning processes. People suffering from Down syndrome are also at a greater risk of suffering from following health conditions:

- Leukemia
- Obesity
- Chronic constipation
- Sleep apnea (Interrupted breathing during sleep)
- Poor vision
- Cataracts
- Strabismus
- Anemia
- Congenital heart defects
- Hearing loss
- Thyroid diseases.

There are three different types of chromosomal patterns resulting in Down syndrome.

2.1   Trisomy 21

Trisomy 21 is a condition where a person has forty-seven chromosomes instead of normal forty-six, i.e., (more than twenty-three sets of chromosomes), which is caused by a faulty cell division. Chromosome 21 is one amongst the given twenty-three pairs of chromosomes and is also the smallest human autosome (not a sex chromosome). A normal person has two copies of this chromosome but a person suffering from Down syndrome has three copies of chromosome 21, because at the time of fertilization the 21st chromosomes of the parent are unable to separate, thus forming a group containing three chromosomes. The extra chromosome is simulated within each cell present in the body through cell division. Trisomy 21 is the most common amongst people suffering from Down syndrome. It accounts for 95% of all cases.

2.2   Mosaicism

Mosaicism is the condition in which the individual with Down syndrome has two different types of cells. The first with 46 chromosomes (twenty-three sets) whereas the second ones are with forty-seven chromosomes which contains an extra copy of 21st chromosome. This is the rarest of all three as it constitutes 1% of all Down syndrome cases.

2.3   Translocation

It constitutes 3–4% of the total cases of Down syndrome. This condition arises when the part of chromosome 21st pair merges with the chromosome of 14th pair. While the number of chromosomes in the cell remains 46, the presence of an extra part of chromosome 21 causes Down syndrome.

3   Related Work

The literature review on the previous works on recognizing Down syndrome has been presented in this section. Saraydemir et al. [13] proposed a model to detect Down syndrome that relied on Gabor wavelets for representation. This was followed by the principle component analysis and LDA and then obtaining results by applying SVM (Support Vector Machine) and K-nearest neighbor on 15 healthy individuals and 15 individuals with Down syndrome. The model was able to achieve an overall recognition rate of 97.3 and 96%. A model made with the help of LBP and template matching for Down syndrome detection from cropped images was then proposed

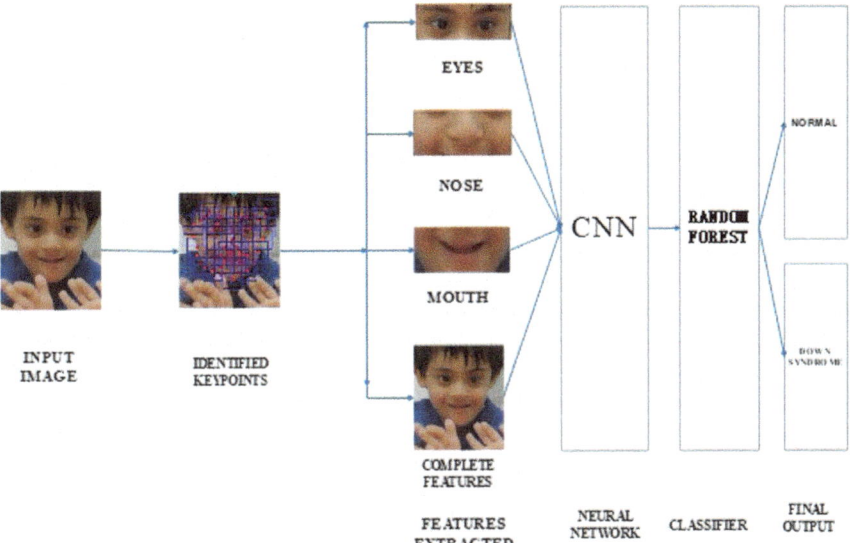

Fig. 1 A block diagram representation of the proposed approach

by Bruçin and Vasif [1]. The concept of cropped images helped us perceive the important features of the human face with the help of local binary pattern. Further, a hierarchical constraint model was proposed by Zhao et al. [15]. The model resulted in an accuracy of 0.97 when tested on 80 healthy subjects and 50 subjects with Down syndrome. There have also been studies that have focused on detecting a spectrum of disorders that also include Down syndrome. Kuru et al. [10] proposed a model to detect five different developmental disorders and the model was able to achieve an accuracy of 79% when tested on a dataset of 25 different students. Ferry et al. [6] proposed a framework for recognizing eight different developmental disorders with the aid of active appearance models [4] on a dataset of 1363 images. Shukla et al. [14] proposed a model to identify different disorders but they included seven different disorders. The proposed model is different as it detects Down syndrome only and the model has much higher accuracy in detecting Down syndrome, as it uses Random Forest as the classifier, than the mentioned model (Fig. 1).

4 Proposed Architecture

Given the distinct differences in the eye and mouth patterns of subjects with Down syndrome, the proposed framework tries to capture essential information from the respective parts of the faces. Since the dataset used for the project was small, therefore to train the neural network better, different parts of the face namely, the region near the eyes, lips, and cheeks were extracted. All three cropped regions along with the full image were then provided as input for the CNN for extracting essential features, which were then combined to form a single feature vector. The entire framework can be divided into three parts: preprocessing, extraction of facial features, and classification.

4.1 Preprocessing and Region Extraction

A mixture of trees with a shared pool of parts approach as described in [16] is applied for facial key point extraction. The extracted facial key points were then used for extracting different parts of the face. The approach is highly robust with the ability to extract faces under varying poses and illumination F_i. The face is represented by a set of key points. The process of key point detection and region extraction have been shown in Fig. 2. Each tree $T_m = (V_m, E_m)$ is represented using a linearly parameterized structure where V_m represents the mixture of a shared pool of parts.

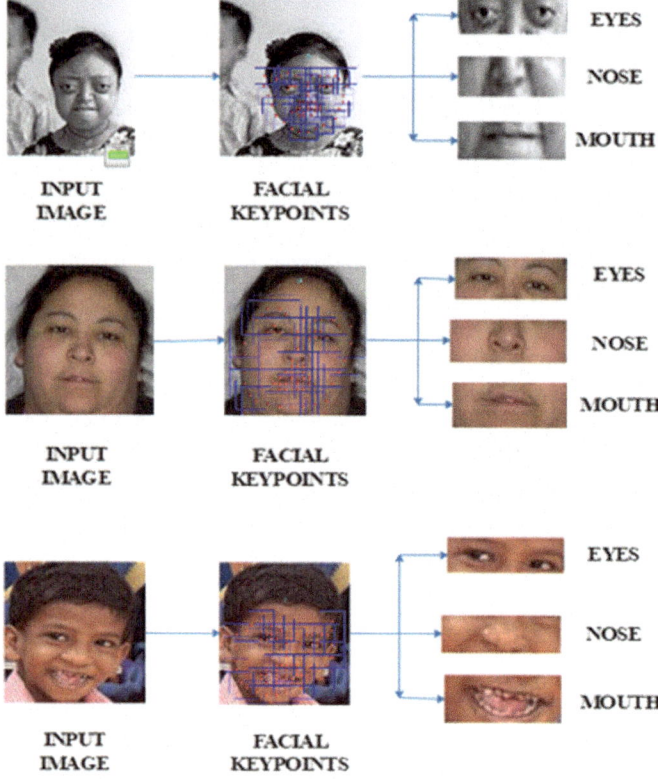

Fig. 2 Facial key points are extracted initially from the faces of subjects. They are followed by extracting different parts of the face

A configuration of parts $L = I(i): iV$ is used for defining a given location l_i in an image I. Each configuration of parts has the following components:

$$S(I, L, m) = A_m(I, L) + Y_M(L) + \alpha^m \tag{1}$$

$$A_m(I, L) = \Sigma w_i^m \cdot \phi(I, l_i) \tag{2}$$

$$Y_m(L) = \Sigma \left(a_{ij}^m dx^2 + b_{ij}^m dx + c_{ij}^m dx^2 + d_{ij}^m dy \right) \tag{3}$$

In the aforementioned equations, $\phi(I, l_i)$ represents the feature vector calculated using feature descriptor like Histogram of Oriented Gradients (HOG). Equation 3 is the summation of the spatial arrangements of L parts where dx and dy are the relative displacements of the i-th part to the j-th part and are given as

$$d_x = x_i - x_j \tag{4}$$

The parameters a, b, c, and d are used for specifying the rigidity and rest location of each spring and every mixture m has an associated prior or scalar bias α^m. The term S(I, L, m) in Eq. 1 is maximized by enumerating all the mixtures and finding the most desired arrangement of parts. The inner maximization for each tree is achieved by dynamic programming and is given as

$$S^*(l) = max_m[max_L\{S(I, L, m)\}] \tag{5}$$

Positive images containing labeled faces along with a mixture of facial landmarks and negative samples are used for training the model in a supervised manner. A structured prediction framework is deployed for learning both the appearance and shape parameters. The maximum likelihood structure for representing the location of landmarks is calculated using the Chow and Liu [2] algorithm. The structured prediction function is defined with the help of labeled positive and negative examples that are represented by I_n, L_n, m_n, and I_n, respectively. Let $I_n = \{L_n, m_n\}$. Now Eq. 1 can be written as

$$S(I, z) = \beta \cdot \phi(I, z) \tag{6}$$

where β is a concatenation of α, ω and (a, b, c, d).

4.2 Deep CNN's for Extracting Representations

The recent researches [8, 12] have proved the superiority of CNN over other methods like HOG [5] and LBP [11] for facial representation. However, CNN requires a lot of data for its implementation. Therefore, for our purpose, we used pretrained models for efficient training of algorithm as our dataset was relatively small. It is based on the pretrained Alex-Net model [9]. All the cropped images were resized to 227 × 227 as it is the requirement of Alex-Net model. The network is calibrated on the LFW database [7]. This database consists of 5749 people providing 13,233 images out of which 1680 people had more than one image. The calibration does not cause any structural change in Alex-Net model.

The initial learning rate for the final layer was equal to 0.001 and it was 0.005 for the remaining layers. The original network was trained for 50000 iterations with a dropout of 0.5 and the momentum of 0.75. In this model, we extracted different features using different Convolution Neural Networks and finally, all the features were concatenated together into a single feature vector that was used to train and test the model.

4.3 Classification Using Support Vector Machines

Support Vector Machines generally tend to perform well with deep representation. Support Vector Machines work by identifying an optimal hyperplane that divides the data into different classes. Since we had only two classes, we used SVM as a binary classifier. The feature vector obtained from CNN was then classified using SVM classifier (Figs. 3 and 4).

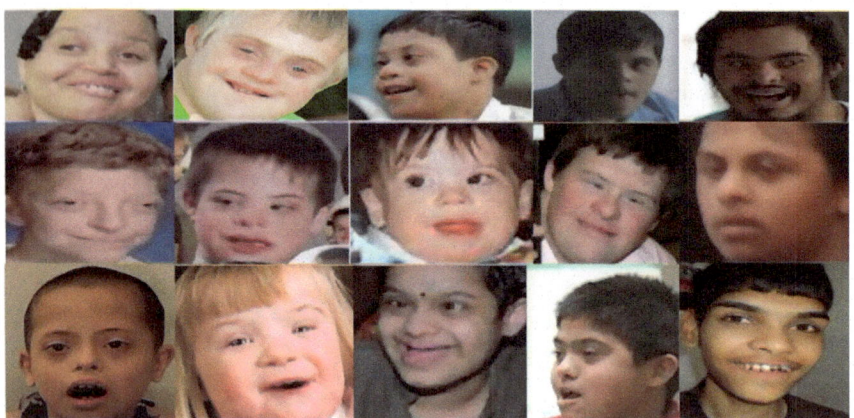

Fig. 3 Samples of subjects with Down syndrome

Fig. 4 Samples of normal subjects from the LFW face database

5 Experiments

This section provides details of the dataset as well as the two experimental scenarios performed on the dataset.

5.1 Dataset Description

The dataset comprises of images of 853 subjects with Down Syndrome and 853 images of normal people. The data was collected from various sources across the Internet. Most of the images were obtained from publicly available sources, while the rest of the images were the part of the dataset of Shukla et al. [14]. The training to testing ratio was 7:3. The data is annotated based on gender as well for detailed study. Faces of normal subjects were a part of the LFW face dataset. Both the classes had an equal number of samples for the training as well as the testing data.

5.2 Experimental Scenarios

Two different experimental scenarios were considered for our study.

- Scenario 1: In this scenario, the model was trained and tested on 800 photos of different people suffering from Down syndrome as well as on 800 photos of normal people. The purpose of this experiment was to differentiate people suffering from Down syndrome from normal healthy people.

- Scenario 2: In this scenario, the entire data was bifurcated based on gender and then the model was trained and tested to differentiate normal people from those suffering from Down syndrome. The results of both the gender were recorded separately (Tables 1, 2, and 3).

Experimental Scenario 1. A dataset comprising of 800 pictures of both normal people as well as of those suffering from Down syndrome was used for training the model and an equal number of images of both the categories were used for testing the model. The training and testing data were divided in the ratio of 1:1. The model was able to correctly distinguish faces with an accuracy of 98.47%. Table 1 gives a statistical comparison of the proposed method with other methods available. It was seen that the model was able to outperform the other models for

Table 1 A comparison of the proposed approach with other sets of features and classifier

Method	Accuracy (%)
HOG+SVM	60.42
LBP+SVM [1]	91.28
GIST+SVM	92.86
GIST+KNN	87.17
GIST+Random Forests	94.12
Proposed approach	98.47

Table 2 A comparison of the proposed approach and other approaches when applied to male subjects only

Method	Accuracy (%)
HOG+SVM	78.14
LBP+SVM [1]	92.03
GIST+SVM	92.32
GIST+KNN	84.73
GIST+Random Forests	94.06
Proposed approach	96.95

Table 3 A comparison of the proposed approach and other approaches when applied to female subjects only

Method	Accuracy (%)
HOG+SVM	66.47
LBP+SVM [1]	91.37
GIST+SVM	90.78
GIST+KNN	86.86
GIST+Random Forests	95.10
Proposed approach	99.07

Down syndrome classification. Overall accuracy of 98.47% was achieved by our model thereby outperforming all other models.

Experimental Scenario 2. Images of subjects of different genders were separated and were tested differently. The results obtained from both cases are tabulated below. The results obtained for female subjects were slightly better than their male counterpart.

5.3 Comparison with Other Existing Approaches

Although not much work has been done in applying computer vision to identify different people suffering from Down syndrome,t this section presents a comparison of the proposed model with three other existing approaches in this field. The proposed architecture performed better than many other existing approaches. Bruçin and Vasif [1] proposed using LBP for classifying and detecting people with Down syndrome. The greater advantage of the proposed model is the use of CNN for detecting Down syndrome. The final accuracy of the model proposed by Bruçin and Vasif [1] was around 90% clearly less than 98.47% accuracy of the proposed model. Saraydemir et al. [13] proposed detection of Down syndrome based on Gabor wavelet transform. Their model was able to achieve an accuracy of 97.3% when used with SVM while it achieved accuracy of 96% using K-nearest neighbors. Our model was able to correctly identify the subjects with an overall accuracy of 98.47% which is better than the accuracy achieved by the model proposed by Saraydemir et al. [13]. Shukla et al. [14] also proposed detection of different disorders using Convolutional Neural Network. Though the overall accuracy of their approach was 98.80%, their model's accuracy for Down syndrome subjects was only around 95%. Focusing on only one disease gave our model an advantage over this model and hence it did better than the model proposed by Shukla et al. [14].

6 Conclusion

The proposed framework uses the Convolution Neural Network for both local and global feature extraction. A comparison was made with other proposed models of image feature extraction, but this model outperformed all the other models. The experiment was conducted in two parts: one was with all the images and the second part was conducted after dividing the dataset based on gender. The results clearly signify the importance and future of deep learning frameworks in the field of detecting Down syndrome. Such models can be used quite cost-effectively in our daily lives to identify people suffering from Down syndrome and improving their lives. This model used relatively small dataset; therefore, it can be improved using a larger dataset. Also, AlexNet and VGGNet can be used to better train the model using a relatively larger dataset, which will surely enhance the accuracy of the model.

References

1. Bruçin, K., Vasif, N.V.: Down syndrome recognition using local binary patterns and statistical evaluation of the system. Expert. Syst. Appl. **38**(7), 8690–8695 (2011)
2. Chow, C., Liu, C.: Approximating discrete probability distributions with dependence trees. IEEE Trans. Inf. Theory **14**(3), 462–467 (1968)
3. Cohen, M.M., Winer, R.A.: Dental and facial characteristics in down's syndrome (moglism). J. Dent. Res. **44**(1), 197–208 (1965)
4. Cootes, T.F., Edwards, G.J., Taylor, C.J.: Active appearance models. IEEE Trans. Pattern Anal. Mach. Intell. **23**(6), 681–685 (2001)
5. Dalal, N., Triggs, B.: Histograms of oriented gradients for human detection. In: IEEE Computer Society Conference on Computer Vision and Pattern Recognition, 2005. CVPR 2005, vol. 1, pp. 886–893. IEEE (2005)
6. Ferry, Q., Steinberg, J., Webber, C., FitzPatrick, D.R., Ponting, C.P., Zisserman, A., Nellaker, C.: Diagnostically relevant facial gestalt information from ordinary photos. Elife **3**, e02020 (2014)
7. Huang, G.B., Ramesh, M., Berg, T., Learned-Miller, E.: Labeled faces in the wild: a database for studying face recognition in unconstrained environments. Technical report, Technical Report 07-49, University of Massachusetts, Amherst (2007)
8. Karpathy, A., Toderici, G., Shetty, S., Leung, T., Sukthankar, R., Fei-Fei, L.: Large-scale video classification with convolutional neural networks. In: Proceedings of the IEEE conference on Computer Vision and Pattern Recognition, pp. 1725–1732 (2014)
9. Krizhevsky, A., Sutskever, I., Hinton, G.E.: Imagenet classification with deep convolutional neural networks. In: Advances in Neural Information Processing Systems, pp. 1097–1105 (2012)
10. Kuru, K., Niranjan, M., Tunca, Y., Osvank, E., Azim, T.: Biomedical visual data analysis to build an intelligent diagnostic decision support system in medical genetics. Artif. Intell. Med. **62**(2), 105–118 (2014)
11. Ojala, T., Pietikainen, M., Maenpaa, T.: Multiresolution gray-scale and rotation invariant texture classification with local binary patterns. IEEE Trans. Pattern Anal. Mach. Intell. **24**(7), 971–987 (2002)
12. Parkhi, O.M., Vedaldi, A., Zisserman, A., et al.: Deep face recognition. In: BMVC, vol. 1, p. 6 (2015)
13. Saraydemir, S., Taşpınar, N., Eroğul, O., Kayserili, H., Dinçan, N.: Down syndrome diagnosis based on gabor wavelet transform. J. Med. Syst. **36**(5), 3205–3213 (2012)
14. Shukla, P., Gupta, T., Saini, A., Singh, P., Balasubramanian, R.: A deep learning frame-work for recognizing developmental disorders. In: 2017 IEEE Winter Conference on Applications of Computer Vision (WACV), March, pp. 705–714. IEEE (2017)
15. Zhao, Q., Okada, K., Rosenbaum, K., Kehoe, L., Zand, D.J., Sze, R., Summar, M., Linguraru, M.G.: Digital facial dysmorphology for genetic screening: hierarchical constrained local model using ica. Med. Image Anal. **18**(5), 699–710 (2014)
16. Zhu, X., Ramanan, D.: Face detection pose estimation, and landmark localization in the wild. In: 2012 IEEE Conference on Computer Vision and Pattern Recognition (CVPR), pp. 2879–2886. IEEE (2012)

Fused Spectral Features in Kernel Weighted Collaborative Representation for Gender Classification Using Ocular Images

Kiran B. Raja, R. Raghavendra and Christoph Busch

Abstract Ocular images have been used to supplement and complement the face-based biometrics. Ocular images are further investigated for identifying the gender of a person such that the soft label can be used to boost biometric performance of the system. Although there are number of works in visible spectrum and Near-Infrared spectrum for gender classification using ocular images, there are limited works in spectral imaging which explore ocular images for gender classification. Considering the advantages of spectral imaging, we explore the problem of gender identification using ocular images obtained using spectral imaging. To this end, we have employed a recent database of 104 unique ocular instances across 2 different sessions and 5 different attempts in each session with a spectral imaging camera capable of capturing 8 different images corresponding to different bands. Further, we present a new framework of using *fused feature descriptors* in kernalized space to fully leverage the number of spectral images for robust gender classification. With the set of experiments, we obtain an average classification accuracy of 81% with the proposed approach of using *fused GIST* features along with the weighted kernel representation of features in collaborative space.

1 Introduction

Biometrics has emerged ubiquitous way of establishing identity in number of applications such as unconstrained surveillance to controlled border crossing. The ease in capturing the biometric face data in unobtrusive nature from a long standoff distance

K. B. Raja (✉)
University of South-Eastern Norway (USN), Kongsberg, Norway
e-mail: kiran.raja@usn.no; kiran.raja@ntnu.no

K. B. Raja · R. Raghavendra · C. Busch
Norwegian University of Science and Technology (NTNU), Gjøvik, Norway

R. Raghavendra
e-mail: raghavendra.ramachandra@ntnu.no

C. Busch
e-mail: christoph.busch@ntnu.no

© Springer Nature Singapore Pte Ltd. 2020
B. B. Chaudhuri et al. (eds.), *Proceedings of 3rd International Conference on Computer Vision and Image Processing*, Advances in Intelligent Systems and Computing 1022, https://doi.org/10.1007/978-981-32-9088-4_12

131

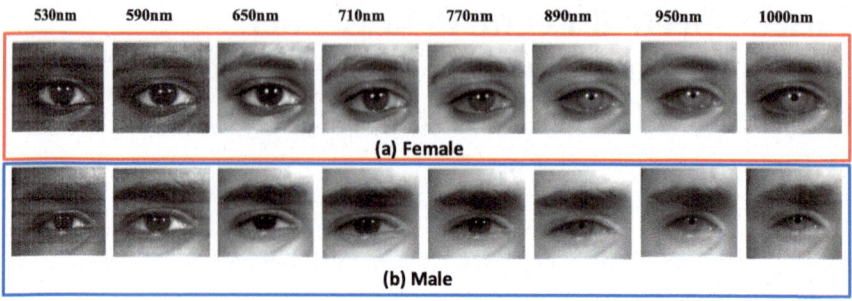

Fig. 1 Sample spectral ocular images for male and female subject collected using *eight* narrow bands

has proven the advantage in not just secure access control but also in surveillance applications for public safety, private access control among many others. The challenges in the present day face surveillance from at-a-distance standoff comes from a major drawback where the face imaging suffers directly as a result of illumination quality. Earlier works have suggested the use of alternative imaging to overcome the problems posed by the illumination and major works advocate thermal imaging [14], Near-Infrared imaging [5] and very recently, spectral/extended-spectral imaging [22]. Along with the different types of imaging, any biometric system can be improved for accuracy by employing soft biometric traits to eliminate false matches and optimize identification [9, 14]. One such soft biometric trait is gender which unlike other soft biometric traits such as weight, height, and skin color, does not change over a period of time unless altered surgically [2].

Further, face recognition becomes challenging under the scenarios where face is partially covered owing to certain way of clothing due to religious practices or to evade the face identification intentionally. Even under such ways of clothing, often people tend to keep the eye region open and thereby, any face biometric system can easily be adapted to also employ ocular region for identification. While there is a series of works that have investigated the ocular characteristics for biometric performance, another set of works focus on identifying the soft information such as gender from ocular region. The major limitation of the current works is that they either consider full face or ocular images captured either in visible spectrum or NIR spectrum for identifying the soft biometric information such as gender, even for spectral imaging in the recent days. There are no works that have explored the gender prediction by using periocular region alone in the spectral imaging to the best of our knowledge. Motivated by such a factor, in this work, we explore the problem of gender prediction using ocular images captured from spectral imaging (Fig. 1).

2   Related Works and Contributions

A set of works have demonstrated the ability to predict gender using face in visible spectrum [8–10]. In the similar lines, another set of works have investigated holistic- and part-based approaches for gender prediction [12, 13]. Most of these works further employ local and global texture descriptors represented in linear space [1, 8, 13]. Recent works have also explored the applicability of Deep Convolutional Neural Network (D-CNN) on large-scale visible spectrum face images for two-class gender prediction problem [10, 12, 14]. However, very few works have been reported to employ Near-Infrared (NIR) [4, 18], Thermal spectrum [14] and spectral imaging [22] for gender classification. It has to be further noted that these works have employed full face for predicting the gender and none of these works have used ocular region alone. Nonetheless, there are a set of works that have explored gender classification using ocular images alone in NIR and visible spectrum domain [19, 20]. Deriving the motivation from the existing works, this work presents a new paradigm of predicting the gender using ocular images from spectral imaging. The key set of contributions from this work are listed below

1. Presents a new approach exploring the ocular images captured in the spectral imaging sensor for predicting the gender.
2. Proposes a new approach of employing *fused GIST features* in a discriminative weighted kernel representation which leverages the performance of GIST features in a highly discriminative manner for gender classification.
3. Presents experiments on a recent database of ocular images collected using spectral imaging with eight different spectral wavelengths. The database consists of 104 unique periocular instances captured across 8 different spectrum. The database is captured in 2 different sessions with 5 samples in each session.
4. Presents quantitative experimental evaluation with respect to other state-of-the-art approaches based on feature descriptors in linear space and Support Vector Machines (SVM) [6] to demonstrate the significance of our proposed framework leveraging the kernel space. The extensive set of results for gender classification is validated newly collected dataset. We further demonstrate the superiority of the proposed approach by validating the results of cross-session gender classification through ocular images captured after a span of 3–4 weeks.

In the rest of this paper, Sect. 4.1 presents an detailed description of spectral ocular database. Section 3 presents the intuition and approach for the proposed framework for gender classification using ocular images. Section 4 details the experimental evaluation protocol and results which is followed by Sect. 5 listing the key remarks from this work.

3 Proposed Approach For Gender Classification

We describe the proposed framework employed in this work for gender classification using the spectral ocular images. Figure 2 illustrates the proposed framework which first extracts the feature descriptors based on the GIST feature extraction [17] for each of the independent spectral ocular image. Further, for the sake of computational efficiency, we represent them in histogram representation. The set of all the histograms for eight of the spectral ocular images are concatenated and further represented in kernel space. The classifier is then learnt using the weighted approach to optimize the collaborative space such that discriminant feature space contributes in modeling a robust classifier. The classifier is further employed to identify the gender for the ocular images in the testing phase. The details of feature extraction and subsequently, weighted kernel approach for collaborative representation is presented in the section below.

3.1 GIST Features

The feature extraction technique to obtain GIST features [17] is based on the multi-scale extraction corresponding to early-stage human visual features which essentially captures the gist/summary of the scene into a low-dimensional feature vector. Compared to the set of approaches based on extracting texture using handcrafted filters or naturally learnt filters, multi-scale processing in GIST feature extraction captures the biologically plausible and computational noncomplex low-level features with a model for visual attention. Our intuition in applying GIST feature extraction is to process the ocular images in human-vision-based approach such that the features correspond to features observed by a human to distinguish the features of different genders in ocular region.

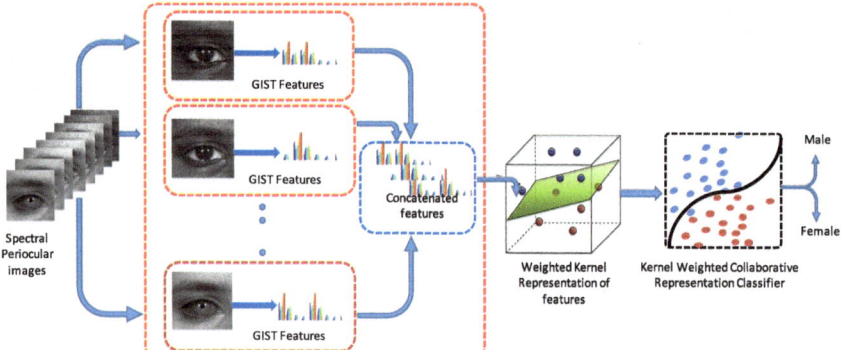

Fig. 2 Proposed framework based on spectral feature fusion for gender classification from spectral ocular images

In order to extract the GIST features, we first remove low-level spatial frequencies such that any impact of frequency due to spectral imaging itself is removed and thereby the data is filtered to be represented in a noise-free manner. Specifically, we carry out filtering and whitening followed by local contrast normalization such that the image is artifact free or the artifacts are reduced to a minimal level. Further, each of the image is filtered using Gabor-like filters in different scales and orientation starting from higher to lower orientation and scale respectively. In this work, we employ filters in eight different scales and eight orientations in each scale. The input image which is convoluted with the filters results in obtaining the set of global features within the image. Further, to obtain the features in an efficient manner, the feature vectors are extracted in a fixed 8×8 grid subregions over the image to account for the set of local features as well.

3.2 Fused GIST Features

Considering eight different bands in our work corresponding to 530, 590, 650, 710, 770, 890, 950, and 1000 nm—the spectral image set can be represented as S where

$$S = \{S_1, S_2, S_2 \ldots S_9\}$$

$for\ bands = 1, 2, \ldots, 8$ corresponding to 530, 590, 650, 710, 770, 890, 950, and 1000 nm.

The GIST features for each spectral band can be represented as G^f where

$$G = \{G_1, G_2, G_2 \ldots G_8\}$$

$for\ bands = 1, 2, \ldots 8$

where the features are extracted using the GIST. Further, the fused feature can be represented as

$$\mathcal{G}^f = \{G_1 \cup G_2 \cup G_3 \cup G_4 \cup G_5 \cup G_6 \cup G_7 \cup G_8\}$$

where $\cup$ operator represents the feature level fusion of the individual GIST histogram features obtained for 8 different spectral band images. For the sake of simplicity, we further represent the fused GIST features $\mathcal{G}^f$ as X.

3.3 Kernel Weighted Collaborative Representation Classification (KWCRC)

The commonly used classifiers have explored the linear spaces to represent the samples of classification through a combination of linear representation such that the coefficients minimize the residual between a new sample of a class and it's linear reconstruction [3]. Collaborative Representation space is one such space where the optimal space combination can be obtained with a reasonable efficiency for classification. However, in many real classification tasks, the training samples are not equally discriminative to devise a classifier such as in our case of ocular characteristic discrimination for male and female gender. Such problems can be easily handled if one can determine the coefficients of the samples of classes that correlate closely and thereby identify the corresponding weights. While these factors of determining the correlation of coefficients and the weights lead to good classification accuracy, it is very often difficult to obtain good performance when the feature space is not well separated in the linear space. This argument holds specifically good for ocular images captured in the spectral imaging where the ocular images are influenced by number of factors such as absorption of spectrum by skin in ocular region and absorption of spectrum by scleral region. These factors result in images that are typically not separable in the linear space as can be seen for certain spectral images as shown in the Fig. 3. Although extracting robust features using descriptors can aid to separate the ocular samples, tasks such as gender classification may not yield optimal performance as the features tend to be common for both the genders to a certain extent. Considering the principles of Support Vector Machines, it can be intuitive to assume a suboptimal performance in linear space.

The nonlinear separability problem can be easily converted to a linear problem using the kernel representation of the feature space [3, 21]. Based on the motivation and arguments provided above, we simply adopt the Kernel Weighted Collaborative Representation Classification for efficient gender classification using the ocular images from spectral imaging [21]. The general collaborative representation classifier can be provided by

$$\hat{\beta} = (X^T X + \lambda)^{-1} X^T y \tag{1}$$

which can be further simplified as $\hat{\beta} = Py$ where $P = (X^T X + \lambda)^{-1} X^T$ is the projection matrix computed based on input features X and λ is the regularization parameter (X is the set of all fused GIST features from 8 spectral bands). The idea of classification is further to find the residual such that the new testing sample y is assigned to the closest class.

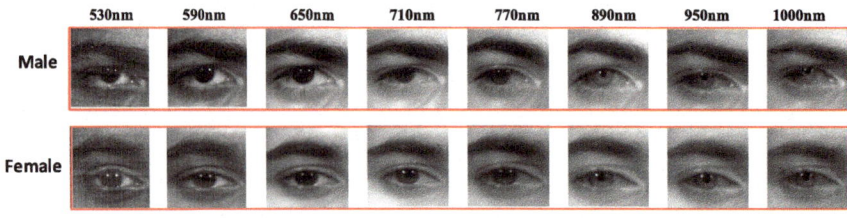

Fig. 3 Highly similar ocular images which cannot be distinguished in linear space of representation for each corresponding bands

Table 1 Database division for the experiments

Details	Session 1		Session 2	
	Training set	Testing set	Training set	Testing set
Total ocular instances	104		104	
Unique ocular instances	60	44	60	44
Samples per session	5	5	5	5
Spectral images for capture	8	8	8	8
Total images	2400	1760	2400	1760

We further provide the analogy of kernel space formulation for the convenience of the reader in this section. Assume that there exist m samples in feature space X such that $x_1, x_2, x_m \in \Re^{n \times n}$ belonging to C classes ($C = 2$ for male and female gender in our work) and F is a feature space induced by a nonlinear mapping $\varphi : R_n \to F$. We can now represent the feature space as $\kappa(, y)$ representing the kernel space mappings for the samples y such that

$$\kappa(, y) = [\kappa(x_1, y), \kappa(x_2, y), , \kappa(x_m, y)]^T \in \Re^{m \times 1} \qquad (2)$$

where ($\Re^n$) represents an Euclidean space. Considering the K representing Gram matrix and I representing the identity matrix. The kernelized version of the collaborative representation classifier can be presented as

$$\hat{\beta}_K = (K + \lambda K(\kappa_1 I + \kappa_2 \Gamma_K^T \Gamma_K))^{-1} \kappa(, y) \qquad (3)$$

where the projection matrix P in Eq. 1 is equivalently represented by $K + \lambda K(\kappa_1 I + \kappa_2 \Gamma_K^T \Gamma_K))^{-1}$ in the Kernalized version and the sample y is represented as $\kappa(, y)$ in kernel space. Further, Γ is the Tikhonov matrix in the kernel space that corresponds to the determined weights for the elements in the kernel. Thus, the task of learning the classifier is to learn the *Kernel Weighted Collaborative Representation Classifier (KWCRC)* such that the residual of the sample is computed using simple *Regularized Least Square* method and the class is assigned in the lines similar to CRC [3]. We simply employ Gaussian kernel for representing the features extracted using various feature extraction techniques. For the ease of understanding of reader, we have adopted the notations from previous articles [3, 21] and the reader is further referred to [3, 21].

4 Experiments and Results

This section presents the experimental protocols and the results obtained for different protocols. To provide a comparison to state-of-the-art techniques, we employ the standard feature descriptors such as Local Binary Pattern (LBP) [15], Local Phase Quantization (LPQ) [16], Binarized Statistical Image Features (BSIF) [11], Histogram of Orientation (HoG) [7] along with GIST descriptors [17]. As current state-of-the-art methods in gender classification employ Support Vector Machines (SVM) for classification, we employ the SVM in conjunction with all the feature descriptors mentioned above. Further, the SVM itself was employed from Matlab 2014b.[1] To demonstrate the applicability of kernelized feature space for all independent feature descriptors for each image captured across eight different spectrums, we employ KWCRC. The results are presented in terms of the classification accuracy in this work.

4.1 Spectral Ocular Database

This section of presents the details of the ocular image database employed for the experimental evaluation. The spectral ocular image database is captured using in-house custom-built spectral imaging sensor operating in $eight$ narrow spectral bands corresponds to 530, 590, 650, 710, 770, 890, 950, and 1000 nm wavelength which covers the spectrum from Visible (VIS) and Near-Infrared (NIR) range. The ocular data collected using spectral imaging sensor is captured from 33 male and 19 female subjects amounting to the total of 104 unique periocular instances. Further, the protocols are carefully designed to consider the variation in terms of sessions and thereby, the duration between two sessions varies from 3 to 4 weeks.

The database consists of a total of 104 ocular $\times 8$ spectrum $\times 2$ sessions $\times 5$ samples $= 8320\ images$. Further, each of the image corresponds to a size of 1280×1024 pixels of spatial resolution. In order to further eliminate the non-ocular region within the captured image, we have processed the database manually and cropped the ocular region alone which is resized to a smaller size of 256×256 pixels for the sake of computational efficiency. Figure 1 illustrates the sample ocular images in spectral bands corresponding to different spectral wavelengths for a sample male and female subject. It has to be further noted that the ocular images are captured from both male and female subjects when the subjects have not used any external make-up such as mascara, eyeliners or additional eyelashes. This practice of collecting the database has introduced no external bias in classifying the gender of the subject. The ocular database in this work is further captured in a cooperative manner such that the subjects are in a standoff distance of approximately $1 - meter$.

---

[1]LIBSVM—Provided similar and equivalent performance to SVM from MATLAB from our experimental trials.

Table 2 Same session gender classification

Methods	Session-1 versus Session-1		Session-2 versus Session-2	
	SVM	KWCRC	SVM	KWCRC
LBP	65.90 ± 3.71	63.40 ± 6.98	64.31 ± 3.87	59.31 ± 6.37
LPQ	58.86 ± 1.29	65.22 ± 5.03	59.54 ± 2.34	58.63 ± 2.34
BSIF	66.36 ± 5.00	72.95 ± 4.07	60.90 ± 4.25	72.50 ± 5.71
HOG	59.09 ± 0.00	74.09 ± 5.98	59.09 ± 0.00	73.40 ± 4.15
Proposed	61.59 ± 2.50	**81.59 ± 5.51**	66.36 ± 5.00	**81.13 ± 4.67**

4.2 Protocols

As the database consists of images captured in two sessions (referred hereafter as *Session*-1 and *Session*-2), we design the corresponding protocols where (1) the first set of experiments corresponds to training and testing samples captured in same session (2) the training samples are obtained from different sessions than the session from which the testing samples are chosen. However, for both the set of experiments, the database is separated into two disjoint sets of training and testing set where the samples are from five different captures in a single session. The database is divided further to have 60 unique ocular instances for training and 44 unique ocular instances for testing in each session. The complete split of the training and the testing set for each session is provided in the Table 1. Further, to account for the number of variations and the biases introduced by the classifier itself, we divide the database into 10 different partitions of independent and disjoint ocular instances while keeping the number of training and testing samples constant. The trials are conducted on the 10 different partitions for all the techniques and average results along with the standard deviation is presented.

4.3 Protocol 1—Same Session Evaluation

The main motivation for this protocol is to evaluate the proposed approach for the proof of concept and evaluate the intention of kernalized representation of the feature space. Thus, for this set of experiments, we conduct two independent evaluation of gender classification with standard SVM classification and further with proposed kernel space exploration for images in *Session*-1 and *Session*-2. Table 2 and Fig. 4 present the results obtained for each session independently with SVM and proposed approach of GIST in kernel space.

- As observed from the Table 2, there is an average accuracy gain starting from a gain of 7% to a maximum gain of 15% for data in *Session*-1.
- Along the similar lines, we can observe a maximum gain of ≈15% in classification accuracy for data in *Session*-2 as compared to rest of the techniques based on SVM.

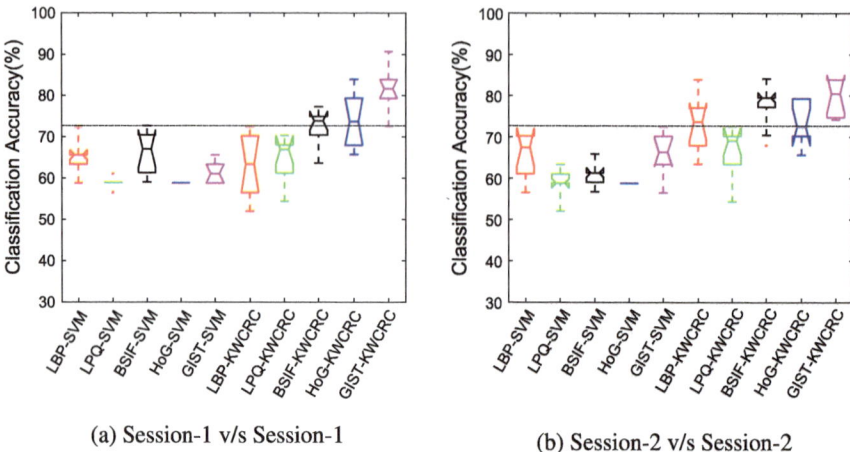

(a) Session-1 v/s Session-1   (b) Session-2 v/s Session-2

Fig. 4 Gender classification accuracy using ocular images captured in the same session for different techniques

– The key point to be noted here is that the kernelized representation of the features for all set of descriptors has resulted in a gain of classification accuracy when the weighted collaborative representation is employed against the SVM-based classification which justifies the motivation behind proposed framework.

It can also be observed from the box plot in Fig. 4 that the average classification accuracy with proposed approach (GIST+KWCRC) outperforms all the other approaches even under number of different trials.

4.4 Protocol 2—Cross-Session Evaluation

This protocol is mainly designed to evaluate the robustness of the proposed approach when the factors such as appearance, illumination of *Session*-2 are different from the factors of *Session*-1 and thereby introduces number of variation in data. As a result of varied factors, the classifiers trained on a particular session may result in suboptimal performance when tested on data collected in the exact session. In this set of evaluation, we determine the classification accuracy of the proposed approach when the classifier is trained using a session different from session data employed in testing. Table 3 and Fig. 5 present the results obtained from this set of experiments.

– As it can be noted from the Table 3, the proposed approach of kernelized representation with weighted representation of feature descriptors results in higher accuracy as compared to SVM-based classification for cross-session evaluation.
– When the classifier is trained with data from *Session*-1 and tested on data from *Session*-2, an maximum gain of 17% is obtained using proposed approach of $GIST + KWCRC$ while a gain of $\approx$13% is achieved in alternative case.

Table 3 Cross-session-based gender classification

Methods	Session-1 versus Session-2		Session-2 versus Session-1	
	SVM	KWCRC	SVM	KWCRC
LBP	60.22 ± 4.04	62.95 ± 5.97	65.55 ± 5.11	72.95 ± 6.10
LPQ	58.18 ± 1.58	65.22 ± 4.01	59.09 ± 3.03	67.04 ± 5.79
BSIF	61.86 ± 3.83	74.31 ± 6.78	60.90 ± 3.17	77.72 ± 4.88
HOG	59.09 ± 0.00	70.22 ± 5.29	59.09 ± 0.00	73.40 ± 4.67
Proposed	61.81 ± 2.58	$\mathbf{78.40 \pm 5.79}$	66.04 ± 4.81	$\mathbf{79.71 \pm 4.00}$

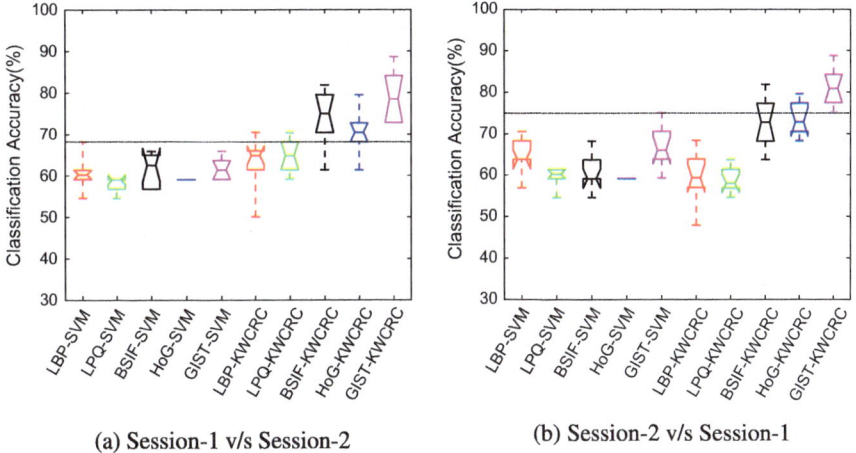

(a) Session-1 v/s Session-2   (b) Session-2 v/s Session-1

Fig. 5 Gender classification accuracy using ocular images captured in the same session for different techniques

– The key factor to be noticed here is that the average gain achieved through kernelized representation is higher than the linear representation of SVM classification. The kernel space is effective in considering the variation across sessions and still obtain better performance than linear representation based SVM.

Similar observations can be noted from the box plot in Fig. 5 where the average classification accuracy with proposed approach (GIST+KWCRC) outperforms rest of the approaches indicating the robustness of the proposed approach.

5 Conclusions

Gender classification is explored to boost the performance of biometric systems by using the gender label as a soft information. Identification and classification of the gender are also important in surveillance scenarios where the face may be partially visible and the systems are typically operated in visible or NIR spectrum. In order to make the better use of partial face in surveillance scenarios, some works have investigated ocular images as complementary or supplementary modality. Further, recent works have explored spectral imaging to address some of the key concerns confronted in visible and NIR spectrum for both face and ocular imaging. While there is an advancement for face recognition and gender classification using spectral imaging in very recent works, there are no works that have explored ocular images from spectral imaging itself for gender classification. This work has presented a new paradigm of detecting gender using ocular characteristics captured using spectral imaging. Further, we have presented a new approach of exploring the kernel space with GIST features to achieve efficient gender classification. With the set of experiments, we have demonstrated the applicability of kernel space along with weighted collaborative representation of GIST features to be efficient compared to linear representation and SVM. Specifically, with an average classification accuracy of 81% on the spectral ocular database, it is an evident illustration of applicability of kernel space representation of spectral images across eight different bands. Future works in this direction shall include feature selection and dimension reduction for efficient feature representation of spectral images across different bands to boost the performance even further.

Acknowledgements This work was carried out under the funding of the Research Council of Norway under Grant No. IKTPLUSS 248030/O70. Ethical and privacy perspectives for the study were taken into consideration according to the national guidelines for this work.

References

1. Ahmad, F., Ahmed, Z., Najam, A.: Soft biometric gender classification using face for real time surveillance in cross dataset environment. In: INMIC, pp. 131–135 (Dec 2013)
2. Bekios-Calfa, J., Buenaposada, J.M., Baumela, L.: Revisiting linear discriminant techniques in gender recognition. IEEE Trans. Pattern Anal. Mach. Intell. **33**(4), 858–864 (April 2011)
3. Cai, S., Zhang, L., Zuo, W., Feng, X.: A probabilistic collaborative representation based approach for pattern classification. In: Proceedings of the IEEE Conference on Computer Vision and Pattern Recognition, pp. 2950–2959 (2016)
4. Chen, C., Ross, A.: Evaluation of gender classification methods on thermal and near-infrared face images. In: 2011 International Joint Conference on Biometrics (IJCB), pp. 1–8 (Oct 2011)
5. Chen, X., Flynn, P.J., Bowyer, K.W.: Ir and visible light face recognition. J. Comput. Vis. Image Underst **99**, 332–358 (2005)
6. Cortes, C., Vapnik, V.: Support vector machine. Mach. Learn. **20**(3), 273–297 (1995)
7. Dalal, N., Triggs, B.: Histograms of oriented gradients for human detection. In: IEEE Computer Society Conference on Computer Vision and Pattern Recognition, 2005. CVPR 2005, vol. 1, pp. 886–893. IEEE (2005)
8. Hadid, A., Ylioinas, J., Bengherabi, M., Ghahramani, M., Taleb-Ahmed, A.: Gender and texture classification: a comparative analysis using 13 variants of local binary patterns. Pattern Recognit. Lett. **68, Part 2**, 231–238 (2015), special Issue on Soft Biometrics
9. Jain, A.K., Dass, S.C., Nandakumar, K.: Can Soft Biometric Traits Assist User Recognition? (2004)

10. Jia, S., Cristianini, N.: Learning to classify gender from four million images. Pattern Recognit. Lett. **58**, 35–41 (2015)
11. Kannala, J., Rahtu, E.: Bsif: binarized statistical image features. In: Proceedings of the 21st International Conference on Pattern Recognition (ICPR2012), pp. 1363–1366 (2012)
12. Levi, G., Hassner, T.: Age and gender classification using convolutional neural networks. In: IEEE Conference on Computer Vision and Pattern Recognition (CVPR) Workshops (June 2015). http://www.openu.ac.il/home/hassner/projects/cnn_agegender
13. Moeini, H., Mozaffari, S.: Gender dictionary learning for gender classification. J. Vis. Commun. Image Represent. **42**, 1–13 (2017)
14. Narang, N., Bourlai, T.: Gender and ethnicity classification using deep learning in heterogeneous face recognition. In: 2016 International Conference on Biometrics (ICB), pp. 1–8 (June 2016)
15. Ojala, T., Pietikainen, M., Maenpaa, T.: Multiresolution gray-scale and rotation invariant texture classification with local binary patterns. IEEE Trans. Pattern Anal. Mach. Intell. **24**(7), 971–987 (2002)
16. Ojansivu, V., Heikkilä, J.: Blur insensitive texture classification using local phase quantization. In: International conference on image and signal processing, pp. 236–243. Springer (2008)
17. Oliva, A., Torralba, A.: Modeling the shape of the scene: a holistic representation of the spatial envelope. Int. J. Comput. Vis. **42**(3), 145–175 (2001)
18. Ross, A., Chen, C.: Can gender be predicted from near-infrared face images? In: Proceedings of the 8th International Conference on Image Analysis and Recognition, vol. Part II, pp. 120–129. ICIAR'11 (2011)
19. Singh, M., Nagpal, S., Vatsa, M., Singh, R., Noore, A., Majumdar, A.: Gender and ethnicity classification of iris images using deep class-encoder (2017). arXiv preprint arXiv:1710.02856
20. Tapia, J.E., Perez, C.A., Bowyer, K.W.: Gender classification from the same iris code used for recognition. IEEE Trans. Inf. Forensics Secur. **11**(8), 1760–1770 (2016)
21. Timofte, R., Van Gool, L.: Adaptive and weighted collaborative representations for image classification. Pattern Recognit. Lett. **43**, 127–135 (2014)
22. Vetrekar, N., Raghavendra, R., Raja, K.B., Gad, R., Busch, C.: Robust gender classification using extended multi-spectral imaging by exploring the spectral angle mapper. In: IEEE International Conference on Identity, Security and Behavior Analysis (ISBA), pp. 1–8 (Jan 2018)

Face Anti-spoofing Based on Specular Feature Projections

Balaji Rao Katika and Kannan Karthik

Abstract The need for facial anti-spoofing has emerged to counter the usage of facial prosthetics and other forms of spoofing at unmanned surveillance stations. While some part of literature recognizes the difference in texture associated with a prosthetic in comparison with a genuine face, the solutions presented are largely prosthetic model specific and rely on two-sided calibration and training. In this paper, we focus on the specular component associated with genuine faces and claim that on account of the natural depth variation, its feature diversity is expected to be much larger as compared to prosthetics or even printed photo impersonations. In our work concerning one-sided calibration, we first characterize the specular feature space corresponding to genuine images and learn the projections of genuine and spoof data onto this basis. The trained SVM corresponding to genuine projections, 3D mask projections, and printed photo projections is then used as an anti-spoofing model for detecting impersonations.

Keywords Face Anti-spoofing · Specular feature · Eigenspace · Printed photos · $3D$ Mask · Low-rank component · Sparse component · SVM classifier

1 Introduction

Given the widespread deployment of unmanned surveillance units, there arises a need to scan and recognize the faces of individuals on the move, especially in large-scale organizations. To prevent any form of impersonation of legitimate people through prosthetic masks or even plain printed photographs, there is a need for an algorithmic

B. R. Katika (✉) · K. Karthik
Department of Electronics and Electrical Engineering, Indian Institute of Technology Guwahati, Guwahati 781039, Assam, India
e-mail: k.balaji@iitg.ac.in

K. Karthik
e-mail: k.karthik@iitg.ac.in

© Springer Nature Singapore Pte Ltd. 2020
B. B. Chaudhuri et al. (eds.), *Proceedings of 3rd International Conference on Computer Vision and Image Processing*, Advances in Intelligent Systems and Computing 1022, https://doi.org/10.1007/978-981-32-9088-4_13

layer which scans, analyzes, and attests the genuineness or the naturalness of the facial image presented to the unmanned camera.

The first thing to recognize in this problem space is that while the class of genuine facial images is available at our disposal, the spoof class is likely to have diverse modalities and formats. In a typical impersonation attack, some individual (either a legitimate part of an organization or an outsider), passes on as another individual who is from the same organization. Because of this genuine concern, that one legitimate individual should not pass on as another genuine individual, this anti-spoofing problem is a *Identity- Independent Detection problem* [7]. In this frame, the anti-spoofing algorithm does not concern itself with the actual biometric feature associated with the individual, but rather focusses on a layer which can be be used to establish the genuineness of the feature. In many ways, this anti-spoofing problem draws several parallels from ideas connected with blind image quality assessment algorithms [3, 8]. In the present literature survey, we restrict our analysis and scope to static facial images as opposed to videos.

Much of the earlier was driven toward the development of spoof model-specific anti-spoofing solutions, wherein images were captured on both fronts: genuine/natural and also with masks/prosthetics. These images were used to train suitable classifiers, with the objective of minimizing the misclassification probability [2, 9]. Some papers [2], have assumed even prior knowledge of the imposter set, to account for mask contortions coming from the mismatch in facial profiles, between the imposter and the person being impersonated.

When the spoof model is regularized and it becomes possible to recreate the spoof settings to a certain degree of precision, then the corresponding anti-spoofing solutions can be optimized and diversified. This is largely true for spoofing through printed photographs [9, 10], wherein the line of anti-spoofing solutions encompass physical and imaging model-based solutions involving physical constraints. In Gao et al. [4], it was surmised that photographs of planar images tend to have a more homogeneous specular illumination component as compared to natural faces.

The same printed photograph was treated as spatial resampling problem by Garcia et al. [5], leading to the introduction of a few artifacts in the image domain, termed as Moire patterns, which can be treated as noise. Since this noise is wideband, its power can be estimated by filtering the query images and then subtracting the original image from the filtered image. The drawback of this frame is that this Moire model assumes a zooming in effect with respect to the printed photo, otherwise the model would change to an aliasing one and the same analysis cannot be repeated.

In Karthik et al. [7], the recapturing phenomenon has been shown physically to be imperfect registration process if the original object distance is unknown. This results in a cumulative blur, which supersedes the natural blur resulting from depth variation, lowering the mean patch sharpness profile in the printed photo. If the original settings can be recreated, then there would be no cumulative blur, but this is unlikely. This frame works for both zooming in and zooming out operations while recapturing the photo.

Furthermore with a planar spoofing model, there is a lack of depth information, which is normally present in natural scenes. Light field cameras [6], tend to accumulate information from different angles and formulate a depth map of the scene or the facial frame. This aspect has been used to detect planar spoofing in Ji et al. [6].

We position our work as a model-specific solution which hinges on certain assumptions or physical constraints. Since 3D masks and printed photos are largely smooth in nature, they tend to have a larger specular component, albeit more homogeneously distributed in the case of the printed photo [4]. However, in the case of natural facial photos, the specular component is expected to be more diverse, mainly because of the depth variations and the self-shadowing effects. Two printed photos belonging to two different individuals are expected to show some coherence in terms of their specular profiles, however, the specular profiles from two natural faces from two different individuals, are unlikely to exhibit much similarity, because of the depth variation associated with the facial profiles. Since the specular component distribution is a function of the surface geometry, the association between two natural photos is expected to be more complex diverse in relation to the association between a natural photograph and the printed photo or between two printed photos. This philosophy has led to the following contributions:

- One-sided characterization of the specular feature eigenspace corresponding to natural photos.
- Learning of the conditional density functions associated with the specular projections of natural photos onto natural photos and projections of spoofed images onto natural photos. Using a Maximal Likelihood inferencing frame by building an SVM to classify the query images into natural and spoof classes.
- Of less significance is the application of specular component isolation algorithm by Candes et al. [1] to extract features from both natural and spoofed faces.

The rest of the paper is organized as follows: The proposed architecture is discussed in detail in Sect. 2. The database selection, training, and testing modes along with the performance evaluation are presented in Sect. 3.

2 Proposed Architecture

The proposed architecture has the following layers:

- Extracting the specular feature vectors from a set of natural faces (including illumination variations and partial pose variations).
- Using these specular feature vectors to characterize the space of natural faces, through an eigen-decomposition procedure.
- Extracting specular feature vectors from training sets corresponding to natural and spoof classes.
- Projecting these specular vectors onto the eigenspace of natural faces to learn the associativity profiles of the two classes (natural and spoof).
- Formulating an inferencing policy based on these learnt conditional densities.
- Deploying test images from both the classes and checking the misclassification rate.

2.1 Specular Feature Extraction

Let $U = \{1, 2, 3, \ldots, n\}$ correspond to n legitimate subjects whose natural facial space needs to be characterized. Each subject i has M pose and illumination variations which are scanned into column vectors and concatenated into an $N^2 \times M$ sized matrix $D_i, i \in \{1, 2, 3, \ldots n\}$. Each subject-specific data matrix D_i is decomposed into diffused component $\overline{I}_i$ and a specular component $\overline{S}_i$. The diffused component is richer and contains more visual information than the specular component which to some extent is an abstraction of the geometry of the surface and its orientation with respect to the illuminating source. While the diffused component is largely contiguous, the sparse components are more patchy but highly correlated within that curved patch. Hence, the diffused component is expected to exhibit significant spatial structure and visual meaning while the specular component is expected to be noisy, patchy, and to carry a certain amount of depth information. When several partial poses and illumination variations of the same subject i, are presented in the form of the data matrix D_i, the column space of D_i corresponds to the common diffused component which is approximately the centroid over all the variations, while the row space of D_i carries information regarding the geometry of the surface based on pixel intensity correlation as a function of the relative positions of the pixels. The latter forms the specular component and is isolated using the accelerated proximal gradient descent algorithm described in Algorithm 1.

Algorithm 1 Accelerated proximal gradient descent algorithm (APG)

Require: $D_i \in \Re^{\{N^2 \times M\}}, \lambda$

1: $\overline{I}_0 \leftarrow 0; \overline{S}_0 \leftarrow 0; t_0 \leftarrow 1\overline{\mu} \leftarrow \delta\mu_0$

2: **while** not converged **do**

3: $\quad Y_k^{\overline{I}} \leftarrow \overline{I}_k + \frac{t_{k-1}-1}{t_k}(\overline{I}_k - \overline{I}_{k-1})$

4: $\quad Y_k^{\overline{S}} \leftarrow \overline{S}_k + \frac{t_{k-1}-1}{t_k}(\overline{S}_k - \overline{S}_{k-1})$

5: $\quad G_k^{\overline{I}} \leftarrow Y_k^{\overline{I}} - \frac{1}{L_{f=2}}(Y_k^{\overline{I}} + Y_k^{\overline{S}} - D_i)$

6: $\quad (U, \Sigma, V) \leftarrow svd(G_k^{\overline{I}}); \overline{I}_{k+1} = U\beta_{\mu_k * 0.5}[\Sigma]V^T$

7: $\quad G_k^{\overline{S}} \leftarrow Y_k^{\overline{S}} - \frac{1}{2}(Y_k^{\overline{I}} + Y_k^{\overline{S}} - D_i)$

8: $\quad \overline{S}_{k+1} = \beta_{\frac{\lambda\mu_k}{2}}[G_k^{\overline{S}}]$

9: $\quad t_{k+1} \leftarrow \frac{1+\sqrt{4t_k^2+1}}{2}$

10: $\quad \mu_{k+1} \leftarrow max(\mu_k, \overline{\mu}); k \leftarrow k+1$

11: **end while**

Parameters of $\lambda = \frac{1}{\sqrt{M}}$, where M is number of pose varying samples. Thresholding function $\mu_0 = 0.99\|D\|_2$ convergence parameter $\delta = 10^{-5}$.

2.2 Genuine Space Characterization

In the earlier section, given a subject specific data matrix D_i, the sparse component is isolated and scanned as a t-point real valued column vector $\overline{S}_i$, $i \in \{1, 2, \ldots n\}$ using Algorithm 1. Let $\mathbf{S}_{GEN}$ be the covariance matrix associated with specular feature set corresponding to the natural images. This is computed as

$$\mathbf{S} = \sum_{r=1}^{n} (\overline{S}_r - \overline{\mu})(\overline{S}_r - \overline{\mu})^T \tag{1}$$

where $\overline{\mu}$ is defined as the centroid over all the specular features from the natural photo set.

$$\mu = \frac{1}{n} \sum_{r=1}^{n} \overline{S}_r \tag{2}$$

The eigen-decomposition of this specular covariance matrix from the natural set, $\mathbf{S}$ is performed as

$$\mathbf{S} = UDU^H \tag{3}$$

where the eigenvector matrix is given by, $\mathbf{U} = [\bar{u}_1, \bar{u}_2, \ldots \bar{u}_t]$ which is a $t \times t$ matrix with t being the length of the specular feature vector,

$$\bar{u}_i^H \bar{u}_j = \begin{cases} 1 \text{ IF } i = j \\ 0 \text{ IF } i \neq j \end{cases} \tag{4}$$

and D is a diagonal matrix comprising of t eigenvalues $\gamma_1, \gamma_2, \ldots, \gamma_t$. Let $t_{sig} \leq t$ be the number of significant eigenvalues detected by first sorting all the eigenvalues and taking the modulus of successive eigenvalues. Thus, the specular eigenspace is eventually characterized by t_{sig} eigenvectors $\bar{u}_{s1}, \bar{u}_{s2}, \ldots, \bar{u}_{s(t_{sig})}$ with $\bar{u}_{s(j)} \in columns(U)$. The truncated eigenvector matrix is U_{sig}. This important matrix represents the EIGENSPACE of the natural face set in the specular feature space.

2.3 Learning the Conditional Densities

The eigenspace is characterized by the reduced eigenset, U_{sig} which is $t \times t_{sig}$ matrix comprising of t_{sig} significant eigenvectors in the descending order of significance $1, 2, 3 \ldots, t_{sig}$. The training operation is performed by collecting samples from three different classes of images:

- Natural/Genuine facial photos (N_G images).
- Printed photographs representing one form of spoofing (N_{PP} images).
- Photographs of people wearing 3D masks build using paper folds (N_M images).

The corresponding specular feature sets derived using the illumination/pose variations for every subject using Algorithm 1, can be defined as

$$\mathbf{TR}_G = \left\{ \overline{S}_{G_1}, \overline{S}_{G_2}, ..., \overline{S}_{G_{N_G}} \right\}$$
$$\mathbf{TR}_{PP} = \left\{ \overline{S}_{PP_1}, \overline{S}_{PP_2}, ..., \overline{S}_{PP_{N_{PP}}} \right\}$$
$$\mathbf{TR}_M = \left\{ \overline{S}_{M_1}, \overline{S}_{M_2}, ..., \overline{S}_{M_{N_M}} \right\}$$

It is important to know what form of information the diffused and sparse/specular components provide for not just the genuine images but also for the spoof models including both the printed photographs [9] (Fig. 1a–d) and also the 3D masks [2] (Fig. 2a–d).

In case of the genuine versus printed photographs, there is a significant similarity in the diffused components derived from genuine samples and also printed photos Fig. 1a, b since these images are concerned NOT with the facial topography or curvature, but rather with the intensity profile. On the other hand, the specular component shows considerable complexity and diversity in the case of genuine faces as compared to spoofed ones (Fig. 1c, d). There is more depth information and complexity in the specular map of genuine images as compared to the spoofed ones.

The results are much closer for 3D masks owing to the presence of depth information in the spoofing operation. Again the diffused components are very similar, Fig. 2a, b However the richness of the genuine specular component is greater than that of the 3D mask specular component, despite the depth since the mask has to be smooth enough to fit multiple subjects to facilitate a many to one type of impersonation (Fig. 2c, d). Masks are prepared in such a manner so that they have to fit all potential clients. To ensure that the face cut of individual-A does not emboss onto the impersonation mask of individual-B, the mask of individual-B ends up being constructed, as a smoothened version of the actual face cut of individual-B. This allows several individuals of type-A to wear a singular mask derived from individual-B.

These specular vector sets are projected onto the eigenspace corresponding to natural images to obtain

$$\overline{P}_{G_i} = U_{sig}^H \overline{S}_{G_i}$$
$$\overline{P}_{PP_i} = U_{sig}^H \overline{S}_{PP_i}$$
$$\overline{P}_{M_i} = U_{sig}^H \overline{S}_{M_i}$$

for $\forall i$ in the respective training sets. The energies of the respective projections are computed as

$$E_{G_i} = \| \overline{P}_{G_i} \|_2$$
$$E_{PP_i} = \| \overline{P}_{PP_i} \|_2$$
$$E_{M_i} = \| \overline{P}_{M_i} \|_2$$

Fig. 1 **a** Low rank/Diffused component of attack face especially printed photo attack faces; **b** Low rank/Diffused component of real legitimate faces; **c** Sparse/Specular component of printed photo attack faces; **d** Sparse/Specular component of real genuine faces. Images are taken from Wen et al. [9] printed photo database

for $\forall i$. A Gaussian fit is done to the Histograms emerging from the energies of the projected components corresponding to Genuine photos, printed photos, and 3D mask images. The Gaussian for Genuine in relation to the conditional distribution for printed photos is shown in Fig. 3a while the comparison with the 3D mask distribution is shown in Fig. 3b.

Since the specular feature, in essence, is an abstraction of the facial surface topography and the reflectivity pattern, the conditional distribution connected with the projections of a genuine set onto the calibrated genuine set has a large variance (σ_G^2 large)indicating a complex association between the two topographies (the blue colored conditional density function in both Fig. 3a, b). The association between the genuine space and the printed photos is a simple one as the latter is a planar representation and this is described by a Gaussian having a relatively small mean and very small variance (μ_{PP} small and σ_{PP} small) as can be witnessed in Fig. 3a (red-colored graph). The conditional distributions (Fig. 3b) get closer for the 3D mask case and $\sigma_M^2 < \sigma_G^2$ (simpler association as compared to genuine-genuine), since the masks essentially designed to fit multiple individuals leading to a spatial smoothing effect, however the means μ_M and μ_G are close. Overall, these conditional distributions

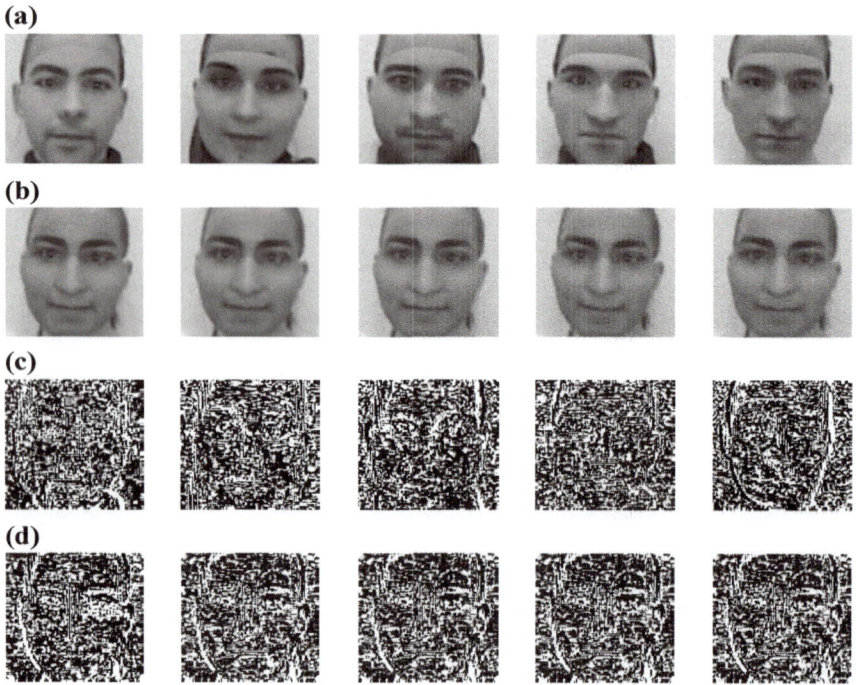

Fig. 2 **a** Low rank/Diffused component of spoofed faces wearing $3D$ masks; **b** Low rank/Diffused component of real legitimate faces; **c** Sparse/Specular component of $3D$ mask faces; **d** Sparse/Specular component of real genuine faces

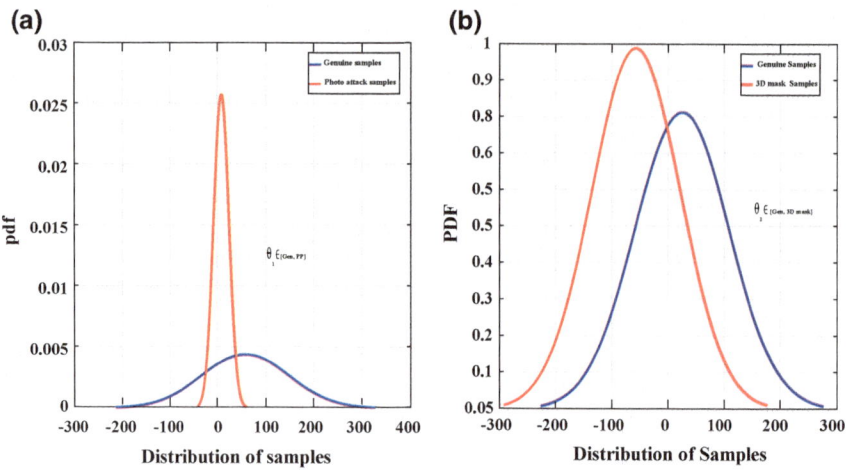

Fig. 3 **a** Gaussian Distribution of Genuine and photo attack samples computed by extracting energy from eigenspace projected features. **b** Gaussian Distribution of Genuine and $3D$ Mask samples computed by extracting energy from eigenspace projected features

Table 1 Performance of $SVM_{GENandPP}$ SVM classifier evaluated over eigenspace projected features across each fold for printed photo attack face detection

Measure	Accuracy @FAR	FAR
K_1	0.9523	0.035
K_2	0.9560	0.042
K_3	0.9584	0.0438
K_4	0.9658	0.036
K_5	0.9650	0.0330
Avg.	0.8651	0.1121

Table 2 Performance of SVM $SVM_{GENand3D}$ classifier evaluated over eigenspace projected features across each fold for $3D$ mask face detection

Measure	Accuracy @FAR	FAR
K_1	0.8972	0.070
K_2	0.8920	0.0583
K_3	0.8930	0.0507
K_4	0.8987	0.0537
K_5	0.8925	0.0470
Avg.	0.7603	0.1353

clearly indicate that there is substantial variability across genuine and both spoof models, which can be used for constructing a suitable classifier.

3 Performance Evaluation

Two databases were deployed for testing the proposed architecture: (i) Printed photo database [9] and (ii) 3D mask database [2] based on paper craft masks. Two SVM models are constructed: (i) $SVM_{GENandPP}$: SVM for Genuine versus Printed photos and (ii) $SVM_{GENand3D}$: SVM for Genuine versus 3D Masks. The training and testing arrangement for $SVM_{GENandPP}$ was

- Training: CLASS-GENUINE: 17 subjects (Genuine) and 150 variations per subject; CLASS-SPOOF: 17 subjects (printed photos) and 150 variations per subject;
- Testing: CLASS-GENUINE: 17 subjects (Genuine) and 150 variations per subject; CLASS-SPOOF: 17 subjects (printed photos) and 150 variations per subject;

The training and testing arrangement for $SVM_{GENand3D}$ was,

- Training: CLASS-GENUINE: 15 subjects (Genuine) and 100 variations per subject; CLASS-SPOOF: 15 subjects ($3D$ Masks) and 100 variations per subject;

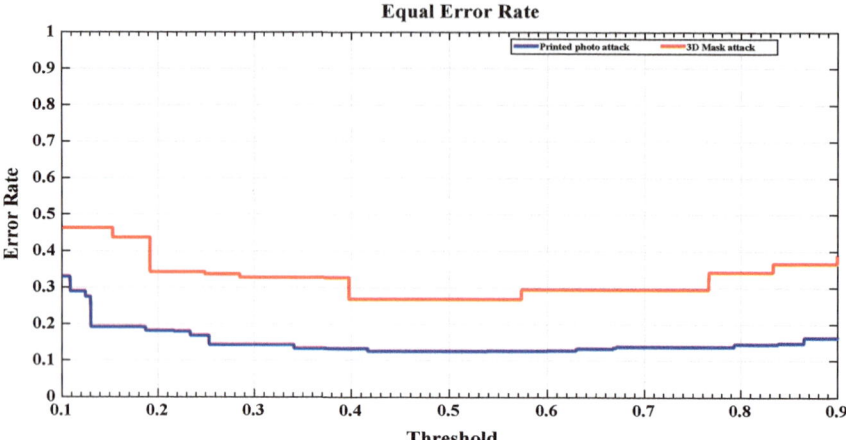

Fig. 4 Expected performance curve (EPC) with RBF kernel of SVM classifier. Error as a function of the threshold for SVM model $SVM_{GENandPP}$ and SVM model $SVM_{GENand3D}$. The optimal threshold is the point of minima in the two curves

– Testing: CLASS-GENUINE: 15 subjects (Genuine) and 100 variations per subject; CLASS-SPOOF: 15 subjects ($3D$ Masks) and 100 variations per subject;

The training and testing data were split into fivefold K_1, K_2, K_3, K_4, $and K_5$ (data rotation to ensure there is no bias in the formation of the SVM models). Within each fold, 80% of the data was used for training and 20% for testing. The accuracies and false alarm rates for SVM models $SVM_{GENandPP}$ and $SVM_{GENand3D}$ are shown in Tables 1 and 2 respectively. For the printed photograph based SVM model, the optimal classification accuracy (corresponding to the Equal Error Rate—EER) was found to be 86.5% while for the 3D mask based SVM model the classification accuracy was 76.0%. The corresponding false alarm rates were 11.21% and 13.53, respectively. The classification error rates as a function of the threshold are shown in Fig. 4 and the minimal error happens to be the point of minima in both the curves. 3D masks do influence depth profile capturing, hence their spoof detection accuracy is on the lower side as compared to printed photos. Their projection variability, however, is small in comparison with genuine faces, which is reflected in a moderate detection rate as far as accuracy is concerned.

4 Conclusions

In this paper, we begin with a hypothesis that since the specular feature contains information pertaining to the topography and depth variation, in a face or a mask, the association between the specular sets from two genuine image sets, will exhibit significant complexity and diversity. On the other hand, the specular set from a genuine image set, is expected to have little similarity with that of a printed photo set and a moderate similarity with a 3D mask set (because of the introduction of depth information). This hypothesis has been ratified by testing the two spoof models (printed photo and 3D mask) onto two different SVM models. The classification accuracies for genuine versus printed photo and genuine versus 3D mask were 85.5% and 76.0%, respectively.

References

1. Candès, E.J., Li, X., Ma, Y., Wright, J.: Robust principal component analysis? J. ACM (JACM) **58**(3), 11 (2011)
2. Erdogmus, N., Marcel, S.: Spoofing face recognition with 3d masks. IEEE Trans. Inf. Forensics Secur. **9**(7), 1084–1097 (2014)
3. Galbally, J., Marcel, S.: Face anti-spoofing based on general image quality assessment. In: 2014 22nd International Conference on Pattern Recognition (ICPR), pp. 1173–1178. IEEE (2014)
4. Gao, X., Ng, T.T., Qiu, B., Chang, S.F.: Single-view recaptured image detection based on physics-based features. In: 2010 IEEE International Conference on Multimedia and Expo (ICME), pp. 1469–1474. IEEE (2010)
5. Garcia, D.C., de Queiroz, R.L.: Face-spoofing 2d-detection based on moiré-pattern analysis. IEEE Trans. Inf. Forensics Secur. **10**(4), 778–786 (2015)
6. Ji, Z., Zhu, H., Wang, Q.: Lfhog: a discriminative descriptor for live face detection from light field image. In: 2016 IEEE International Conference on Image Processing (ICIP), pp. 1474–1478. IEEE (2016)
7. Karthik, K., Katika, B.R.: Face anti-spoofing based on sharpness profiles. In: 2017 IEEE International Conference on Industrial and Information Systems (ICIIS), pp. 1–6. IEEE (2017)
8. Karthik, K., Katika, B.R.: Image quality assessment based outlier detection for face anti-spoofing. In: 2017 2nd International Conference on Communication Systems, Computing and IT Applications (CSCITA), pp. 72–77. IEEE (2017)
9. Wen, D., Han, H., Jain, A.K.: Face spoof detection with image distortion analysis. IEEE Trans. Inf. Forensics Secur. **10**(4), 746–761 (2015)
10. Zhang, Z., Yan, J., Liu, S., Lei, Z., Yi, D., Li, S.Z.: A face antispoofing database with diverse attacks. In: 2012 5th IAPR International Conference on Biometrics (ICB), pp. 26–31. IEEE (2012)

Efficient Sparse to Dense Stereo Matching Technique

**Piyush Bhandari, Meiqing Wu, Nazia Aslam, Siew-Kei Lam
and Maheshkumar Kolekar**

Abstract Acquiring accurate dense depth maps with low computational complexity is crucial for real-time applications that require 3D reconstruction. The current sensors capable of generating dense maps are expensive and bulky, while compact low-cost sensors can only generate the sparse map measurements reliably. To overcome this predicament, we propose an efficient stereo analysis algorithm that constructs a dense disparity map from the sparse measurements. Our approach generates a dense disparity map with low computational complexity using local methods. The algorithm has much less computation time than the existing dense stereo matching techniques and has a high visual accuracy. Experiments results performed on KITTI and Middlebury datasets show that our algorithm has much less running time while providing accurate disparity maps.

Keywords Depth estimation · Sparse to dense · Stereo matching · Interpolation

1 Introduction

In recent years, the rapid growth of 3D imaging technology has brought about a new era of depth sensing and signal processing. A dramatic rise in-depth sensors can be seen. Current high-quality depth sensors which include time-of-flight cameras

---

P. Bhandari (✉) · N. Aslam · M. Kolekar
Indian Institute of Technology, Patna, Bihar, India
e-mail: piyush.ee13@iitp.ac.in

N. Aslam
e-mail: 1821ee10@iitp.ac.in

M. Kolekar
e-mail: mahesh@iitp.ac.in

M. Wu · S.-K. Lam
Nanyang Technological University, Singapore, Singapore
e-mail: meiqingwu@ntu.edu.sg

S.-K. Lam
e-mail: siewkei_lam@pmail.ntu.edu.sg

© Springer Nature Singapore Pte Ltd. 2020
B. B. Chaudhuri et al. (eds.), *Proceedings of 3rd International Conference on Computer Vision and Image Processing*, Advances in Intelligent Systems and Computing 1022,
https://doi.org/10.1007/978-981-32-9088-4_14

and high-resolution structured light scanners which are expensive and bulky. While innovations, such as Kinect (Microsoft Corp.), have reduced the size and the cost, there is a need for even more compact and less expensive sensors. In this paper, we use a stereo camera which, unlike Kinect, can be used in outdoors environment and is compact with low cost.

3D imaging technology has a vast number of applications ranging from building reconstruction to navigation of self driven cars [1, 2]. Applications such as pedestrian tracking [3, 4] and human reidentification [5, 6] also use depth informations (RGBD data) since it provides easier extraction of human-distinguishable features. Previously, many solutions for complex vision problems such as [7–9] has been proposed without the use of depth informations. Depth information could boost up the performance of certain tasks such as [10–12].

Miniature robots and other small maneuvering or autonomous systems do not only have limited processing power but also require less power consumption. In such cases, stereo vision is usually a low- level task and should not consume much energy, leaving enough processing power for high-level operations. Therefore, a need for efficient stereo matching algorithms that are able to satisfy the needs of such applications is called into play. One of the ways to achieve faster processing is to incorporate sparse- or feature-based algorithms. Such algorithms have been much explored in recent years, though the information provided by them is usually not sufficient for implementing them in real time. Sparse maps, by definition are unable to provide us with a larger view of the picture and obtaining the depth values of a few points is hardly any help to the application systems. Little research is conducted on sparse to dense map generation techniques, but they too suffer from a few drawbacks. Most of the techniques are based on transforms and compressive sensing [13, 14] which themselves require higher computation than most of the dense map algorithms. The algorithms are difficult to implement on low energy hardware and the running time of the algorithms are high, for example, [15]. Other sparse to dense techniques are based on interpolation techniques, which have low computational time but are unable to bring an accuracy comparable to the dense map techniques. Hence, we try to close the gap brought about by various sparse to dense techniques and try to promote the idea of exploring such techniques.

In this paper, we propose an efficient way to generate a dense disparity map. Working on a dense map is computationally complex, therefore we initially generate a sparse disparity map using local methods and then convert it into a dense map using the proposed algorithm. The main contribution of the paper is to introduce an efficient sparse to dense disparity map conversion technique using stereophotogrammetric systems. The method uses the concept of super pixels through which we interpolate the values of pixels with less distinguishable features using the pixels whose correspondence can be found with much accuracy. Using such an approach, we can generate dense disparity maps much faster than the current approaches and not compromise on the accuracy. Our experiments with this novel sparse to dense conversion technique on the KITTI [16], Middlebury (2001, 2003, 2014) datasets [17–20] show that the algorithm is much faster and considerably more accurate.

The organization of the paper is as follows: in Sect. 2, we describe the proposed depth estimation algorithm. Following that is Sect. 3, which explains the experiments that we run to evaluate the our method and results are presented in Sect. 4. Section 5 concludes the paper.

2 Methodology

The motivation behind our approach is to reduce the computational complexity by incorporating a sparse to dense conversion algorithm. We segment our algorithm into parts and then explore ways to make each part of the algorithm efficient and compatible with our sparse to dense approach. The subdivisions of our algorithm are

- Generation of Sparse Map

    - Preprocessing of Images
    - Similarity Computation
    - Cost aggregation
    - Sparse Post-processing

- Generation of Dense Map

    - Sparse To Dense Interpolation
    - Dense Post-processing

2.1 Generation of Sparse Map

Preprocessing of Images—Since first we are trying to obtain a sparse disparity map of a given stereo pair, it is of utmost importance that the accuracy of disparity values of the sparse set is particularly high. If pixels with less accuracy are mapped into a dense map, the effectiveness of the disparity map obtained would itself be extremely low. Also, the number of sparse pixels obtained must be significant in quantity and must be distributed across the whole image instead of just a small segment, as a few post-processing techniques (proposed in the upcoming subsections) will remove some of the less accurate points leaving us with even less points in the sparse map. Therefore, highly distinguishable pixels like edges, corners are the most likely choice as the sparse feature set for our algorithm. Keeping all of the above in mind, we use the concept of super pixels as our input feature. The reason for using super pixels are as follows:

- When converting our sparse map into a dense disparity map, the values of the pixels are interpolated using the values of the pixels comprising the boundary of super pixel. If the number of super pixels in an image are large each super pixel can to some extent belong to the subpart of the same or nearby object and using

Fig. 1 SEEDS output

the interpolation technique specified all the pixels of an object (similar in texture and closer in spatial distance) would have similar disparities.
– We can tune the parameters of the super pixels so that the highly distinguishable features of an object (edges, corners, etc.) comprise the sparse set of our system, therefore a fairly accurate sparse depth map will be obtained (Fig. 1).

For our algorithm we use SEEDS [21] (Super pixels Extracted via Energy-Driven Sampling) Fig. 2, the algorithm is computationally less expensive compared to the other Super Pixel algorithms. Using SEEDS, the percentage of pixels obtained in the sparse set are between 6 and 13%, therefore the time complexity of the overall system reduces much significantly.

Similarity Computation—Stereo correspondence greatly depends on the matching costs for determining the similarity in left and right image locations. Few of the common methods to compute costs are the sum of absolute or squared differences (SAD, SSC), normalized cross correlation (NCC), Rank and Census Transforms [22]. Research conducted by Heiko Hirschmuller and Daniel Scharstein [23] provides us with a detailed evaluation on stereo matching cost with radiometric differences. Their results display greater performance for Census Transform as a measure of

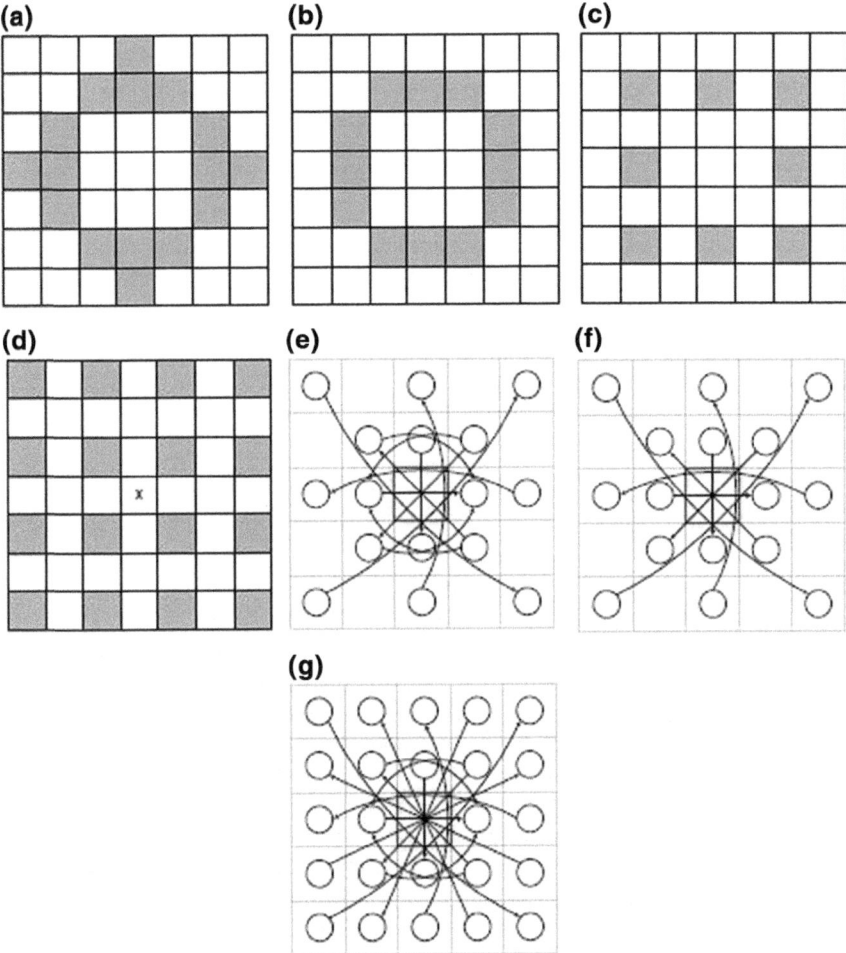

Fig. 2 Sparse Windows **a** 16-Point Diamond **b** 8-Point Diamond **c** 8-Point Alternate **d** 16-Point Alternate **e** 8 Edge (Comparisons) **f** 12 Edge (Comparisons) **g** 16 Edge (Comparisons)

matching cost. Since Census Transform is a nonparametric cost which relies on the relative ordering of pixels values, they are invariant to all the radiometric differences that preserve such an order. Therefore, we implement Census Transform as a filter followed by a dissimilarity measure using Hamming's Distance.

Census Transform is usually implemented in an $N \times N$ square window size, but we try to adapt a more sparse census transform window which significantly reduces the computational complexity of our stereo matching algorithm. For example, [24] introduces a comparison between the computational complexity and the accuracy of the maps produced by a large variety of sparse and generalized census transform windows. The paper also described a hardware architecture for efficiently implementing high-performance census-based correlation systems. Therefore, we use a sparse census window which seeks to achieve a balance between the accuracy of the stereo system and the resource requirements of its implementation. The amount of logic required to implement the census stereo method is directly related to the length of the census vector and it is reduced to a significantly large amount by reducing the number of comparisons to compute the transform.

The results for the best sparse census transform window (specified in Fig. 2) for each dataset are presented in Table 1. Each dataset provides particularly high accuracy for a different sparse window. Therefore instead of using a generalized window, we find the best window type for each datatype and run the algorithm separately.

Cost Aggregation—The cost correlation between any two points in the corresponding left and right images are computed using Hamming Distance. In order to make the computations faster, we use SSE instructions to determine the cost (similarity measure) for any two points.

Instead of using a 5×5 aggregation window, we implement a rectangular aggregation window of 6×4. The window is adjusted to a rectangular form based on the research by Nils Einecke and Julian Eggert [25] (Multi-Block Matching, MBM) and the comparison outputs we ran. They specify an approach of in which they use a number of blocks (61×1, 1×61, 9×9, 3×3) and find the best match using a probabilistic fashion of multiplying the cost of all the blocks. The computations for each aggregation vastly increases (212 comaprisons compared to our 24), though they reduce the computation time by using IPP routines and optimizing their algorithm. From our analysis, we found that increasing a small percentage of block size horizontally can provide better results than a normal square window based aggregation while keeping the complexity of computations same. This behavior is mainly because the horizontal component of a block has more similarity when moving to the right image than the vertical shifts. The change in the accuracy of pixels due to the above discussed block sizes are shown in Table 1.

Post-Processing—In the sparse map stage, after finding the correspondences of the sparse feature set, we incorporate left–right consistency check and sub-pixel estimation techniques. In consistency check, we filter out the matches with high uncertainty by imposing a uniqueness constraint, i.e., the matching cost for a selected feature pair has to be smaller than the cost for the next best match times a uniqueness factor γ. For our algorithm the value of $\gamma = 0.7$ for all datasets. For sub-pixel estimation, we fit a quadratic curve through the neighboring cost and find the minimum cost position.

Table 1 Percentage error for different block sizes

Sparse window	16P Dia	8P Dia	8P Alter	8 Edge
Agg.Block 5×5	14.5456	15.0605	16.74	18.2824
Agg Block 6×4	13.53	14.04	15.14	16.47

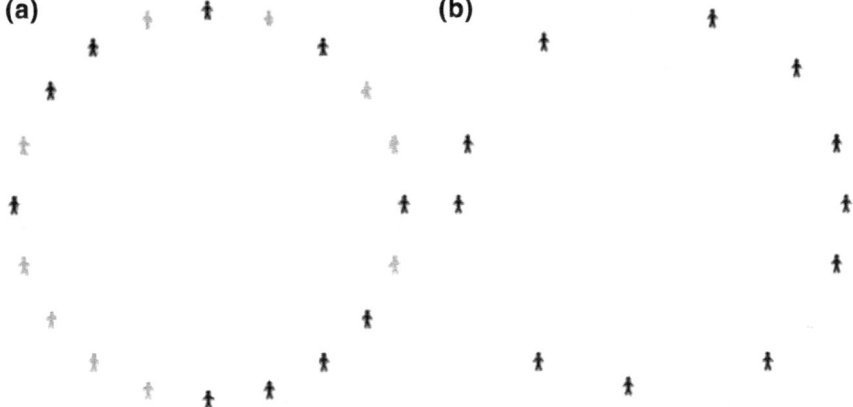

Fig. 3 Single cluster analysis. **a** Before post-processing **b** After post-processing

2.2 Generation of Dense Map

Sparse To Dense Map—Some of the common ways of converting a sparse disparity map into a dense one are compressive sensing using wavelet coefficients [14], applying contour-let transforms [13], multilayer conditional random field [26] or basic interpolation techniques such as nearest-neighbor interpolation. The above approaches are not time efficient as the running time can sometimes reach a few minutes and the accuracy of disparity maps obtained are not comparable to the other state-of-the-art algorithms. Also, the common approaches take a random or predefined set of sparse features for conversion and assume that the sparse set is of high accuracy. Instead of taking the ideal case of highly accurate sparse set, we use the above defined routines to get the sparse feature set and then convert the sparse map into a dense one.

Using SEEDS super pixels, we obtain a set of clusters. To better understand our approach, we take our analysis to a single cluster (shown in Fig. 3a), the points (shapes) in the image corresponds to the pixels at the boundary of the cluster. The result obtained from the sparse map generation routines will give us a few pixels which are highly accurate (black in color) and a few of them which are less accurate (gray in color). After dense consistency, check the cluster's sparse set reduces to only a few accurate pixels (shown in Fig. 3b).

Using this model, we try to predict the plausible values of the pixels inside the cluster (disparity unknown) based on the pixels on the boundary (disparity known) of the cluster. The problem could be seen as a two-dimensional approach of plane fitting. However, the huge differences in the disparity measure of the accurate and less-accurate pixels create a plane prone to inaccuracies. Plane fitting is a suitable method when small error in disparity (e.g., 3 px) do not vary the depth to a large extent, however, in majority changes, even small errors can lead to huge differences

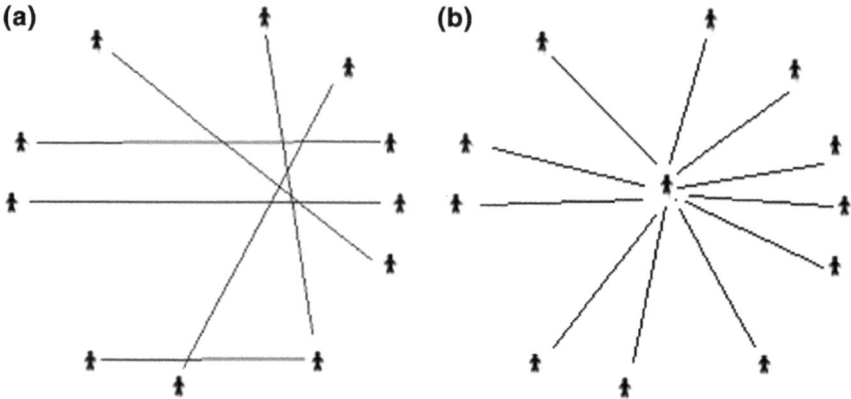

Fig. 4 Interpolation technique. **a** Cosine interpolation **b** Weighted interpolation

in the depth obtained for a particular object. Therefore, we use one-dimensional approach in which we predict the values of pixels lying on a line joining the two extreme boundary pixels (Fig. 4a). This approach when applied using super pixels can lead to much better results in any scenario and does not involve the computation of eigenvectors or SVD. Common one-dimensional approaches are linear interpolation, cubic interpolation, cosine interpolation, spline. Cubic interpolation is the best among all since it incorporates double derivatives as well as single derivatives. However, for cubic interpolation, we require 4 points in order to fit the curve between them. Unfortunately, we will not always have the luxury of choosing 4 points which we can easily tell by looking at the sparse map for a single cluster received after sparse post-processing routines. Therefore, we use cosine interpolation whose results are close to that of cubic interpolation and also, a simple and efficient look-up table can be implemented to speed up cosine calculations.

For our algorithm, we use the following formula to interpolate the value of the pixel based on the two extreme pixels using cosine interpolation:

$$w1 = \frac{distance(p, i)}{distance(p, i) + distance(p, j)} \tag{1}$$

$$w2 = \frac{(1 - cos\,(w1 * \pi))}{2} \tag{2}$$

$$D(p) = D(i)(1 - w2) + D(j)(w2) \tag{3}$$

where p, i, j are pixels and the value of pixel p is to be determined using pixel extremities i and j, all of which lie on the same line. Variable $w1$ is the weight based on the spatial distance of the pixel p with pixel i. $D(p)$ corresponds to the disparity value of the pixel p.

The 1D interpolation explained above depends on the fact that for every pixel inside the super pixel cluster, we are able to extract two extremities in the same horizontal line, which might not always be plausible. To tackle the problem we use adaptive weight-based interpolation technique, all the pixels for whom the extremities are not located in a predefined range of the line, we interpolated their disparity values using normalized weighted aggregation based on color similarity and spatial distance between the unknown pixel and all the known pixels (for each cluster). The formula for the above technique represented in Fig. 4b is described below

$$\omega_i(p, q) = exp \left\{ -(\frac{color(p, q)}{\gamma_c} + \frac{spatial(p, q)}{\gamma_s}) \right\} \tag{4}$$

$$D(p) = \frac{\sum_{i=0}^n w_i(p, q)D(q)}{\sum_{i=0}^n w_i(p, q)} \tag{5}$$

where p is current pixel(disparity unknown), q is superPixel boundary(disparity known), ω corresponds to the weight based on color similarity and spatial distance and n is the number of Super Pixel boundary points.

The results portrayed that the computational time for normalized adaptive weight was higher than cosine, therefore we used a combination of cosine and weighted interpolations described in Algorithm 1. The function closestExtremeties(superPixel(p),p) will provide us the with the closest two extremities based on the vertical distance of pixel p with the Super Pixel boundary points.

Algorithm 1 Interpolation Algorithm

1: **procedure** INTERPOL(p, $super Pixel$)
2:    $super Pixel(p)$;   ▷ Super Pixels of p
3:    $Point\ i,\ j=closest Extremeties(super Pixel(p), p)$
4:    **if** $abs(i.y - p.y) < 4$ && $abs(j.y - p.y) < 4$ **then**
5:        $D(p) = cosine Interpol(p, i, j)$
6:    **else**
7:        $D(p) = weighted Interpol(p, super Pixel(p))$
8:    **end if**
9: **end procedure**

Dense Post-Processing—or the dense stage, after converting our sparse map into a dense map we use standard block matching post-processing steps:- speckle filtering (<120 pixels), background fill in and median filtering using kernel size (3×3). Each of the process is used to remove noise or fill the holes the disparity image generated.

3 Experimental Results and Discussion

For evaluating our sparse to dense stereo approach, we used the two popular datasets available online : : KITTI and Middlebury. We used the color images as input and the ground truth of the maps for comparing the results of the algorithm. The tests were conducted on an Intel Core i3-4010U Processor with clock frequency of 1.70 GHz We use the results given in the benchmark tables and compare the timing of our algorithm compared to that of SGM(Semi-global matching) and MBM (Multi-Block Matching).

It should be noted that the comparison is made with algorithms (on the benchmark tables) such as ELAS [27], SGBM whose results are specified on an i7-6700K processor with clock frequency @4GHz and GTX TITAN X graphics card and MBM results are specified on 1 core processor with clock frequency @ 3.0 GHz with IPP routine optimization. GCSF algorithm [28] on the benchmark has higher accuracy that our algorithm but our algorithm is at least seven times faster even on a much less powerful hardware. The results for various alogrithms on the given benchmark datasets are uploaded online using different processors, also we were unable to produce the exact same results (as specified in the benchmark tables) using the same processor for different algorithms. Therefore, we tried to compare the experimental results while taking into consideration the processor type used. It can be easily verified from the results in Tables 3 and 4 that even using a less powerful CPU and without IPP optimization the timing of our program is close to that of the MBM and SGM. This way the algorithm proposed is much more suitable for real-time applications and implementing it on an FPGA or stand-alone processors will be much simpler. Since semi-global and global methods utilize heavy optimization methods (Dynamic Programming, Graph Cuts, etc.), it would be much more difficult to implement then in real-time scenarios with much simpler hardwares. The memory footprint of our algorithm is also small since we work on a small part of an image at a time, whereas the global methods tend to work using the whole images, which results in a much larger memory footprint. Table 2 displays the accuracy of our algorithm obtained for the respective datasets. The best result with respect to Census windows on our algorithm is specified in the table, the difference in the Census windows for each dataset presents because of the difference in the texture and type of images presented in the different datasets. The percentage errors are defined for certain thresholds. If

Table 2 Percentage error for different datasets using sparse to dense algorithm

Error threshold	3.0 px	2.0 px	1.0 px	Census window
Midd. 2001	6.9	11.8	18.8	8 Point alternate
Midd. 2003	13.3	18.9	29.2	12 edge
Midd. 2014	28.7	35.5	47.7	12 edge
KITTI	27.3	46.5	54.5	8 edge

Table 3 Percentage error (Threshold 3px) for KITTI dataset

Algo	Error-Percentage	Run-time (s)	Processor
GCSF	12.05	2.4	1 core, i5@ 2.5 Ghz
SGBM	12.22	1.1	1 core, i7@ 2.5 Ghz
Our Algo	27.3	0.35	2 core, i3 @1.7 GHz

Table 4 Percentage error (Threshold 2 px) For Middlebury 2014 Dataset

Algo	Error-Percentage	Run-Time(s)	Processor
SGBM1	36.4	0.23	1 core, i7@3.3 GHz
ELAS	34.6	3	1 core, i7@3.6 Ghz
Our Algo	35.5	0.31	2 core, i3@1.7 GHz

the difference in the computed disparity and ground truth for a pixel is greater than the threshold, the pixel is then accounted as an error.

"Error Threshold" specified in the table refers to the number of pixels with true disparity difference of less than x (where x can be 1, 2, or 3) when compared to the ground truth values of the respective datasets. The approach for sparse to dense produced high inaccuracy in KITTI dataset mainly because local search methods are unable to match the stereo images with much confidence. More than one matches of a particular pixel are obtained and the correct matching pixel sometimes has higher cost than an incorrect match. Such irregularities can be addressed by using multiple blocks as done by MBM(multi-block) matching, though this may affect the computing time many folds. Heavy optimization routines can enable the use of local methods in such environments.

References

1. Scharwchter, T., Schuler, M., Franke, U.: Visual guard rail detection for advanced highway assistance systems. In: 2014 IEEE Intelligent Vehicles Symposium Proceedings, June 2014, pp. 900–905 (2014)
2. Hne, C., Sattler, T., Pollefeys, M.: Obstacle detection for self-driving cars using only monocular cameras and wheel odometry. In: 2015 IEEE/RSJ International Conference on Intelligent Robots and Systems (IROS), Sept 2015, pp. 5101–5108 (2015)
3. Pallauf, J., Wagner, J., Len, F.P.: Evaluation of state-dependent pedestrian tracking based on finite sets. IEEE Trans. Instrument. Measurem. **64**(5), 1276–1284 (2015)
4. Gao, S., Han, Z., Li, C., Ye, Q., Jiao, J.: Real-time multipedestrian tracking in traffic scenes via an rgb-d-based layered graph model. IEEE Trans. Intell. Transp. Syst. **16**(5), 2814–2825 (2015)
5. John, V., Englebienne, G., Krose, B.: Person re-identification using height-based gait in colour depth camera. In: 2013 IEEE International Conference on Image Processing, Sept 2013, pp. 3345–3349 (2013)

6. Wu, A., Zheng, W.S., Lai, J.H.: Depth-based person reidentification. In: 2015 3rd IAPR Asian Conference on Pattern Recognition (ACPR), Nov 2015, pp. 026–030 (2015)
7. Kolekar, M.H., Sengupta, S.: Bayesian network-based customized highlight generation for broadcast soccer videos. IEEE Trans. on Broadcast. **61**(2), 195–209 (2015)
8. Kolekar, M.H.: Bayesian belief network based broadcast sports video indexing. Int. Springer J. Multimedia Tools Appl. **54**, 27–54 (2011)
9. Kolekar, M.H., Palaniappan, K., Sengupta, S., Seetharaman, G.: Semantic concept mining based on hierarchical event detection for soccer video indexing. Int. J. Multimedia **4**(5), 298–312 (2009)
10. Kolekar, M.H., Talbar, S.N., Sontakke, T.R.: Texture segmentation using fractal signature. IETE J. Res. India **46**(5), 319–323 (2000)
11. Kolekar, M.H., Dash, D.P.: A nonlinear feature based epileptic seizure detection using least square support vector machine classifier. IEEE Region 10 Conference TENCON 2015
12. Chatterjee, S., Bhandari, P., Kolekar, M.H.: Feature extraction and segmentation techniques in static hand gesture recognition system. In: Book Chapter of Book on Hybrid Intelligence for Image Analysis and Understanding. John Wiley Publication, UK (2017)
13. Liu, L.K., Nguyen, T.D.: Sparse reconstruction for disparity maps using combined wavelet and contourlet transforms. In: 2014 IEEE International Conference on Acoustics, Speech and Signal Processing (ICASSP), May 2014, pp. 3553–3557 (2014)
14. Hawe, S., Diepold, K., Kleinsteuber, M.: Dense disparity maps from sparse disparity measurements. In: 2011 IEEE International Conference on Computer Vision (ICCV 2011), vol. 00, no. undefined, pp. 2126–2133 (2011)
15. Gehrig, S.K., Eberli, F., Meyer, T.: A Real-Time Low-Power Stereo Vision Engine Using Semi-Global Matching, pp. 134–143. Berlin, Heidelberg: Springer (2009)
16. Geiger, A., Lenz, P., Urtasun, R.: Are we ready for autonomous driving? In: The Kitti Vision Benchmark Suite Conference on Computer Vision and Pattern Recognition (CVPR) (2012)
17. Scharstein, D., Szeliski, R., Zabih, R.: A taxonomy and evaluation of dense two-frame stereo correspondence algorithms. In: Proceedings of IEEE Workshop on Stereo and Multi-baseline Vision (SMBV 2001), pp. 131–140 (2001)
18. Scharstein, D., Szeliski, R.: High-accuracy stereo depth maps using structured light. In: Proceedings of 2003 IEEE Computer Society Conference on Computer Vision and Pattern Recognition, vol. 1, June 2003, pp. I195–I202 (2003)
19. Scharstein, D., Pal, C.: Learning conditional random fields for stereo. In: 2007 IEEE Conference on Computer Vision and Pattern Recognition, June 2007, pp. 1–8 (2007)
20. Scharstein, D., Hirschmuller, H., Kitajima, Y., Krathwohl, G., Nesic, N., Wang, X., Westling, P.: High-Resolution Stereo Datasets with Subpixel-Accurate Ground Truth, pp. 31–42. Springer International Publishing, Cham (2014)
21. Van den Bergh, M., Boix, X., Roig, G., Van Gool, L.: Seeds: superpixels extracted via energy-driven sampling. Int. J. Comput. Vision **111**(3), 298–314 (2015)
22. Zabih, R., Woodfill, J.: Non-parametric Local Transforms for Computing Visual Correspondence. Springer, Berlin, Heidelberg, pp. 151–158 (1994)
23. Hirschmuller, H., Scharstein, D.: Evaluation of stereo matching costs on images with radiometric differences. IEEE Trans. Pattern Anal. Mach. Intell. **31**(9), 1582–1599 (2009)
24. Fife, W.S., Archibald, J.K.: Improved census transforms for resource-optimized stereo vision. IEEE Trans. Circ. Syst. Video Technol. **23**(1), 60–73 (2013)
25. Einecke, N., Eggert, J.: A multi-block-matching approach for stereo. In: 2015 IEEE Intelligent Vehicles Symposium (IV), June 2015, pp. 585–592
26. Li, F., Li, E., Shafiee, M.J., Wong, A., Zelek, J.: Dense depth map reconstruction from sparse measurements using a multilayer conditional random field model. In: 2015 12th Conference on Computer and Robot Vision (CRV), June 2015, pp. 86–93 (2015)
27. Geiger, A., Roser, M., Urtasun, R.: Efficient Large-scale Stereo Matching, pp. 25–38. Springer, Berlin, Heidelberg (2011)
28. Čech, J., Sanchez-Riera, J., Horaud, R.: Scene flow estimation by growing correspondence seeds. In: CVPR 2011, pp. 3129–3136, June 2011

Handwritten Numeral Recognition Using Polar Histogram of Low-Level Stroke Features

Krishna A. Parekh, Mukesh M. Goswami and Suman K. Mitra

Abstract The paper focuses on the handwritten numeral recognition of an Indian script, Gujarati, with the reduced dimensions of features. The proposed method employees the low-level stroke (LLS) for feature extraction and the polar histogram method for feature vector generation that enables the reduced size representation of features. The baseline experiments were performed using k-nearest neighbor (k-NN) classifier and the result was improved further using support vector machine (SVM) classifier with radial basis function (RBF) kernel. The method of the Polar histogram of LLS features was also tested on Devanagari and English handwritten numeral datasets. The accuracy of classification for Gujarati, Devanagari, and English are on par with the state-of-the-art methodologies. The experiments were also performed for mixed dataset Gujarati-English, Gujarati-Devanagari, English-Devanagari, and Gujarati-English-Devanagari. In all experiments, the feature vector size is significantly less while the accuracy is not compromised much.

Keywords Handwritten numerals · Polar histogram · Low-level stroke features

1 Introduction

Optical character recognition (OCR) is a technology that converts handwritten and printed text into the digital documents. In developing country like India, digitization has reached many of the fields but the use of handwritten paperwork is unavoidable in many circumstances. An OCR is required for the digitization of such handwritten

K. A. Parekh (✉) · S. K. Mitra
DA-IICT, Gandhinagar 382007, Gujarat, India
e-mail: 201611037@daiict.ac.in

S. K. Mitra
e-mail: suman_mitra@daiict.ac.in

M. M. Goswami
Dharmsinh Desai University, Nadiad 387001, Gujarat, India
e-mail: mgoswami.it@ddu.ac.in

© Springer Nature Singapore Pte Ltd. 2020
B. B. Chaudhuri et al. (eds.), *Proceedings of 3rd International Conference on Computer Vision and Image Processing*, Advances in Intelligent Systems and Computing 1022,
https://doi.org/10.1007/978-981-32-9088-4_15

paperwork. Moreover, majority of people in India uses native language and scripts for oral and written communications. It is also necessary to convert the handwritten literature of regional script into the digital form for automated processing and world-wide dissemination. Hence, handwritten character recognition for regional scripts is the need of the hour. The development of OCR for many Indian scripts, including Gujarati, is still in progress. The diversity of Indian scripts, cursive nature of writing, and complex character set consisting of conjuncts, base characters, modifiers, and special symbols are some of the major challenges [9]. The OCR for handwritten symbols is considered even more difficult than printed because of the variations in writing styles.

In this paper, we are addressing the problem of offline handwritten Gujarati numeral recognition. The problem also has many direct applications like automated bank cheque processing, postal mail sorting, and automated form processing. However, the work may be extended with some modifications for alphabets and alphanumeric symbols for other wide range of real-life applications. Handwritten numeral recognition can be done either by the online or offline way. The online method includes the recognition of character directly from input on the digital writing pad. The offline method deals with the image obtained by scanning the handwritten document. Here, we deal specifically with the offline handwritten numeral recognition. A considerable amount of work has been done for handwritten symbol recognition in various scripts like English, Chinese, and Arabic. Also, a noticeable amount of work can be found in the same area for dominating Indian scripts like Devanagari, Bengali, and Tamil. However, for other regional scripts like Gujarati, it is still in the primary stage. Therefore, this paper focuses on handwritten numeral recognition for Gujarati numerals. The experiments were also extended to mixed numeral recognition considering the multilingual scenario in Indian context.

Traditionally, the features like chain code, directional element features (DEF), stroke orientation, and gradient-based features are used for handwritten character or numeral recognition. The proposed method uses a recently developed technique for feature extraction called low-level stroke (LLS) feature (explained in Sect. 3.1). LLS features are template matching-based features and give compact representation. The extracted features are represented in the reduced mathematical form using some type of feature vector generation method such that they do not lose their discriminating power. Block histogram method is generally used for feature vector generation. Small-sized feature vectors are efficient in terms of disc space and classification computation cost. However, the size of the feature vector in different existing methods is considerably large. So, efforts should be made to reduce the size of features. So, the feature vector generation is accomplished using Polar histogram method with LLS features in order to achieve the reduced size representation.

The proposed method is tested on the individual as well as mixed handwritten numeral samples from Gujarati, English, and Devanagari scripts. The performance of proposed technique is found to be on par with that of recent techniques. However, the main contribution of the proposal is evident from the size of the feature vector which is significantly less than that of all recent works.

The organization of rest of the paper is as follows: Sect. 2 provides a quick overview of related existing work. Section 3 gives details of feature generation process. It explains of feature extraction by low-level stroke (LLS) features and the proposed method for feature vector generation. Section 4 discusses about the experimental setup. It gives an insight of dataset, classifier, and results. Finally, the work is concluded in Sect. 5.

2 Related Works

There have been many works reported of handwritten numeral recognition for English, Devanagari and Bengali languages. For Indian script, first handwritten symbol recognition can be dated back to 1977 for Devanagari symbols by Sethi and Chatterjee [14]. Pal et al. [13] performed recognition of handwritten numerals from six popular Indian scripts like Devanagari, Bengali, Telugu, Oriya, Kannada, and Tamil. Directional features with Modified Quadratic Discriminant Function (MQDF) is used and accuracy of 99.56%, 98.99%, 99.37%, 98.40%, 98.71%, and 98.51%, respectively, has been reported. Bhattacharya et al. [4] worked on Devanagari and Bangla handwritten numeral recognition and had also built a dataset for the same. They also worked on mixed handwritten numerals of Devanagari, Bangla, and English. Multistage Recognition scheme using MLP classifier is used in it.

Considering Gujarati numeral recognition, Desai [5] in 2010 used four different profiles, horizontal, vertical, and two diagonal as features. Multi-layer feedforward backpropagation neural network was used for classification. The accuracy achieved was 81.66%. Maloo et al. [11] used affine invariant moments with SVM for Gujarati handwritten numeral recognition. The accuracy of 90.55% was achieved. Baheti et al. [1] used kNN and PCA on the feature set from affine invariant moments. Accuracy reported in it was 90.04% for kNN and 84.1% by PCA.

More recently, Nagar and Mitra [12] have used stroke orientation-based features with SVM classifier for recognition of Devanagari and Gujarati handwritten numerals. The accuracy claimed for Devanagari and Gujarati script numerals were 99.14% and 98.97%, respectively, with feature size 1296. The accuracy reported is highest so far for handwritten Gujarati numeral classification.

Goswami and Mitra [8] worked on printed Gujarati character classification using low-level stroke (LLS) features extraction method. Accuracy of 96.5% was reported for the feature vector of size 300. The same author has used low-level stroke (LLS) feature extraction method and Block histogram method for feature vector generation for Gujarati handwritten numerals [7]. The average accuracy obtained for it is 98.49% with feature vector of size 588 using SVM classifier with RBF kernel. They also tested the method for Devanagari handwritten numerals as well as for Gujarati-English and Gujarati-Devanagari mixed dataset. The accuracy of 98.65% with feature size 588 for Devanagari and 96.54 and 97.88% with feature vector size 192 for Gujarati-English and Gujarati-Devanagari was obtained for mixed numeral dataset.

Table 1 Recognition accuracy achieved in various related work in different languages

References	Languages	Accuracy (%)
Pal et al. [13]	Devanagari	99.56
	Bangla	98.99
	Telugu	99.37
	Oriya	98.40
	Kannada	98.71
	Tamil	98.51
Bhattacharya et al. [4]	Devanagari	99.04
	Bangla	98.01
Desai [5]	Gujarati	81.66
Dhandra et al. [6]	Kannada	96.80
	Devanagari	96.00
	Telugu	98.80
Singh et al. [15]	Devanagari	98.49
Maloo et al. [11]	Gujarati	91.00
Baheti et al. [1]	Gujarati	92.28
Nagar and Mitra [12]	Devanagari	99.14
	Gujarati	98.97
Goswami and Mitra [7]	Devanagari	98.65
	Gujarati	98.49

All related works in various Indian scripts along with the accuracy obtained are listed in Table 1. It is evident from the comparison that fewer works are reported for Gujarati handwritten symbols recognition as compared to other Indian scripts. However, a fairly large dataset for Gujarati handwritten numerals is developed in recent time. Most of the works, either in Gujarati or in other Indian scripts, reported fairly large feature vector size. However, not more than 50% of the total pixels from the numeral images are informative. Thus, there is a need for exploring the idea of achieving similar accuracy with much smaller feature vector size. The present work is an attempt to reduce the feature size while achieving similar accuracy for Gujarati numerals.

3 Feature Generation

The process of feature generation consists of three parts; (i) Preprocessing of data, (ii) Extraction of features, and (iii) Feature vector generation.

The first step for character recognition is data preprocessing, so that we get data fit for feature extraction. The following preprocessing is done on image dataset:

- *Noise removal*: Binary images may contain salt and pepper noise due to scanning process. So to remove the salt and pepper noise, 2D median filtering is used.
- *Resizing*: Bilinear interpolation is used to normalize the size of numeral images in case they differ in size due to writer's style.
- *Thinning*: Images are thinned to single pixel stroke using ZW algorithm [16] for better stroke based feature extraction.

Note that similar preprocessing techniques are used in [7]. The next step is feature extraction.

3.1 Extraction of Low-Level Stroke (LLS) Features

After the preprocessing of data, feature extraction is to be performed. The extraction of LLS features provides detailed representation of binary numeric images containing only 0 and 1. Current proposal uses a recently developed technique called low-level strokes for feature extraction. The details of LLS features is described in [7], we are furnishing here the salient points for the better understanding of readers. Low-level stroke (LLS) method of feature extraction is an extension of directional element features (DEF) and it has been proposed by Goswami and Mitra in [7] and [8]. Low-level strokes represent the basic elements of geometry like endpoints, junction points, line elements, and curve elements. Line and curve elements are further classified based on orientation and slope. In line elements, we get Horizontal, Vertical, Right, and Left Slant, whereas in Curve elements, we get Right Flat and Deep curve and Left Flat and Deep curve. The junction points can also be classified as "Cross", "Y", or "T" junctions. There are total of 12 different LLS pattern codes, which include endpoint, four line, and curve elements each and three for junction points.

The character comprises of high-level strokes which are in turn described as sequence of low-level strokes as shown in Fig. 1. A 3×3 mask is used to get LLS from single-pixel thinned character image by the method of template matching.

LLS categorization can be done by number of pixels. Let N_I be the number of on pixels in the 3×3 neighborhood of an object pixel I.

- $N_I = 1$; I is an isolated object pixel. It is likely to be a noisy pixel, so ignore it.
- $N_I = 2$; I is an endpoint.
- $N_I = 3$; I is either a line or a curve segment.
- $N_I = 4$; I is either a "T" junction or a "Y" junction.
- $N_I = 5$; I is a cross junction.
- $N_I \geq 6$; I is an unknown LLS. It is denoted by pattern code (-1) and is handled specially.

The 12 different pattern codes incorporate 34 different template patterns shown in Fig. 2. So using LLS feature extraction method, a matrix of pattern codes is generated

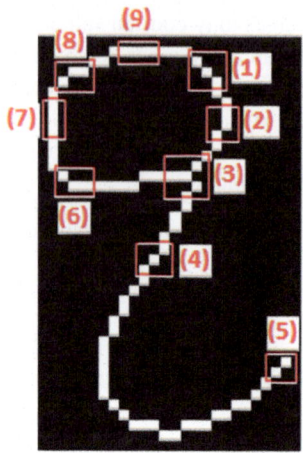

Fig. 1 Representation of character as HLS and HLS as sequence of LLS; (1) Right Slant; (2) Deep Left Curve; (3) "Y" Junction; (4) Left Slant; (5) End Point; (6) Flat Right Curve; (7) Vertical Line; (8) Flat Left Curve; (9) Horizontal Line

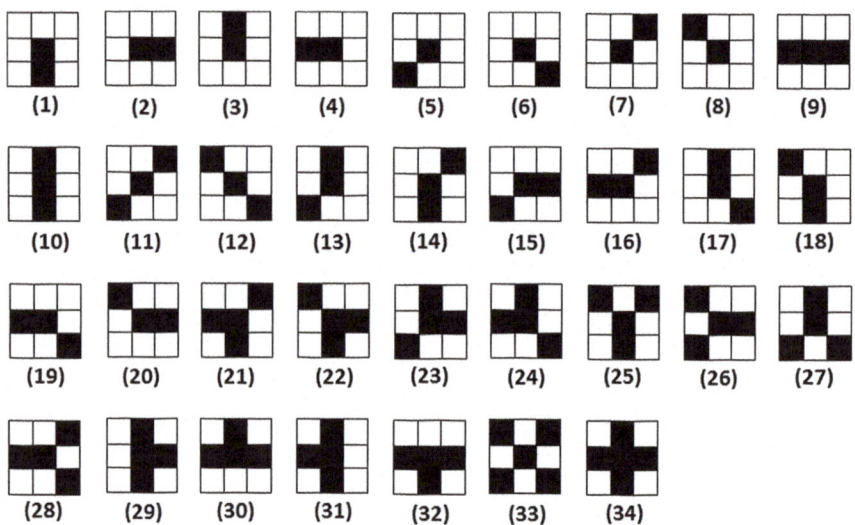

Fig. 2 LLS template patterns for 3 × 3 mask; (1) to (8)—endpoint with different orientations; (9) to (12)—Line segment; (13) to (20)—Flat and Deep curve segments ; (21) to (28)—"Y" Junction; (29) to (32)—"T" Junction; (33) and (34)—Cross Junction

so we get histogram with bin size 12 for the binary numeric images by incorporating the pattern codes of LLS features. This matrix is further used in feature vector generation.

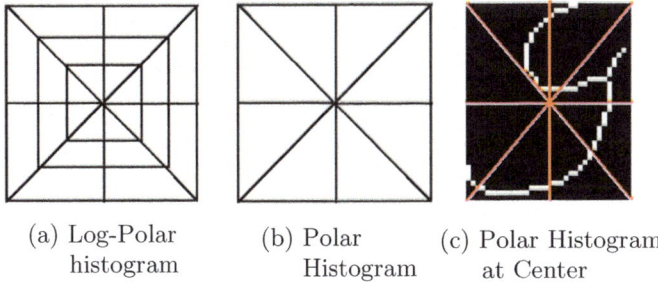

(a) Log-Polar histogram   (b) Polar Histogram   (c) Polar Histogram at Center

Fig. 3 Feature vector generation

3.2 Feature Vector Generation

The feature vector is a reduced mathematical representation of extracted features. Reduced dimensional feature vectors are better in terms of low memory space requirement and computation cost of classification. The proposed method of the Polar histogram for feature vector generation is capable of generating low dimensional feature vectors without much compromising the accuracy of classification.

Prior to using Polar histogram, experiments were performed with Log-Polar histogram [2] and [3] for feature vector generation. Figure 3 shows the diagram of how images are divided to get log-polar and polar histogram. For each subdivision feature, histogram is generated and finally concatenated.

The following experiments were performed for feature vector generation. However, polar histogram at the center of image seems to be best as far as recognition is considered.

i Log-Polar histogram at endpoints and junction points without incorporating location information.
ii Log-Polar histogram at endpoints and junction points by incorporating location information of end points and junction points. Image is divided into four quadrants for inclusion of location information.
iii Log-Polar histogram at the center of image with three distance divisions (3×8 bins of histogram) and two distance divisions (2×8 bins of histogram).
iv Polar histogram at the center of the image (8 bins of histogram).

Experiments show that Polar histogram is showing better accuracy compared to Log-polar histogram. The results of Log-polar histogram method were not up to the mark as the feature matrix of pattern codes is highly sparse in nature. One of the reasons can be the small-sized image is divided into many divisions for feature generation. Experiments showed the improvement in the result while moving from log-polar histogram with many divisions to log-polar histogram with fewer divisions. Hence, specificity of the feature is reduced. So, Polar histogram is taken at the center of pattern code matrix for feature vector generation.

In Polar histogram method, the matrix of pattern codes for each character image is divided into eight angular divisions. As discussed in Sect. 3.1, there are 12 different pattern codes of LLS features. So the histogram bin size is 12. The feature vector of size 8×12 (i.e., **96**) is generated for each image since eight angular divisions are considered. For this, the angle of each nonzero pixel is computed with respect to the center of the matrix of the character image and assigned to one of the eight polar division based on the angle. Thus, each thinned character image is first represented by LLS pattern code matrix (Sect. 3.1) and then polar histogram (at the center Fig. 3c) is used to get the histogram of LLS features. The generated histograms are concatenated to form the final feature vector of size 96 (i.e., 12(LLS pattern) $\times$ 8(Angular division)).

4   Experimental Setup

The previous section described the generation of the feature vector of size 96 from the thinned input image. The feature vector, in turn, is used as an input to the classifier for recognition purpose. In this section, we explain the dataset and the classifiers used in the experiments followed by some discussion on the results obtained.

4.1   Dataset

The dataset for **Gujarati** numeral consists of 12851 samples of binarized scanned images proposed by Goswami et al. [7]. The dataset contains wide variety of different Gujarati digits. Dataset for **Devanagari** numerals consists of 18783 samples of binarized images used by Goswami et al. [7]. MNIST dataset by LeCun et al. [10] is used for **English** digits.

Gujarati-English dataset contains 22809 samples and total of 17 classes. Numerals 0, 2, and 3 are similar in Gujarati and English. So they are combined to form one class. **Gujarati-Devanagari** dataset contains 24989 samples. Numerals 0, 1, 2, 4, and 7 are similar in Gujarati and Devanagari. So they are merged together and so we get total of 15 classes. **English-Devanagari** dataset contains 9830 samples and 18 classes as 0 and 2 are similar in English and Devanagari. Pooled dataset for **Gujarati-English-Devanagari** contains 11802 samples. 0 and 2 are similar in all the three scripts. 1, 4, and 7 are similar in Gujarati and Devanagari while 3 is common in Gujarati and English. So we get 22 classes in total.

4.2   Classifier

Two classifiers, namely, k-nearest neighbor (k-NN) and support vector machine (SVM) are used in the experiments. First, the baseline experiments are performed by using k-NN and the results are improved further by using SVM classifier.

K-NN: k-nearest neighbor (k-NN) computes the distance of an unknown sample with respect to all the samples in the template database, and estimates the class label of an unknown sample from the class labels of most similar or closest samples in the template database, called nearest neighbors. The value of k is determined experimentally. Appropriate distance measure can be used depending on the application. Here, Frobenius norm is used as distance measure. Frobenius norm is a matrix norm or vector norm of an $m \times n$ matrix A, defined as the square root of the sum of the absolute squares of its elements.

$$\|A\|_F = \sqrt{\sum_{i=1}^{m} \sum_{j=1}^{n} |a_{ij}|^2} \tag{1}$$

It is also equal to the square root of the trace of matrix AA^H, where A^H is the conjugate transpose.

$$\|A\|_F = \sqrt{Tr\left(AA^H\right)} \tag{2}$$

SVM: Support vector machine (SVM) is a supervised Machine Learning classification technique. It is primarily a two class classifier. But it can also be extended for multiclass classification problem. SVM resolves the classification problems by separating the data into two categories by using an n-dimensional hyperplane that maximizes the margin between classes. Kernel functions can be used for nonlinear boundary. Different kernel functions are linear, polynomial, RBF, sigmoid, etc. Different parameters are associated with different kernel. Here SVM with radial basis function (RBF) kernel is used.

4.3 Experimental Results

Three different experimental setups were considered, namely, k-NN, SVM-RBF with default parameter, and SVM-RBF with optimized parameters. In the first case, baseline experiments were performed using k-NN by taking different values of k. The results were further improved by using SVM-RBF with default as well as optimized parameters obtained using grid search algorithm. In all the experiments, random validation technique is used and the average result obtained after performing the experiment five times is reported. Results of the k-NN classifier are shown in Table 2, whereas the results obtained by SVM-RBF classifier with default and optimized parameters are shown in Table 3.

It is observed during experiments that the frequently misclassified samples in Gujarati numeral dataset are digit 7 as 0, digit 9 as 8 ,and digit 1 as 2 due to the similarity in writing styles. In a similar manner, for Devanagari misclassification is found between digits 3 and 5 as 2, digit 6 as 9, and digit 7 as 0.

Table 2 Results of classification using k-NN classifier (LLS features with Polar histogram). Here, Accuracy (%) is in the form of ($\mu \pm \sigma$) is shown

Numeral database	K = 1	K = 3	K = 5	K = 7
Gujarati	96.01 ± 0.27	96.34 ± 0.22	96.03 ± 0.12	95.76 ± 0.23
Devanagari	95.63 ± 0.09	95.78 ± 0.06	95.95 ± 0.28	95.29 ± 0.39
English	88.73 ± 0.47	89.53 ± 0.35	89.25 ± 0.65	89.75 ± 0.86
Gujarati-English	92.36 ± 0.29	92.08 ± 0.34	92.37 ± 0.53	92.09 ± 0.27
Gujarati-Devanagari	93.58 ± 0.22	93.88 ± 0.33	93.74 ± 0.34	93.90 ± 0.09
English-Devanagari	83.86 ± 0.25	85.37 ± 0.56	85.64 ± 0.55	85.22 ± 0.31
Gujarati-English-Devanagari	88.01 ± 0.30	88.10 ± 0.50	87.46 ± 0.31	87.55 ± 0.28

Table 3 Results of classification using SVM-RBF classifier (LLS features with Polar histogram). Here, Accuracy (%) is in the form of ($\mu \pm \sigma$) is shown

Numeral Database	SVM-RBF with default *parameters*[a]	SVM-RBF with optimized parameters	Parameters (C, γ)
Gujarati	97.00 ± 0.38	98.30 ± 0.12	(8, 0.0025)
Devanagari	97.54 ± 0.11	98.35 ± 0.10	(5, 0.005)
English	94.39 ± 0.42	95.31 ± 0.36	(7, 0.015)
Gujarati-English	95.88 ± 0.18	96.51 ± 0.05	(7, 0.005)
Gujarati-Devanagari	96.21 ± 0.18	97.49 ± 0.07	(7, 0.005)
English-Devanagari	90.68 ± 0.5	92.00 ± 0.51	(1, 0.006)
Gujarati-English-Devanagari	93.36 ± 0.37	94.88 ± 0.48	(5, 0.005)

Notes [a] $C = 1$ and $\gamma = 1 \div no. of features$

The proposed method is compared with state-of-the-art methodologies. The comparison of the proposed method with existing work on handwritten Gujarati and Devanagari numeral recognition is shown in Tables 4 and 5 respectively. The results of proposed method are the average of five runs of random validation of data. It is evident from the results of Table 4 that for Gujarati numerals by compromising accuracy of 0.67%, we have achieved the reduction in feature size of about 92.5%. For Devanagari, 76% reduction in feature size is achieved by compromising accuracy of 1.21%. This is evident from Table 5. The reduction in the feature vector size implies faster processing of recognition.

Table 4 Comparison of proposed method with existing work in Gujarati handwritten numeral recognition

Reference	Feature type	Feature size	Classifier	Accuracy (%)
Nagar and Mitra [12]	Stroke orientation features	1,296	SVM	98.97
Goswami and Mitra [7]	Block wise histogram of low-level stroke features	588	SVM-RBF	98.49
Proposed method	Polar histogram of low-level stroke features	96	SVM	98.30
Desai [5]	Profiles of digits taken blockwise	94	Multi layer Feed Forward Neural Network	81.66

Table 5 Comparison of proposed method with existing work in Devanagari handwritten numeral recognition

Reference	Feature type	Feature size	Classifier	Accuracy (%)
Pal et al. [13]	Block wise histogram of directional feature	400	MQDM	99.56
Nagar and Mitra [12]	Stroke orientation features	1,296	SVM	99.14
Bhattacharya and Chaudhuri [4]	Multi-resolution wavelet features	256	MLP	99.04
Goswami and Mitra [7]	Block wise histogram of low-level stroke features	588	SVM-RBF	98.65
Jangid et al. [15]	Zone density and directional distribution	144	SVM-RBF	98.49
Proposed method	Polar histogram of low-level stroke features	96	SVM	98.35

Goswami and Mitra [7] have used LLS features with Block histogram method. They have worked on Gujarati and Devanagari handwritten numerals as well as on Gujarati-English and Gujarati-Devanagari mixed dataset. The comparison of results reported in [7] along with the feature size is shown in Table 6.

Table 6 Comparison of proposed method with results reported in [7]

Reference	Feature size	Accuracy using SVM for Gujarati numerals	Accuracy using SVM for Devanagari numerals (%)	Accuracy using SVM for Guj-Eng dataset (%)	Accuracy using SVM for Guj-Dev dataset (%)
LLS features with	48	94.15	91.30	90.42	91.27
Block histogram	108	97.98	97.14	94.18	96.52
(M. Goswami and	192	98.37	98.26	95.73	97.57
S.K. Mitra [7])	300	98.45	98.53	96.27	97.76
	432	98.41	98.64	96.51	97.88
	588	98.49	98.65	96.54	97.77
LLS features with Polar histogram (Proposed method)	96	98.30	98.35	96.51	97.49

5   Conclusion

Low-Level Stroke feature is a recently developed feature which is based on local stroke pattern. The same is utilized well for numeral and character recognition [7] and [8]. However, large feature vector size makes it a little costly. In this work, we addressed the problem of reducing the feature vector size considerably while keeping the accuracy at par.

Considerably, large size of features are generated using different feature vector generation methods like block-based histogram. Reduced size of feature are efficient in terms of cost of computation of classifier and memory space. With the help of Polar histogram method with low-level stroke features, a considerable amount of feature vector size reduction is achieved with almost the same accuracy. The reduction of 50% to 92.5% has been achieved by using Polar histogram method for computing the feature vector. Thus, reduced feature size is advantageous for any real-time applications.

References

1. Baheti, M., Kale, K., Jadhav, M.: Comparison of classifiers for Gujarati numeral recognition. Int. J. Machine Intell. **3**(3) (2011)
2. Belongie, S., Malik, J., Puzicha, J.: Matching shapes. In: Eighth IEEE International Conference on Proceedings of Computer Vision, 2001. ICCV 2001 . vol. 1, pp. 454–461. IEEE (2001)

3. Belongie, S., Malik, J., Puzicha, J.: Shape context: A new descriptor for shape matching and object recognition. In: Advances in Neural Information Processing Systems, pp. 831–837 (2001)
4. Bhattacharya, U., Chaudhuri, B.B.: Handwritten numeral databases of indian scripts and multi-stage recognition of mixed numerals. IEEE Trans. Pattern Anal. Machine Intell. **31**(3), 444–457 (2009)
5. Desai, A.A.: Gujarati handwritten numeral optical character reorganization through neural network. Pattern Recogn. **43**(7), 2582–2589 (2010)
6. Dhandra, B., Benne, R., Hangarge, M.: Kannada, Telugu and Devanagari handwritten numeral recognition with probabilistic neural network: a novel approach. Int. J. Comput. Appl. **26**(9), 83–88 (2010)
7. Goswami, M.M., Mitra, S.K.: Offline handwritten gujarati numeral recognition using low-level strokes. Int. J. Appl. Pattern Recogn. **2**(4), 353–379 (2015)
8. Goswami, M.M., Mitra, S.K.: Classification of printed Gujarati characters using low-level stroke features. ACM Trans. Asian Low-Resource Lang. Informat. Process. (TALLIP) **15**(4), 25 (2016)
9. Kompalli, S., Nayak, S., Setlur, S., Govindaraju, V.: Challenges in OCR of Devanagari documents. In: Proceedings of Eighth International Conference on Document Analysis and Recognition, 2005, pp. 327–331. IEEE (2005)
10. LeCun, Y., Bottou, L., Bengio, Y., Haffner, P.: Gradient-based learning applied to document recognition. Proc. IEEE **86**(11), 2278–2324 (1998)
11. Maloo, M., Kale, K.: Support vector machine based gujarati numeral recognition. Int. J. Comput. Sci. Eng. **3**(7), 2595–2600 (2011)
12. Nagar, R., Mitra, S.K.: Feature extraction based on stroke orientation estimation technique for handwritten numeral. In: 2015 Eighth International Conference on Advances in Pattern Recognition (ICAPR), pp. 1–6. IEEE (2015)
13. Pal, U., Sharma, N., Wakabayashi, T., Kimura, F.: Handwritten numeral recognition of six popular Indian scripts. In: Ninth International Conference on Document Analysis and Recognition, 2007, ICDAR 2007, vol. 2, pp. 749–753. IEEE (2007)
14. Sethi, I.K., Chatterjee, B.: Machine recognition of constrained hand printed Devanagari. Pattern Recogn. **9**(2), 69–75 (1977)
15. Singh, M.J.K., Dhir, R., Rani, R.: Performance comparison of Devanagari handwritten numerals recognition. Int. J. Comput. Appl **22** (2011)
16. Zhang, Y., Wang, P.S.P.: A parallel thinning algorithm with two-subiteration that generates one-pixel-wide skeletons. In: Proceedings of the 13th International Conference on Pattern Recognition, vol. 4, pp. 457–461. IEEE (1996)

Integrated Semi-Supervised Model for Learning and Classification

Vandna Bhalla and Santanu Chaudhury

Abstract Labelled data are not only time consuming but often expensive and difficult to procure as it involves skilful inputs by humans to tag and annotate. Contrary to this unlabelled data is comparatively easier to procure but fewer methods exist to optimally use them. Semi-Supervised Learning overcomes this problem and assists to build better classifiers by using unlabelled data along with sufficient labelled data and may actually yield higher accuracy with considerably less human input effort. But if the labelled data set is inadequate in size then the Semi-Supervised techniques are also stuck. We propose a novel framework where the small labelled dataset is appropriately augmented using the intelligent learning mechanisms of artificial immune systems to train the proposed model. The model retrains with the unlabelled data to fortify the learning mechanism. We show that the generative deep framework utilizing artificial immune system principles provides a highly competitive approach for learning in the semi-supervised environment.

Keywords Semi-supervised · Self-training · Clonal data · Integrated

1 Introduction

A key complication in Conventional Classifiers is that enormous quantities of labelled samples are required for accurate training and learning. 'Labels are hard to obtain while unlabelled data are abundant, therefore semi-supervised learning is a good idea to reduce human labour and improve accuracy' [24]. In our modern world, the data set sizes are for ever increasing but acquiring the label information for these data

V. Bhalla (✉) · S. Chaudhury
Indian Institute of Technology Delhi, Hauz Khas, Delhi, India
e-mail: vbhalla.du@gmail.com

S. Chaudhury
CEERI PILANI, Pilani, India
e-mail: schaudhury@gmail.com

© Springer Nature Singapore Pte Ltd. 2020   183
B. B. Chaudhuri et al. (eds.), *Proceedings of 3rd International Conference on Computer Vision and Image Processing*, Advances in Intelligent Systems and Computing 1022,
https://doi.org/10.1007/978-981-32-9088-4_16

is a demanding and complicated task. This led to Semi-Supervised Learning gain consequential practical significance.

Automatic classification of personal data is of significant relevance in today's scenario. The problem is that classification of such data is a challenge as the various categories desired by an individual may not have sufficient labelled instances for training and moreover, the user has to hand label the training data repository. The hand labelling will increasingly become infeasible when the numbers approach millions. We present a novel deep CNN Semi- Supervised learning architecture, using clonal selection techniques for such applications with limited labelled data. The high-level complex features learnt by Deep Models are more resilient and eloquent when compared to shallow classical methods. We harness the potential of Deep Learning in our model and represent the data features as deep features. We have thus developed an innovative generative model, which gives appreciable results when working with unlabelled data along with small-sized labelled data specifically in the domain of personal photo collections.

2 Review

A vast number of semi-supervised techniques for clustering like Nonnegative matrix factorization via constraint propagation [19], Active learning [20], Hierarchical clustering [23], Linear discriminant clustering [10], Kernel mean shift clustering [17], Maximum margin clustering [21] and more [7] are found in literature. A well-structured semi-supervised learning technique was proposed by Fergus et al. [5] and Liu et al. [11] put forth a proposal that clusters billions of images using map reduce.

Many semi-supervised learning methods in literature [25] include generative models like the Self-Training and Expectation–Maximization with mixture models and the Discriminative models like graph-based methods, Gaussian processes and Support vector machines. Expectation–Maximization is prone to local maxima. Unlabelled instances can be detrimental to learning in some cases with these methods like for instance where a local maxima is away from the global maxima. Normally a method is chosen based on its assumptions that best fit the structure of the problem. Self-training is a common and popular approach for semi- supervised learning. Initially, a small quantity of labelled data is used to train a classifier. The trained classifier helps to categorize or label the unlabelled data. Now, the most confident unlabelled data points along with the newly learnt labels are subjoined to the original labelled data set. The new enhanced dataset retrains the classifier yet again. This technique has been successfully applied to many natural language processing problems. Subjective nouns were identified by Riloff et al. [15]. Classification of dialogues with two classifiers was accomplished by Maereizo et al. [12] in 2004. Yet again in 2005 Rosenberg et al. [16] achieved object detection in images using self-training. Though self-training is an algorithm which is hard to analyse, Culp and

Michailidis [3] have analysed the convergence of algorithms in this setting. We have used self-training in our work.

Deep hybrid Architectures in semi-supervised environs have been successfully implemented for a multitude of recognition problems. Deep models have surpassed popular shallow architectures especially in image [18] and language [6] domains. Most of these architectures use the greedy approach to pretraining and undergo a multistage generative learning. Various auxiliary approaches have been used to help deep models in early learning and also tackle recalcitrant input variations [22]. Two interesting hybrid semi-supervised deep architecture [13] combine multi-objective learning with efficient layer wise greedy approach for text categorization optical character recognition. But auxiliary free parameters got added introducing additional challenges. These hybrid semi-supervised deep models are promising with good results but it is an undeniable fact that such architectures have their limitations. Further, the test images used so far have been very small in size with neither change in illumination nor background clutter or any other such problems that are inevitable in many natural personal datasets [9]. The kernel methods, comparatively, disregard not only the structure of the input data but also its dimensionality. The flexibility and scalability are also inadequate besides needing large amount of training data.

The Architecture of our model is discussed in the next section. Experiments conducted and Results are in the subsequent sections. Discussion and future work concludes this paper.

3   Architecture of the Integrated Semi-Supervised Learning and Classification Model

We have designed and realized a semi-supervised artificial immune hybrid classifying framework, an SS-AIHC model, presented in Fig. 1. The model consists of series of Convolution and Subsampling layers constituting a deep Convolutional Neural Network (CNN) architecture integrated with Clonal Selection (CS). The softmax layer of our earlier supervised model, the CNN-AIHC [1], is replaced with the Artificial Immune System inspired classifier, AIHC [2]. The complete training of the novel SS-AIHC model resulting in the memory cell maturation process can be divided into three modules as shown in Fig. 2.

The model is first and foremost trained with the completely labelled data in the Supervised Convolutional-AIS module. The model parameters are further fine-tuned with additional artificially generated data for each class constituting module 2. The Supervised Classification is performed with Artificial test data produced using clonal selection algorithm. Finally, the unlabelled data is used in the module 3, called the semi-supervised stage, to benefit the system further in the training and learning process. All the stages assimilate to mature the memory cells of the novel SS-AIHC framework. A trained SS-AIHC classifier will consist of matured memory cells corresponding to each class obtained from the labelled, clonal and unlabelled data. These

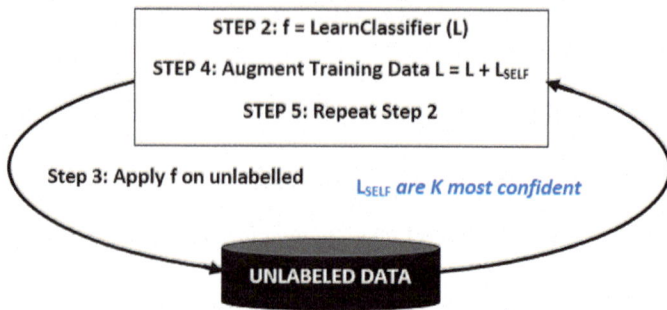

Fig. 1 Semi-supervised learning

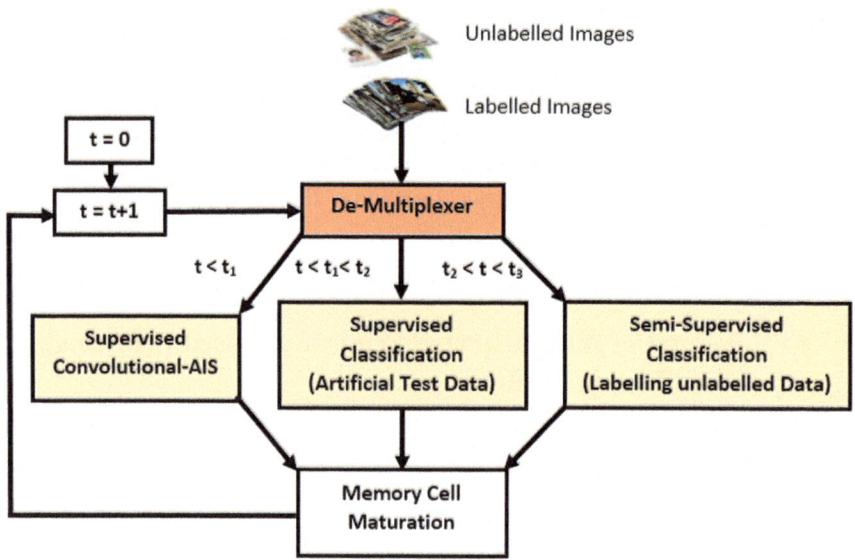

Fig. 2 Semi-supervised AIHC model: learning framework

memory cells are the set of antibodies representing each class. The memory cell for all class are initialized randomly. The model automatically matures and enhances these memory cells. The CNN generates a distinct pattern for each input sample and the Clonal Selection Algorithm [4] inspires optimal additional data generation. We have used Inner Product to ascertain the affinity between two samples amongst the many measures available like Euclidean Distance, Relative Distance, Manhattan Distance as this measure resulted in best performance.

A deep CNN architecture is used to learn data features and is realized by alternately stacking convolution and sampling layers. Each input sample is convolved with a linear filter, a bias term is added and passed through a non linear function repeatedly to generate its feature map.

$$n_{ij}^k = f\left(\sum_v \sum_{x=0}^{X_i-1} \Omega_{ijv}^x \eta_{(i-1)v}^{k+x} + b_{ij}\right) \tag{1}$$

where n_{ij}^k is the neuron value in the ith layer of the jth map at kth position. In (1) v is the index of the previous layer, i.e. the layer $(i-1)$ and Ω_{ijv}^x represents the weight at position x in the vth feature map. X_i is the kernel width, b_{ij} the bias of the current map in the current layer and f is a non linear function like for instance $tanh$.

The k kernels of the convolutional layer produce k feature maps of size $m-n+1$ where $m \times m$ is the dimension of the input sample and $n \times n$ is the size of the kernel. Each map is subsampled with max pooling which provides invariance.

$$p_j = \max_{M \times 1} \left(p_i^{mx1} w(m, 1)\right) \tag{2}$$

where p_j is the maxima and $w(m, 1)$ is the window function. The pooling layer neuron combines a $M \times 1$ patch of the convolutional layer. The entire CNN is trained using the back propagation algorithm. We explain each stage of the learning process in detail in the next sections.

3.1 Supervised Convolutional AIS: Module 1

Since the memory cells of the classifier are initialized randomly hence the initial epoch 1 to t_1 of Fig. 2 uses only the original labelled data to train the deep CNN network. This helps to optimize the population of memory cells toward the best representation of its class. The entire dataset is divided into batches. A batch consists of a number of images. The batch is fed to multiple convolutional and subsampling layers of a deep CNN resulting in the generation of feature map for all images. These feature maps are converted to one-dimensional feature vectors. Finally, an N × D vector size where the number of the images taken is N and the dimension of each is D is generated. Borrowing our terminology from the Artificial Immune Systems, we name this the antibody set. From this set one antibody is chosen at a time and is termed antigen. For each antibody in the set execute the following:

{label of $picked_{antigen}$= label (antigen (i));
Pick the class corresponding to label of the antigen. Let class is $class_i$
Do {

– Perform the affinity measure, i.e. inner product of chosen antigen with the predetermined antibody set (N_i) of the class and store the value in the local array.
– Choose the best n_1 antibody from the antibody set having highest affinity value.

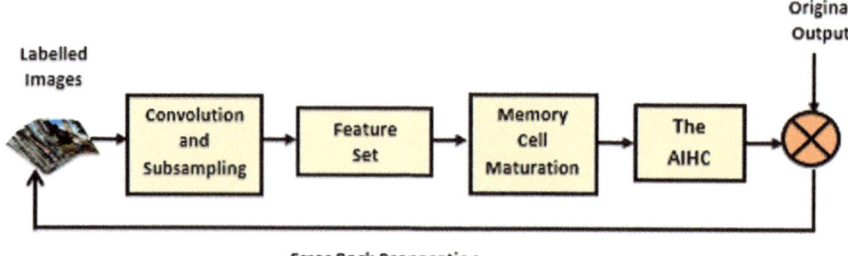

Fig. 3 Supervised convolutional-AIHC model

– Generate additional features using principles of clonal selection. This process
  yields n_2 number of new antibody.
– Choose the best N_i from total of old (N_i) and new (n_2).

  }
  }

This process leads to optimal maturation of antibodies in the memory cells of each
class using the labelled data only. This is the clonal selection process to optimally
augment labelled data. Once the above process of training the classifier is accomplished data can be passed through the AIHC to ascertain their labels (classes). The
class which shows maximum affinity is chosen as the output class. The output is
compared with the original output. Error is calculated and back propagated. The
entire process is presented in Fig. 3. This way both the memory cells, which are the
trained representative antibodies of each class, and the kernels at the Convolutional
layer gets trained for each class.

3.2 Enhancement of the model with misclassification error: module 2

This process is from epoch no. t_1 to t_2 in Fig. 2. The further training is done using
the now somewhat trained and optimally populated supervised CNN-AIHC accomplished in the previous stage. This module is explained in Fig. 4. The Misclassification at the first (Original) output layer is used to produce the additional data to
train further at the feature level. The property of convergence is directly related to
misclassification.

The entire dataset will have images from each class. The feature set is divided
into blocks of the classes. So, one class has n_1 feature vectors each of size d and
another class may have n_2 feature vectors of size d. Each feature vector from the
feature set are fed to the semi-trained model and misclassification is calculated. This
step provides the error corresponding to each feature vector in feature set. Now, we
have error corresponding to each feature vector for each class. Based on error, clonal

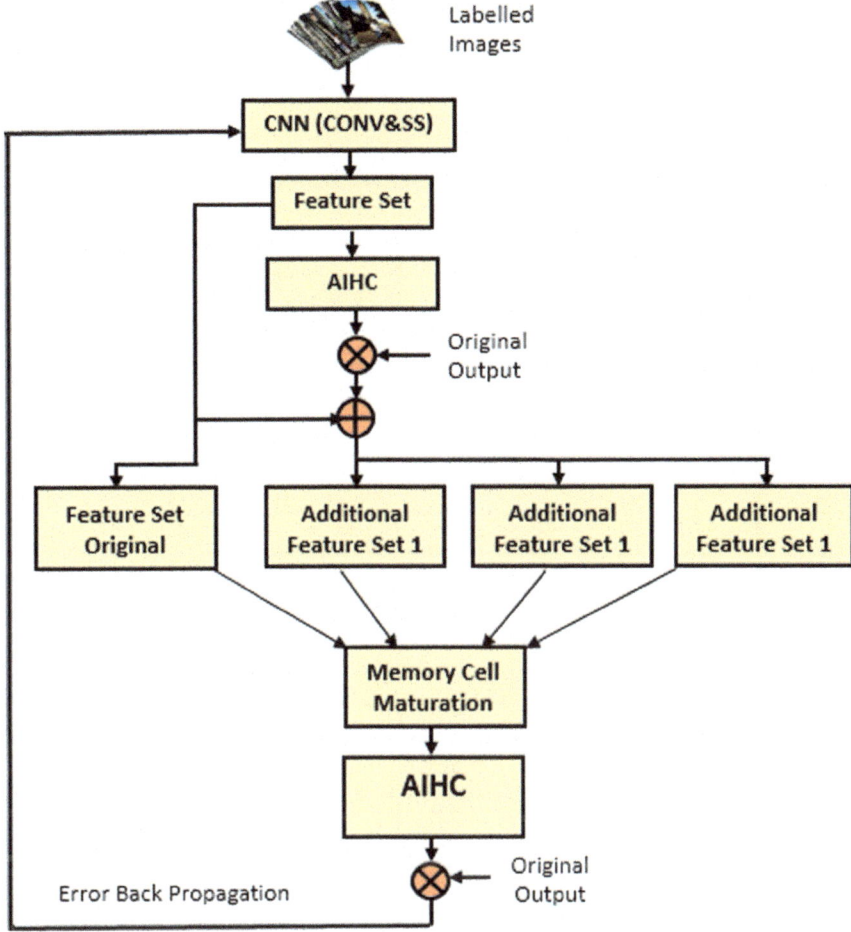

Fig. 4 Enhancement of the Model with Misclassification Error

selection and mutation process is performed in each class leading to creation of new additional data.

Clonal Rate $\propto$ 1/error and

Mutation Rate $\propto$ 1/error.

The process results in generation of artificial clonal training data based on misclassification. The SS-AIHC model now has new batches along with the original batch. The entire set of data is used to mature the memory cells exactly as explained in the module 1. All batches are given to the model and the error generated is back-propagated as usual. The newly generated training data strengthens the model in its learning and hence improves the accuracy of overall system. Figure 5 illustrates the memory-maturation process.

The model now progresses to its final stage.

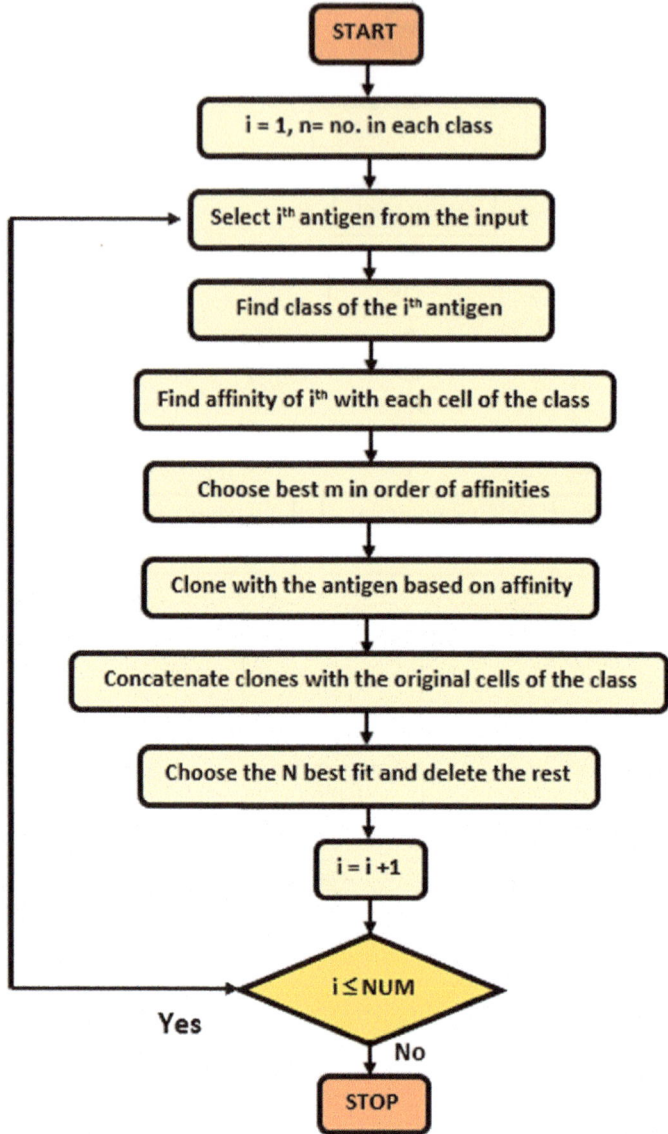

Fig. 5 Memory-maturation process in semi-supervised convolution—AIS model

3.3 Semi-Supervised Convolution-AIS: Module 3

The module 1 and module 2 are the pretraining stages before the actual semi-supervised stage. The above two modules result in a trained Supervised integrated Convolutional-AIS Classifier using the labelled data and the misclassification error.

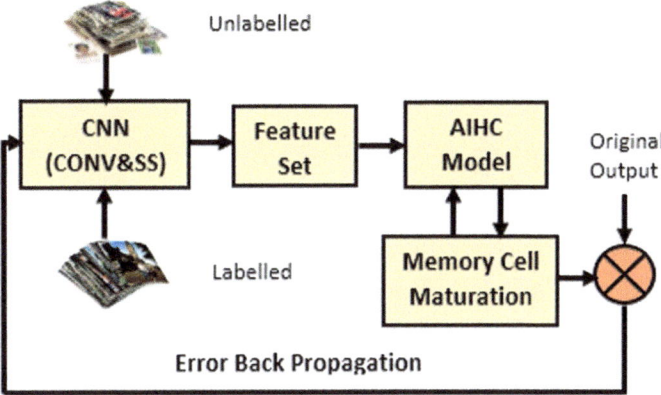

Fig. 6 Semi-supervised convolution-AIS module

To further strengthen the learning system, the model now uses the unlabelled data. The module is shown in Fig. 6.

The following steps are undertaken:

1. Use the Model that has trained with the labelled data and the misclassification error, a task accomplished in the first two modules.
2. Apply this semi-trained model on the unlabelled data and learn their labels.
3. After ascertaining the labels of the unlabelled data, mature the memory cell population of each class using this newly labelled data and the initial labelled data.
4. The framework is now retrained using the entire data.
5. Repeat steps 2–4 till the convergence condition is achieved.

The unlabelled data is hence helping in the learning and training process of model which subsequently results in improving the accuracy. This novel approach of using Convolutional-AIS can address the small data problem in a semi-supervised environment. The model now is a trained SS-AIHC (Semi-Supervised Artificial Immune Hybrid Classifier) which can be used to learn and classify test data.

4 Experiments and Results

We tested the trained SS-AIHC model with the data from personal data collections as well as on standard datasets.

For each test sample do

{

– Extract the feature of the image using the now trained convolutional and subsampling layer.

Table 1 Error attained with different techniques on MNIST data

S. no	Algorithms/Techniques	$N = 100$ (%)	$N = 500$ (%)	$N = 1000$ (%)
1	NN [8]	25.81	11.44	10.7
2	CNN [8]	22.98	7.68	6.45
3	TSVM [8]	16.81	6.16	5.38
4	CAE [8]	13.47	6.3	4.77
5	MTC [8]	12.03	5.13	3.64
6	**SS-AIHC**	**10.8**	**6.9**	**3.99**
7	AtlasRBF [8, 14]	9	–	3.7

- The extracted feature is compared against the pre populated memory cell of each class following the two layer classification process of our AIHC.
- Affinity calculation is done with each memory cell.
- Class having maximum affinity value is chosen as the output class.
- The photo is rightly classified if the output class matches with its true label.

}

Experiment 1: Results on MNIST dataset: We compared our results in SS learning with the existing results of Pitepis's Atals RBF, MTC (manifold tangent classifier) using CAE (contractive auto encoders), TSVM (Transductive SVM), NN (nearest neighbours), CNN (convolutional Neural Networks) [8]. The semi- supervised data set for learning was built by dividing the 50,000 samples between unlabelled and labelled set. The size of the labelled was 100, 500 and 1000, respectively. For higher number of labelled data, in thousands, the accuracy was expectedly higher. Unlike all these models our architecture performs well with smaller number of labelled data and would be having similar order cost as these alternatives. Table 1 tabulates errors of these standard semi-supervised techniques on MNIST data. The Atals RBF follows a two-step approach and is specifically for high-dimensional data. The manifold of data is approximated on the original space using the small dimensional affine charts, completely unsupervised. The second step uses SVM-based supervised learning. The data points are given soft allotments to the affine charts which are low dimensional. The unlabelled data is used to understand the detailed shape of the manifolds underlying, which helps improve the accuracy of the classifier trained with minimal labelled data. Though the method has recorded better results but its ability and accuracy in personal data collections remains unexplored.

Experiment 2: Results on SVHN dataset: We compare the performance of classification on the far more complex image dataset of SVHN with some other techniques [8] from literature. Classification on SVHN dataset for techniques in literature is with 1000 labels. Table 2 shows that our model works well with SVHN data too.

The optimistic results on the two standard datasets prove the efficacy of our model. Our model records superior performance on the standard datasets using smaller number of labelled data.

Table 2 Accuracy achieved with different techniques on SVHN data

S. no	Algorithms/Techniques	Accuracy (%)
1	KNN [8]	77.93
2	TSVM [8]	66.55
3	M1+KNN [8]	65.63
4	M1+TSVM [8]	54.33
5	M1+M2 [8]	36.02
6	**SS-AIHC**	**89**

Table 3 Accuracy achieved on personal photo album using **SS-AIHC**

S. no	Class	Accuracy (%)
1	Wedding	75–78
2	Picnic	78–82
3	Conference	80–86

Experiment 3: Results on Personal Photos: We have performed experiments on our dataset which is uploaded at https://github.com/vandnabhalla/Database. Table 3 presents the results of the Semi-Supervised AIHC model.

5 Discussions and Conclusions

The Deep CNN-AIS Semi-Supervised model, our contribution in this work, shows a definite improvement in accuracy on the standard as well as our application datasets. Tables 1 and 2 show the superiority of our model on the standard datasets. Table 3 shows consistency in the model's performance for a unique data comprising of personal photos. We observe that with sufficiently large amounts of unlabelled data a better classifier can be realized than with just labelled data by itself. As observed our model is able to perform better than most previous methodologies implemented for these environments. There are many applications with abundant availability of the unlabelled data while labelled data is scarce. The Semi-Supervised Hybrid Deep Convolutional–Artificial Immune System Architecture (SS-AIHC) combines the knowledge of a small annotated dataset to build a larger database integrating principles of Clonal Selection from Artificial Immune System with Deep CNN. The learning is subsequently enhanced using unlabelled data too. We need to somehow use the properties of the existent available data to enhance the boundaries of accurate classification decisions. We explore the generative models with semi-supervised learning approach and have developed a new hybrid model that results in efficacious generalization starting from small size hand labelled data. The model is self-learning and the labelled data is augmented from the most confident prognosis. Our experiments show that after augmenting data with Artificial Immune System techniques, deep generative models can bring about considerable enhancement under semi-supervised settings. Our problem is exciting yet exacting for the following reasons:

- It is arduous to manually hand label any dataset. We have used a personal photo collection as an example dataset in addition to the standard MNIST and SVHN datasets.
- Clustering such similar datasets with many classes is challenging especially with few labelled instances and large unlabelled data.

References

1. Bhalla, V., Chaudhury, S.: Artificial immune hybrid photo album classifier. In: Proceedings of International Conference on Computer Vision and Image Processing CVIP 2016, vol. 1 (2016)
2. Bhalla, V., Chaudhury, S., Jain, A.: A novel hybrid cnn-ais visual pattern recognition engine, pp. 215–224. Springer International Publishing, Cham (2015)
3. Culp, M., Michailidis, G.: An iterative algorithm for extending learners to a semi-supervised setting. J. Comput. Graph. Stat. **17**(3), 545–571 (2008)
4. De Castro, L.N., Von Zuben, F.J.: Learning and optimization using the clonal selection principle. IEEE Trans. Evol. Comput. **6**(3), 239–251 (2002)
5. Fergus, R., Weiss, Y., Torralba, A.: Semi-supervised learning in gigantic image collections. In: Bengio, Y., Schuurmans, D., Lafferty, J.D., Williams, C.K.I., Culotta, A. (eds.) Advances in Neural Information Processing Systems 22, Curran Associates, Inc., pp. 522–530 (2009)
6. Glorot, X., Bordes, A., Bengio, Y.: Domain adaptation for large-scale sentiment classification: a deep learning approach. In: Getoor, L., Scheffer, T. (eds.) ICML, Omnipress, pp. 513–520 (2011)
7. Jiao, L.C., Shang, F., Wang, F., Liu, Y.: Fast semi-supervised clustering with enhanced spectral embedding. Pattern Recogn. **45**(12), 4358–4369 (2012)
8. Kingma, D.P., Mohamed, S., Rezende, D.J., Welling, M.: Semi-supervised learning with deep generative models. In: Ghahramani, Z., Welling, M., Cortes, C., Lawrence, N.D., Weinberger, K.Q. (eds.) Advances in Neural Information Processing Systems 27, Curran Associates, Inc., pp. 3581–3589 (2014)
9. Krizhevsky, A., Sutskever, I., Hinton, G.E.: Imagenet classification with deep convolutional neural networks. In: Advances in neural information processing systems, pp. 1097–1105 (2012)
10. Lee, C.H., Liu, C.L., Hsaio, W.H., Gou, F.S.: Semi-supervised linear discriminant clustering. IEEE Trans. Cybern. **44**(7), 9891000 (July 2014)
11. Liu, T., Rosenberg, C., Rowley, H.A.: Clustering billions of images with large scale nearest neighbor search (2007)
12. Maeireizo, B., Litman, D., Hwa, R.: Co-training for predicting emotions with spoken dialogue data. In: Proceedings of the ACL 2004 on Interactive Poster and Demonstration Sessions (Stroudsburg, PA, USA), ACLdemo '04, Association for Computational Linguistics (2004)
13. Ororbia II, A.G., Reitter, D., Wu, J., Lee Giles, C.: Online learning of deep hybrid architectures for semi-supervised categorization. In: Machine Learning and Knowledge Discovery in Databases—European Conference, ECML PKDD, Porto, Portugal, September 7–11, 2015. Proceedings, Part I, 2015, pp. 516–532 (2015)
14. Pitelis, N., Russell, C., Agapito,L.: Semi-supervised Learning Using an Unsupervised Atlas, pp. 565–580. Springer, Berlin, Heidelberg (2014)

15. Riloff, E., Wiebe, J., Wilson, T.: Learning subjective nouns using extraction pattern bootstrapping. In: Proceedings of the Seventh Conference on Natural Language Learning at HLT-NAACL 2003, vol. 4 (Stroudsburg, PA, USA), CONLL '03, Association for Computational Linguistics, pp. 25–32 (2003)
16. Rosenberg, C., Hebert, M., Schneiderman, H.: Semi-supervised self-training of object detection models. In: WACV/MOTION, pp. 29–36. IEEE Computer Society (2005)
17. Tuzel, O., Anand, S., Mittal, S., Meer, P.: Semi-supervised kernel mean shift clustering. IEEE Trans. Pattern Anal. Mach. Intell. **36**(6), 1201–1215 (June 2014)
18. Vincent, P., Larochelle, H., Lajoie, I., Bengio, Y., Manzagol, P.-A.: Stacked denoising autoencoders: learning useful representations in a deep network with a local denoising criterion. J. Mach. Learn. Res. **11**, 3371–3408 (2010)
19. Wang, D., Gao, X., Wang, X.: Semi-supervised nonnegative matrix factorization via constraint propagation. IEEE Trans. Cybern. **46**(1), 233–244 (2016)
20. Xiong, S., Azimi, J., Fern, X.Z.: Active learning of constraints for semi-supervised clustering. IEEE Trans. Knowl. Data Eng. **26**(1), 43–54 (2013)
21. Zeng, H., Cheung, Y.-M.: Semi-supervised maximum margin clustering with pairwise constraints. IEEE Trans. Knowl. Data Eng. **24**(5), 926–939 (2012)
22. Zhang, J., Tian, G., Mu, Y., Fan, W.: Supervised deep learning with auxiliary networks. In: Proceedings of the 20th ACM SIGKDD International Conference on Knowledge Discovery and Data Mining (New York, NY, USA), KDD '14, pp. 353–361. ACM (2014)
23. Zheng, L., Li, T.: Semi-supervised hierarchical clustering. In: Proceedings of the IEEE 11th International Conference on Data Mining, p. 982991 (2011)
24. Zhu, X.: Semi-supervised learning literature survey. Technical Report 1530, Computer Sciences, University of Wisconsin-Madison (2005)
25. Zhu, X., Goldberg, A.B., Brachman, R., Dietterich, T.: Introduction to Semi-supervised Learning. Morgan and Claypool Publishers (2009)

Mushroom Classification Using Feature-Based Machine Learning Approach

Pranjal Maurya and Nagendra Pratap Singh

Abstract Mushroom is an important fungus which contains a good source of vitamin B and a large amount of protein when compared to all other vegetables. It helps to prevent cancer, useful in weight loss and increases the immunity power of human. On the other hand, some mushrooms are toxic and can prove dangerous if we eat them. Therefore, it is a prominent task to differentiate, the edible and poisonous mushrooms. This paper focuses on developing a method for classification of mushroom using its texture feature, which is based on the machine learning approach. The performance of the proposed approach is 76.6% by using SVM classifier, which is found better with respect to the other classifiers like KNN, Logistic Regression, Linear Discriminant, Decision Tree, and Ensemble classifiers.

1 Introduction

The organisms belong to the kingdom of fungi. It contains yeasts, puffballs, molds, truffles, smuts, and mushrooms. There are thousands of species of fungi found in nature. Mushroom is an important fungus used by a human. Unlike animals and plants they don't contain chlorophyll and digest their food externally through mycelium, they absorb nutrients from the constituent of organic material where they live. Mushrooms are a good source of vitamin B, which admits niacin, riboflavin, and pantothenic acid. They also have beta-glucans, minerals such as copper, potassium, ergothioneine, selenium, etc. Moreover, they prevent cancer, diabetes, weight loss, and immunity power. Therefore, it comes on top rank for containing protein among all vegetables. On the other hand, some mushrooms are toxic and can prove dangerous if eat them. Also, they have side effects like food poisoning, headache, allergy, anxiety, and also fatal. Therefore, it is very important to differentiate, which mushrooms are

P. Maurya (✉) · N. P. Singh
Department of CSE, M.M.M. University of Technology, Gorakhpur 273010, UP, India
e-mail: pranjalmaurya1996@gmail.com

N. P. Singh
e-mail: npscs@mmmut.ac.in

© Springer Nature Singapore Pte Ltd. 2020
B. B. Chaudhuri et al. (eds.), *Proceedings of 3rd International Conference on Computer Vision and Image Processing*, Advances in Intelligent Systems and Computing 1022, https://doi.org/10.1007/978-981-32-9088-4_17

edible and which are poisonous. There are many violent properties of mushrooms such as Cap, Stalk, Scales on the cap, Gills under the cap, Rings, Mycelial threads, and many others. The main focus of this paper is to develop a precautionary methodology that distinguishes between edible and poisonous mushroom on the basis of their texture features. These features are extracted from the image. There are many researchers, who already working on the classification of mushrooms using different approaches. Authors Raji and Alamutu proposed an automated sorting machine for mushroom recognition, which is used in food industry [1]. However, in this case, the program is specific for one task and is not used for general applications. Because of this technique, an author focuses on some color features, which are not efficient in general and used an image classification method to classify the edible and poisonous mushroom. An author Subramaniam and Byung-Joo proposed a Principal Component Analysis (PCA) based algorithm for recognition of mushrooms [2]. In the paper [3], the authors Beniwal and Bishan Das used data mining approach and to classify the mushrooms by using WEKA (Waikato Environment for Knowledge Analysis) tool and found that the Bayesian network classification technique performs better among all other classifiers. Mushrooms are recognized by general classification technique according to its features for a mobile application by the author Matti [4]. It is described by Maftoun et al., mushrooms are mostly edible, which belong to Pleurotus species. These types of mushrooms are more diversified and distributed worldwide and contain almost all nutrients [5].

Rest of the paper is organized as follows: Sect. 2 explains the proposed method in detail. Experimental results and discussion are given in Sect. 3. Finally, Sect. 4 contains the conclusion of the proposed work.

2 The Proposed Method

The proposed method is the combination of the preprocessing, feature extraction, and classification on the basis of selected features using machine learning classification techniques. In the preprocessing step, first resize (equal size) all the mushroom images taken from various datasets then convert the color image into grayscale image by using PCA-based color to grayscale conversion method followed by contrast-limited adaptive histogram equalization (CLAHE) [6, 7], because these steps preserve both color discriminability and the texture by using simple linear computations in subspaces with low computational complexity [8]. The next step, extract the color features as well as grayscale features of the mushroom image. Feature extraction is a process of deriving new features for describing and further processing of an image and also it is related to dimensionality reduction. Whereas features are the significant properties of an image which are helpful in feature extraction such as color, shape, edge, texture etc. In this paper we worked on a statistical feature of the image such as GLCM features, Tamura's features, Wavelet features, Law's Texture Energy (LTE)-based features, and HOG features which are described as follows:

Table 1 Notations and their meaning

Notation	Meaning	Notation	Meaning
$P(i, j)$	Element i, j in GLCM	N_g	Number of gray levels
μ_x	$\mu_x = \sum_{i=1}^{N_g} i\, P_x(i)$	μ_y	$\sum_{j=1}^{N_g} j\, P_y(j)$
$(\sigma_x)^2$	$\sum_{i=1}^{N_g} (i - \mu_x)^2 P_x(i)$	$(\sigma_y)^2$	$\sum_{j=1}^{N_g} (j - \mu_y)^2 P_y(j)$
$P_x(i)$	$\sum_{j=1}^{N_g} P(i.j)$	$P_y(j)$	$\sum_{i=1}^{N_g} P(i, j)$

Gray-Level Co-occurrence Matrix (GLCM): Gray-Level Co-variance Matrix is a method of examining second order texture features from the image. GLCM is a matrix of frequencies at which a pair of the pixel is separated by a certain vector, occur in the image. The distribution of the matrix depends on the displacement and angular relationship between pixels. The various vectors are used for the capturing of dissimilar texture properties. First, a gray-level co-occurrence matrix is created and by using this matrix, GLCM features such as entropy, variance, dissimilarity, cluster prominence, cluster shade, maximum probability, energy, contrast, correlation, homogeneity are extracted. These features are evaluated by using Eqs. (1)–(10), respectively, and the notations and their meaning used in equations are mentioned in Table 1.

$$Entropy = \sum_{i,j=1}^{N_g} -P(i, j) * log\, P(i, j) \tag{1}$$

$$Variance = \sum_{i,j=1}^{N_g} (i - \mu)^2 P(i, j) \tag{2}$$

$$Dissimilarity = \sum_{i,j=1}^{N_g} |i - j| \cdot P(i, j) \tag{3}$$

$$Cluster\, Prominence = \sum_{i,j=1}^{2N_g} (i + j - 2\mu)^3 P(i, j) \tag{4}$$

$$Cluster\, Shade = \sum_{i,j=1}^{2N_g} (i + j - 2\mu)^4 P(i, j) \tag{5}$$

$$Maximum\, Probability = maxi, j\, P(i, j) \tag{6}$$

$$Energy = \sum_{i,j=1}^{N_g} (P(i, j))^2 \tag{7}$$

$$Contrast = \sum_{i,j=1}^{N_g} (i - j)^2 P(i, j) \tag{8}$$

$$Correlation = \frac{\sum_{i,j=1}^{N_g}(i, j)P(i, j) - \mu_x\mu_y}{\sigma_x\sigma_y} \tag{9}$$

$$Homogeneity = \frac{\sum_{i,j=1}^{N_g}(1)P(i, j)}{(1 + (i - j)^2} \tag{10}$$

In the above equations, Eqs. (7)–(10) are based on two parameters: relative distance measured in pixel numbers which is assumed to be one and their relative orientation (θ) which varies from $0°$ to $135°$ with common difference $45°$. Therefore, totally 22 GLCM features are extracted from GLCM matrices and all features are used for the classification of edible and poisonous mushroom [9].

Tamura: Tamura texture features are computed according to the human visual perception system and are often used for image recovery. Six textual features namely coarseness, contrast, directionality, line-likeness, regularity, and roughness are described by Tamura feature and these features are compared with psychological measurements for human subjects. Coarseness detects the largest size texture among smaller micro textures. There is a direct relationship between coarseness to scale and repetition rates. The dynamic range of gray level in an image is identified, by contrast, the polarization of distribution of black and white, repetition, edge sharpness. In an image, directionality is measured in the presence of orientation. If only orientations of images are different, then their degree of the property will be the same. There are 3 different Tamura features such as coarseness, directionality, and contrast existing in the literature [10], and all 3 features are used for the classification of edible and poisonous mushroom.

Wavelet: It is a statistically dependent feature. Wavelet transform provides the capability of decomposition of the image into sub-bands having a different frequency. This characteristic of wavelet makes it more desirable for the segmentation and classification of texture images. Energy is the most important wavelet feature for each one sub-band. In feature selection procedure, the wavelet feature selection approach uses the tree body structure of the wavelet decomposition. Wavelet is used when frequency and time both are importantly considered. There are 32 different wavelet features that are used for the classification of edible and poisonous mushrooms.

Law's Texture Energy (LTE): LTE is a method of giving texture feature using local masking for detection of many texture feature. LTE method introduced a texture-energy approach that computes the total fluctuation within a rigid window. In the LTE method, for extracting texture feature from a digital image energy is transformed. For finding local texture feature, images are convoluted with micro-texture filters. Texture energy image is obtained after translation of micro-texture feature, by displacing-window to macro-texture properties. After that, similar texture model is clustered into regions, by aggregating macro-texture feature planes. Totally 15 different LTE features are extracted and all of them are used for the classification of edible and poisonous mushroom.

Histogram of Oriented Gradients (HOG): The histogram of oriented gradients (HOG) is an element descriptor utilized as a part of PC vision and image preparing with the end goal of question identification. The strategy includes events of inclination introduction limited bits of an image. This strategy is like that of edge introduction histograms, scale-invariant element change descriptors, and shape settings, however, it varies in that it is figured on a thick matrix of consistently dispersed cells and utilizations covering neighborhood differentiate standardization for enhanced exactness. There are 36 different HOG features that are extracted and used for the classification of edible and poisonous mushroom.

Classifiers: In this paper, various machine learning based classifiers such as SVM, KNN, Decision tree, Ensemble training, and Discriminant analysis are used for the classification of edible and poisonous mushroom. A short discussion of various machine learning based classifiers are as follows:

Support Vector Machine: In machine learning, support vector machines are supervised by learning patterns with related understanding algorithms that analyze information applied for regression and classification study. It uses a method referred to as the kernel trick to remodel our information to support these renovations. It discovers the associated optimum border between the attainable results. Basically placed, it will bring some very advanced information conversions, then find out the way to isolate your information supported the labels or results you have outlined.

K-Nearest Neighbors (KNN): In KNN classifier, it contains all classes and on the basis of similarity classifies a new class. In it, unlabeled data are assigned to the most similar labeled class and classification is performed. Classes are classified by the majority votes of its neighbors. Features of data are contained in both training and test dataset.

Decision Tree: It is a tree-like model used for the decision. Decision tree classifies the working area into subarea by identifying lines. In this classifier, it reaches a general conclusion through specific examples. This classifier predicts unseen circumstances by training and learning of decision tree themselves.

Ensemble Tree: An ensemble is an arrangement of classifiers that learn a target, and their individual expectations are combined to characterize new illustrations. It improves the generalization performance of an arrangement of classifiers on a domain. Ensemble method is well established for obtaining more accurate classifiers by combining less accurate ones.

Discriminant Analysis: Discriminant analysis is a statistical tool used to figure out categorically dependent variable by one or more predictor variables that are also called a binary independent variable with the objective to access the sufficiency of a classification. In other words, it analyzes the accuracy of classification.

Table 2 All 108 selected features of first image of Fig. 1

Image	Features					
e_1.jpg	f_1	f_2	f_3	f_4	f_5	f_6
	0.0365	18.2906	2.8134	0.5524	0.5524	267.2451
	f_7	f_8	f_9	f_{10}	f_{11}	f_{12}
	10.3072	1.1579	3.6297	0.6047	0.565	0.0964
	f_{13}	f_{14}	f_{15}	f_{16}	f_{17}	f_{18}
	19.5019	8.1367	41.4418	2.5089	2.8134	1.4174
	f_{19}	f_{20}	f_{21}	f_{22}	f_{23}	f_{24}
	−0.1296	0.6276	0.887	0.9631	478.295	985.044
	f_{25}	f_{26}	f_{27}	f_{28}	f_{29}	f_{30}
	58.8956	157.701	4.6451	17.285	112.0618	239.1125
	f_{31}	f_{32}	f_{33}	f_{34}	f_{35}	f_{36}
	58.0389	11.2404	15.1812	6.1126	6.7103	14.7466
	f_{37}	f_{38}	f_{39}	f_{40}	f_{41}	f_{42}
	16.6528	1.9631	4.074	6.0163	6.5302	4.9654
	f_{43}	f_{44}	f_{45}	f_{46}	f_{47}	f_{48}
	3.4707	3.8122	4.2132	4.0636	4.5177	4.4055
	f_{49}	f_{50}	f_{51}	f_{52}	f_{53}	f_{54}
	4.2212	10.4717	3.5483	3.3384	1.3731	1.1521
	f_{55}	f_{56}	f_{57}	f_{58}	f_{59}	f_{60}
	3.05317	3.94172	3.56277	2.13722	0.87782	1.762
	f_{61}	f_{62}	f_{63}	f_{64}	f_{65}	f_{66}
	2.0569	1.73665	3.43123	3.19395	1.48021	4.48591
	f_{67}	f_{68}	f_{69}	f_{70}	f_{71}	f_{72}
	0.84525	0.42777	0.83711	3.7356	5.22598	0.03159
	f_{73}	f_{74}	f_{75}	f_{76}	f_{77}	f_{78}
	0.10632	0.126	0.11	0.13444	0.24303	0.12565
	f_{79}	f_{80}	f_{81}	f_{82}	f_{83}	f_{84}
	0.07318	0.08107	0.07848	0.0564	0.0936	0.1047
	f_{85}	f_{86}	f_{87}	f_{88}	f_{89}	f_{90}
	0.1578	0.3709	0.245	0.0638	0.0762	0.0607
	f_{91}	f_{92}	f_{93}	f_{94}	f_{95}	f_{96}
	0.136	0.1138	0.0882	0.1333	0.3173	0.1459
	f_{97}	f_{98}	f_{99}	f_{100}	f_{101}	f_{102}
	0.1038	0.1622	0.125	0.0653	0.0797	0.0877
	f_{103}	f_{104}	f_{105}	f_{106}	f_{107}	f_{108}
	0.1556	0.3709	0.2695	0.0313	0.045	0.0422

3 Experimental Results and Discussion

The proposed approach has been implemented on 250 different mushroom images, which is publicly available on the Internet sources [11–14]. Out of 250 images, 118 edible and 132 are poisonous images of different species of mushrooms such as Agaricus, Chanterelle, Crimini, Shiitake, Oyster, Enoki, Portabello, Porcini, Morel, Reishi, Shiitake, White Button, Maitake, Turkey Tail, Giant Puffball, Black Trumpet, etc. For experimental analysis, 108 features are extracted from every 250 images. For example, all 108 features from f_1 to f_{108} of first image of Fig. 1 (e.g., "e1.jpg") are given in Table 2. MATLAB 2017a is used for feature extraction and classification. After feature extraction, apply the different machine learning based approaches together with K-fold cross-validation technique. The performance of the proposed approach is measured by evaluating an accuracy (ACC) by using Eq. (6) as given below.

$$Accuracy = \frac{TP + TN}{TP + TN + FP + FN} \qquad (11)$$

The performance (accuracy) of SVM, KNN, Logistic Regression, Linear Discriminant, Decision Tree, and Ensemble classifiers are mentioned in Tables 3 and 4. The performance of classifiers is evaluated in two ways. First, evaluate the performance by selecting all extracted features and then apply k-fold by using the values of k are 5, 10, and 15 (i.e., 5-fold, 10-fold, and 15-fold) and hold-out (i.e., 20, 25, and 45%) validation techniques. The results are mentioned in Table 3. In next way, evaluate the performance by apply PCA-based feature selection technique to select the suitable features and then apply k-fold by using the values of k are 5, 10, and 15 (i.e., 5-fold, 10-fold, and 15-fold) and hold-out (i.e., 20, 25, and 45%) validation techniques on selected features. The results are mentioned in Table 4.

Table 3 Performance of classifiers by selecting all extracted features with k-fold and hold-out validation technique

	Accuracy					
	5-fold	10-fold	15-fold	20% hold-out	25% hold-out	45% hold-out
SVM	67.6	67.6	66.0	76.6	66.8	61.6
KNN	63.6	71.6	67.6	71.7	62.9	66.1
Logistic regression	58.0	53.6	54.8	54.0	54.8	49.1
Linear discriminant	56.8	58.8	56.4	54.0	58.1	58.0
Decision tree	50.8	58.8	49.2	48.0	48.4	57.1
Ensemble	64.4	65.2	63.6	60.0	61.3	65.2

Table 4 Performance of classifiers by applying PCA-based feature selection technique with k-fold and hold-out validation technique

	Accuracy					
	5-fold	10-fold	15-fold	20% hold-out	25% hold-out	45% hold-out
SVM	54.0	66.8	55.6	56.0	53.2	54.5
KNN	52.8	67.6	58.8	56.0	59.7	57.1
Logistic regression	54.0	58.0	54.0	56.0	53.2	54.5
Linear discriminant	54.0	56.8	54.0	56.0	53.2	54.5
Decision tree	53.2	53.2	58.4	56.0	64.6	55.4
Ensemble	54.0	62.0	59.6	52.0	64.5	58.0

Fig. 1 Some sample of edible mushroom images out of 118 edible mushroom images taken from [11–14] for feature extraction and classification

After result analysis, it is observed that the overall performance of SVM classifier is better with respect to the other classifiers as mentioned above. The performance of SVM classifier is 76.6%, when 20% hold-out validation technique is used and PCA-based feature selection technique are disable. The performance of our proposed approach is improved in the future. Reason behind that the used mushroom images in the proposed approach contain different backgrounds as shown in Figs. 1 and 2, and therefore the values of extracted features contain the features of backgrounds as well as the mushroom images. If the background of used mushroom images is removed by applying some suitable segmentation approach then performance of our proposed approach must be improved because in this case, the values of extracted features contain only mushroom image feature.

Fig. 2 Some sample of poisonous mushroom images out of 132 poisonous mushroom images taken from [11–14] for feature extraction and classification

4 Conclusion

Mushroom has good source of vitamin B and helps to prevent cancer, weight loss, and immunity power of human. But some mushrooms are toxic and can be dangerous if we eat them. Therefore to differentiate the edible and poisonous mushrooms is a prominent task. This paper focused on developing a method for classification of mushroom using feature-based machine learning approach. The performance of the proposed approach was achieved 76.6%, which is better but it can be improved in future by applying some suitable segmentation approach to remove the backgrounds from the mushroom images.

References

1. Raji, A.O., Alamutu, A.O.: Prospects of computer vision automated sorting systems in agricultural process operations in Nigeria. CIGR J. Agric. Eng. Int. (2005)
2. Subramaniam, A., Byung-Joo, O.: Mushroom recognition using PCA algorithm. Int. J. Softw. Eng. Appl. **10**(1), 43–50 (2016)
3. Beniwal, S., Das, B.: Mushroom classification using data mining techniques
4. Matti, D.: Mushroom recognition. Master semester project, Ecolepolytechniquefederale de Lausanne (2010)
5. Maftoun, P., Johari, H., Soltani, M., Malik, R., Othman, N.Z., El Enshasy, H.A.: The edible mushroom Pleurotus spp.: I. Biodiversity and nutritional values. Int. J. Biotechnol. Wellness Ind. **4**(2), 67–83 (2015)
6. Singh, N.P., Srivastava, R.: Retinal blood vessels segmentation by using Gumbel probability distribution function based matched filter. Comput. Methods Programs Biomed. **129**, 40–50 (2016)
7. Singh, N.P., Srivastava, R.: Segmentation of retinal blood vessels by using a matched filter based on second derivative of Gaussian. Int. J. Biomed. Eng. Technol. **21**(3), 229–246 (2016)
8. Seo, J.-W., Kim, S.D.: Novel PCA-based color-to-gray image conversion. In: 2013 20th IEEE International Conference on Image Processing (ICIP), pp. 2279–2283. IEEE (2013)

9. Cen, H., Renfu, L., Zhu, Q., Mendoza, F.: Nondestructive detection of chilling injury in cucumber fruit using hyperspectral imaging with feature selection and supervised classification. Postharvest Biol. Technol. **111**, 352–361 (2016)
10. Howarth, P., Rüger, S.: Evaluation of texture features for content-based image retrieval. In: International Conference on Image and Video Retrieval, pp. 326–334. Springer (2004)
11. Jana Bohdalov Ji Bohdal. Fungi, mushroom photo gallery. http://www.naturephoto-cz.com/mushrooms.html. Accessed 29 Dec 2017
12. Publitek Inc. dba Fotosearch USA. Poisonous mushroom stock photos and images. http://www.fotosearch.com/photos-images/poisonous-mushroom.html. Accessed 29 Dec 2017
13. Frank. List of edible mushroom with photo's. http://www.michiganmorels.com/funtalk/showthread.php?t=7617. Accessed 29 Dec 2017
14. CrystalGraphics Inc. USA. Poisonous mushroom images, pictures & photos. http://www.crystalgraphics.com/powerpictures/images.photos.asp?ss=poisonous+mushroom. Accessed 29 Dec 2017

Pose-Invariant Face Recognition in Surveillance Scenarios Using Extreme Learning Machine Based Domain Adaptation

Avishek Bhattacharjee

Abstract Face Recognition (FR) under adversarial conditions has been a big challenge for researchers in the computer vision community. FR performance deteriorates in surveillance condition due to poor illumination, blur, noise, and pose variation in test samples (probe), when compared to training samples (gallery). Even recent deep learning methods fail to perform well in such conditions. This paper proposes a novel framework called PIFR-EDA (Pose-Invariant Face Recognition using Extreme learning machine based Domain Adaptation) that performs pose-invariant face recognition (PIFR) in cross-domain settings. It consists of two stages where the first stage performs face frontalization using a single unmodified 3D facial model and the second stage performs the task of robust domain adaptation by simultaneously learning a category transformation matrix and an $\ell_{1,1}$-regularized sparse extreme learning machine classifier. The proposed method outperforms state-of-the-art shallow and deep methods (in terms of rank-1 recognition rates) when experimented on three real-world face datasets captured using surveillance cameras.

Keywords Face recognition · $\ell_{1,1}$-regularized sparse extreme learning machine · Face frontalization · Domain adaptation

1 Introduction

The success of automated FR has improved steadily in the past decade, since the development of robust feature representation and classification techniques [35]. The performance of FR degrades in surveillance scenarios due to change in illumination conditions, occlusion, blur, noise, pose variation, and low resolution of test (probe) samples compared to that of training (gallery) samples. Conventional as well as deep machine learning based FR techniques fail to provide satisfactory results (accuracy)

A. Bhattacharjee (✉)
Department of CS&E, IIT Madras, Chennai, India
e-mail: avi@cse.iitm.ac.in

© Springer Nature Singapore Pte Ltd. 2020
B. B. Chaudhuri et al. (eds.), *Proceedings of 3rd International Conference on Computer Vision and Image Processing*, Advances in Intelligent Systems and Computing 1022,
https://doi.org/10.1007/978-981-32-9088-4_18

on datasets which mimic surveillance scenarios as their working principle assumes that training data and test data belong to the same distribution.

The proposed method PIFR-EDA can be divided into 2 stages: (i) face frontalization (FF) and (ii) $\ell_{1,1}$-regularized sparse extreme learning machine based domain adaptation (referred to as $\ell_{1,1}$-sparse-EDA). The first stage ensures pose invariance in the proposed framework by generating frontal views from off-frontal faces by making use of a single unmodified 3D reference facial surface (which approximates the shape of all query faces). The second stage performs domain adaptation by learning a category transformation matrix and an $\ell_{1,1}$-regularized sparse extreme learning machine (referred to as $\ell_{1,1}$-sparse-ELM) classifier by simultaneously minimizing $\ell_{1,1}$-norm of the output weights of ELM and classifier loss.

The contributions of this paper are as follows: (1) $\ell_{1,1}$-sparse-EDA loss obtained by incorporating the $\ell_{1,1}$-sparse-ELM in the EDA [33] framework and (2) the PIFR-EDA framework which unites the face frontalization method with the proposed $\ell_{1,1}$-sparse-EDA method.

In the rest of the paper, Sect. 2 lists recent advances in the field of face detection (FD), domain adaptation (DA), face frontalization, and surveillance face datasets. Section 3 provides a detailed analysis of the two stages of the proposed method. Section 4 describes the experimental procedure and provides performance analysis through quantitative results (using rank-1 recognition rates). Finally, Sect. 5 concludes the paper.

2 Related Work

The most widely used method for the task of face detection (FD) proposed by Viola et al. [28] (referred to as VJFD, alias Viola Jones Face Detector, subsequently) uses the AdaBoost algorithm for selecting a subset of visual features from a bigger set of HAAR features for the task. Another widely prevalent method for face detection is Chehra, proposed by Asthana et al. [2], which detects 49 fiducial landmark points to estimate a tight bounding box on the detected face. The proposed work uses VJFD for the purpose of FD.

Domain adaptation (DA) is a popular machine learning technique, which performs domain invariant classification by either feature transformation or classifier transformation or both, primarily having applications in speech, computer vision and natural language processing [5, 21]. Figure 1 gives a pictorial representation of the DA problem using 3 classes of 2 different domains. Instance weighting is an approach to solve DA where every source domain data instance is weighted appropriately and used for training with the objective of minimizing the expected loss [10, 26]. Another approach to solve the problem is to project source and target data into multiple intermediate subspaces such that the difference between the two distributions are minimized [11, 18, 24].

The method proposed by Banerjee et al. named "Soft-Margin Learning for Multiple Feature-Kernel Combinations" (SML-MFKC) employ a learning method based

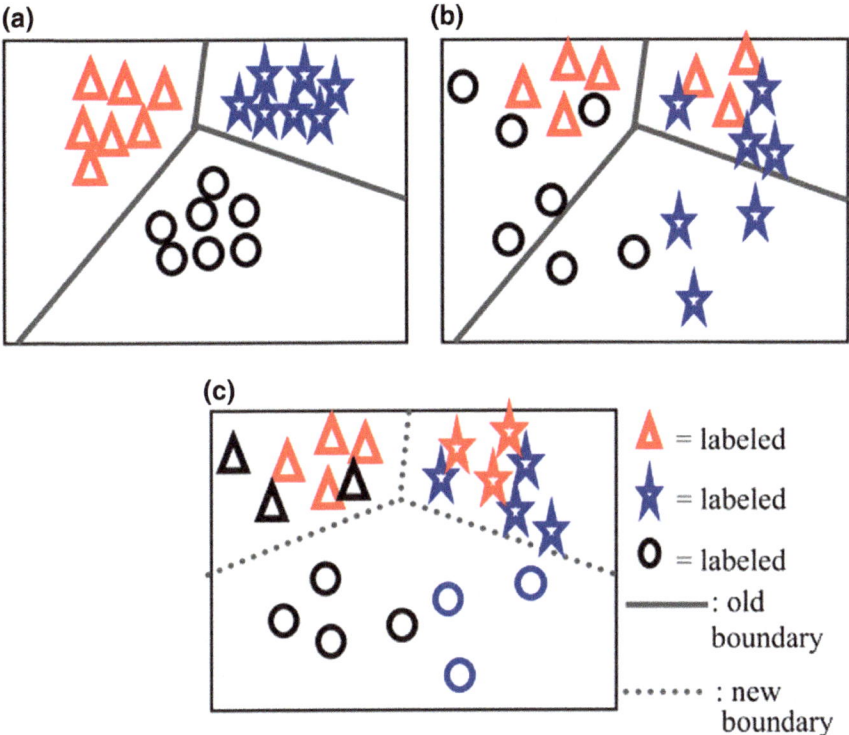

Fig. 1 The problem defining necessity of DA: **a** source domain data of three classes classified using linear boundaries; **b** target domain data misclassification using same boundaries; **c** transformed target distribution and classifier boundaries using DA [33]

on soft-margin to find multiple feature-kernel combinations and then transform the features by DA. Chen et al. proposed a method named "Fisher vector encoded deep convolutional features" (FV_DCNN) [8], which combines Fisher vector representations with deep-CNN features for robust face recognition and verification tasks. "Landmarks-based Kernelized Subspace Alignment for Unsupervised Domain Adaptation" (LSSA) proposed by Aljundi et al. [1] jointly performs subspace alignment and landmark selection in two domains to reduce domain discrepancy. FaceNet by Schroff et al. [25] utilizes a deep-ConvNet to perform embedding optimization by minimizing a triplet loss designed by triplet mining method. "Deep Adaptation Network" (DAN) proposed by Long et al. [20] performs DA by diminishing the separation between the means of the two domain. "Extreme Learning Machine (ELM) based Domain Adaptation" (EDA) proposed by Zhang et al. [33] learns an $\ell_{2,1}$-regularized-ELM classifier along with a category transformation matrix to make the data more separable and improve classification accuracy. It acts as the motivation for our proposed work. The proposed method utilizes an $\ell_{1,1}$-sparse-ELM classifier which, unlike EDA [33], is capable of selecting different sets of features for different categories which enhances classification accuracy (further explained in Sect. 3.2). It also performs PIFR in DA settings, which has not been attempted in any of the works mentioned above.

The face frontalization (FF) task has been previously approached with morphable model-based methods [6, 7, 27, 31] which utilize multiple aligned 3D face models to learn possible facial geometries. Shape from shading models [17] also perform the task of face frontalization with great detail but are not robust to occlusion. The FF method proposed by Hassner et al. [14] approximates the shape of all query faces to an unmodified 3D facial surface which makes the implementation efficient and simple. The incorporation of soft-symmetry also makes the implementation robust to occlusion. This work has been used as the first stage of the proposed method and is briefly explained in Sect. 3.1.

Some of the existing face datasets characterizing surveillance scenarios are FR_SURV [23], SCface [12] and ChokePoint [29] which are described in detail in Sect. 4.1.

3 Proposed Method

This section comprises of three parts, with the first two parts explaining the task of face frontalization and proposed $\ell_{1,1}$-sparse-EDA and the last part explaining the working of the proposed method using the previously mentioned two tasks.

3.1 Face Frontalization

The face frontalization method, proposed in [14], utilizes a single unmodified 3D facial reference surface for approximating the shape of all input face images. Figure 2 shows the result of FF task on ChokePoint dataset [29]. It involves performing face detection on gallery and probe images using VJFD [28] to obtain tightly cropped face images, followed by localization of fiducial landmark points which are used to align the images with respect to a textured 3D facial reference model. A reference coordinate system is obtained from the rendered frontal view of the aforementioned face.

The back-projection of appearance of the input face onto the previously obtained reference coordinate system (by using the 3D facial model) generates an intermediate frontalized face. Further fine-tuning is done on this face by borrowing appearance from the symmetric side of the face to reduce the effect of pose. Details of these steps are as follows:

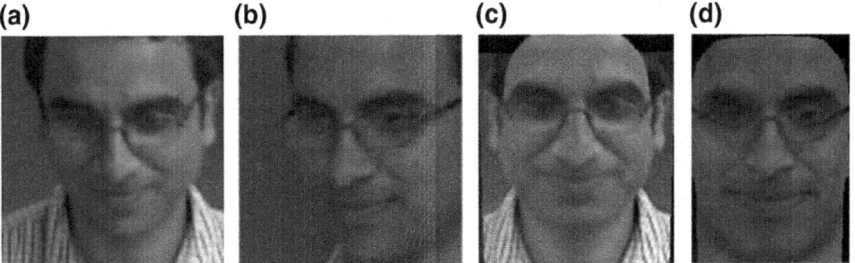

Fig. 2 Results of face frontalization: **a** and **b** are samples from gallery and probe set of ChokePoint dataset [29]; **c** and **d** are corresponding frontalized counterparts

Frontal View Generation

Initially, a 3×4 projection matrix is computed, approximating the one used for capturing the input image. This requires localizing fiducial landmark points on both query image and rendered frontal view of the 3D facial reference model and also finding the 2D–3D correspondences between them. Facial feature detection is performed with the help of Supervised Descent Method (SDM) [30], which finds 49 fiducial landmark points on a query image. SDM does not localize landmark points along the jawline, thus ensuring that all the landmark points are close to the 3D plane in front of the face.

Next step involves pose estimation where a reference projection matrix $\mathbf{C}_M = A_M[R_M \ t_M]$ is used for finding the synthetic rendered view of the textured 3D facial model (A_M is the intrinsic matrix and $[R_M \ t_M]$ is the extrinsic matrix with R_M as rotation matrix and t_M as translation vector). The 3D coordinates $\mathbf{P}^T$ located on the 3D facial surface is stored for every pixel $\mathbf{p}'$ of the generated reference view I_R which satisfies the following condition:

$$\mathbf{p}' \sim \mathbf{C}_M \mathbf{P} \tag{1}$$

Considering $\mathbf{p}_i$ to be the set of fiducial landmark points on the query image I_Q and $\mathbf{p}'_i$ to be the same fiducial landmark points from the generated reference view, the projection matrix $\mathbf{M}_Q = A_Q[R_Q \ t_Q]$ (approximating camera matrix $\mathbf{C}_Q$ of the query image I_Q) can be estimated using the correspondences ($\mathbf{p}_i^T$, $\mathbf{P}_i^T$) by standard techniques described in [13].

In the next step, frontal pose synthesis is performed by using (1) to obtain the 3D coordinates $\mathbf{P}$ on the reference facial surface which was projected onto $\mathbf{q}'$ (pixel coordinates in I_R) using $\mathbf{C}_M$. The following equation is used to estimate coordinates of those same facial features in I_Q (denoted by $\mathbf{p}$):

$$\mathbf{p} \sim \mathbf{C}_Q \mathbf{P} \tag{2}$$

The intensities at coordinate $\mathbf{p}$ in I_Q is obtained using bilinear interpolation, which is then assigned to the corresponding $\mathbf{q}'$ in the fresh frontalized image.

Soft Symmetry for Self-occlusions

In this step, visibility of a face is estimated by an approach similar to multi-view 3D reconstruction methods as in [19, 32]. A single view (I_R) and a reference face (3D geometry approximation) is used to determine visibility in the query image (I_Q). As the yaw angle is directly proportional to the angle between the camera plane and poorly visible facial features, an increase in yaw angle results in a large number of points in the 3D surface projecting onto the same pixel in the photo. The sampling-rate visibility measure can be formulated using the above property.

For each pixel $\mathbf{q}'$ in the reference view I_R, (2) is used to estimate the location of its corresponding pixel $\mathbf{q}$ (quantized integer value rather than the estimated non-integer value of $\mathbf{q}$, thus referring to nearest neighboring pixel) in the query photo. The visibility score metric described in (3) is calculated for every pixel in the frontalized view $\mathbf{q}'$ as

$$v(\mathbf{q}') = 1 - \exp(-\#\mathbf{q}) \tag{3}$$

where $\#\mathbf{q}$ refers to the number of times $\mathbf{q}$ corresponded with *any* frontalized pixel. Finally, the less visible pixel intensities, having low visibility score in (3), are substituted with a weighted average of itself and the corresponding symmetric pixel's intensity weighted by the visibility score.

3.2 $\ell_{1,1}$-*Sparse-ELM Based Domain Adaptation* ($\ell_{1,1}$-*Sparse-EDA*)

Overview of Extreme Learning Machines (ELM)

ELMs are "generalized" feed-forward neural networks with a single hidden layer, trained by estimating the output weights β using Moore–Penrose generalized inverse, while keeping the input weights (randomly initialized) fixed. In comparison with other single-layer feed-forward neural networks, the ELMs perform better in regression and classification tasks [15, 16]. The hidden layer representation can be generated with random input weights W, bias B and activation function $\mathcal{H}(\cdot)$. Thus, ELM training process becomes identical to solving regularized least squares. The conventional $\ell_{2,1}$-regularized ELM model is formulated as

$$\min_{\beta \in \mathbb{R}^{L \times c}} \frac{1}{2} \|\beta\|_{2,1} + \frac{C}{2} \sum_{i=1}^{N} \|e_i\|^2 \tag{4}$$
$$st. \ e_i^T = t_i^T - \mathcal{H}(x_i^T W + B^T)\beta$$

where $t_i, e_i \in \mathbb{R}^c$ denotes the label and error vector for the ith training data $x_i \in \mathbb{R}^d$, d is the input dimension, c represents number of categories, C is the training error penalty coefficient, N denotes number of training samples, and $\mathcal{H}(\cdot)$ is the hidden layer activation function. $\beta \in \mathbb{R}^{L \times c}$, $\|\beta\|_{a,b} = (\sum_{i=1}^{L} (\sum_{j=1}^{c} |\beta_{ij}|^a)^{b/a})^{1/b}$, $W \in \mathbb{R}^{d \times L}$ and $B \in \mathbb{R}^L$ and L is the number of hidden layer nodes. A closed-form solution for β exists in this case as discussed in [33].

Such a closed form solution does not exist for β in case of $\ell_{1,1}$-sparse-ELM and solution is obtained using iterative methods such as FISTA (Fast Iterative Shrinkage-Thresholding Algorithm) [4]. The motivation behind using $\ell_{1,1}$-regularization comes from the hypothesis that a certain set of features may be useful for describing a particular category whereas some other set of features may better describe a different category. The $\ell_{1,1}$-regularization, unlike $\ell_{2,1}$-regularization used in [33], makes the proposed framework capable of selecting different sets of features for different categories, thus providing more flexibility to the learning algorithm to enhance its performance. The $\ell_{1,1}$-sparse-ELM optimization function is formulated as

$$\min_{\beta \in \mathbb{R}^{L \times c}} \|\beta\|_{1,1} + \frac{C}{2} \sum_{i=1}^{N} \|e_i\|^2 \tag{5}$$
$$st. \ e_i^T = t_i^T - \mathcal{H}(x_i^T W + B^T)\beta$$

where the notations have their usual meaning.

Summary of Notations for $\ell_{1,1}$-Sparse-EDA

$X_S \in \mathbb{R}^{d \times N_S}$ is used to denote source domain (S) data matrix. $T_S = (t_S^1 \dots t_S^{N_S})^T \in \mathbb{R}^{N_S \times c}$ denotes label matrix of S, c denotes the number of categories, $X_T \in \mathbb{R}^{d \times (N_{Tl} + N_{Tu})}$ refers to the target domain (T) data matrix, $T_T = (t_{Tl}^1 \dots t_{Tl}^{N_{Tl}})^T \in \mathbb{R}^{N_{Tl} \times c}$ defines label matrix of labeled target data and $\mathbf{H_S} \in \mathbb{R}^{N_S \times L}$, $\mathbf{H_{Tu}} \in \mathbb{R}^{N_{Tu} \times L}$, $\mathbf{H_T} \in \mathbb{R}^{N_{Tl} \times L}$ and $\mathbf{H} \in \mathbb{R}^{(N_{Tl} + N_{Tu}) \times L}$ refers to the hidden layer output matrix for source, unlabeled target, labeled target, and all target data, respectively. L is the number of hidden layer nodes. $\beta \in \mathbb{R}^{L \times c}$ is defined as the learned classifier and $\Theta \in \mathbb{R}^{c \times c}$ as output category transformation matrix. The predicted label for ith labeled sample from S is defined by $y_S^i|_\beta$, whereas the predicted label for jth labeled sample from T is defined by $y_{Tl}^j|_\beta$. $y_{Tu}^k|_\beta$ represents predicted label for kth unlabeled sample from T. ϕ_ρ refers to a base classifier trained on X_S (pre-learned classifier) and $\phi_{\rho,Tu} = (\phi_{\rho,Tu}^1 \dots \phi_{\rho,Tu}^{N_{Tu}})$ defines the target label matrix of the unlabeled target data predicted by the pre-learned classifier.

$\ell_{1,1}$-Sparse-EDA Optimization Function

The proposed $\ell_{1,1}$-sparse-EDA simultaneously learns a category transformation matrix and a classifier in DA settings. The proposed loss function for $\ell_{1,1}$-sparse-EDA is defined as follows:

$$\min_{\beta,\Theta} R_{S(\ell_{1,1})}(\beta) + V_{Tl}(\beta, \Theta) + V_{Tu}(\beta, \phi_\rho) \qquad (6)$$

The first term $R_{S(\ell_{1,1})}(\beta)$ is the proposed loss defining the $\ell_{1,1}$-sparse-ELM, which utilizes labeled source data to train the classifier. It is formulated as

$$R_{S(\ell_{1,1})}(\beta) = ||\beta||_{1,1} + C_S \sum_{i=1}^{N_S} ||\xi_S^i||^2 \qquad (7)$$

where C_S refers to a penalty coefficient. $\xi_S^i = t_S^i - y_S^i|_\beta$ is the classification error. $t_S^{i,j} = 1$ if x_i corresponds to the jth category and -1 otherwise.

The second term $V_{Tl}(\beta, \Theta)$, as in [33], utilizes a small number of labeled target data for learning a cross-domain classifier β and a category transformation matrix Θ. It is formulated as

$$V_{Tl}(\beta, \Theta) = C_T \sum_{j=1}^{N_{Tl}} ||\xi_T^j|_{\beta,\Theta}||^2 + \gamma||\Theta - I||_F^2 \qquad (8)$$

where $\xi_T^j|_{\beta,\Theta} = (t_{Tl}^j)^T \circ \Theta - y_{Tl}^j|_\beta$, I is an identity matrix and $\circ$ is the multiplication operator of category transformation via Θ as defined in [34]. The category distortion during transformation is controlled by $||\Theta - I||_F^2$. The category transformation matrix performs output adaptation in label space. C_T and γ are trade-off parameters.

The third term $V_{Tu}(\beta, \phi_\rho)$, as described in [33], provides stability and generalization by utilizing the unlabeled target data. The systematic perturbation error between β and ϕ_ρ (for same inputs) is minimized to achieve this. Mathematically, it can be formulated as

$$V_{Tu}(\beta, \phi_\rho) = \tau \sum_{k=1}^{N_{Tu}} ||\xi_{Tu}^k|_{\beta,\phi_\rho}||^2 + \Omega_M(\beta) \qquad (9)$$

where $\xi_{Tu}^k|_{\beta,\phi_\rho} = \phi_{\rho,Tu}^k - y_{Tu}^k|_\beta$, τ is a penalty coefficient, and $\Omega_M(\beta)$ represents manifold regularization. Manifold regularization maintains that if two points x_i and x_j are close in the feature space, then their conditional probabilities $P(y|x_i)$ and $P(y|x_j)$ should also be similar. The manifold regularization framework is mathematically formulated as

$$\min \frac{1}{2} \sum_{i,j} \mathcal{A}_{i,j} ||y_i - y_j||^2 \qquad (10)$$

where y_i and y_j are predicted labels for given inputs x_i and x_j. $\mathcal{A}$ (referred to as the similarity matrix) is sparse where every element $\mathcal{A}_{i,j}$ is a nonzero value only if $x_i \in \mathcal{N}_k(x_j)$ ($\mathcal{N}_k(x_j)$ represents k-nearest neighbors of x_j) or $x_j \in \mathcal{N}_k(x_i)$. Equation (10) can be expanded in matrix trace form resulting in the following formulation of the manifold regularization term:

$$\Omega_M(\beta) = \lambda \cdot tr(F^T \mathcal{L} F) \tag{11}$$

where D (diagonal matrix) is defined as $D_{ii} = \sum_j \mathcal{A}_{i,j}$, which in turn defines the Laplacian graph matrix $\mathcal{L} = D - \mathcal{A}$. $F = \mathbf{H}\beta$ is the target output matrix.

By substituting (7), (8), (9) and (11) into (6), and substituting the constraints, the proposed $\ell_{1,1}$-sparse-EDA optimization function is obtained as

$$\min_{\beta, \Theta} \mathcal{J}(\beta, \Theta) = ||\beta||_{1,1} + C_S ||\mathbf{H_S}\beta - T_S||_F^2 + C_T ||\mathbf{H_T}\beta - T_T \circ \Theta||_F^2$$
$$+ \gamma ||\Theta - I||_F^2 + \tau ||\mathbf{H_{Tu}}\beta - \phi_{\rho,Tu}||_F^2 + \lambda \cdot tr(\beta^T \mathbf{H}^T \mathcal{L} \mathbf{H}\beta) \tag{12}$$

Learning Algorithm

As the loss function is minimized *with respect to* two parameters β and Θ, it is done individually *with respect to* both, in an alternating iterative manner.

In the first step, the objective function in (12) is minimized *with respect to* β with FISTA by initially setting $\Theta = I$ to obtain an intermediate optimal β.

In the second step, (12) is minimized *with respect to* Θ by differentiating it *with respect to* Θ and setting to 0 to obtain a closed-form solution as

$$\Theta = (C_T T_T^T T_T + \gamma I)^{-1}(C_T T_T^T \mathbf{H_T}\beta + \gamma I) \tag{13}$$

where β is obtained from the previous step. The above two steps are iterated alternately to find the optimal β and Θ.

3.3 PIFR-EDA

The block diagram in Fig. 3 explains the working procedure of the proposed PIFR-EDA method. The tightly cropped face images obtained by VJFD [28] from both gallery and probe images are frontalized by the method described in Sect. 3.1. Contrast enhancement is performed only on frontalized probe samples with separate gamma values for different datasets as specified in [3]. The preprocessing step consists of face detection, face frontalization, and contrast enhancement (only on probes). The preprocessed gallery and probe images are made compatible with the input specifications of the VGG-Face model [22] by normalizing and resizing to 224×224 pixels and are then passed through a pretrained VGG-Face model. The

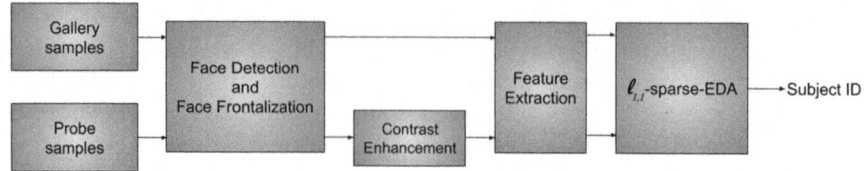

Fig. 3 The different stages of the proposed PIFR-EDA method

4096-dimensional embeddings obtained for each image from fc7-layer (layer preceding the softmax classifier) of the model are used as the corresponding feature vectors, which are then used as input to the $\ell_{1,1}$-sparse-EDA framework described in Sect. 3.2.

4 Experiments

In the following section, the datasets are described first, followed by details of the experimentation protocols. The experimental results and the observations drawn from the results are discussed next.

4.1 Datasets and Experimental Setup

The performance of PIFR-EDA is determined through experimentation on 3 real world surveillance face datasets namely, FR_SURV [23], ChokePoint [29] and SCface [12]. Only FR_SURV probe samples are captured outdoor while ChokePoint and SCface samples are captured indoor. Figure 4 shows gallery and probe samples of all 3 datasets cropped using VJFD [28]. The FR_SURV dataset consists of 51 subjects with 20 samples per subject for gallery as well as probe having average resolution of 150×150 (gallery) and 33×33 (probe) pixels. SCface dataset consists of images from 130 subjects with the gallery images (cropped using VJFD [28]) having a resolution of 800×600 pixels. The images from cams 1–5 at *Distances 1,2 and 3*, also cropped using VJFD, are considered as probe samples with average resolutions 40×40, 60×60 and 100×100 pixels respectively. The ChokePoint dataset contain 25 and 29 subjects in portals 1 and 2, respectively, with 64, 204 face images in total. For experimentation, samples from camera $C1$ are considered as gallery whereas samples from $C2$ and $C3$ cameras are considered as probes. The feature vectors of gallery and probes of each dataset are obtained by following the steps specified in Sect. 3.3.

The performance of PIFR-EDA is compared with state-of-the-art shallow methods (EDA [33], LSSA [1], SML-MFKC [3]) as well as deep methods (VGG-Face [22], DAN [20], FV_DCNN [8], FaceNet [25]). A Naïve method (RBF-Kernel SVM [9])

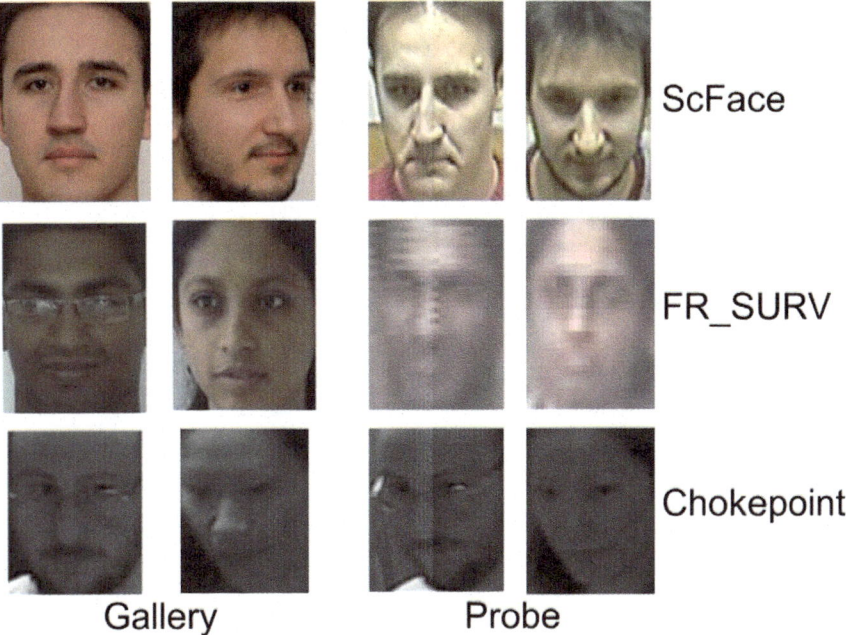

Fig. 4 Gallery and corresponding probe samples (cropped using VJFD [28]) from SCface [12], FR_SURV [23] and ChokePoint [29] datasets

is used to provide baseline performance. PIFR-EDA, Naïve and shallow methods (barring SML-MFKC) use source domain data (gallery VGG-features) and randomly split 20% target domain data (probe VGG-features) for training while keeping the rest 80% target data for testing purposes. The deep methods, which use the preprocessed gallery and probe images as source and target data, respectively, also follow the same target data splitting policy as shallow methods. For SML-MFKC, the procedures followed for preprocessing of gallery and probe images, feature extraction and target data splitting are as described in [3].

4.2 Experimental Results and Observations

The rank-1 recognition rates of PIFR-EDA along with recent methods such as SML-MFKC [3], LSSA [1], FV_DCNN [8], EDA [33], VGG-Face [22], DAN [20], FaceNet [25] as well as Naïve [9] are reported in Table 1. It is observed that PIFR-EDA outperforms all other methods on SCface and ChokePoint datasets while performing marginally below SML-MFKC [3] for FR_SURV dataset. The second best performance is shown by SML-MFKC. The Naïve method performs the worst among all other methods. The overall performance of deep methods is low compared to shal-

Table 1 Rank-1 recognition rates of various FR methods. Results in bold indicate best performances

Sl.	Algorithm	SCface (%) [12]	ChokePoint (%) [29]	FR_SURV (%) [23]
1	LSSA [1]	62.79	69.37	36.61
2	FV_DCNN [8]	56.82	68.37	38.31
3	Naïve [9]	35.17	61.73	18.91
4	SML-MFKC [3]	79.51	85.49	**56.76**
5	VGG-Face-19 [22]	46.38	69.74	32.58
6	DAN [20]	58.62	70.41	39.67
7	FaceNet [25]	48.31	69.83	36.43
8	EDA [33]	74.37	81.06	49.64
9	Proposed	**80.76**	**86.91**	56.59

low methods. The proposed PIFR-EDA method outperforms EDA (base method) on all datasets for 2 reasons: (1) the face frontalization step makes the FR task robust to pose variations, thus improving the performance; (2) $\ell_{1,1}$-regularization used in PIFR-EDA allows the proposed framework to find different set of optimal features for different categories which helps to maximally enhance its classification performance.

A careful row-wise study of Table 1 also reveals that performance of all methods on FR_SURV dataset is far below compared to other datasets. This is because the gallery samples of FR_SURV dataset are captured indoor, whereas the probe samples are captured outdoor. This is in contrast with SCface [12] and ChokePoint [29] where both gallery and probe samples are captured indoor. This causes a larger domain shift between the gallery and probes in FR_SURV compared to others, making it a difficult dataset to evaluate on and also showing scope for further performance improvement of FR in surveillance scenarios.

5 Conclusion

The problem of FR in surveillance scenario still remains an open problem due to real-world perturbation present in probes (testing conditions) compared to gallery (training conditions). Deep learning methods, which have enhanced the performance of various tasks in the field of computer vision, also fail to perform in surveillance conditions. This paper proposes a novel loss function ($\ell_{1,1}$-sparse-EDA loss) and also introduces a novel framework for face recognition, called the PIFR-EDA (Pose Invariant Face Recognition using Extreme learning machine based Domain Adaptation), which tackles the problem of pose variation along with other degradations such as poor illumination, blur and noise. Rigorous experimentation of the proposed framework on 3 surveillance datasets show its superiority over recent state-of-the-art methods.

Acknowledgements We would like to thank the faculty and members of the Visualization & Perception Lab, Dept. of CS&E, IIT Madras for their insight, expertise, and support that greatly assisted the research.

References

1. Aljundi, R., Emonet, R., Muselet, D., Sebban, M.: Landmarks-based kernelized subspace alignment for unsupervised domain adaptation. In: Proceedings of the IEEE Conference on Computer Vision and Pattern Recognition, pp. 56–63 (2015)
2. Asthana, A., Zafeiriou, S., Cheng, S., Pantic, M.: Incremental face alignment in the wild. In: Proceedings of the IEEE Conference on Computer Vision and Pattern Recognition, pp. 1859–1866 (2014)
3. Banerjee, S., Das, S.: Soft-margin learning for multiple feature-kernel combinations with domain adaptation, for recognition in surveillance face dataset. In: Proceedings of the IEEE Conference on Computer Vision and Pattern Recognition Workshops, pp. 169–174 (2016)
4. Beck, A., Teboulle, M.: A fast iterative shrinkage-thresholding algorithm for linear inverse problems. SIAM J. Imaging Sci. **2**(1), 183–202 (2009)
5. Beijbom, O.: Domain adaptations for computer vision applications. Technical report, University of California, San Diego (2012)
6. Blanz, V., Scherbaum, K., Vetter, T., Seidel, H.P.: Exchanging faces in images. In: Computer Graphics Forum, vol. 23, pp. 669–676 (2004)
7. Blanz, V., Vetter, T.: A morphable model for the synthesis of 3d faces. In: Proceedings of the Annual Conference on Computer Graphics and Interactive Techniques, pp. 187–194 (1999)
8. Chen, J.C., Zheng, J., Patel, V.M., Chellappa, R.: Fisher vector encoded deep convolutional features for unconstrained face verification. In: IEEE International Conference on Image Processing, pp. 2981–2985 (2016)
9. Cortes, C., Vapnik, V.: Support vector machine. Mach. Learn. **20**(3), 273–297 (1995)
10. Dai, W., Yang, Q., Xue, G.R., Yu, Y.: Boosting for transfer learning. In: Proceedings of the International Conference on Machine Learning, pp. 193–200 (2007)
11. Gopalan, R., Li, R., Chellappa, R.: Domain adaptation for object recognition: an unsupervised approach. In: Proceedings of the IEEE International Conference on Computer Vision, pp. 999–1006 (2011)
12. Grgic, M., Delac, K., Grgic, S.: Scface-surveillance cameras face database. Multimed. Tools Appl. **51**(3), 863–879 (2011)
13. Hartley, R., Zisserman, A.: Multiple View Geometry in Computer Vision. Cambridge University Press (2003)
14. Hassner, T., Harel, S., Paz, E., Enbar, R.: Effective face frontalization in unconstrained images. In: Proceedings of the IEEE Conference on Computer Vision and Pattern Recognition, pp. 4295–4304 (2015)
15. Huang, G.B., Zhou, H., Ding, X., Zhang, R.: Extreme learning machine for regression and multiclass classification. IEEE Trans. Syst. Man Cybern. Part B (Cybern.) **42**(2), 513–529 (2012)
16. Huang, G.B., Zhu, Q.Y., Siew, C.K.: Extreme learning machine: theory and applications. Neurocomputing **70**(1), 489–501 (2006)
17. Kemelmacher-Shlizerman, I., Basri, R.: 3d face reconstruction from a single image using a single reference face shape. IEEE Trans. Pattern Anal. Mach. Intell. **33**(2), 394–405 (2011)

18. Kulis, B., Saenko, K., Darrell, T.: What you saw is not what you get: domain adaptation using asymmetric kernel transforms. In: Proceedings of the IEEE Conference on Computer Vision and Pattern Recognition, pp. 1785–1792 (2011)
19. Kutulakos, K.N., Seitz, S.M.: A theory of shape by space carving. Int. J. Comput. Vis. **38**(3), 199–218 (2000)
20. Long, M., Cao, Y., Wang, J., Jordan, M.I.: Learning transferable features with deep adaptation networks. In: Proceedings of the International Conference on Machine Learning (2015)
21. Pan, S.J., Yang, Q.: A survey on transfer learning. IEEE Trans. Knowl. Data Eng. **22**(10), 1345–1359 (2010)
22. Parkhi, O.M., Vedaldi, A., Zisserman, A., et al.: Deep face recognition. In: British Machine Vision Conference, vol. 1, p. 6 (2015)
23. Rudrani, S., Das, S.: Face recognition on low quality surveillance images, by compensating degradation. In: International Conference on Image Analysis and Recognition, pp. 212–221 (2011)
24. Saenko, K., Kulis, B., Fritz, M., Darrell, T.: Adapting visual category models to new domains. In: European Conference on Computer Vision, pp. 213–226 (2010)
25. Schroff, F., Kalenichenko, D., Philbin, J.: Facenet: A unified embedding for face recognition and clustering. In: Proceedings of the IEEE Conference on Computer Vision and Pattern Recognition, pp. 815–823 (2015)
26. Sugiyama, M., Nakajima, S., Kashima, H., Buenau, P.V., Kawanabe, M.: Direct importance estimation with model selection and its application to covariate shift adaptation. In: Advances in Neural Information Processing Systems, pp. 1433–1440 (2008)
27. Tang, H., Hu, Y., Fu, Y., Hasegawa-Johnson, M., Huang, T.S.: Real-time conversion from a single 2d face image to a 3d text-driven emotive audio-visual avatar. In: Proceedings of the IEEE International Conference on Multimedia and Expo, pp. 1205–1208 (2008)
28. Viola, P., Jones, M.J.: Robust real-time face detection. Int. J. Comput. Vis. **57**(2), 137–154 (2004)
29. Wong, Y., Chen, S., Mau, S., Sanderson, C., Lovell, B.C.: Patch-based probabilistic image quality assessment for face selection and improved video-based face recognition. In: IEEE Biometrics Workshop, Computer Vision and Pattern Recognition Workshops, pp. 81–88 (2011)
30. Xiong, X., De la Torre, F.: Supervised descent method and its applications to face alignment. In: Proceedings of the IEEE Conference on Computer Vision and Pattern Recognition, pp. 532–539 (2013)
31. Yang, F., Wang, J., Shechtman, E., Bourdev, L., Metaxas, D.: Expression flow for 3d-aware face component transfer. ACM Trans. Graph. **30**(4), 60 (2011)
32. Zeng, G., Paris, S., Quan, L., Sillion, F.: Progressive surface reconstruction from images using a local prior. In: Proceedings of the IEEE International Conference on Computer Vision, vol. 2, pp. 1230–1237 (2005)
33. Zhang, L., Zhang, D.: Robust visual knowledge transfer via extreme learning machine-based domain adaptation. IEEE Trans. Image Process. **25**(10), 4959–4973 (2016)
34. Zhang, L., Zuo, W., Zhang, D.: LSDT: latent sparse domain transfer learning for visual adaptation. IEEE Trans. Image Process. **25**(3), 1177–1191 (2016)
35. Zhao, W., Chellappa, R., Phillips, P.J., Rosenfeld, A.: Face recognition: a literature survey. ACM Comput. Surv. (CSUR) **35**(4), 399–458 (2003)

Classification of Abandoned and Unattended Objects, Identification of Their Owner with Threat Assessment for Visual Surveillance

Harsh Agarwal⬤, Gursimar Singh⬤ and Mohammed Arshad Siddiqui⬤

Abstract Terrorism is on an ever-increasing rise and is one of the major threats the world is facing today. Terrorist attacks mostly take place in crowded areas such as railway stations and airports. They involve the use of explosives which are placed inside suspicious abandoned objects like bags, suitcases, etc. In this paper, we are proposing a model that can classify abandoned and unattended objects separately and backtrack to identify the owner as well as find the last known location of the owner in a social environment using visual surveillance feed in real time for rapid alert and action.

Keywords Video surveillance · Human detection and tracking · Background modeling · Foreground analysis · Scene perception

1 Introduction

The world is witnessing terrorism and security risk at a global level. There has been a significant rise in such attacks in recent years. According to [1], in the last five years, India alone has been a victim of over 1400 attacks with over 300 of them in 2016 itself. A large number of these attacks occur in a crowded social setting such as bus stand, railway station or airport resulting in a substantial number of casualties. Such attacks use explosives that are hidden inside bags and suitcases which are left at public places and go unnoticed.

Video surveillance is used in such crowded settings. The video surveillance market was valued at USD 30.37 Billion in 2016 and is projected to reach USD 75.64 Billion by 2022 according to [2]. These video feeds require manual monitoring by authorities at every instant. This is not very effective as abandoned objects often go unnoticed

H. Agarwal · G. Singh (✉) · M. A. Siddiqui
PDPM Indian Institute of Information Technology Design & Manufacturing Jabalpur, Jabalpur, India
e-mail: gursimarsingh@iiitdmj.ac.in

H. Agarwal
e-mail: harshagarwal@iiitdmj.ac.in

© Springer Nature Singapore Pte Ltd. 2020
B. B. Chaudhuri et al. (eds.), *Proceedings of 3rd International Conference on Computer Vision and Image Processing*, Advances in Intelligent Systems and Computing 1022, https://doi.org/10.1007/978-981-32-9088-4_19

by authorities or get considered as unattended baggage with no means to distinguish between the two of them.

Some solutions have been proposed in the past but they have certain limitations. These systems often mistakenly detect still humans as unattended or abandoned objects. Though some systems detect the baggage, they are unable to distinguish between unattended or abandoned baggage. These models require state of the art hardware to give results in an effective period which is costly. These problems lead to the need of having a low-cost, effective, and robust system for tackling security risks.

Our proposed system not only outperforms the existing systems, but it also distinguishes between unattended and abandoned objects. Along with this, our model can detect the real owner of the abandoned object and trace the last known location of that person in the video feed automatically. This model can also be implemented on Single Board Computers.

Apart from this, our model differentiates between multiple threat levels associated with the time for which the object was left abandoned. The abandoned object is flagged as soon as it is discovered and subsequently, the threat level is raised once a certain predefined time has passed without any action. The proposed system, with the help of low-cost SBCs can be integrated easily with the existing surveillance systems.

2 Our Model

Our model follows a set of procedural steps pipelined to get the results it needs to outperform existing systems utilizing limited computational power. A detailed explanation of these steps has been stated below.

2.1 Marking Region of Interest (ROI)

An interactive GUI is developed for easy selection of the ROI, thus excluding unwanted regions like train tracks, roads, etc. In Fig. 1a, the input background image is displayed on the GUI. Using mouse events such as double click, ROI can be selected as shown in the Fig. 1b.

(a)   **(b)**

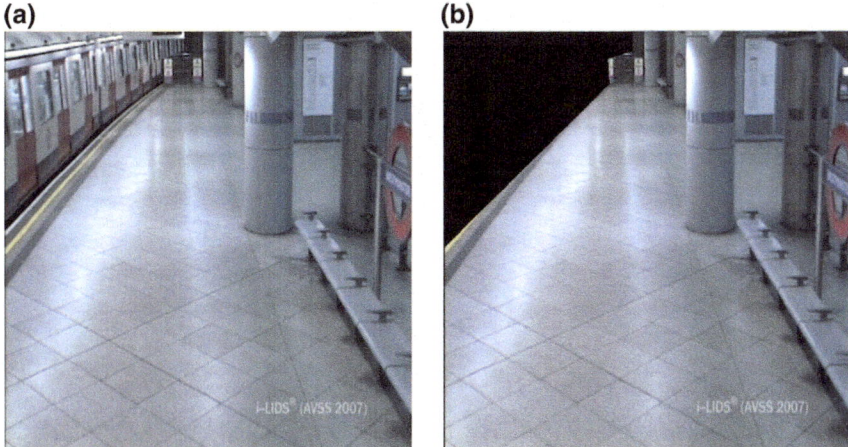

Fig. 1 GUI for selecting region of interest (ROI)

2.2 Long-Term and Short-Term Model for Background Subtraction

We have used the approach of Gaussian mixture model, proposed by Zivkovic and Van Der Heijden [3] as a part of our background subtraction model.

The long-term and short-term background model is used for background subtraction. This method comprises of two models running at different learning rates. The short-term model having a higher learning rate compared to the long-term model. Higher rate signifies that the model is updated frequently.

The model converts the input RGB frame into a binary image. This deletes the background and outputs only the foreground pixel. The output is a binary code, where 1 signifies a foreground element represented in white while 0 represents the background in black.

Long-Term Model (*denoted by* F_{LT}):
Here, the learning rate for long-term model is taken as zero. The model is initialized with an image containing only the background with no person or moving object in the view of the camera. The long-term model considers all-new elements, static as well as moving and label it as foreground. The working of long-term model is depicted in Fig. 2e.

Short-Term Model (*denoted by* F_{ST}):
The learning rate for short-term model is taken as 0.004. The time taken by the model to detect a static object depends on this learning rate. This means that the model is updated frequently as compared to the long-term model. A static object, after a certain period of time, will be considered as background while only the moving objects are considered as foreground. The working of short-term model is depicted in Fig. 2d.

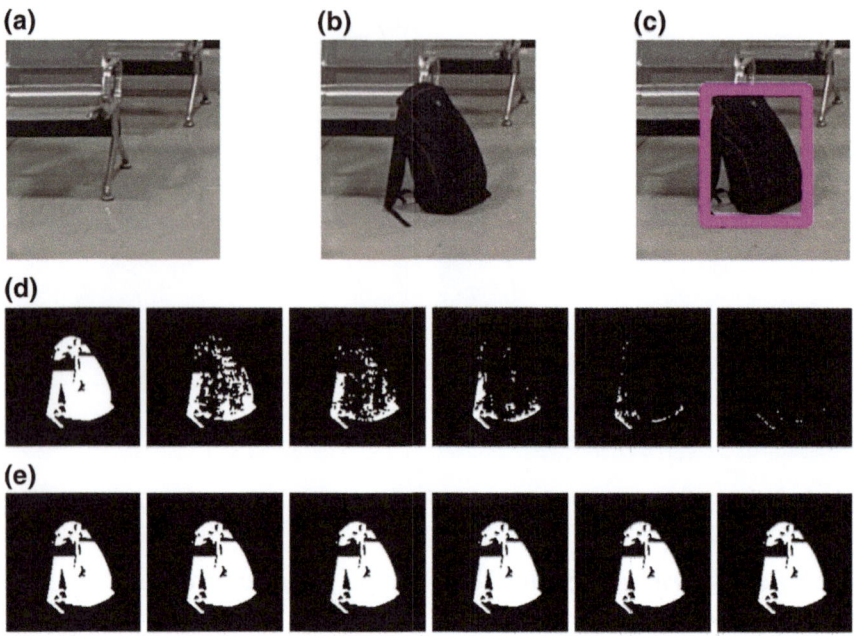

Fig. 2 Long-term and short-term model detecting static object

We then define a two-bit state function S(i) for ith pixel by combining the long-term and short-term pixel values in the frame as explained in Table 1.

$$S(i) = F_{ST}(i) \, F_{LT}(i) \tag{1}$$

Figure 2 demonstrates the long-term and short-term model in our dataset to detect the static object. Figure 2a is the background. As the learning rate is zero for long-term model, any object other than this image will be considered as foreground by the long-term model. Figure 2b has a static object in this environment and Fig. 2c has the final detection of the object surrounded by a bounding box. Figure 2d demonstrates the short-term model, where we can see that the static object with time becomes a part of the background while in the long-term model, Fig. 2e we can see that this object stays as a foreground object. Thus, the output of the static pixel after T_{ST} time results to 10.

Table 1 State and its hypothesis for pixel i	

S(i)	Hypothesis for pixel i
00	Pixel is part of the background
01	Uncovered background pixel
10	Static foreground pixel
11	Moving foreground pixel

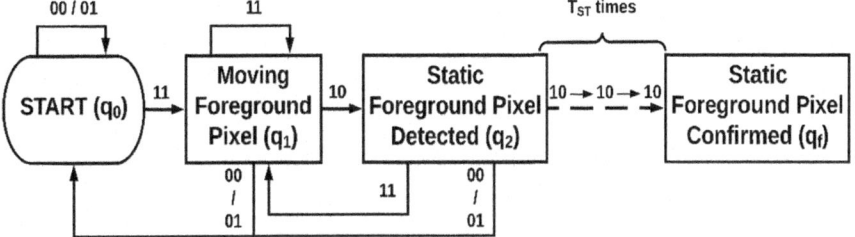

Fig. 3 Pixel-based finite state machine used in our model

2.3 Pixel-Based Finite State Machine

Using the state definitions from the above model we can detect all the foreground static pixels.

For this, we have constructed a finite state machine which is depicted in Fig. 3. The finite state machine moves from one state to another based on the intensity value of a pixel in subsequent frames.

The Pixel-Based Finite State Machine or PFSM, remains on the initial state, q_0, if it sees a background pixel or an uncovered background pixel. The state of the finite state machine moves to the next state, q_2, on seeing a moving foreground pixel and on seeing subsequent foreground pixel, the state remains the same. The other possibility for this state is that it encounters a background or uncovered background pixel which changes the state back to q_0. The third state, q_3 is reached when the pixel is a static foreground pixel. If at q_3 the state of pixel changes to 00 or 01, that is, if it changes to background or uncovered background image, the state changes back to q_0.

Now, at state q_3 we have encountered the first static pixel. We now check for T_{ST} times that the pixel state remains the same as the static pixel, we classify this pixel as a static object. We check for T_{ST} so that noise originating from imperfect modeling can be removed.

2.4 Back-Tracing to Find Owner

Detecting Humans in a Frame.
Our model uses a combination of two different models to obtain state of the art accuracy for detecting humans in each frame. MobileNet-SSD (Single Shot Multi-Box Detector) and HOG (Histogram of Oriented Gradients) with SVM (Support Vector Machines) are used to detect the humans.

MobileNet-SSD, depicted in Fig. 4a uses an RCNN or region-based convolutional neural network for identifying humans in an image. It is different from a CNN model as the object detection and localization occurs in a single forward pass, making it a

(a)   **(b)**   **(c)**

Fig. 4 Detecting humans in a frame

better model for real time systems, scoring over 74% mAP (mean Average Precision) at 59 frames per second as stated by Liu et al. [4].

HOG is used with linear SVM, as proposed by Dalal and Triggs [5] in their approach to detect humans. Dalal and Triggs [5] have shown that grids of Histograms of Oriented Gradient (HOG) descriptors for human detection significantly outperform existing feature sets. The results have been depicted in Fig. 4b.

The result from these two models is then combined and depicted in Fig. 4c. The overlapping detection is considered as a single human while those which are identified by a single model is also considered, increasing the accuracy for prediction, taking the result of both models together.

Tracking Multiple Humans in Video Feed

We create a new queue for every new human that enters the scene and as soon as he/she is detected. The queue stores the coordinates of the location of the user for 20 frames. If the human is in the frame for more than 20 frames, the coordinates of the oldest frame are removed, and the new frame coordinates are added to this queue.

Scale-invariant feature transform or SIFT method as presented by Lowe [6], is used to extract features and descriptors of all humans in a frame. The features are stored for all the humans of subsequent frames. FLANN or Fast Library for Approximate Nearest Neighbors proposed by Muja and Lowe [7], is used to match the stored features in consecutive frames and thus track them. FLANN is a collection of algorithms used for nearest neighbor search for high dimensional features. When compared to the brute force method, both provide comparable results, but FLANN is much faster, helping us in detecting the humans in all the frames of a real time surveillance feed. This process may still be noisy and fail when a human pass from behind an object or when human detection fails in a frame. Thus, in order to increase the robustness of our approach Kalman filter [8] is used for predicting the future location of the human in the frame.

Performing this, we get all the humans and their position in subsequent frames. This allows us to track the human in a surveillance video feed. In Fig. 5, the trajectory of the human is marked in yellow which leads all the way up to him in each frame.

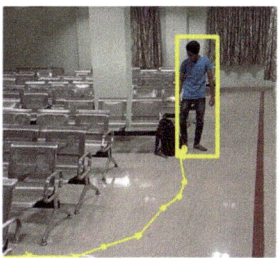

Fig. 5 Tracking of human in video feed

Classifying Static Object as Abandoned or Unattended

According to the current security standards, if the owner of an object is within 2 m distance of the baggage, then it is considered as unattended and is not flagged as abandoned. But, if the owner of the objects moves more than 3 m away, then the object is flagged as abandoned.

It is impossible to calculate 3D distances from a 2D image using perspective transformation. A well calibrated camera or more than one camera system is required to obtain high precision, real-world locations of the 3D objects. Both of these systems are not used for surveillance in public places. Therefore, we use the camera model proposed by Baur et al. [9]. In this approach, the necessary conditions are that the object whose distance is to be calculated, is on the ground plane and the camera parameters such as the camera tilt, focal length, and camera height should be known.

In video surveillance, these parameters are known, and all the conditions are satisfied. Therefore, once a static object is detected, we backtrack to the frame where the static object was first detected. The distances of humans from this object is calculated for this frame using this approach. The person closest to this object in the frame where this object was first detected is identified as the owner. This approach is then applied to the current frame to calculate the distance between the static baggage and its owner to classify it as abandoned or unattended.

In Eqs. (2) and (3), camera tilt (θ_x), camera height (y_c), and the focal length (f) are known, (u_c, v_c) are the coordinates of the optical center of the camera and (u_b, v_b) are the bottom coordinates of the bounding box of the object.

Equation (2) gives us the depth in 3D plane of the object.

$$Z = \frac{f y_c}{v_b - v_c \cos\theta_x + f \sin\theta_x} \tag{2}$$

Equation (3) gives the horizontal distance of the object from the camera optical center.

$$\chi = \frac{z(u_b - u_c \cos\theta_x)}{f} \tag{3}$$

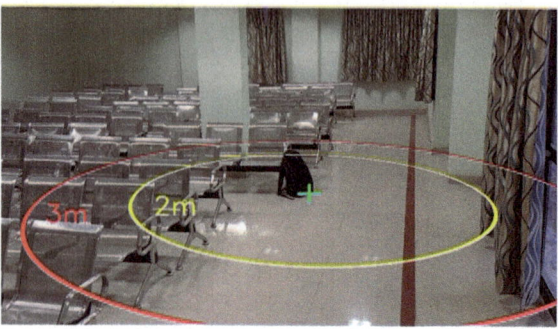

Fig. 6 Distances in 2D image for different threat levels

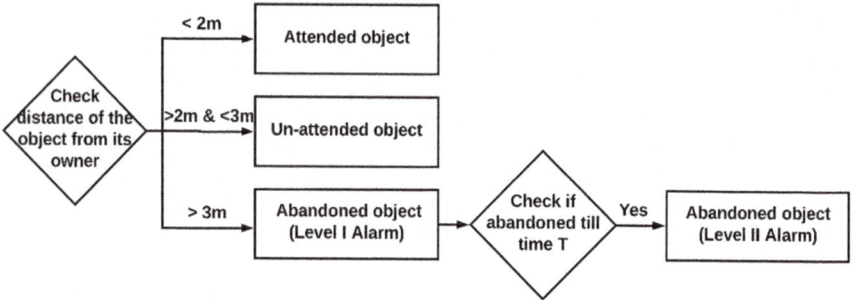

Fig. 7 Algorithm for classification as abandoned object and different threat level

Though this model gives accurate distances when the camera tilt is low, the error tends to increase with the increase in the tilt angle. However, the error is comparatively very small than the distances required in surveillance and hence can be ignored (Fig. 6).

Threat Levels

There are two different threat levels. First, an unattended baggage is identified using PFSM. Then after calculating the distance of the owner from the baggage, the object is marked as abandoned. This is the first security flag. As soon as the owner is found to be more than 3 m away from the baggage. The next level alarm is raised when after a certain time *T*, the object is still abandoned. Rapid action can be taken as soon as this flag is raised. Figure 7 depicts the flow of this process in our approach.

3 Datasets

The solution was tested on three datasets. First two datasets are open-source datasets and the videos for the third dataset were shot by us.

PETS 2006:

- Performance Evaluation of Tracking and Surveillance dataset for 2006 [10]
- Total of 28 videos
- Comprises of 7 distinct events shot from 4 different viewpoints.

AVSS 2007:

- Dataset provided by Advanced Video and Signal based Surveillance [11]
- Comprises of 3 videos: Easy, Medium, and Hard
- In medium and hard, the number of people around the objects is quite high.

Our Dataset:

- The dataset created by us to test and evaluate our model
- Comprises of total 18 videos shot from 3 different viewpoints
- The authors and their colleagues are featured in the dataset.

4 Results and Experiments

The model was evaluated on the benchmark datasets available open-source [10, 11], and also, on a dataset created by us described above. Testing configuration for testing the dataset video feeds are Intel i5 dual-core processor with 4 GB RAM. The video feed was processed at 10 frames per second.

The PETS 2006 dataset comprises of 28 videos. The system failed to detect the abandoned object in three cases. In two cases, the owner was detected incorrectly. Also, in two cases, a static bystander was detected as an object while in three cases light reflection blobs were detected as an unattended object.

In the AVSS 2007 dataset, a total of three cases are presented as easy, medium, and hard. The system performed perfectly for the easy and medium category, while for the hard category, the owner is identified incorrectly.

We also evaluated the model on our dataset. A total of 18 videos, out of which the model detected blobs generated by changes in light as a static object in three cases and the owner is identified incorrectly in one case.

Traditionally, most of the previous models have been tested using a regular confusion matrix and calculating the corresponding F-score. Table 2 presents the comparison of our proposed model with another state of the art models in ascending order of their F-scores. Table 2(a) shows the comparison on the PETS 2006 dataset, while Table 2(b) shows the comparison of different models on AVSS 2007 dataset. However, it is observed that some models have only shown their results on a very small number of sequences to increase the F-score value, whereas we have used the complete dataset for the evaluation. Also, [12] and [13] were evaluated on a single dataset which is AVSS 2007 and PETS 2006 respectively.

Table 2 Comparison of different state of the art models

(a)

	[15]	[13]	[16]	Proposed model	[14]
Precision	0.03	1.0	0.75	0.83	0.85
Recall	1.0	0.71	1.0	0.89	1.0
F-score	0.05	0.83	0.86	0.86	0.92

(b)

	[15]	[17]	[14]	[16]	[13]	Proposed model
Precision	0.05	0.21	0.35	1.0	1.0	1.0
Recall	1.0	1.0	1.0	1.0	1.0	1.0
F-score	0.09	0.35	0.52	1.0	1.0	1.0

Though the only model that surpassed our result was [14] in PETS 2006 dataset, but it did miserably on AVSS 2007 dataset, showing us that their model might have possible overfitting for the PETS 2006 dataset.

Although the model has been evaluated above using the conventional F-Score metric, it does not give a good evaluation of the overall steps involved and the complexity and various cases involved in this condition. Therefore, we define our own rate of accuracy. It is equivalent to the number of successful cases by the total number of cases.

The metric A is defined as,

$$A = \frac{N - (N_0 + N_1 \times 0.50 + N_2 \times 0.50 + N_3 \times 0.25)}{N}$$

where,

N Total number of cases tested
N_0 No abandoned object detected
N_1 Abandoned object is detected but owner is incorrectly identified
N_2 Abandoned object labelled correctly but human is also detected as object
N_3 Abandoned object labelled correctly but incorrect blob also detected

The equation can be interpreted as a penalizing metric. If the object is not identified, the case is penalized completely. If the owner is detected incorrectly, a 50% penalization is done and if extra incorrect object detections occur along with the correct abandoned object, then a penalty of 50% for static human as an object and a penalty of 25% in case of some blob as an object is made to that case. Following are the values obtained for our model on PETS 2006, AVSS 2007 and our dataset.

$$N = 49, N_0 = 3, N_1 = 4, N_2 = 2 \text{ and } N_3 = 6$$

Therefore, evaluating our model based on these results: $A = \mathbf{0.857}$.

5 Conclusion

The paper presents a model for detecting abandoned objects in video surveillance. It is a robust and efficient framework to detect abandoned objects and distinguish them from unattended objects in a complex and crowded environment from real time video surveillance. The framework uses a dual background segmentation method, a short-term and long-term model to effectively detect static objects in any background. The model also detects and localizes humans in the video feed using SSD and HOG with SVM. The combined model for detecting and localizing humans has an accuracy of over 90% in our datasets. Features were extracted from individual humans detected in each frame using SIFT and matched in subsequent frames using FLANN followed by filtering using Kalman filter. Once a static foreground object is detected, we backtrace to identify the owner and using the threat level model, we classify the object as attended, unattended or abandoned.

The model outperforms the existing models. In Table 2(a), a comparison of the current models on the PETS dataset is shown. We can see that [13, 15, 16] get a lower score compared to this model while [14] has a slightly better F-score. On the other hand, in Table 2(b), we can see that [14] has a substantially lower F-score compared to [13, 16], the proposed model.

Acknowledgements We thank Mr. Anuj Khare, Mr. Chinmay Swaroop Saini, and Mr. Harshit Choubey (Undergraduate Students at IIITDM Jabalpur) for helping us in creating our own dataset. The videos in the dataset were shot at IIITDM Jabalpur campus not violating any ethical obligation and feature the above mentioned volunteers and authors with their consent.

References

1. Gurung, S.: India witness highest bomb blasts in world in past two years (2018). https://economictimes.indiatimes.com/news/defence/india-witnessed-highest-number-of-bomb-blasts-in-world-in-past-two-years/articleshow/57082541.cms
2. Video Surveillance Market by System, Offering, Vertical, and Geography—Global Forecast to 2023. https://www.marketsandmarkets.com/Market-Reports/video-surveillance-market-645.html
3. Zivkovic, Z., Van Der Heijden, F.: Efficient adaptive density estimation per image pixel for the task of background subtraction. Pattern Recogn. Lett. **27**(7), 773–780 (2006)
4. Liu, W., Anguelov, D., Erhan, D., Szegedy, C., Reed, S., Fu, C.Y., Berg, A.C.: SSD: single shot multibox detector. In: European Conference on Computer Vision, pp. 21–37. Springer, Cham (2016)
5. Dalal, N., Triggs, B.: Histograms of oriented gradients for human detection. In: IEEE Computer Society Conference on Proceedings of CVPR 2005, vol. 1, pp. 886–893. IEEE (2005)
6. Lowe, D.G.: Distinctive image features from scale-invariant keypoints. Int. J. Comput. Vis. **60**(2), 91–110 (2004)
7. Muja, M., Lowe, D.G.: Fast approximate nearest neighbors with automatic algorithm configuration. In: VISAPP (1), vol. 2(331–340), p. 2 (2009)
8. Bishop, G., Welch, G.: An introduction to the Kalman filter. In: Proceedings of SIGGRAPH, Course, vol. 8(27599-3175), p. 59 (2001)

9. Baur, R., Efros, A., Hebert, M.: Statistics of 3d object locations in images (2008)
10. In: Ninth IEEE International Workshop on Performance Evaluation of Tracking and Surveillance 2006, PETS 2006 dataset. http://www.cvg.reading.ac.uk/PETS2006/data.html
11. i-Lids dataset for AVSS 2007. http://www.eecs.qmul.ac.uk/~andrea/avss2007_d.html
12. Pan, J., Fan, Q., Pankanti, S.: Robust abandoned object detection using region-level analysis. In: 2011 18th IEEE International Conference on Image Processing (ICIP), pp. 3597–3600. IEEE (2011)
13. Li, L., Luo, R., Ma, R., Huang, W., Leman, K.: Evaluation of an IVS system for abandoned object detection on PETS 2006 datasets. In: Proceedings of the IEEE Workshop PETS, pp. 91–98 (2006)
14. Tian, Y., Feris, R.S., Liu, H., Hampapur, A., Sun, M.T.: Robust detection of abandoned and removed objects in complex surveillance videos. IEEE Trans. Syst. Man Cybern. Part C (Appl. Rev.) 41(5), 565–576 (2011)
15. Porikli, F., Ivanov, Y., Haga, T.: Robust abandoned object detection using dual foregrounds. EURASIP J. Adv. Signal Process. 2008, 30 (2008)
16. Liao, H.H., Chang, J.Y., Chen, L.G.: A localized approach to abandoned luggage detection with foreground-mask sampling. In: IEEE Fifth International Conference on Advanced Video and Signal Based Surveillance, 2008, AVSS'08, pp. 132–139. IEEE (2008)
17. Evangelio, R.H., Senst, T., Sikora, T.: Detection of static objects for the task of video surveillance. In: 2011 IEEE Workshop on Applications of Computer Vision (WACV), pp. 534–540. IEEE (2011)

Enabling Text-Line Segmentation in Run-Length Encoded Handwritten Document Image Using Entropy-Driven Incremental Learning

R. Amarnath, P. Nagabhushan and Mohammed Javed

Abstract In today's digital era, archival and transmission of document images are generally carried out in a compressed form in order to avoid wastage of storage space and bandwidth. In the case of CCITT Group 3 and Group 4, the compressed representation is a stream of white and black pixel intensity values called runs, correspondingly indicating background and foreground regions of the document image. In this research paper, we propose a novel entropy-driven incremental learning technique that directly works on the compressed stream of runs, and subsequently facilitates text-line segmentation in handwritten document images using entropy and connected component analysis. Spatial Entropy Quantifier (SEQ) is extracted from the stream of runs based on a suitable window. Further, incremental entropy and connected component analysis are carried out thus separating text and non-text regions leading to automatic text-line segmentation. The proposed method is validated with the compressed dataset of handwritten document images and performance is reported.

Keywords Compressed document · Compressed stream · CCITT Group 3 Group 4 · Entropy · Incremental learning · Text-line segmentation

1 Introduction

The CCITT (International Telegraph and Telephone Consultative Committee) [1, 2] standards, Group-3 and Group 4, facilitate the development of communication protocols for document image compression and facsimile transmission particularly of binary images over telephone lines, data networks, and digital libraries. RLE (Run-Length Encoding) is the backbone of CCITT compression standards which is also widely supported in TIFF [4] and JPEG [4] formats. In real time, scanned version of printed or handwritten documents generally occupy a large storage space in terms

R. Amarnath (✉) · P. Nagabhushan
Department of Studies in Computer Science, University of Mysore, Mysore, India
e-mail: amarnathresearch@gmail.com

P. Nagabhushan · M. Javed
Department of IT, Indian Institute of Information Technology, Allahabad, India

© Springer Nature Singapore Pte Ltd. 2020
B. B. Chaudhuri et al. (eds.), *Proceedings of 3rd International Conference on Computer Vision and Image Processing*, Advances in Intelligent Systems and Computing 1022,
https://doi.org/10.1007/978-981-32-9088-4_20

233

of Mega Bytes (MBs), and through applications like e-governance and AADHAR, a huge volume of such documents are being accumulated leading to a Big data problem [3–5]. Therefore, in order to overcome the volume aspects of big data, compression techniques are used in the current digital world for space and time efficacy. Although compression overcomes the issue of volume, its processing becomes expensive as it requires decompression and recompression operations as many times the data need to be processed for archival, segmentation and OCR [5]. Also, in case of Document Image Analysis (DIA) [3], decompression of a document is inevitable supposing that the operations like feature extraction, segmentation, skew detection and OCR [5] have to be carried out in the compressed document format. Therefore, processing directly in the compressed representation of a document could be an ideal solution for space and time trade-offs [3]. Recent literature [4, 5, 7] have reported operations like feature extraction, segmentation and word spotting processing of compressed printed document images. However, the challenges involved in processing compressed unconstrained handwritten documents with oscillatory variations, inclined orientations and frequent touching of text-lines [3] are yet to be addressed.

The motivation of the proposed research work is to demonstrate efficient direct processing on the compressed stream of handwritten document images and subsequently carry out text-line segmentation. A novel idea of incremental learning using entropy as a feature is demonstrated for solving the compressed handwritten text-line segmentation. The proposed model is illustrated in Fig. 1. In a compressed stream of data consisting of runs, when data are received as batches, in regular intervals, instant processing could improve the efficacy of space, time and accuracy [3]. Just maintaining the metadata (temporal understanding) from every packet could miti-

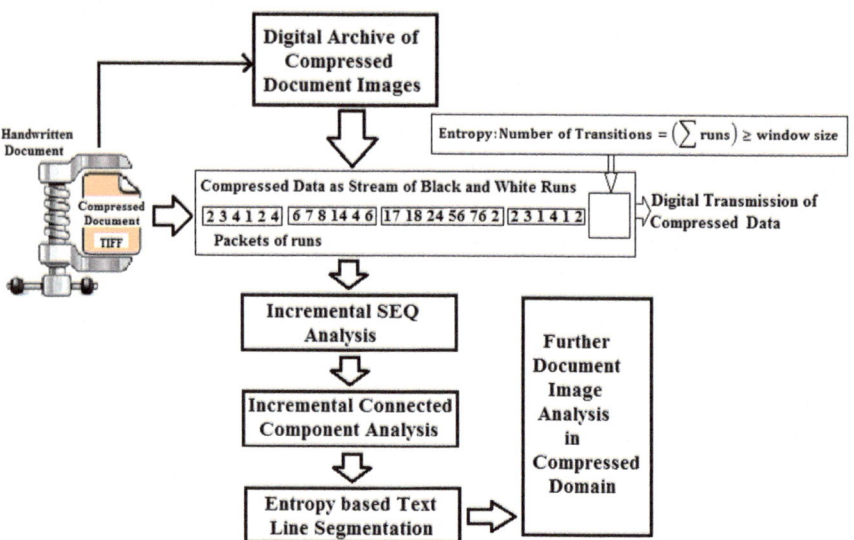

Fig. 1 Proposed entropy-driven incremental learning model for text-line segmentation

gate future packet chunks that lead to distinguish text and non-text regions. However, metadata could be an entropy feature [8–10] which is widely applicable for DIA [8].

Spatial Entropy Quantifier (SEQ) [6, 8] provides the position of transition (pos) that takes an advantage over Conventional Entropy Quantifier (CEQ) [8] for analysis of both compressed and uncompressed documents. The application of SEQ has been reported [8] just for printed documents. However, the proposal using SEQ works equally well for compressed handwritten document images.

The SEQ for the transition between two runs $C_{\beta 1}$ and $C_{\beta 2}$ in the stream with respect to an uncompressed document width (row) r_α is written as

$$E(\beta) = \frac{r_\alpha}{m} \left(\frac{pos}{n} \times \log \frac{n}{pos} + \left(m - \frac{pos}{n} \right) \times \log \frac{m}{m \times n - pos} \right)$$

where $\beta = 1, \ldots, m$. The variables m and n represent the number of rows and columns respectively. Total entropy for each row is computed as the summation of entropy at even and odd columns represented as $E^+(\beta)$ and $E^-(\beta)$ respectively given as $E(\beta) = E^+(\beta) + E^-(\beta)$.

A detailed procedure is provided showing the text-line segmentation in compressed representation of a document image using incremental learning. An illustration is also presented with an example. Developing such a system provides computational benefit over the voluminous documents (big data) [3, 5, 9] which is applicable for cloud computing, digital libraries and network technologies arising from text document images [3]. Rest of the paper is organized as follows: Sect. 2 introduces compressed representation and related terminologies, Sect. 3 discusses the proposed model, Sect. 4 reports experimental results and discussion, Sect. 5 summarizes the entire research work.

2 Compressed Representation and Related Terminologies

Figure 2 shows a sample binary document image. RLE is a technique primarily used for compressing binary document images. In RLE, a run is a sequence of similar pixel values, and the RLE representation indicates the alternating white (background) and black (foreground) runs [3, 5]. Here, four streaming orders such as top-down, bottom-up, left-to-right, and right-to-left are considered for text-line segmentation though CCITT Group 3 and Group 4 recommends top-down streaming order. Figure 3 shows row-wise and column-wise orders with respect to the example shown in Fig. 2.

top-down (row-wise ordering) = {14, 2, 13, 5, 7, 4, 2, 1, 1, 3, 8, 3, 4, 3, 7, 4, 6, 1, 7, 5, 5, 1, 5, 4, 1, 2, 4, 2, 6, 2, 2, 2, 5, 1, 7, 1, 4, 1, 4, 1, 15, 5}
bottom-up (row-wise ordering) = {5, 15, 1, 4, 1, 4, 1, 7, 1, 5, 2, 2, 2, 6, 2, 4, 2, 1, 4, 5, 1, 5, 5, 7, 1, 6, 4, 7, 3, 4, 3, 8, 3, 1, 1, 2, 4, 7, 5, 13, 2, 14}
left-to-right (column-wise ordering) = {36, 1, 9, 2, 4, 1, 1, 5, 3, 5, 5, 4, 6, 6, 7, 3, 10, 1, 2, 2, 8, 1, 9, 3, 5, 5, 2, 1, 2, 11, 9, 1, 9, 1}

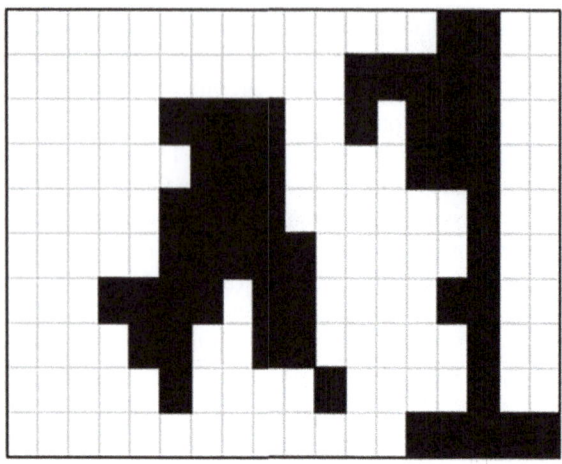

Fig. 2 A sample binary image of size 10×18 (*Height* $\times$ *Width*)

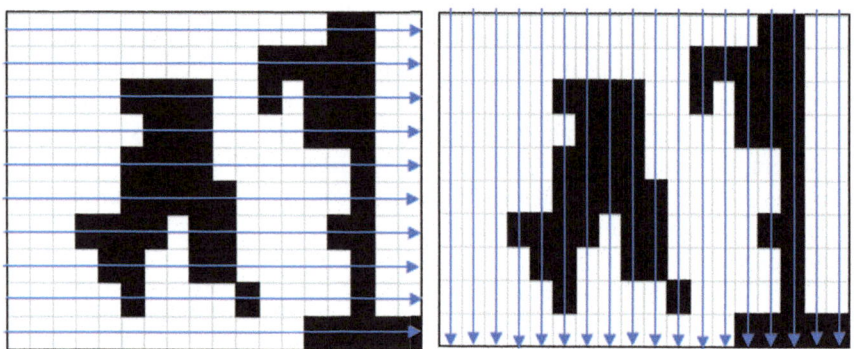

Fig. 3 Row-wise and Column-wise ordering

right-to-left (column-wise ordering) $= \{1, 9, 1, 9, 11, 2, 1, 2, 5, 5, 3, 9, 1, 8, 2, 2, 1,$
$10, 3, 7, 6, 6, 4, 5, 5, 3, 5, 1, 1, 4, 2, 9, 1, 36\}$

Suppose, h and w represent the height and width of an uncompressed document image, respectively. If w is a four-digit number, then 2^4 bits memory space is adequate for representing a row. Obviously, a run for a non-text document (blank document) requires $2^{(h \times w)}$ bits in its worst-case scenario with regard to memory space, given that the run is of $h \times w$ digits. On the other hand, network protocols [10] determine a packet size. A notion is that, a packet may contain k number of runs [3]. In brief, the objective is to determine the text and non-text region from every packet containing k number of runs. Also, k varies across packet chunks, which is studied in the proposed model.

Table 1 SEQ for compressed data

Scanning directions	Total runs (n)	SEQ (n − 1)
top-down (Row-wise ordering)	42	41
bottom-up (Row-wise ordering)	42	41
left-to-right (Column-wise ordering)	34	33
right-to-left (Column-wise ordering)	34	33

In convention [7], SEQ is computed for every row to find the horizontal entropy which is introduced in the previous section. In the proposed model, the total horizontal entropy is computed by identifying the total number of runs. If n represents the total number of runs, then $n - 1$ would be the entropy value. The SEQ for the example (Fig. 2) is tabulated (Table 1).

There are 'm' packet chunks. Number of runs ki varies from one packet chunk to other. One less than the summation of ki would be the SEQ for the total packet chunks.

$$SEQ\,for\,compressed\,data = \left(\sum_{i=1}^{m} ki \right) - 1$$

where, $k_1, k_2, \ldots k_m$ represent the number of runs in packets $p_1, p_2 \ldots p_m$ respectively.

$$\sum_{i=1}^{m} ki = n \,where,\, n\,is\,the\,total\,runs.$$

3 Model Building

The proposed methodology, in fact, makes use of a window for estimating entropy values irrespective of directions and number of packet chunks. The size of the window t determines a portion of the document image specifying the spatial coordinates. Obviously, in the proposed method t would be of one dimension, because RLE would be streamed either row-wise or column-wise.

For instance, considering t to be a row size w of a document image, the entropy value would be e_1. For example (In Fig. 1, top-down approach), the entropy value estimated for a very first row is 2 which is calculated based on the number of transitions [3, 7] (background to foreground and vice versa), provided t be a row size of a document image.

3.1 Top-Down/Bottom-Up Approach: Entropy Computation

A notion is that, odd and even positioned runs represent the background and foreground of a document image respectively. With this ground truth, if the first runs rc_1 in packet p_1 is greater than that of multiples of w (row width), then it is considered as a *top_margin*. Otherwise, there exist a text. The same is applicable for the final odd positioned run rc_n, if exist, in the last packet p_m and obviously it is considered as *bottom_margin*.

$$top\_margin = rc_1 > c \times w, \text{ where } c \geq 1$$

$$bottom\_margin = rc_n > c \times w, \text{ where } c \geq 1$$

Also the left indent [3] in the starting line and right indent [3] in the ending line can be computed as

$$left\_indent = \frac{rc_1}{w}$$

$$right\_indent = \frac{rc_n}{w}$$

If other odd positioned runs are found to be greater than or equal to that of twice the quantity of w, then it is indicating a non-text-line.

$$nontext\ line = rc_i > c \times w, \text{ where } c \geq 2 \text{ and } mod\,(i, 2) \neq 0 \text{ or } i \text{ is odd}$$

Selection of a window size t is explained in the literature [11]. Here, entropy could be calculated for a particular region or portion. For example (Fig. 2: top-down approach), variations in t with respect to w, $w/2$ and $w/4$ result in 2, 0 and 0 entropy values respectively. Similarly, the entropies shall be estimated for all the runs in every packet. Here, a portion where entropy is greater than zero (0) determines the text content. Statistically, linear summation of runs rc_i provide the spatial coordinates.

$$Spatial\ Coordinates\ of\ an\ document\ image = \sum_{i=1}^{n} rc_i$$

where, $r_1, r_2, \ldots r_n$ represent the run counts.

The aim is to find the entropy for every portion of an image with reference to t. For instance, if $t = 6$ (Fig. 1: top-down approach), then the number of portions would be $(w \times h)/t$, which is equal to 30 as depicted in Fig. 4. In this case, the entropy value would be zero (0) for the portions $p_1, p_2, p_4, p_{10}, p_{28}$ and p_{29}, because no occurrences of transition could be witnessed.

Fig. 4 Selection of window size t results with 30 portions (From Fig. 1)

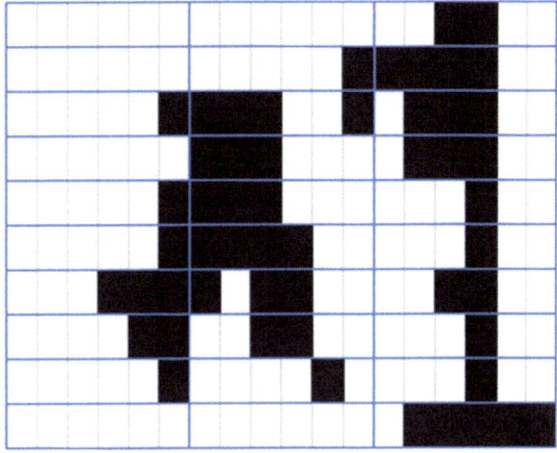

3.2 Hypothesis for the Top-Down Approach: An Example

With reference to the examples illustrated (Figs. 2 and 5) in previous section considering window size $t = 6$ and the first run $rc_1 = 14$, here $t < rc_1$ and $2 < \left(\frac{rc_1}{t}\right) < 3$. This infers that the entropy values for the first two portions namely p_1 and p_2 remain zero. It also infers a transition takes place in the portion p_3. Next, consider $rc_2 = 2$, where $rc_1 + rc_2 = 16$, and $2 < \frac{(rc_1 + rc_2)}{t} < 3$. In this case, there is one more transition in the portion p_3 could be found. Therefore, total entropy for the portion p3 would be 2. In the sequence, considering $rc_3 = 13$ the cumulated runs result with $rc_1 + rc_2 + rc_3 = 29$ and $4 < \frac{(rc_1 + rc_2 + rc_3)}{t} < 5$. Therefore, no transition in portion p_4 could be witnessed and there exists one transition in portion p_5. Next, consider $rc_4 = 5$, where $5 < \frac{(rc_1 + rc_2 + rc_3 + rc_4)}{t} < 6$ and therefore a transition takes place in the

Fig. 5 Selection of window size t results in 36 portions (From Fig. 2)

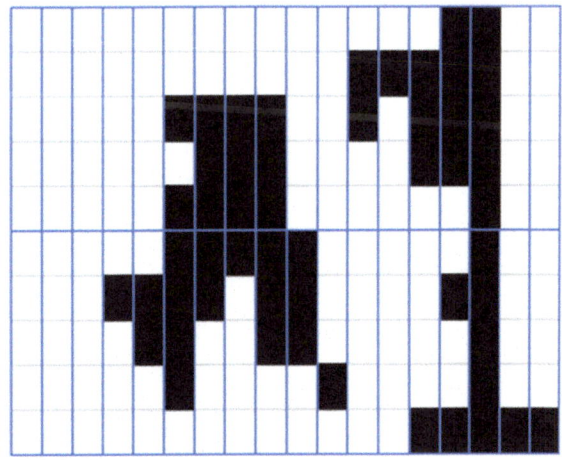

portion p_6. Similarly, entropy values can be calculated for every portion taking a run count at a time. It can be generalized as follows

$$Entropy\ in\ Portion\ p_{j+1} = p_j - f, where, j < \frac{\sum_{i=1}^{f \le n} rc_i}{t} < (j+1)$$

where $p_0 = 0, f = 0$.

$$Where\ p_0 = 0, f = 0$$

The bottom-up approach uses the same model for entropy analysis irrespective of orders.

3.3 Column-Wise Approach: Entropy Computation

CCITT Group3 and Group4 streamed runs adhere row-wise ordering. However, to show the adaptability of the proposed model, column-wise ordering is considered for experimentation. Consider odd and even positioned run counts as background and foreground. Here, left and right margins can be computed as

$$left\_margin = rc_1 > c \times w, where\ c \ge 1$$

$$right\_margin = rc_n > c \times w, where\ c \ge 1$$

A non-text-line cannot be determined directly like row-wise approach. However, distribution of the text content could be estimated for every divided portion. For instance, $t = 6$ leads to 36 portions. Heuristically, t could be an average font-size [6] of the document image. Here, $t = h/f$, where $f > 2$. Figure 5 shows an example of selection of t with respect to h which results to 36 portions. Here, portions between p_1 and p_7, both inclusive hold an entropy value zero (0) and thus those indicate the left-margins. Subsequently, the entropy values for the portions p_9, p_{19}, p_{21}, p_{24}, p_{26}, p_{34} and p_{36} also remains zero (0). The technique used in previous section for estimating the entropy for every region can be extended for column-wise approach. Since, $f = 2$, the relative portions with zero entropy values are considered for text-line segmentation. In this case, portions p_{19}, p_{21}, p_{24} and p_{26} are connected with 8-neighborhood relationship. A sequence of such connected portions would result in a sequence of non-text region thus segmenting two adjacent text-lines. The same applies for row-wise ordering.

$p_{i-(w/t)-1}$	$p_{i-(w/t)}$	$p_{i-(w/t)+1}$	$p_{i-(h/t)-1}$	p_{i-1}	$p_{i+(h/t)-1}$
p_{i-1}	p_i	p_{i+1}	$p_{i-(h/t)}$	p_i	$p_{i+(h/t)}$
$p_{i+(w/t)-1}$	$p_{i+(w/t)}$	$p_{i+(w/t)+1}$	$p_{i-(h/t)+1}$	p_{i+1}	$p_{i+(h/t)+1}$

Fig. 6 8-connectivity for row-wise column-wise ordering

3.4 Establishment of Connected Component Between Regions for Text-Line Segmentation

There are $(w \times h)/t$ portions in an image irrespective of ordering directions. Entropy values are estimated for each portion p_i, that is $e_i = f(p_i)$. Figure 6 shows the 8-connectivity relationships with respect to the portion p_i for row-wise and column-wise orders. Whenever $e_i = 0$, the connectivity can be established with regard to its 8-neighbors. A monotonically increasing sequence of such connectivity between the portions results in text-line segmentation.

4 Illustration

An algorithm for entropy computation is given below

Algorithm 1: Entropy Computation
Input: $RLE\ Sequence = \{rc_1, rc_2, rc_3 \dots rc_n\}$
$Given\ Window\ Size\ t\ and\ document\ image\ size = w \times h$
Output: Entropy values e_i for all portions
    1. Initiate $e_i = 0, \forall i$
    2. For all run counts, $i = 1 \to n$
        $k = \left\lceil \frac{rc_i}{t} \right\rceil$ // Ceil
        $e_k = e_k + 1$
    3. Stop

Algorithm 2: Connected Component
Input: $Entropies\ with\ respect\ to\ portions = \{e_1, e_2, e_3 \dots e_n\}$
Output: Connected component
    1. For all $e_i>0\ \ i = 1 \to n$
        Compute 8-connectivity relationships
    2. Stop

The proposed algorithms such as Algorithm 1 and Algorithm 2 require $O(n)+O(n)$ in its worst-case scenario. Here, n represents the number of runs. In comparison, the literatures [3, 7, 8] require $O(h \times w')$ where w' and h represent width and height of an compressed document image (RLE). The proposed method claims that $O(n) < O(h \times w')$ and $\theta(n) < \theta\left(\frac{h \times w'}{2}\right)$ in its worst and average cases, respectively, for enabling text-line segmentation.

4.1 An Example

For the purpose of illustration, a sample document image portion from the ICDAR dataset [11] is considered. Figure 7 shows an example of a document image portion and its corresponding connected component analysis enabling text-line segmentation. However, for implementation purpose, the window size t is modified accordingly to accomplish the requirements. t is chosen empirically as shown in Fig. 8. Further, Fig. 9 shows the comparative study of the compressed and uncompressed version of images (from ICDAR13 dataset).

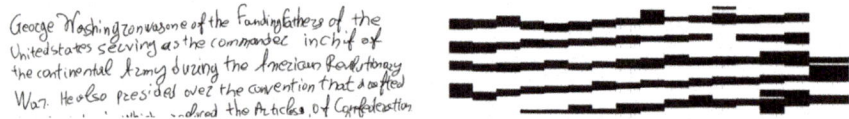

Fig. 7 A portion of a document image [11] and its corresponding segmented portion

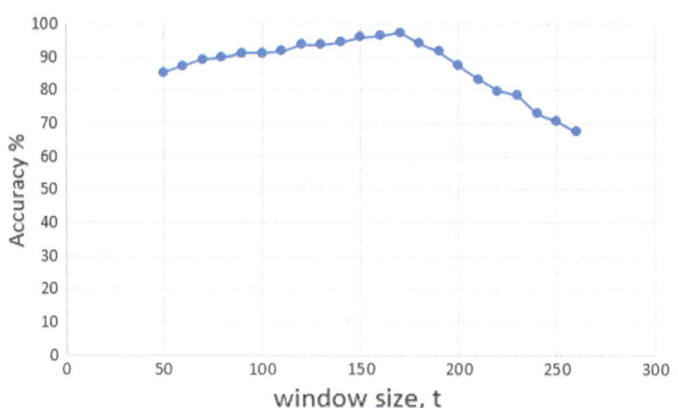

Fig. 8 Accuracy rate by varying t with respect to row-wise ordering (From Fig. 7)

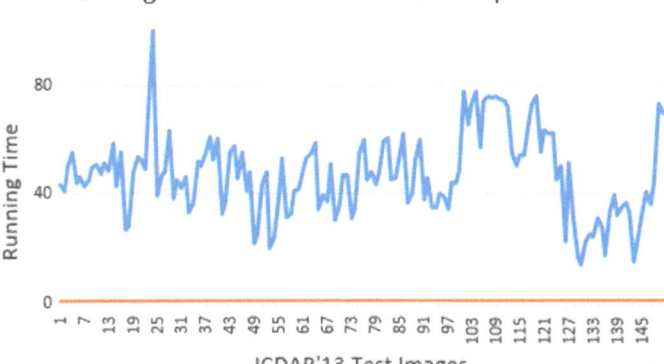

Fig. 9 Time comparison between RLE and Uncompressed document image

4.2   Accuracy and Comparative Analysis

As the proposed work is carried out directly using the RLE of a document image for text-line segmentation, the accuracy rate cannot be compared with the conventional document image analysis. Although, few attempts have been reported for enabling text-line segmentation directly in RLE. Table 2 depicts the efficacy of the proposed model.

5   Discussion and Conclusion

Enabling text-line segmentation directly in run-length encoded document image of unconstrained handwritten document images is presented. Though CCITT Group3 and Group4 prefer horizontal streaming orders, the proposed method could enable segmentation irrespective of streaming orders. Advantages and limitations of the proposed model are listed as follows:

Advantages of the proposed system:

1. The proposed method requires a minimum buffer space and less time for enabling text-line segmentation.
2. Decompression of a document image is completely avoided.
3. Demonstrated the application of both entropy quantifier and incremental learning directly in compressed representation for the text-line segmentation.
4. Proposed connected component analysis in RLE document with respect to divided regions.
5. Proposed algorithm could handle the packets irrespective of streaming orders. However, the packet size depends on the transmission protocols.

Table 2 Comparative Analysis

Authors	Contributions	Accuracy (%)	Worst-case—time	Worst-case—memory
Jeevad et al. [7] (**MARG dataset**)	Extraction of line word character segments directly from run-length compressed **printed text documents**	100	$h \times w'$	$2^{h \times w'}$
Amarnath and Nagabhushan [3] (**ICDAR13 dataset**)	Text line segmentation in the compressed representation of handwritten document using tunneling algorithm	89	$h \times w'$	$2^{h \times w'}$
Proposed model (**ICDAR13 dataset**)	Entropy-driven incremental learning based text-line segmentation model	94.31	n	$2^{f \leq 4}$

Limitations of the proposed system:

1. The proposed system is modeled to enable text-line segmentation in compressed document image.
2. The accuracy rate for enabling text-line segmentation is limited to compressed RLE documents.
3. The window size or packet size is heuristically chosen with experimentation.
4. The packet streaming orders are limited to column-wise or row-wise.

Compressed data processing particularly using RLE should find applications in document analysis and content-based image retrieval in Cloud Computing environment. However, the future scope is to improve the accuracy rate, and then to perform actual text-line segmentation.

References

1. T.4-Recommedation Standardization of group 3 facsimile apparatus for document transmission, terminal equipments and protocols for telematic services, vol. vii, fascicle, vii. 3, Geneva. Technical report (1985)
2. T.6-Recommendation Standardization of group 4 facsimile apparatus for document transmission, terminal equipments and protocols for telematic services, vol. vii, fascicle, vii. 3, Geneva. Technical report (1985)

3. Amarnath, R., Nagabhushan, P.: Spotting separator points at line terminals in compressed document images for text-line segmentation. Int. J. Comput. Appl. **172**(4) (2017)
4. Javed, M., Krishnanand, S.H., Nagabhushan, P., Chaudhuri, B.B.: Visualizing CCITT Group 3 and Group 4 TIFF Documents and Transforming to Run-Length Compressed Format Enabling Direct Processing in Compressed Domain International Conference on Computational Modeling and Security (CMS 2016) Procedia Computer Science 85 213 – 221. Elsevier. (2016)
5. Javed, M., Nagabhushan, P.: A review on document image analysis techniques directly in the compressed domain. Artif Intell Rev. s10462-017-9551-9. Springer Science+Business Media Dordrecht (2017)
6. Gowda, S.D., Nagabhushan, P.: Entropy Quantifiers Useful for Establishing Equivalence between Text Document Images International Conference on Computational Intelligence and Multimedia Applications (ICCIMA 2007)
7. Javed, M., Nagabhushan, P., Chaudhuri, B.B.: Entropy computations of document images in run-length compressed domain. In: Fifth International Conference on Signal and Image Processing (2014)
8. Sindhushree, G.S., Amarnath, R., Nagabhushan, P.: Entropy based approach for enabling text line segmentation in handwritten documents. In: First International Conference on Data Analytics and Learning (DAL), Mysore (2018). In Press Springer, LNNS
9. Preeti M., P. Nagabhushan, P.: Incremental feature transformation for temporal space. Int. J. Comput. **145**(8), Appl. 0975–8887 (2016)
10. https://en.wikipedia.org/wiki/Transmission_Control_Protocol. Accessed from 31 Mar 2018
11. Alaei, A., Pal, U., Nagabhushan, P., Kimura, F.: Painting based technique for skew estimation of scanned documents. In: International Conference on Document Analysis and Recognition (2011)

Diagnosis of Prostate Cancer with Support Vector Machine Using Multiwavelength Photoacoustic Images

Aniket Borkar, Saugata Sinha, Nikhil Dhengre, Bhargava Chinni, Vikram Dogra and Navalgund Rao

Abstract Photoacoustic (PA) imaging is an emerging soft tissue imaging modality which can be potentially used for the detection of prostate cancer. Computer-aided diagnosis tools help in further enhancing the detection process by assisting the physiologist in the interpretation of medical data. In this study, we aim to classify the malignant and nonmalignant prostate tissue using a support vector machine algorithm applied to the multiwavelength PA data obtained from human patients. The performance comparison between two feature sets, one consisting of multiwavelength PA image pixel values and the other consisting of chromophore concentration values are reported. While chromophore concentration values detected malignant prostate cancer more efficiently, the PA image pixels detected the nonmalignant prostate specimens with higher accuracy. This study shows that multiwavelength PA image data can be efficiently used with the support vector machine algorithm for prostate cancer detection.

Keywords Photoacoustic imaging · Prostate cancer · Support vector machine

1 Introduction

Prostate Cancer causes the second-highest number of cancer-induced deaths in the USA. It is the most common cancer after skin cancer. American cancer society estimated that there would be about 164,690 new cases and 29,430 deaths from prostate cancer in 2018 [1]. Although it is a serious disease, the prior discovery of prostate cancer can prevent the death of the patient [2]. Typical screening procedures for prostate cancer consist of prostate-specific antigen (PSA) level testing and digital rectal examination [3]. If any abnormality is found in these preliminary tests, the

A. Borkar (✉) · S. Sinha · N. Dhengre
Department of Electronics and Communication Engineering, Visvesvaraya National Institute of Technology, Nagpur, India
e-mail: aniket996633@gmail.com

B. Chinni · V. Dogra · N. Rao
Department of Imaging Science, University of Rochester Medical Center, Rochester, USA

© Springer Nature Singapore Pte Ltd. 2020
B. B. Chaudhuri et al. (eds.), *Proceedings of 3rd International Conference on Computer Vision and Image Processing*, Advances in Intelligent Systems and Computing 1022,
https://doi.org/10.1007/978-981-32-9088-4_21

patient is referred to transrectal ultrasound (TRUS) guided biopsy procedure. Due to the low sensitivity of ultrasound (US) imaging technique for identifying this cancer, sometimes misleading diagnosis occurs in TRUS guided biopsy [4]. Different US based techniques have been incorporated in the existing US imaging technology to improve the sensitivity metric, but these techniques have not yielded any significant improvement in the detection of prostate cancer [5, 6]. There is a need of a new imaging technology which can detect prostate cancer at an early stage with high sensitivity.

1.1 Photoacoustic Imaging

Photoacoustic (PA) imaging is a hybrid noninvasive soft tissue imaging modality. It is based on the PA effect in which high-frequency US waves are emitted by the soft tissues after their exposure to nanosecond pulse laser radiation [7]. The spatial variation of the optical absorption property is the source of contrast in a PA image of a tissue specimen. PA imaging can perform soft tissue characterization more efficiently that US imaging owing to the dominance of variation of optical property over echogenicity. At the same time, since emitted US waves in PA imaging do not suffer from excessive scattering inside soft tissue like optical waves, the spatial resolution of PA imaging is better than that of pure optical imaging inside soft tissue [8]. Using multiwavelength PA imaging of a soft tissue specimen, important functional information like the concentration of different light-absorbing tissue elements or chromophores, oxygen saturation, etc., can be recovered which can be further used for efficient tissue characterization, e.g., differentiating between malignant and nonmalignant tissue [9, 10]. In multiwavelength PA imaging, PA images of the same tissue specimen are acquired at different wavelengths. These wavelengths are chosen corresponding to the absorption peak of a particular chromophore, suspected to be present in the tissue specimen.

1.2 Computer-Aided Diagnosis for Prostate Cancer

Literature consists of a number of instances of computer-aided diagnosis assisted prostate cancer detection. Mehdi Moradi et al. proposed a new feature based on ultrasound data for computer-aided diagnosis of prostate cancer [11]. Islam Reda et al. diagnosed prostate cancer from diffusion-weighted magnetic resonance imaging (DW-MRI) [12]. Lee et al. did their study with different algorithms such as logistic regression, SVM and ANN classifiers for classification and found that SVM gave the best results among them [13]. Puspanjali Mohapatra et al. used SVM, Naive Bayesian and k-Nearest neighbor classifiers for classification of publicly available biomedical microarray datasets (Prostate cancer). They found that SVM outperformed other classifiers [14]. Botoca, C. et al. found that ANN outperformed logistic regression for predicting prostatic capsule penetration [15]. Dheeb Albashish et al. proposed a new features selection method using SVM recursive feature elimination (RFE) and conditional mutual information (CMI) [16]. Chiu et al. used an ANN model for predicting skeletal metastasis in prostate cancer suspected patients [17]. Chuan-Yu Chang et al. proposed a prostate cancer detection system in dynamic MRIs using SVM classifier to distinguish between malignant and nonmalignant tumor [18]. Çınar et al. used SVM and ANN classifiers for prostate cancer detection with prostate volume, PSA, density and age information as input variables [19].

In this study, we applied the SVM algorithm on multiwavelength PA features for distinguishing malignant prostate tissues from nonmalignant tissues. In Sect. 2, a detailed experimental procedure is provided followed by Sects. 3 and 4 with experimental results and discussion.

1.3 Support Vector Machine

The two-class classification problem using linear model can be expressed as

$$y(x) = W^T \Phi(x) + b \tag{1}$$

where b is a bias parameter, W is a weight vector and y(x) represent the output of model. $\Phi(x)$ represents a fixed feature-space transformation. Let $t_i = \pm 1$ be the ground truth labels corresponding to the training inputs x_i, where the value of $t_i = +1$ denotes the class 1 while $t_i = -1$ denotes the class 2. The labels are assigned to the new input x depending on the sign of discriminant function y(x). If y(x) corresponding to any input x, turns out to be positive, the input x is assigned to class 1, otherwise, it is assigned to class 2. The boundary between regions corresponding to class 1 and class 2 is called the decision boundary of the classifier. The distance of any point x_n and the decision boundary is given by

$$\frac{y(x_n)}{||W||} = \frac{W^T \Phi(x_n) + b}{||W||} \tag{2}$$

The SVM algorithm works toward maximizing this distance along with the constraint of $t_i(W^T \Phi(x) + b)$ being always positive through the optimization of the values of W and b.

2   Method

In the photoacoustic imaging laboratory (Dogra Lab) at University of Rochester Medical Center, a detailed multiwavelength PA imaging study was performed in which 3D PA data from freshly excised human prostate tissue specimens were acquired at five different wavelengths. In the digital images of histology slides, 53 different region of interest (ROI) corresponding to 19 malignant, 8 benign prostatic hyperplasia (BPH) and 26 normal prostate were marked. A total of 30 patients took part in the study. All the details about the PA imaging system, imaging protocol and different wavelengths used can be found in our earlier reports [20–22].

From the 3D PA data generated from a prostate specimen, 2D grayscale C-scan PA images were formed. A C-scan PA image represents the cross-sectional plane of the tissue specimen. For each tissue specimen, five C-scan PA images were generated corresponding to five different wavelengths. Using C-scan PA image pixel values corresponding to different ROIs at five different wavelengths, one category of PA dataset was formed which will be referred to as the category 1 PA dataset in this report. Using the C-scan PA pixel values and absorption spectra of different chromophores, chromophore concentration images were formed. The pixel values of chromophore concentration images represent qualitative values of chromophore concentration in the cross-sectional plane of the tissue specimen. The four different chromophores considered here were deoxyhemoglobin, oxyhemoglobin, lipid and water. Using chromophore concentration image pixel values corresponding to different ROIs for four different chromophores, one category of PA dataset was formed which will be referred as the category 2 PA dataset. The details of the chromophore concentration algorithm can be found in one of our earlier report [23].

The 53 ROIs, identified from the PA data, generated a total of 807 data vectors of which 398 data vectors belonged to malignant ROIs, 133 belonged to normal ROIs and the rest 276 belonged to BPH ROIs. Thus, leading to the dimensionality of category 1 PA and category 2 PA datasets as 807×5 and 807×4, respectively. The classification performance of SVM algorithm was evaluated on the above mentioned two PA datasets. For each PA data category, we divided the input dataset into two separate and independent groups namely training set and testing set. 70% of total data was assigned to the training set and the remaining 30% was assigned to the testing set. Assignment of individual data vector to training and testing set was completely random. To select the most efficient kernel for our dataset, we implemented SVM with different kernel functions (Gaussian, polynomial and radial basis function (RBF)). RBF kernel was found to outperform other kernels for our training dataset. We used the RBF kernel with C value equal to 100 and gamma value equal to 1.5. Once training was done, the performance of the trained SVM algorithm was tested using the testing set. Different performance metrics like sensitivity, specificity, and accuracy were evaluated from the confusion matrices computed after testing as well as a training procedure for each data category. Receiver operating characteristic (ROC) curves and area under ROC (AUROC) were computed for malignant prostate class after testing for both the categories.

3 Results

Tables 1 and 2 depicts the quantitative evaluation of classification performance for category 1 and category 2 PA dataset, respectively. Figure 1 shows the ROC plot for the classification of malignant prostate cancer for category 1 PA data. The AUROC obtained for category 1 PA data was 0.7197. Figure 2 shows the ROC plot for the classification of malignant prostate cancer for category 2 PA data. The AUROC obtained for this category was 0.8296.

Table 1 Classification performance of the category 1 PA dataset

Metrics	After training (%)	After testing (%)
Sensitivity	96.53	66.45
Specificity	90.84	81.39
Accuracy	93.46	71.78

Table 2 Classification performance of the category 2 PA dataset

Metrics	After training (%)	After testing (%)
Sensitivity	97.00	84.21
Specificity	93.31	81.88
Accuracy	95.05	82.98

Fig. 1 ROC plot for classification (category 1 PA data). Obtained AUROC = 0.7197

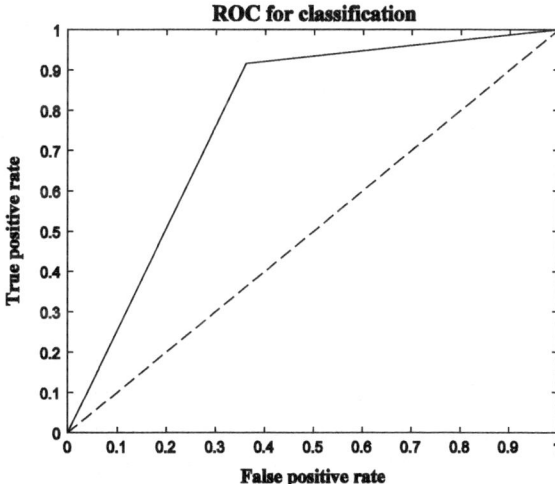

Fig. 2 ROC plot for classification (category 2 PA data). Obtained AUROC = 0.8296

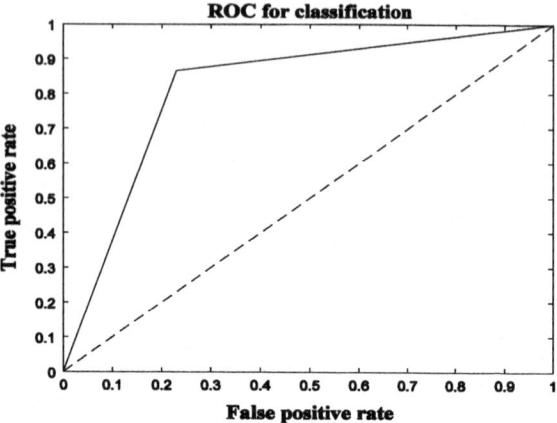

4 Discussion

From the results, it is evident that category 2 outperformed category 1 in terms of AUROC and sensitivity, which means that category 2 was better in malignant prostate cancer detection. Specificity for category 1 after testing is 81.39% while for category 2 this value reads 81.88%. Thus, both categories were equally efficient in terms of specificity. Accuracy for category 1 and category 2 after testing was 71.78% and 82.98%, respectively. Thus, in terms of accuracy category 2 outperformed category 1. As mentioned earlier, category 1 PA data contain the grayscale PA image pixel values corresponding to five different wavelengths while category 2 PA data contain pixel values of the parametric images formed using the spatially varying concentration of four different chromophores. While the category 1 PA data represent the qualitative variation of optical absorption property inside a tissue specimen, category 2 PA data represents the qualitative variation of chromophore concentration. As the correlation between chromophore concentration and tissue pathology is greater than that of optical absorption property and tissue pathology, category 2 PA data are expected to provide more efficient cancer detection than category 1 PA data.

5 Conclusion

The initial results of CAD for prostate cancer detection using photoacoustic imaging suggests that PA imaging can be successfully employed for prostate cancer detection. Fine-tuning of the SVM hyperparameters may yield better accuracy in prostate cancer detection.

References

1. https://www.cancer.org/cancer/prostate-cancer/about/key-statistics.html. Accessed 7 Mar 2018
2. http://www.cancer.org/acs/groups/cid/documents/webcontent/003182-pdf.pdf. Accessed 7 Mar 2018
3. Halpern, E.J.: Contrast-enhanced ultrasound imaging of prostate cancer. Rev. Urol. **8**(Suppl 1), S29 (2006)
4. Presti Jr., J.C.: Prostate biopsy: current status and limitations. Rev. Urol. **9**(3), 93 (2007)
5. Brock, M., von Bodman, C., Palisaar, R.J., Löppenberg, B., Sommerer, F., Deix, T., Noldus, J., Eggert, T.: The impact of real-time elastography guiding a systematic prostate biopsy to improve cancer detection rate: a prospective study of 353 patients. J. Urol. **187**(6), 2039–2043 (2012)
6. Yi, A., Kim, J.K., Park, S.H., Kim, K.W., Kim, H.S., Kim, J.H., Eun, H.W., Cho, K.S.: Contrast-enhanced sonography for prostate cancer detection in patients with indeterminate clinical findings. Am. J. Roentgenol. **186**(5), 1431–1435 (2006)
7. Beard, P.: Biomedical photoacoustic imaging. Interface Focus, pp. 602–631 (2011)
8. Valluru, K.S., Chinni, B.K., Rao, N.A., Bhatt, S., Dogra, V.S.: Basics and clinical applications of photoacoustic imaging. Ultrasound Clin. **4**(3), 403–429 (2009)
9. Wang, L.V.: Photoacoustic Imaging and Spectroscopy. CRC Press, Boca Raton (2009)
10. Hu, S., Wang, L.V.: Photoacoustic imaging and characterization of the microvasculature. J. Biomed. Opt. **15**(1), 011101 (2010)
11. Moradi, M., Abolmaesumi, P., Isotalo, P.A., Siemens, D.R., Sauerbrei, E.E., Mousavi, P.: Detection of prostate cancer from RF ultrasound echo signals using fractal analysis. In: 28th Annual International Conference of the IEEE on Engineering in Medicine and Biology Society, 2006. EMBS'06, pp. 2400–2403. IEEE (2006)
12. Reda, I., Shalaby, A., Khalifa, F., Elmogy, M., Aboulfotouh, A., El-Ghar, M.A., Hosseini-Asl, E., Werghi, N., Keynton, R., El-Baz, A.: Computer-aided diagnostic tool for early detection of prostate cancer. In: 2016 IEEE International Conference on Image Processing (ICIP), pp. 2668–2672. IEEE (2016)
13. Lee, H.J., Hwang, S.I., Han, S.m., Park, S.H., Kim, S.H., Cho, J.Y., Seong, C.G., Choe, G.: Image-based clinical decision support for transrectal ultrasound in the diagnosis of prostate cancer: comparison of multiple logistic regression, artificial neural network, and support vector machine. Eur. Radiol. **20**(6), 1476–1484 (2010)
14. Mohapatra, P., Chakravarty, S.: Modified PSO based feature selection for microarray data classification. In: Power, Communication and Information Technology Conference (PCITC), 2015 IEEE, pp. 703–709. IEEE (2015)
15. Botoca, C., Bardan, R., Botoca, M., Alexa, F.: Organ confinement of prostate cancer: neural networks assisted prediction. In: International Conference on Advancements of Medicine and Health Care through Technology, pp. 287–290. Springer (2009)
16. Albbish, D., Sahran, S., Abdullah, A., Adam, A., Shukor, N.A., Pauzi, S.H.M.: Multi-scoring feature selection method based on SVM-RFE for prostate cancer diagnosis. In: 2015 International Conference on Electrical Engineering and Informatics (ICEEI), pp. 682–686. IEEE (2015)
17. Chiu, J.S., Wang, Y.F., Su, Y.C., Wei, L.H., Liao, J.G., Li, Y.C.: Artificial neural network to predict skeletal metastasis in patients with prostate cancer. J. Med. Syst. **33**(2), 91 (2009)
18. Chang, C.Y., Hu, H.Y., Tsai, Y.S.: Prostate cancer detection in dynamic MRIs. In: 2015 IEEE International Conference on Digital Signal Processing (DSP), pp. 1279–1282. IEEE (2015)
19. Çinar, M., Engin, M., Engin, E.Z., Ateşçi, Y.Z.: Early prostate cancer diagnosis by using artificial neural networks and support vector machines. Expert Syst. Appl. **36**(3), 6357–6361 (2009)
20. Sinha, S., Rao, N.A., Chinni, B.K., Dogra, V.S.: Evaluation of frequency domain analysis of a multiwavelength photoacoustic signal for differentiating malignant from benign and normal prostates: ex vivo study with human prostates. J. Ultrasound Med. **35**(10), 2165–2177 (2016)

21. Sinha, S., Rao, N., Chinni, B., Moalem, J., Giampolli, E., Dogra, V.: Differentiation between malignant and normal human thyroid tissue using frequency analysis of multispectral photoacoustic images. In: 2013 IEEE on Image Processing Workshop (WNYIPW), Western New York, pp. 5–8. IEEE (2013)
22. Sinha, S., Rao, N.A., Valluru, K.S., Chinni, B.K., Dogra, V.S., Helguera, M.: Frequency analysis of multispectral photoacoustic images for differentiating malignant region from normal region in excised human prostate. In: Medical Imaging 2014: Ultrasonic Imaging and Tomography, vol. 9040, p. 90400P. International Society for Optics and Photonics (2014)
23. Dogra, V.S., Chinni, B.K., Valluru, K.S., Joseph, J.V., Ghazi, A., Yao, J.L., Evans, K., Messing, E.M., Rao, N.A.: Multispectral photoacoustic imaging of prostate cancer: preliminary ex-vivo results. J. Clin. Imaging Sci. 3 (2013)

Detecting Face Morphing Attacks with Collaborative Representation of Steerable Features

Raghavendra Ramachandra, Sushma Venkatesh, Kiran Raja and Christoph Busch

Abstract Passports have used face characteristics to verify and establish the identity of an individual. Face images provide high accuracy in verification and also present the opportunity of verifying the identity visually against the passport face image if the need arises. Morphed image-based identity attacks are recently shown to exploit the vulnerability of passport issuance and verification systems, where two different identities are morphed into one image to match against both images. The challenge is further increased when the properties in the digital domain are lost after the process of print and scan. This work addresses such a problem of detecting the morphing of face images such that the attacks are detected even after the print and scan process. As the first contribution of this work, we extend an existing database with 693 bonafide and 1202 morphed face images with the newly added of 579 bonafide and 1315 morphed images. We further propose a new approach based on extracting textural features across scale-space and classifying them using collaborative representation. With a set of extensive experiments and benchmarking against the traditional (non-deep-learning methods) and deep-learning methods, we illustrate the applicability of the proposed approach in detecting the morphing attacks. With an obtained Bonafide Presentation Classification Error (BPCER) of 13.12% at Attack Presentation Classification Error Rate (APCER) of 10%, the use of the proposed method can be envisioned for detecting morph attacks even after print and scan process.

Keywords Biometrics · Face morphing · Spoofing attacks

R. Ramachandra (✉) · S. Venkatesh · K. Raja · C. Busch
Norwegian University of Science and Technology (NTNU), Gjøvik, Norway
e-mail: raghavendra.ramachandra@ntnu.no

S. Venkatesh
e-mail: sushma.venkatesh@ntnu.no

K. Raja
e-mail: kiran.raja@ntnu.no

C. Busch
e-mail: christoph.busch@ntnu.no

© Springer Nature Singapore Pte Ltd. 2020
B. B. Chaudhuri et al. (eds.), *Proceedings of 3rd International Conference on Computer Vision and Image Processing*, Advances in Intelligent Systems and Computing 1022,
https://doi.org/10.1007/978-981-32-9088-4_22

1   Introduction

Biometrics-based access to the restricted services is widely deployed in applications that include border control, smartphone unlocking, national identity card management, forensics identification among many others. Amongst several biometric characteristics, face characteristics are widely used due to nonintrusive nature of capture and user convenience. Face biometrics in the border control applications has carved a niche in automatic border crossing gates due to the feasibility of having the face biometric data on passport and match against the image in ABC kiosk. However, the same benefit has been exploited to make the biometric systems vulnerable through the use of seamless morphing attacks [2]. Morphing attacks employ the process of combing the face images from two different subjects to generate one composite image which matched to both subjects visually and computationally. The challenge is pressing due to the fact that passport application process in many countries relies on the photo provided by applicant. The process leaves the loophole through which the applicant can submit morphed image such that a person with a criminal background will be able to avail the passport. The process can be scrutinized by the supervision of qualified professionals, but the recent study has established that a sufficient quality morphed images can challenge the human observers including border guards to detect the subtle differences in morphed image [3, 15].

With the backdrop of studies, the detection of morphed face image has not only gained the momentum from both academia but also from the practitioners in industry who face the challenge in deployed systems. A number of techniques have been proposed in the recent years to detect the morphing attacks which can be widely classified in two main classes: (1) Single image based: given an image, it is classified either as morph or bonafide. (2) Differential image based: given two images, one from kiosk and one document image, the probability of the submitted image being morphed is measured. The former class can be independent of live capture through kiosks, while the latter is dependent of kiosk and therefore suited for Automatic Border Control (ABC) gates, where the captured images can be compared with the image available in a passport to make the final decision.

The techniques further proposed under each of the abovementioned categories can be of two types: (1) Digital images: The digital version of the morphed and bonafide images are used. (Countries like New Zealand, Ireland and Estonia still use the digital photograph to renew the passport, the morphed detection on digital images are pursued in the literature.) (2) Print-Scanned: The scanned version of the printed images (either morphed or bonafide). This represents a large-scale of real-world use-case as the majority of the countries still accept the passport photo which is then scanned and incorporated in passport. It has to be further noted that most of the reported works have focused on detecting the attacks with digital images based on (a) Engineered texture features (b) Deep learning features and (3) Image degradation features. Table 1 presents the overview of the face morphing detection algorithms.

Observing from Table 1, it can be noted that the majority of the reported work is based on the digital images and also on the single image based approach. The

Table 1 State-of-the-art face morphing attack detection algorithms

Authors	Algorithm(s)	Type of database	Type of approach
Raghavendra et al. [12]	Local binary patterns (LBP)	Digital database	Single image based
	Binarised statistical image features (BSIF)		
	Image gradient magnitude (IG)		
	Local phase quantitation (LPQ)		
Makrushin et al. [8]	Quantized DCT coefficients	Digital database	Single image based
Hildebrandt et al. [6]	StirTrace technique	Digital database	Single image based
Neubert [10]	Image degradations	Digital database	Single image based
Seibold et al. [17]	Deep CNN (VGG)	Digital database	Single image based
Asaad et al. [1]	Topological representation of LBP	Digital database	Single image based
Scherhag et al. [16]	Texture and frequency methods	Digital and print-scanned	Single image based
Raghavendra et al. [13]	Deep learning (VGG and AlexNet)	Digital and print-scanned	Single image based
Raghavendra et al. [14]	Color textures	Print-scanned database	Single image based
Ferrara et al. [4]	De-morphing	Digital database	Differential image based

popular feature extraction techniques are based on the texture features that include LBP, BSIF, and LPQ [1, 12, 16]. Further, the image degradation methods based on JPEG compression [10] and DCT features [8] are also explored on the digital images. Further, the methods on using pretraining deep CNN's especially on using the VGG architecture [17] and fusion of features from VGG and Alex CNN architecture [13] are also studied on the digital images. Experimental results described in [13] has indicated better performance when compared to that of the texture based methods, especially with the low-quality (or highly compressed) morph images. The color texture features [14] are explored on the high-quality morphed face images. Recently, the first method on the differential image based approach using face de-morphing technique was proposed in [4]. Experiments are carried out on the digital high-quality face morph shows interesting and promising results. The key findings from the set of available works are

– The majority of the works are focused on detection of a digital version of the face morphed image, especially with the single image based approach.
– Among the available techniques, the texture-based methods are widely employed.

- Deep learning techniques based on the pretrained network are explored only with VGG and AlexNet architecture.
- The differential morphed face detection based on the de-morphing technique in the high-quality digital morphed images are explored. However, the robustness of such methods is to be demonstrated with different resolution data with real-life noise (illumination and shadow) that are commonly encountered with the ABC systems.

In this paper, we present a novel approach to detect the morphed face images by exploring the collaborative representation of the steerable texture features computed from the luminance component of the face image. To this extent, the proposed method first extracts the luminance component of the given face image. Then texture features are extracted by employing the Steerable pyramids with different scales and orientation which is then classified using the Collaborative Representation Classifier (CRC) to make the final decision as either morph or bonafide. Thus, the following are the main contributions of this paper:

- Presents a new method for detecting the morphed face images based on the collaborative representation of the scale-space features from the luminance component of the given face image.
- Face morphing database available in [14] is further extended such that it has 2518 morphed images and 1273 bona fide images. All the images in the database are print-scanned to reflect the real-life passport issuance procedure. Thus, the final database is comprised of 3791 face images that have resulted in the largest morphed face database in the literature.
- Extensive experiments are carried out by comparing the proposed method with 7 different deep learning methods and 5 different hand-crafted feature based methods. The quantitative performance of the proposed method together with 12 different algorithms are presented using the ISO/IEC metrics for Morphing Attack Detection (MAD).

The rest of the paper is organized as follows: Sect. 2 describes the proposed method, Sect. 3 presents the experimental results of the proposed method compared with both deep learning and non-deep-learning (or hand-craft feature) methods on the newly extended face morphing database. Section 4 draws the conclusion and lists out the key remarks.

2 Proposed Method

Figure 1 shows the block diagram of the proposed approach that can be structured into four functional blocks: (1) extract luminance component (2) extract scale-space features (3) extract high-frequency components from scale-space features (4) classification using Collaborative Representation Classifier (CRC). Given the face image,

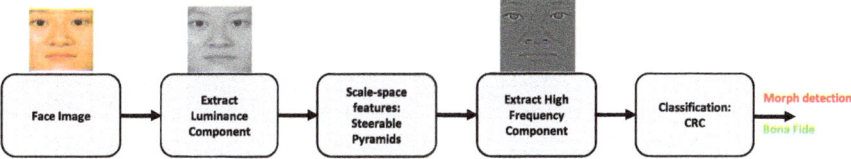

Fig. 1 Block diagram of the proposed method

we first extract the face region using Viola-Jones face detector which is further processed to correct the translation and rotation errors to get the normalized face image I_f of size $250 \times 250 \times 3$ pixels. We then compute the luminance component L_f corresponding to I_f as follows:

$$L_f = K_R.R + K_G.G + K_B.B \tag{1}$$

where, K_R, K_G and K_B denotes three defined constants that are derived from the definition of corresponding RGB space as mentioned in [11].

In the next step, we obtain the scale-space features using Steerable pyramid [5] which are basically a set of oriented filters that are synthesized as a linear combination of the basis function. In this work, we use the steerable pyramid based scale-space features by considering its rotation and translation invariance property which can effectively capture the texture information. It is our assertion that the use of extracted texture features can capture the discontinuity or the degradation in the image in the morphed face image is clearly reflected. Given the luminance component of the face image L_f, corresponding steerable pyramid representation $P_{m,n}(x, y)$ can be obtained as follows:

$$Pm, n(x, y) = \sum_x \sum_y L_f(x, y) D_{m,n}(x - x1, y - y1) \tag{2}$$

where, $D_{m,n}$ denotes the directional bandpass filters at stage $m = 0, 1, 2, \ldots, S1$, and orientation $n = 0, 1, 2, \ldots, K1$.

Figure 2 illustrates features obtained from the steerable pyramid with three different levels, each of which shows the texture features extracted from eight different orientations. In the next step, high-pass residual band components are extracted and they can effectively represent the distortion in the image useful for detecting the morphed face image. Let the high-pass residual band corresponding to the luminance face image L_f be H_f. Figure 3 shows the qualitative results of the proposed method on both bona fide (see Fig. 3a) and morphed face image respectively. It can be observed from Fig. 3 that the high-frequency residual features extracted using steerable pyramids from the luminance component of the image can show the visual differences between the bonafide and morphed face images. This justifies the intuition of the proposed method to effectively capture the useful texture information to aid us in detecting the morphed face image.

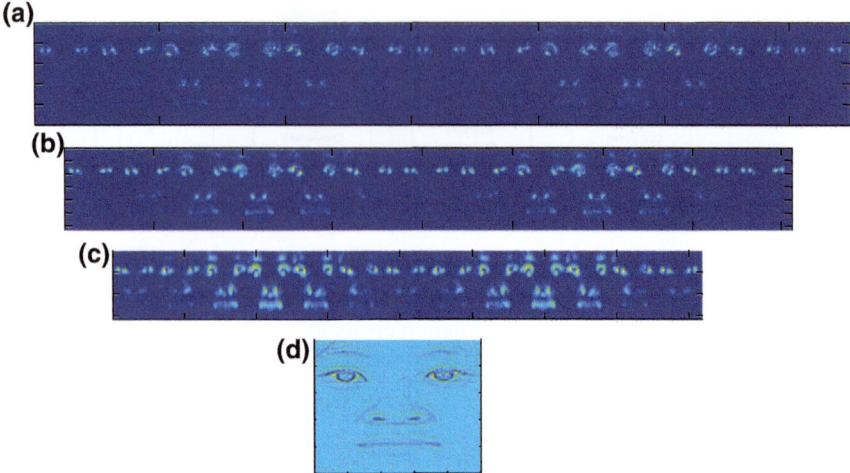

Fig. 2 Qualitative illustration of the scale-space features extracted using steerable pyramid, **a** level 1, **b** level 2, **c** level 3, **d** high pass residual band

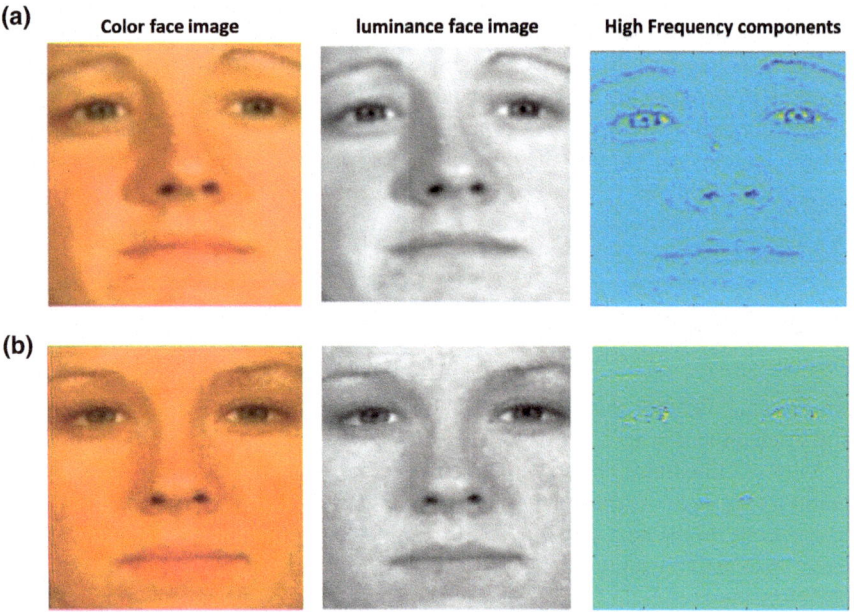

Fig. 3 Qualitative illustration of the proposed method, **a** bona fide face image, **b** morphed face image

To effectively classify the features at various scale-space, in this work, we have employed the Collaborative Representation Classifier (CRC) [18]. The extracted feature from the training set of the database is learned in a collaborative subspace ρ_λ and the final classification scores are obtained using regularized Least Square Regression coefficients on the learned spectral feature vectors against the test sample image, which is explained mathematically as follows:

$$D = argmin_\beta \left\| H_f - \rho_\lambda \beta \right\|_2^2 + \sigma \left\| \beta \right\|_2^2 \qquad (3)$$

3 Experiments and Results

Experiments presented in this work are carried out on the semi-public database available from [14]. In this work, we have extended this database by strictly following the protocol as described in [14]. To make sure that quality of the bonafide and morphed face images is the same so that classier is not biased, we have evaluated the blind image quality measure based on BRISQUE [9]. Figure 4 shows the measured quality values of both bonafide and morphed images are completely overlapped indicating no bias in the image quality of bonafide and morphed images. The extended database comprises of 693 bonafide and 1202 morphed face images in the training set and 579 bonafide and 1315 morphed images in the testing set. Thus, the complete database comprises of 3791 samples.

Experiments are carried out by comparing the performance of the proposed method with 13 different state-of-the-art algorithms that includes both deep learning and non-deep-learning methods. The results are presented using the metrics: *Bonafide Presentation Classification Error Rate (BPCER) and Attack Presentation Classification Error Rate (APCER)* along with the corresponding DET curves (APCER vs. BPCER) as described in ISO/IEC 30107-3 [7]. **BPCER** is defined as proportion of bonafide presentations incorrectly classified as presentation attacks at the attack

Fig. 4 BRISQUE quality distribution on bonafide and morphed face images

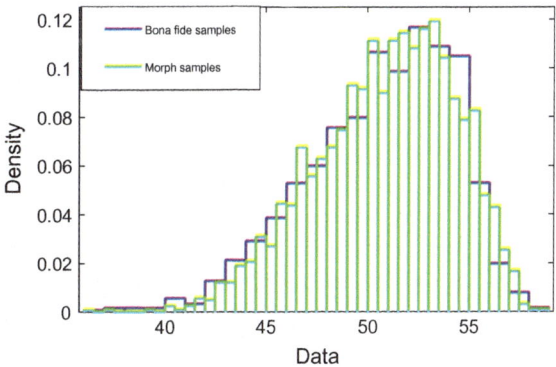

Table 2 Quantitative performance of the proposed method

Algorithm type	Algorithm	BPCER (%) @	
		APCER = 5%	APCER = 10%
Deep learning methods	AlexNet	32.76	23.62
	VGG16	36.38	24.83
	VGG19	33.62	21.55
	ResNet50	32.24	20.17
	ResNet101	30.34	22.59
	GoogleLe Net	41.38	30.86
	Google IceptionV3	73.97	59.83
Non-deep learning methods	LBP-SVM	89.63	75.14
	LPQ-SVM	94.64	82.13
	IG-SVM	68.08	56.47
	BSIF-SVM	96.2	86.87
	Color textures	80.48	51.64
	Proposed method	**45.76**	**13.12**

detection subsystem in a specific scenario while **APCER** is defined as proportion of attack face images incorrectly classified as bonafide images at the attack detection subsystem in a specific scenario. Besides, we also report the performance of the system by reporting the value of BPCER by fixing the APCER to 5 and 10% corresponding to realistic operating values of commercial face recognition systems.

Table 2 indicates the qualitative results of the proposed method together with 13 different state-of-the-art methods. The corresponding DET curves are shown in Fig. 5 for deep learning methods. For the simplicity, we have indicated the DET curves only deep learning methods, however, similar observation in DET curves can also be observed with non-deep learning methods. The deep learning methodology employed in this work is based on fine-tuning the pretrained deep CNN architecture. While performing the fine-tuning, the data augmentation is carried out to avoid the over-fitting of the deep CNN networks. Based on the obtained results, the following can be observed:

- Among the deep CNN architectures, the best performance is noted with the ResNet50 architecture with, $BPCER = 32.24\%@APCER = 5\%$ and $BPCER = 20.17\%@APCER = 10\%$.
- Among non-deep learning methods, the state-of-the-art technique based on the color textures shows the best performance with, $BPCER = 80.48\%@APCER = 5\%$ and $BPCER = 51.64\%@APCER = 10\%$. However, the similar performance is also noted with other techniques such as: BSIF-SVM and LPQ-SVM.

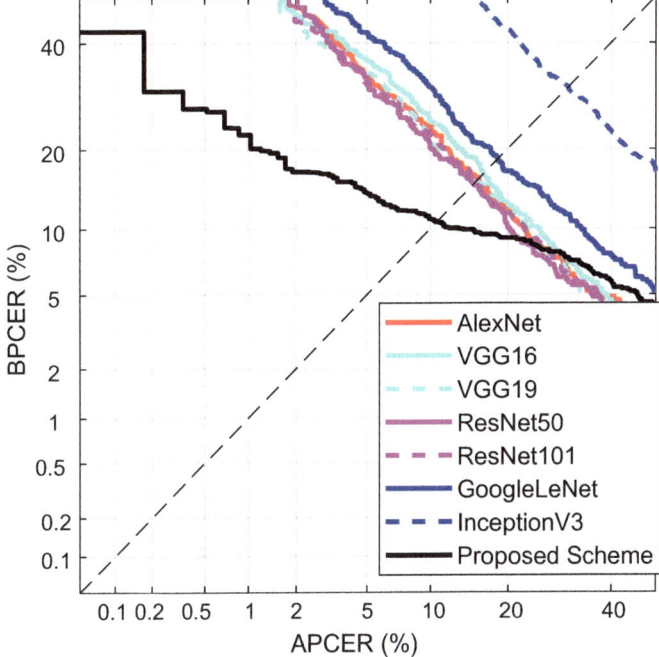

Fig. 5 DET curves indicating the performance of the proposed method with deep learning methods

- Deep learning techniques shows the improved performance when compared to that of non-deep learning techniques.
- The performance of the proposed method shows the best performance with $BPCER = 45.76\%@APCER = 5\%$ and $BPCER = 13.12\%@APCER = 10\%$.

4 Conclusion

The use of face biometrics in passport document for border crossing is known to provide high accuracy while aiding the border guards to verify the identity visually. The recent works have demonstrated the vulnerability of passport systems where a morphed image can be submitted to avail a valid passport that can match with colluding identities. The challenge, therefore, is to detect such morphed images before they are used for passport issuance not only in the digital domain, but also the detecting morphed image after the image has been printed and scanned. In this work, we have presented a new approach to detect such morphed image attacks where the images are first morphed, printed and scanned. The approach has been validated with existing database of 693 bonafide and 1202 morphed face images and with the newly extended database 579 bonafide and 1315 morphed images. The approach based on extracting textural features across scale-space and classifying them using collaborative representation has been experimentally proven effective in detecting the morphing attacks. The proposed approach has provided results with a BPCER of 13.12% at an APCER of 5% exemplifying the applicability in detecting the morphed image attacks after the print and scan process.

Acknowledgements This work was carried out under the funding of the Research Council of Norway under Grant No. IKTPLUSS 248030/O70.

References

1. Asaad, A., Jassim, S.: Topological data analysis for image tampering detection. In: International Workshop on Digital Watermarking, pp. 136–146 (2017)
2. Ferrara, M., Franco, A., Maltoni, D.: The magic passport. In: IEEE International Joint Conference on Biometrics, pp. 1–7 (2014)
3. Ferrara, M., Franco, A., Maltoni, D.: Face recognition across the imaging spectrum. In: On the Effects of Image Alterations on Face Recognition Accuracy, pp. 195–222. Springer International Publishing (2016)
4. Ferrara, M., Franco, A., Maltoni, D.: Face demorphing. IEEE Trans. Inf. Forensics Secur. **13**(4), 1008–1017 (2018)
5. Freeman, W.T., Adelson, E.H., et al.: The design and use of steerable filters. IEEE Trans. Pattern Anal. Mach. Intell. **13**(9), 891–906 (1991)
6. Hildebrandt, M., Neubert, T., Makrushin, A., Dittmann, J.: Benchmarking face morphing forgery detection: application of stirtrace for impact simulation of different processing steps. In: International Workshop on Biometrics and Forensics (IWBF 2017), pp. 1–6 (2017)
7. International Organization for Standardization: Information Technology—Biometric Presentation Attack Detection—Part 3: Testing and Reporting. ISO/IEC DIS 30107-3:2016, JTC 1/SC 37, Geneva, Switzerland (2016)
8. Makrushin, A., Neubert, T., Dittmann, J.: Automatic generation and detection of visually fault-less facial morphs. In: Proceedings of the 12th International Joint Conference on Computer Vision, Imaging and Computer Graphics Theory and Applications—vol. 6: VISAPP, (VISI-GRAPP 2017), pp. 39–50 (2017)
9. Mittal, A., Moorthy, A.K., Bovik, A.C.: No-reference image quality assessment in the spatial domain. IEEE Trans. Image Process. **21**(12), 4695–4708 (2012)
10. Neubert, T.: Face morphing detection: an approach based on image degradation analysis. In: International Workshop on Digital Watermarking, pp. 93–106 (2017)
11. Poynton, C.: Digital Video and HD: Algorithms and Interfaces. Elsevier (2012)
12. Raghavendra, R., Raja, K.B., Busch, C.: Detecting morphed face images. In: 8th IEEE International Conference on Biometrics: Theory, Applications, and Systems (BTAS), pp. 1–8 (2016)
13. Raghavendra, R., Raja, K.B., Venkatesh, S., Busch, C.: Transferable deep-CNN features for detecting digital and print-scanned morphed face images. In: Proceedings of the IEEE Conference on Computer Vision Pattern Recognition Workshops (CVPRW), pp. 1822–1830 (2017)
14. Raghavendra, R., Raja, K., Venkatesh, S., Busch, C.: Face morphing versus face averaging: vulnerability and detection. In: IEEE International Joint Conference on Biometrics (IJCB), pp. 555–563 (2017)
15. Robertson, D., Kramer, R.S., Burton, A.M.: Fraudulent id using face morphs: experiments on human and automatic recognition. PLoS ONE **12**(3), 1–12 (2017)
16. Scherhag, U., Raghavendra, R., Raja, K., Gomez-Barrero, M., Rathgeb, C., Busch, C.: On the vulnerability of face recognition systems towards morphed face attack. In: International Workshop on Biometrics and Forensics (IWBF 2017), pp. 1–6 (2017)

17. Seibold, C., Samek, W., Hilsmann, A., Eisert, P.: Detection of face morphing attacks by deep learning. In: International Workshop on Digital Watermarking, pp. 107–120 (2017)
18. Zhang, L., Yang, M., Feng, X.: Sparse representation or collaborative representation: which helps face recognition? In: IEEE International Conference on Computer Vision (ICCV), pp. 471–478 (2011)

Vehicle Speed Determination and License Plate Localization from Monocular Video Streams

Akash Sonth, Harshavardhan Settibhaktini and Ankush Jahagirdar

Abstract Traffic surveillance is an important aspect of road safety these days as the number of vehicles is increasing rapidly. In this paper, a system is generated for automatic vehicle detection and speed estimation. The vehicle detection and tracking are done through optical flow and centroid tracking using frames of the low-resolution video. The vehicle speed is detected using the relation between the pixel motion and the actual distance. The dataset is taken from the system (Luvizon et al. in A video-based system for vehicle speed measurement in urban roadways. IEEE Trans Intell Transp Syst 1–12, 2016 [1]). The measured speeds have an average error of $+0.63$ km/h, staying inside $[-6, +7]$ km/h in over 90% of the cases. The license plate localization is done on the vehicle detected by extracting the high-frequency information and using the morphological operations. This algorithm can be used to detect the speed of multiple vehicles in the frame. The measured speeds have a standard deviation error of 4.5 km/h, which is higher than that in the system (Luvizon et al. in A video-based system for vehicle speed measurement in urban roadways. IEEE Trans Intell Transp Syst 1–12, 2016 [1]) by 2 km/h. But our algorithm uses a low-resolution video which reduces the processing time of the system.

Keywords Optical flow · License plate localization · Centroid tracking · Speed detection · Morphological operations

A. Sonth (✉) · H. Settibhaktini · A. Jahagirdar
Department of Electrical & Electronics Engineering, BITS Pilani, VidyaVihar Campus, 333031 Pilani, Rajasthan, India
e-mail: f2015265@pilani.bits-pilani.ac.in

H. Settibhaktini
e-mail: s.harsha@pilani.bits-pilani.ac.in

A. Jahagirdar
e-mail: ankush.chandrakant@pilani.bits-pilani.ac.in

© Springer Nature Singapore Pte Ltd. 2020
B. B. Chaudhuri et al. (eds.), *Proceedings of 3rd International Conference on Computer Vision and Image Processing*, Advances in Intelligent Systems and Computing 1022, https://doi.org/10.1007/978-981-32-9088-4_23

1 Introduction

Automatic vehicle speed determination is quite essential due to the number of road accidents occurring every year. This scenario made the researchers think about this problem for a lot of time. In these kinds of problems, the vehicle speed is most often determined by radar guns or speed cameras that work on the concept of Doppler effect [2]. Research has led to several other methods such as stereo vision [3] and inductive loop sensors [1]. Traffic model-based systems have also been developed for highly congested or accident-prone roads [4–6].

Recognizing the license plate is an important part of traffic surveillance. A high-resolution camera is required for this step as we have to detect and recognize text. License plate recognition and localization are mainly done by machine learning techniques. Generally, these machine learning algorithms feed the entire frame as input and the bounding box surrounding the license plate as ground truth to the model. This process is computationally expensive and is dependent on a large amount of training data [7].

In this paper, we present the use of video frames from a low-resolution camera to predict vehicle speed. The dataset used in our model is taken from [1] and is of high resolution. The video frames are therefore resized to lower resolution to increase the processing speed. The dataset used has the camera in the configuration as shown in Fig. 1 [1]. Different motion detection methods such as adaptive background subtraction [8] and optical flow [9] have been tested. Morphological operations [10] are carried out on the detected vehicles to segment them. The detected vehicles are tracked throughout the duration for which they are present in the frame. The speed of these tracked vehicles is estimated by using the frame rate and vehicle duration in the video. License plate localization of the detected vehicles has been done using image processing algorithms [11, 12].

Section 2 of this paper explains the related work which consists of optical flow, vehicle speed detection, and license plate localization. Section 3 discusses our simulations and algorithms used in vehicle detection, tracking, speed estimation, and license plate localization. Section 4 presents the conclusion and future work.

Fig. 1 Camera setup used for recording the video [1]

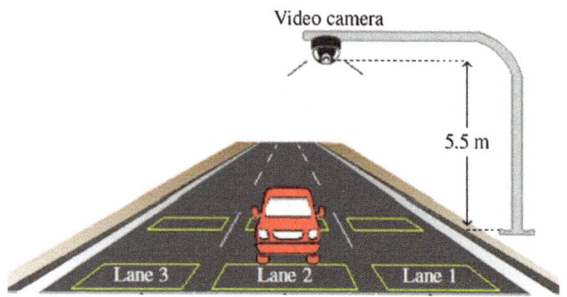

2 Related Work

2.1 Optical Flow

Optical flow is used to detect and measure motion in a video. It gives the magnitude and direction of every pixel in the frame. Optical flow is used when a pixel moves a differential amount of distance in consecutive frames based on the assumption that the brightness and intensity of the moving pixel remain constant. Optical flow has been used in many scenarios such as, to estimate the amount of motion blur [13] from the pixel movement in a frame. It has also been used to predict the future frames with the help of artificial neural networks [14].

The two major optical flow algorithms are Horn–Schunck [15] and Lucas–Kanade [9]. Horn–Schunck method is quite often implemented using a pyramid-based coarse-to-fine scheme to achieve performance similar to the much more recent Lucas–Kanade method. In this paper, we present the use of Lucas–Kanade optical flow algorithm to detect vehicles and segment them.

2.2 Vehicle Speed Determination

Vehicle speed estimation has been accomplished in the past by using several methods such as stereo vision [3], radar-based speed cameras [2], and inductive loop sensors [1]. Stereo cameras are quite expensive and their installation process is quite tedious. Inductive loop sensors-based systems are easily damaged by wear and tear. Radar-based speed cameras are prone to noise and do not differentiate between targets in traffic. There are some traffic model-based systems that are developed for particular road sections and rely on the time of the day to predict vehicle speed [4–6].

Our work is based on a video feed from a low-resolution camera fixed at a particular height and angle above the ground. This method is a stand-alone system and can be used in any situation as long as the camera orientation is the same.

2.3 License Plate Localization

License plate localization is done so that we need not store the entire video frame. Moreover, the localized region can be further used to recognize the text. There are different license plate localization algorithms based on artificial neural networks (ANN) [16] and discrete wavelet transform [17]. ANN based system depends on a large amount of training data. Convolutional neural networks (CNN) based systems such as in [7] have become quite popular due to their high accuracy. The downside of using CNNs is that they require a huge amount of training data and are very expensive computationally.

We present a license plate localization algorithm which is independent of training data and carries out image processing operations to give the candidate regions for the license plate. Our algorithm is independent of the license plate dimensions and orientation.

3 Simulations

Vehicle detection, tracking, and speed estimation have been done with low-resolution video frames, i.e., the original video frames from the dataset [1] were downscaled. This along with license plate localization were performed completely using MAT-LAB. Vehicle speed estimation is a four-step procedure namely, motion detection, vehicle segmentation, vehicle tracking, and speed calculation. For the purpose of license plate localization, we use the original high-resolution video feed. The schematic representation of the proposed work is shown in Fig. 2. We have considered the cases in which vehicle moves in a single lane and the vehicle speed is constant throughout the duration.

3.1 Vehicle Detection

This step demonstrates the advantage of optical flow over background subtraction for localizing the vehicles. Background in a video does not always remain the same. Change in the lighting condition may cause the background to change. In scenarios like if an object is partially present in a frame or remains stationary for a long time, the background changes. Thus, the need for an adaptive background is required. Issues with adaptive background subtraction are vehicles leaving behind a trail in the subtracted image and stationary vehicles taking time to merge into the background. Therefore, we chose to use optical flow.

Lucas–Kanade [9] optical flow method is used on the video frames (see Fig. 3a). This method is based on the assumption that the intensity of a pixel would remain the same after moving a differential amount of distance in a differential amount of time.

The flow vector magnitudes of all the pixels in a frame are averaged to get a dynamic threshold for vehicle detection (see Fig. 3b). Morphological closing is then applied to get a cleaner output [10] (see Fig. 3c). Morphological closing is dilation followed by erosion. Similarly, morphological opening of an image by a structuring element is erosion followed by dilation. The structuring element used is a disk as there are no sharp edges in the silhouette of a vehicle. Bounding boxes are drawn around the candidate regions in the resultant image as in Fig. 4.

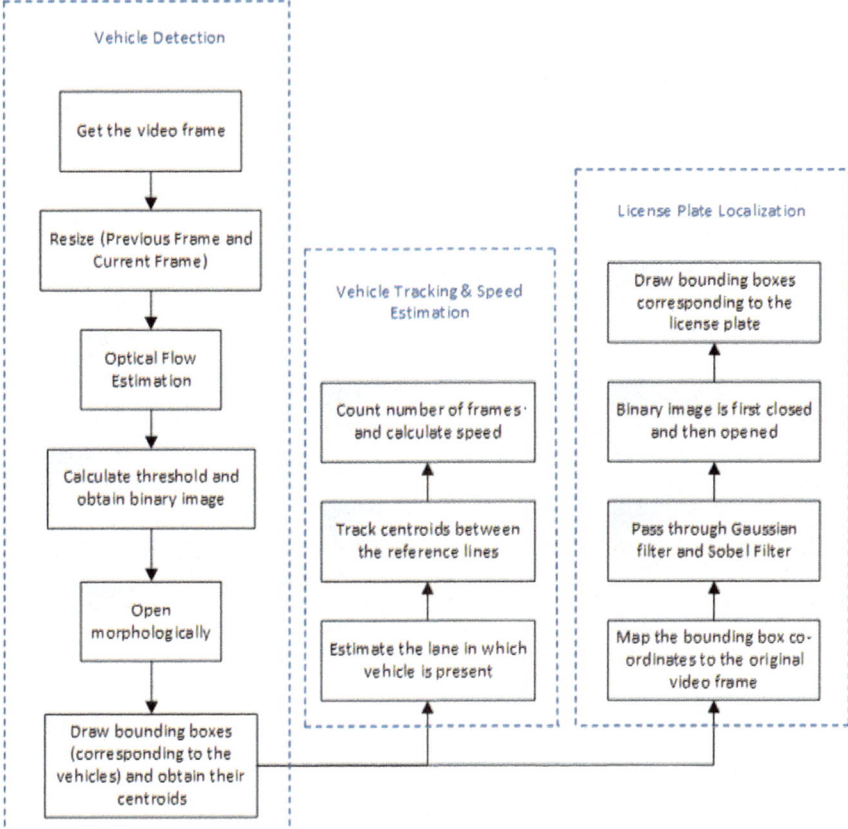

Fig. 2 Schematic diagram of the proposed model

3.2 Vehicle Tracking

Centroids of each bounding box are calculated for the ease of tracking. This means that we just store one centroid coordinate instead of four corner coordinates for every vehicle per frame. This centroid is used to determine the lane in which the vehicle is present. This is done by drawing vertical lines at the midpoint (between the references lines) of every lane marking on the road. In this case, it is considered that there are three lanes as the fourth lane isn't completely visible in the frame. The distance between the centroids of bounding boxes in consecutive frames is calculated for every lane. If the centroid has moved further away from the camera, then the centroid belongs to the same vehicle. Vehicles are tracked in this way and the centroid of each vehicle is stored in a separate matrix for the purpose of speed estimation.

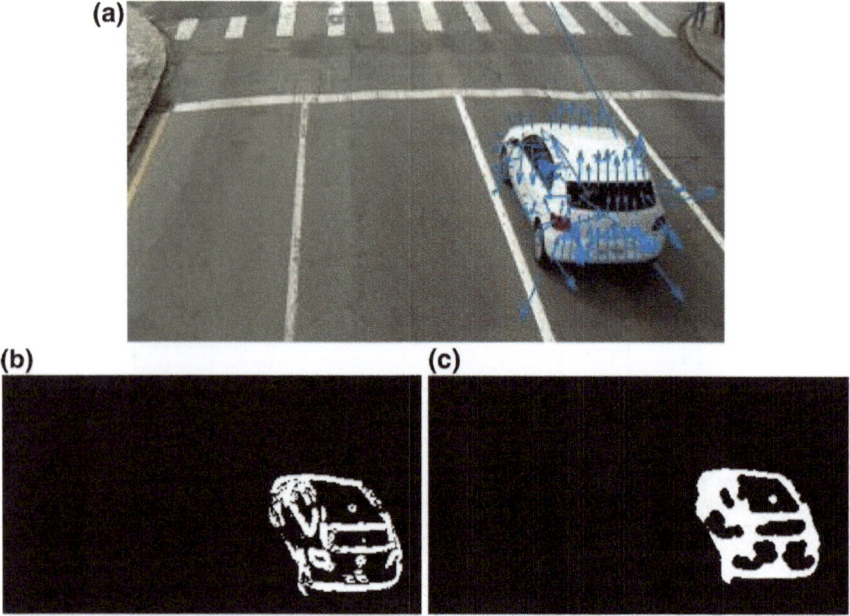

Fig. 3 **a** Results of the Lucas–Kanade optical flow method. **b** Binary image after thresholding. **c** Morphological closed image

Fig. 4 Tracked vehicle is shown in the third lane

3.3 Vehicle Speed Estimation

As the video frame rate is constant, the time between successive frames is fixed. Total time taken by a vehicle in the frame can be calculated by multiplying the number of frames for which it is visible, with the frame rate. The frame rate for the dataset from [1] is 25 frames per second.

$$v = k \times \frac{D}{n} \tag{1}$$

To track the speed, reference start and end lines are marked as shown in Fig. 5. Equation (1) is used for calculating the speed of the vehicle. Here, v is the estimated speed in Km/h, k is a constant factor, D is the Euclidean distance between the first and last centroid present between the reference lines for every vehicle, and n is the number of frames for which a vehicle is present between the reference lines. The constant factor k is unique for every lane, and was calculated by averaging the constant value found for some test vehicles in a lane. The constant factor is different for every lane because the ratio of actual distance in kilometers to that in pixels is different for each lane (as the view of each lane with respect to the camera is different), even if the actual distance moved is same. The resulting formula gives an overall mean error of +0.63 km/h with a standard deviation of 4.5 km/h when compared to the actual speed from the dataset of [1]. Figure 6 shows the error histogram in which the y-axis is the number of vehicles and the x-axis is the error in calculated speed. Although the standard deviation is higher than that in the system [1] by 2 km/h, our algorithm uses a low-resolution video which increases the processing speed of the system. Table 1 presents the tracked centroid coordinates of a vehicle moving in lane 3 with the given ground truth speed as 58 km/h. Our algorithm predicts the vehicle's speed as 60 km/h based on the centroid pixel movement.

Fig. 5 Reference lines for estimating the speed

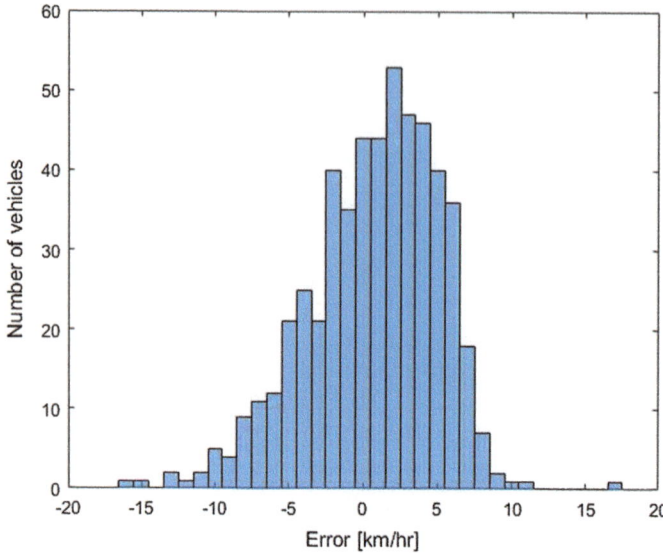

Fig. 6 Error in the speeds calculated. 91% of the error values lie in the range between 6 km/h and 7 km/h

Table 1 The centroid coordinates of a vehicle moving in lane 3 are given below, as they are tracked throughout the duration the centroid is present between the reference lines

S. No	Pixel x-coordinate	Pixel y-coordinate	Speed (pixels/frame)	Speed (pixels/s)
1	264.08011	129.33897	12.2339	305.8480
2	259.56772	117.96764	10.6827	267.0679
3	253.80603	108.97190	10.3623	259.0566
4	248.26517	100.21545	9.8471	246.1778
5	242.72108	92.077347	9.3378	233.4450
6	237.56258	84.293755	8.8759	221.8980
7	233.27232	76.523575	8.6689	216.7220
8	228.68057	69.170662	7.8762	196.9055

3.4 License Plate Localization

License plate localization is required for capturing the information of over-speeding vehicles. License plate localization and recognition is dependent on high-quality video as text recognition comes into the picture. Therefore, bounding box coordinates which we found in the tracking algorithm are mapped to the original video frame. The localization algorithm can be used when a vehicle exceeds the speed limit. Our method is based on the high contrast between the license plate background and text.

The input image is first converted into grayscale and then passed through a high-pass Gaussian filter [11] (see Fig. 7b). This is done so as to preserve the high-frequency content, i.e., the text characters in this case. The resulting image is then passed through a vertical Sobel filter to obtain a binary image (see Fig. 7c). The binary image is then morphologically closed by a 5 × 5 square (see Fig. 7d) and then opened by a 2 × 3 rectangular structural element (see Fig. 7e). Closing was done so that the region between the boundary of text gets filled, and the opening was then done so as to remove lone white pixels and thin lines for better results of localization. Different-sized structural elements were used which worked only for particular license plate dimensions. Bounding boxes are drawn around the final

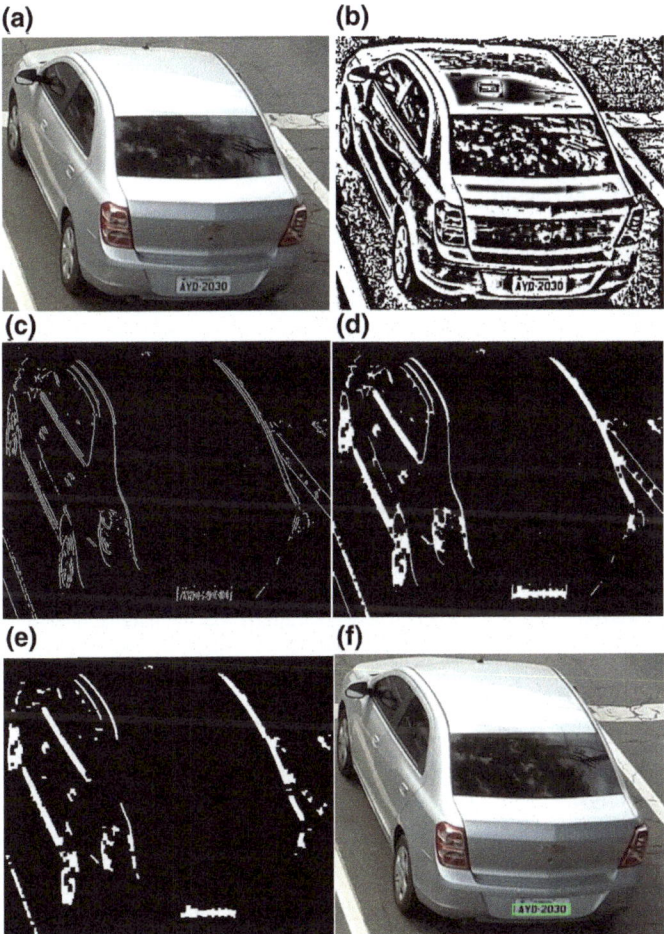

Fig. 7 **a** Input image. **b** Gaussian high-pass filtered output. **c** Edges obtained through vertical Sobel filter. **d** Morphological closing for making the text region denser. **e** Morphological opening for removing thin lines and isolated white pixels. **f** Bounding box drawn around the region of interest

output and their coordinates are mapped on to the original video frame to get the license plate (see Fig. 7f).

4   Conclusion and Future Work

We have shown the advantage of optical flow over background subtraction for motion detection. Although the standard deviation of the error in speed was higher than the dataset [1] by 2 km/h, our model is capable of estimating the speed of multiple vehicles simultaneously from a low-resolution video. Thus, our processing speed compared to them would be higher. We also proposed a license plate localization algorithm which is based on image processing operations.

Future work could involve (1) text recognition on the localized license plate. The localized license plate can be passed through a CNN to improve its resolution and optical character recognition can be used on the output of CNN, (2) alignment of license plates for better text recognition accuracy, and (3) high-speed cameras can be used to get better accuracy for estimated speed.

References

1. Luvizon, D.C., Nassu, B.T., Minetto, R.: A video-based system for vehicle speed measurement in urban roadways. IEEE Trans. Intell. Transp. Syst. 1–12 (2016)
2. Adnan, M.A., Norliana, S., Zainuddin, N.I., Besar, T.B.H.T.: Vehicle speed measurement technique using various speed detection instrumentation. In: IEEE Business Engineering and Industrial Applications Colloquium (BEIAC), pp. 668–672. Langkawi (2013)
3. Jalalat, M., Nejati, M., Majidi, A.: Vehicle detection and speed estimation using cascade classifier and sub-pixel stereo matching. In: International Conference of Signal Processing and Intelligent Systems (ICSPIS), pp. 1–5. Tehran (2016)
4. Park, J, Li, D., Murphey, Y.L., Kristinsson, J., McGee, R., Kuang, M., Phillips, T.: Real time vehicle speed prediction using a neural network traffic model. In: International Joint Conference on Neural Networks, San Jose, California (2011)
5. Jiang, B., Fei, Y.: Traffic and vehicle speed prediction with neural network and hidden Markov model in vehicular networks. In: IEEE Intelligent Vehicles Symposium. COEX, Seoul, Korea (2015)
6. Robert, K.: Video-based traffic monitoring at day and night. Vehicle features detection and tracking. In: International IEEE Conference on Intelligent Transportation Systems. St. Louis, MO, USA (2009)
7. Li, H., Wang P., Shen, C.: Towards end-to-end car license plates detection and recognition with deep neural networks. In: IEEE Transactions on Intelligent Transportation Systems. arXiv preprint arXiv:1709.08828 (2017)
8. Thadagoppula, P.K., Upadhyaya, V.: Speed detection using image processing. In: 2016 International Conference on Computer, Control, Informatics and its Applications (2016)
9. Lucas, B.D., Kanade, T.: An iterative image registration technique with an application to stereo vision. In: Proceedings of the 7th International Joint Conference on Artificial intelligence (IJCAI), vol. 2, pp. 674–679, San Francisco, CA, USA (1981)

10. Kumar, D., Rathour, P.S., Gupta, A.: A model for multiple vehicle speed detection by video processing in MATLAB. Int. J. Res. Appl. Sci. Eng. Technol. (IJRASET) **3**(6) (2015). ISSN: 2321-9653

11. Rashedi, E., Nezamabadi-Pour, H.: A hierarchical algorithm for vehicle license plate localization. Multimed Tools Appl **77**, 2771–2790 (2018)

12. Kim, J.-H., Kim, S.-K., Lee, S.-H., Lee, T.-M., Lim, J.: License plate detection and recognition algorithm for vehicle black box. In: International Automatic Control Conference (CACS), Pingtung, Taiwan (2017)

13. Harshavardhan, S., Gupta, S., Venkatesh, K.S.: Flutter shutter based motion deblurring in complex scenes. In: 2013 Annual IEEE India Conference (INDICON) (2013)

14. Verma, N.K., Gunesh, D.E., Rao, G.S., Mishra, A.: High accuracy optical flow based future image frame predictor model. In: Applied Imagery Pattern Recognition Workshop (AIPR), pp. 1–6. IEEE, Washington, DC (2015)

15. Aslani, S., Mahdavi-Nasab, H.: Optical flow based moving object detection and tracking for traffic surveillance. Int. J. Electr. Comput. Energ. Electron. Commun. Electr. Eng. **7**(9) (2013)

16. Roomi, S.M.M., Anitha, M., Bhargavi, R.: Accurate license plate localization. In: International Conference on Computer, Communication and Electrical Technology (ICCCET), pp. 92–97, Tamil Nadu (2011)

17. Chen, T.: License plate text localization using DWT and neural network. In: International Conference on Granular Computing, pp. 73–77. IEEE, Nanchang (2009)

Retinal-Layer Segmentation Using Dilated Convolutions

T. Guru Pradeep Reddy, Kandiraju Sai Ashritha, T. M. Prajwala,
G. N. Girish, Abhishek R. Kothari, Shashidhar G. Koolagudi
and Jeny Rajan

Abstract Visualization and analysis of Spectral Domain Optical Coherence Tomography (SD-OCT) cross-sectional scans has gained a lot of importance in the diagnosis of several retinal abnormalities. Quantitative analytic techniques like retinal thickness and volumetric analysis are performed on cross-sectional images of the retina for early diagnosis and prognosis of retinal diseases. However, segmentation of retinal layers from OCT images is a complicated task on account of certain factors like speckle noise, low image contrast and low signal-to-noise ratio amongst many others. Owing to the importance of retinal layer segmentation in diagnosing ophthalmic diseases, manual segmentation techniques have been proposed and adopted in clinical practice. Nonetheless, manual segmentations suffer from erroneous boundary detection issues. This paper thus proposes a fully automated semantic segmentation technique that uses an encoder–decoder architecture to accurately segment the prominent retinal layers.

Keywords Retinal layer segmentation · Optical coherence tomography · Dilated convolutions · Deep learning · Retina

1 Introduction

Optical Coherence Tomography (OCT) has gained a lot of medical importance in retinal imaging and diagnosis of several ocular diseases. OCT is a noninvasive technique which uses low coherence infrared light to generate the imagery of retinal structures [10]. The introduction of Spectral Domain OCT (SD-OCT) has led to

T. Guru Pradeep Reddy (✉) · K. S. Ashritha · T. M. Prajwala · G. N. Girish · S. G. Koolagudi
J. Rajan
Department of Computer Science and Engineering, National Institute
of Technology Karnataka, Surathkal, India
e-mail: gurupradeept@gmail.com

A. R. Kothari
Pink City Eye and Retina Center, Jaipur, India

© Springer Nature Singapore Pte Ltd. 2020   279
B. B. Chaudhuri et al. (eds.), *Proceedings of 3rd International Conference on Computer Vision and Image Processing*, Advances in Intelligent Systems and Computing 1022,
https://doi.org/10.1007/978-981-32-9088-4_24

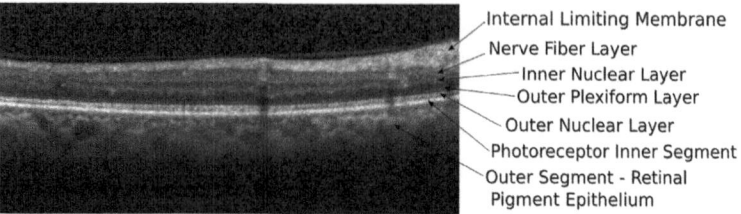

Fig. 1 Sample retinal OCT B-Scan with annotation of different retinal layers

greater imaging speed, greater resolving power and high scan density, which has proven to be beneficial in OCT data acquisition [7].

The SD-OCT cross-sectional scans have a typical resolution of about 6×6 mm and a sufficient depth penetration of about (0.5–2 mm) [1]. These scans have found their applications in skin imaging [16], intra-vascular imaging [2], lumen detection [13], plaque detection [14], and retinal imaging [16]. A few clinical applications of the SD-OCT images in retinal imaging are: detection and identification of retinal fluid, neurosensory detachments, pigment-epithelial detachments, and assessment of choroidal neovascular membranes [6]. Cross-sectional analysis and investigation of OCT scans has significance in the early diagnosis and prognosis of several retinal diseases.

The interpretation of morphological changes of retinal layers from OCT images has significance in detecting several ocular disorders such as age-related macular degeneration, macular edema, and retinal detachment. Figure 1 shows the retinal OCT B-scan with annotation of different layers. Effective segmentation of these layers plays a vital role in diagnosis of retinal diseases.

Segmentation of retinal layers is a challenging task due to constraints posed by the presence of speckle noise, low image contrast, low signal-to-noise ratio, and associated pathologies. Also inherent challenges during the acquisition of OCT images like motion artifacts, blood-vessel shadows and noise make it increasingly difficult to accurately segment the layers and detect the boundaries. But owing to the critical importance of retinal layer segmentation in diagnosing ophthalmic diseases, manual segmentation techniques have been adopted in most of the OCT related studies [3]. However, manual segmentation is exhaustive, tedious and prone to errors. Also, commercial algorithms available for retinal layer segmentation suffer with erroneous boundary and layer detection issues. A lot of effort and time has been invested by several researchers in order to come up with segmentation techniques that are automated and requiring minimal human intervention.

Various approaches to segmenting retinal layer boundaries have been reported with different levels of success. Chiu et al. [4] present an automatic approach for segmenting retinal layers in SD-OCT images using graph theory and dynamic programming (GTDP), which significantly reduced the processing time. The method proposed earlier in [4] for the quantitative analysis of retinal and corneal layer thicknesses is extended in [5] to segment retinal objects like cells and cysts. A quasi-polar domain transform used to map the closed-contour features in the Cartesian domain into lines in the quasi-polar domain, which is used as a basis in this technique. These features of interest are used for segmenting layers using the graph theory approach described in [4].

The authors of [3] have proposed a method based on kernel regression (KR)-classification for the estimation of retinal layer and fluid positions. A fully automated algorithm based on KR and the previously described GTDP framework [4] has been presented for the identification of fluid-filled regions and segmentation of seven retinal layers in SD-OCT images. KR-based classifiers have been created in the initial stages which are then used in the succeeding phases for guiding GTDP to provide more accurate results.

All of the abovementioned algorithms are automatic segmentation approaches. However, they all use hand-designed features. It also needs to be considered that these manually extracted features are not discriminative enough for the differentiation of semantic object boundaries and unpredictable changes in low-level image cues.

Inspired by deep neural networks (DNN), a number of researchers have proposed DNN approaches, which replaced hand-designed features with deep features. Hence, introducing the deep learning techniques into the edge segmentation problem is reasonable and feasible. The authors of [8] present a deep learning based approach to learn discriminative features for layer segmentation. With the learned deep features, they claim that layer segmentation can be implemented by transmitting them into any classifiers. They have employed the computationally efficient random structured forest to classify layers and background. Then, dynamic programming has been performed for the identification of the shortest path from the top pixel of the left column to the bottom pixel of the right column.

Inspired by the results obtained using the deep convolutional neural networks for the purpose of segmentation of retinal layers, the authors of [15] have proposed a novel fully convolutional DNN architecture called ReLayNet, for end-to-end segmentation of retinal layers and fluids from SD-OCT scans. ReLayNet makes use of a contracting path of convolutional blocks (encoders) that learn a hierarchy of contextual features, followed by an expansive path of convolutional blocks (decoders) corresponding to the encoder blocks, for the purpose of semantic segmentation. ReLayNet is trained to optimize on a combined loss function comprising of a weighted logistic loss and Dice loss. It is validated on a publicly available Duke DME dataset and obtained segmentation results were compared against five state-of-the-art retinal layer segmentation techniques. As a result, this is considered as one of the best state-of-the-art approaches in comparison to all the previous approaches used for retinal layer segmentation.

Our paper proposes a fully convolutional network which makes use of dilated convolutions [17] for increasing the receptive field size without compromising on the dimensionality of the images. We have also introduced a weighting scheme that addresses the challenges posed by the segmentation of thin retinal layers, and class imbalance between the retinal layers and the background. The usage of dilated convolutions and the introduction of the weighting scheme has shown significant improvement in performance for the segmentation of the challenging retinal layers in comparison to the ReLayNet model [15].

This paper is organized as follows: Sect. 2 describes the methodology proposed for retinal layer segmentation. Section 3 describes loss functions used in the proposed approach. Section 4 describes how the model has been trained to learn the features using a joint loss and weighting scheme and experimental setup. Section 5 outlines qualitative and quantitative analysis of results. Finally, Sect. 6 concludes the paper.

2 Methodology

The task to be carried out is to assign each pixel of the input OCT Retinal image to a class. Eight classes have been considered for the task of segmentation. This includes seven retinal layer classes and one background class. The proposed architecture is shown in Fig. 2.

The proposed model is symmetric in nature. It is fully convolutional and does not contain any fully connected layers. It is a variant of Encoder–Decoder architecture [15] with the introduction of dilated convolutions and skip connections. The skip connections carry out the task of transfer of features from an encoding block to its corresponding decoding block. The former part of the network is a contracting path consisting of a series of encoder blocks, followed by a series of dilated convolutions.

Dilated convolutions preserve the dimensions of the feature maps. The latter part of the network comprises an expanding path consisting of a series of decoding blocks with skip connections from corresponding encoding blocks. The final classification block computes the probability score of pixels belonging to different classes. The details and significance of each block can be explained as follows.

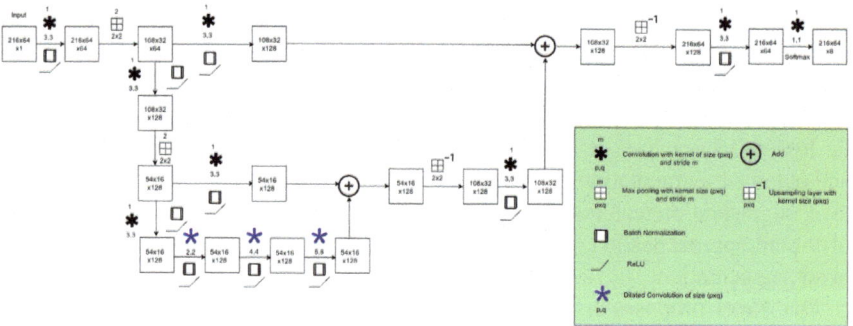

Fig. 2 Dilated retinal layer segmentation network

2.1   Contracting Path

The contracting path comprises of two encoder blocks. Each encoder block consists of 4 components: convolution layer, batch normalization layer, Rectified Linear Unit (ReLU) activation layer and a max-pooling layer. The kernels in the convolution layer are of size [3 × 3]. The number of convolution filters is set to 64 in the first encoder block which is doubled to 128 in the second owing to the reduction in dimensionality of the image as the feature maps are passed from one encoder block to the other. This also ensures a richer encoding of the representations as we move down the hierarchy (the complexity of the features increases as we move down the network).

The feature maps are appropriately zero padded in order to preserve the size of the image after the application of the convolution operation. The next layer is a batch normalization layer which reduces the ill effects of internal co-variate shift, makes learning faster and also provides a regularization effect, which, in turn, helps in prevention of over-fitting the train set [11]. This layer is followed by a ReLU activation layer which introduces non linearity to the network required for the learning of complex mappings. The last layer is a max pooling layer of size [2 × 2], which reduces the dimensionality of the feature maps and increases the field of view.

2.2   Dilation Path

Although the max-pooling layer incorporated in the contracting path increases the field of perception for the kernels, it compromises on the resolution of the feature maps due to the dimensionality reduction. This decrease of resolution is undesirable for the task of semantic segmentation. This is because it requires the usage of a max-pooling layer which in turn, needs a corresponding upsampling layer that learns the complex mapping for an increase in size. Thus, dilated convolutions, also known as Atrous convolution [17] have been introduced to expand the size of the receptive field exponentially without compromising on the resolution. In the proposed model, 3 such dilated convolution layers have been used with dilation rates set to 2, 4, and 8, respectively. This ensures that an exponential increase in the receptive field size is obtained without increasing the number of parameters. The number of filters for all the dilated convolution layers is set to 128.

2.3   Expanding Path

The expanding path consists of two decoder blocks, corresponding to the two encoder blocks. Each decoder block consists of 5 components: skip connection, upsampling layer, convolution layer, batch normalization, and a ReLU activation layer. The skip connections consist of a convolution layer, batch normalization and ReLU activation layer each. These are applied on the output of the encoder and will be added pixel-wise with the output of corresponding decoder. There are two main advantages of using skip connections—they allow transfer of low-level information from lower layers to higher layers in order to simplify the process of learning of the upsampling features and also aid an easier backward flow of gradients, i.e., they provide a direct gateway for the gradient to flow to lower layers which minimizes the problem of vanishing gradient [12]. The next layer is an upsampling layer, which increases the dimensions by repeating the data based on the kernel size, which has been set to 2 in both the blocks. This is followed by a convolution layer with a kernel size of [3 × 3] and zero padding to preserve the dimension of feature maps. The number of convolution filters is set to 128 in the first decoder block and halved to 64 in the second.

2.4 Classification Path

The final part of the network is a classification block, to obtain the probability maps of pixels belonging to eight classes. A convolution layer of kernel size [1 × 1] with 8 filters is applied to the output of the final decoder block followed by a softmax activation layer. Softmax activation layer outputs the probability of a pixel belonging to either of the eight classes (seven different layers and background).

3 Loss Function

The model is optimized on a combined loss function, inspired by [15]. It comprises of the following two loss functions.

3.1 Weighted Multi-class Logistic Loss

In general, cross-entropy measures the similarity between the actual and predicted values in probabilistic terms. Cross-entropy loss, also known as the logistic loss, penalizes the deviation of the predicted value from the actual values. At a pixel location x, if the predicted probability of the pixel belonging to class l is $p_l(x)$, $\omega(x)$ is the weight associated to pixel x and $g_l(x)$ is the actual label of the pixel x, then the weighted multi-class logistic loss $\tau_{logloss}$, is defined as (1):

$$\tau_{logloss} = -\Sigma_{x \epsilon \Omega} \omega(x) g_l(x) log(p_l(x)) \tag{1}$$

The weights associated with the pixels are to compensate for the unbalanced thicknesses of the different retinal layers.

Fig. 3 Histogram portraying class imbalance problem in retinal layer segmentation task from OCT B-scans

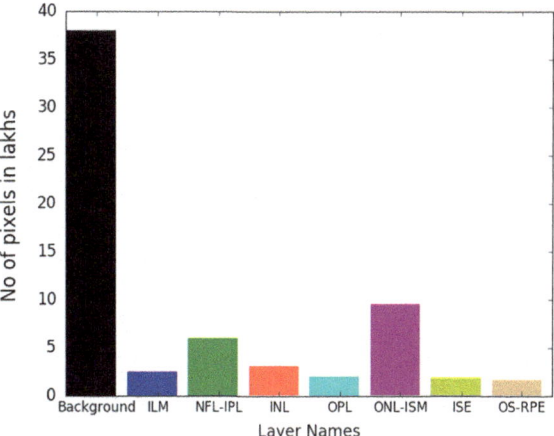

Weighting scheme for the multi-class logistic loss: Class imbalance is one of the major challenges associated with retinal layer segmentation. Figure 3 shows the histogram of the number of pixels belonging to each class. It can be observed that the background class dominates other classes due to the large number of background pixels.

In the absence of any weighting scheme, there are two issues that arise: (i) The boundary pixels in the layer transition regions will be challenging to be segmented, (ii) the model will be biased toward the most dominant background pixels class.

The first issue is resolved by weighting all the pixels belonging to the retinal layer boundaries by a factor of w_1. This ensures that the pixel contributions from the retinal layer transition regions are weighted more relative to the interior retinal layer pixels. The value of w_1 is empirically chosen as 15.

The second issue is addressed with an introduction of weighting factor w_2. This weighs the contributions from the pixels belonging to under-represented classes (including the boundary pixels) and ensures that the convolution kernels are more sensitive to the gradient contributions from these pixels. As is evident from the histogram, the different classes are nonuniformly balanced relative to each other and hence the value of w_2 is varied for different layers using *Median Frequency Method*.

Median frequency method: The value of w_2 for each of the retinal layers is computed using the median frequency method. Here, the number of pixels belonging to each of the layers is computed as n_l. The median of these values is calculated as M. L represents the set of 7 retinal layer classes. The value of w_2 is calculated as (2):

$$w_2 = \begin{cases} M/n_l & \text{when } l(x) \in L \\ 0 & \text{otherwise} \end{cases} \tag{2}$$

Table 1 Computed value of w_2 for different layers

Layer	w_2
ILM	11.459
NFL-IPL	5.63
INL	11.007
OPL	14.368
ONL-ISM	3.336
ISE	13.647
OS-RPE	16.978

Fig. 4 Proposed weighting scheme representation on a sample retinal OCT B-scan

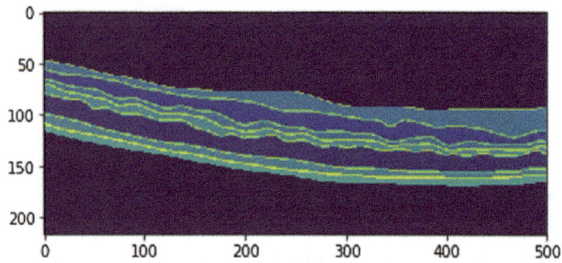

The final weighting scheme is thus derived as (3):

$$w = 1 + w_1 + w_2 \tag{3}$$

The proposed weighting scheme ensures that the boundary pixels receive a weight higher than the maximum value of w_2 assigned to any of the layers. Retina comprises of hypo and hyper-reflective layers and it can be observed from Table 1 that the value of w_2 in hypo reflective layers is lower relative to hyper-reflective layers. Hence, darker pixels are assigned lesser weights and the brighter ones are assigned higher weights, i.e, the weights assigned to the pixels are proportional to the intensity of the pixels. Figure 4 depicts the weighting scheme described.

3.2 Dice Loss

Dice coefficient is used to measure the similarity between the predicted and actual labels, and is used popularly in segmentation to compare the performance of the predicted values with respect to the reference masks in medical applications. Dice loss $\tau_{diceloss}$, defined as $1 - dice\ coefficient$ thus penalizes the deviation of the predicted values from the actual labels. It is defined as (4).

$$\tau_{diceloss} = 1 - \frac{2\Sigma_{x\epsilon\Omega}\, p_l(x)g_l(x)}{\Sigma_{x\epsilon\Omega}\, p_l(x)^2 + \Sigma_{x\epsilon\Omega}\, g_l(x)^2} \tag{4}$$

3.3 Combined Loss

The overall loss function used for optimization of the proposed network is a weighted combination of the weighted multi-class logistic loss ($\tau_{logloss}$) (1) and the dice loss ($\tau_{diceloss}$) (4) along with an additional weight decay term [15]. Overall loss function is defined as (5):

$$\tau_{overall} = \lambda_1 \tau_{logloss} + \lambda_2 \tau_{diceloss} + \lambda_3 \| W^{(\cdot)} \|_F^2 \tag{5}$$

where λ_1, λ_2 and λ_3 are the weight terms and $\| W^{(\cdot)} \|_F^2$ represents the Frobenius norm on the weights W of the model.

4 Training

4.1 Dataset Description

The proposed architecture is evaluated on the Duke DME SD-OCT dataset [3]. The scans of the dataset were obtained from 10 subjects with DME using Spectralis Heidelberg OCT machine. The dataset consists of a total of 110 annotated images (11 B-scans per subject) with a resolution of [496 × 768]. All these scans were annotated for the retinal layers by two expert ophthalmologists.

4.2 Data Preprocessing

The annotations for different patients were of differently sized and hence prepro-cessing was required to obtain a uniformly sized dataset. Different bounds were manually computed for the images of different patients and all the 110 images were sized to [496 × 500]. It was observed that the majority of the pixels in these images comprised of background and cropping a major portion of the background would aid in faster and unbiased model training. On horizontal cropping of certain portions, the images were sized to [216 × 500]. Thus, the dataset comprised of 110 images of size [216 × 500].

The dataset is relatively small in order to train a full scale fully convolutional network. Hence, in order to increase the size of the dataset, to reduce overfitting and to speed up the training process, the images were sliced vertically to obtain 770 images of size [216 × 64] each.

The OCT scans contain speckle noise, which can be approximated with a Gaussian distribution. Thus, the images were denoised using Unbiased Fast Non-local-Means algorithm [9].

4.3 Experimental Settings

The dataset, thus consists of 770 images, out of which 540 were set aside for training and the remaining 230 were used for testing. In order to improve the model's ability to generalize, the train and test images were distributed uniformly across all the patients. The hyper-parameters for the loss function are set as $\lambda_1 = 1$, $\lambda_2 = 0.5$ $\lambda_3 = 0.01$. Every layer of convolution incorporated in the model has a L2 kernel regularizer with a weight decay = 0.01. The optimization was carried out using SGD optimizer with Nesterov Momentum and a batch size of 32 was used. The initial learning rate was set to 0.01, and it is reduced by a factor of 0.5 on plateau where the loss stops decreasing for more than 6 epochs and it is allowed to reach a minimum learning rate of 0.000005. Momentum of 0.9 was used while training. The model is trained for 200 epochs and takes an average of 25 s for training per epoch. All the experiments were performed on Intel Xenon CPU with 64 GB RAM and Nvidia K4O GPU with 10 GB dedicated memory.

5 Results and Analysis

5.1 Qualitative Analysis

Figures 5 and 6 show the sample OCT scans, model predictions and ground truths for qualitative analysis of the results obtained. Figure 5a, b, c show the OCT scan, model prediction and ground truth of a sample image with minimal retinal layer

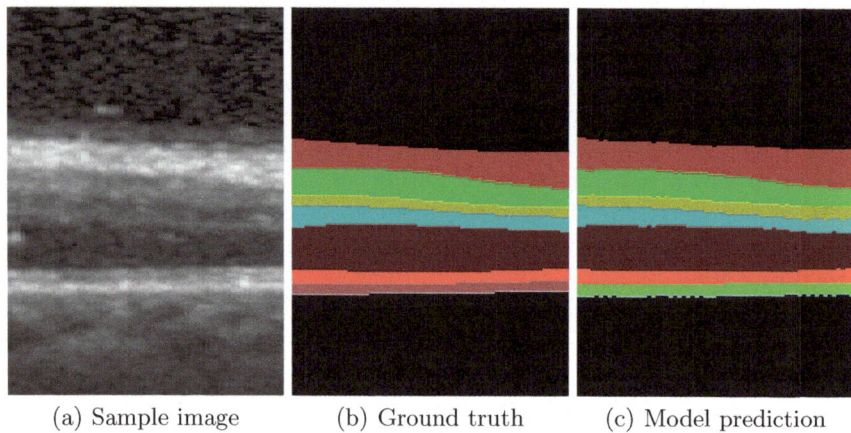

(a) Sample image   (b) Ground truth   (c) Model prediction

Fig. 5 Segmentation result with limited retinal layer distortion

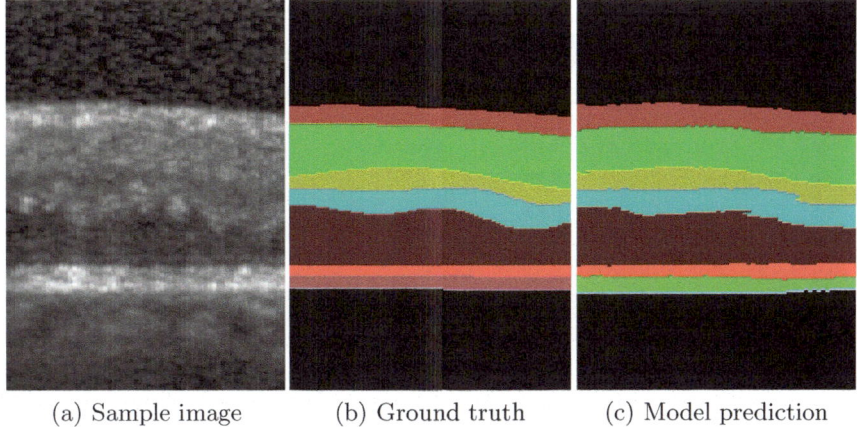

(a) Sample image   (b) Ground truth   (c) Model prediction

Fig. 6 Segmentation result with significant retinal layer distortion

distortion, respectively. Similarly, Fig. 6a, b, c show the OCT scan, model prediction and ground truth of a sample with significant retinal layer distortion, respectively.

It can be seen that the segmentations of the proposed model which makes use of dilated convolutions for retinal layer segmentation are of high quality and the retinal layer boundary demarcations are comparable to human expert clinicians. The shape of the layer boundaries is highly preserved as per visual analysis. This is evident by comparing Fig. 5c to its corresponding groundtruth in Fig. 5b. The model has been able to successfully learn even the complicated edges and fine details as it can be seen in Fig. 6c, b. Also, the predictions are robust against the noise and retinal layer distortions.

5.2 Quantitative Metrics

Dice coefficient has been used as the quantitative metric for evaluation of the performance of the proposed model for retinal layer segmentation. The Dice coefficient score (6) has been computed for every retinal layer for analysis of each of the layer segmentations.

$$DSC = \frac{2|Predicted \cap Actual|}{|Predicted| + |Actual|} \quad (6)$$

Comparative analysis: Quantitative comparison of the Dice score metrics has been carried out against the recently proposed method [15] and they have been tabulated in Table 2.

It can be observed that our proposed method has comparable Dice scores with [15], for most of the retinal layers. The OPL and ONL-ISM show a significant improvement over the corresponding ReLayNet Dice scores. It can be noted that the

Table 2 Comparative analysis of the proposed model with ReLayNet using Dice scores. (The best performance is indicated in **bold**)

Layer	Proposed	ReLayNet
Background	**0.99**	**0.99**
ILM	0.89	**0.90**
NFL-IPL	**0.94**	**0.94**
INL	**0.88**	0.87
OPL	**0.87**	0.84
ONL-ISM	**0.96**	0.93
ISE	**0.94**	0.92
OS-RPE	**0.90**	**0.90**

proposed model has performed well in segmenting intra-retinal layers (IPL to ISE) compared to ReLayNet [15], which are particularly challenging to segment due to lot of undulations and other pathological compartments in these layers. This increase in performance can be owed to the introduction of the novel weighting scheme that weighs the contributions from the thin retinal layers and boundaries.

6   Conclusion

In this paper, we have introduced retinal layer segmentation network using dilated convolutions, and the proposed network is fully convolutional. The proposed model classifies every pixel in the given OCT image to either background or one of the seven prominent retinal layer classes. Combined loss function which is a weighted sum of Dice loss and multi-class logistic loss has been used. A novel pixel-wise weighting scheme has been adopted to solve the class imbalance problem and to accurately segment the pixels in the layer transition regions. Dice coefficient has been used for quantitative performance evaluation and results have been reported for each layer. Our model performs reasonably well in segmenting intra-retinal layers which is crucial in the presence of pathologies and undulations. The proposed model provides a fast response and hence can be incorporated for real time applications. Aiding to its fully convolutional nature, it can be used independently of the test image dimensions.

One major challenge faced is the noise present in the dataset induced during the OCT image acquisition. Thus, devising a method which can denoise the images in real time is an open-ended research problem. Another limitation under concern is that the proposed model is vendor dependent (Spectralis vendor scans) due to the scarcity of data. This can be resolved by the acquisition of a larger annotated dataset from clinical experts. Hence, future works can be directed towards proposing robust denoising methods and validating across cross-vendor OCT scans.

Acknowledgements This work was supported by the Science and Engineering Research Board (Department of Science and Technology, India) through project funding EMR/2016/002677. The authors would like to thank Vision and Image Processing (VIP) Lab, Department of Biomedical Engineering, Duke University, Durham, NC, USA for providing DME dataset.

References

1. Anger, E.M., Unterhuber, A., Hermann, B., Sattmann, H., Schubert, C., Morgan, J.E., Cowey, A., Ahnelt, P.K., Drexler, W.: Ultrahigh resolution optical coherence tomography of the monkey fovea. identification of retinal sublayers by correlation with semithin histology sections. Exp. Eye Res. **78**(6), 1117–1125 (2004)
2. Bouma, B., Tearney, G., Yabushita, H., Shishkov, M., Kauffman, C., Gauthier, D.D., Mac-Neill, B., Houser, S., Aretz, H., Halpern, E.F., et al.: Evaluation of intracoronary stenting by intravascular optical coherence tomography. Heart **89**(3), 317–320 (2003)
3. Chiu, S.J., Allingham, M.J., Mettu, P.S., Cousins, S.W., Izatt, J.A., Farsiu, S.: Kernel regression based segmentation of optical coherence tomography images with diabetic macular edema. Biomed. Opt. Express **6**(4), 1172–1194 (2015)
4. Chiu, S.J., Li, X.T., Nicholas, P., Toth, C.A., Izatt, J.A., Farsiu, S.: Automatic segmentation of seven retinal layers in sdoct images congruent with expert manual segmentation. Opt. Express **18**(18), 19413–19428 (2010)
5. Chiu, S.J., Toth, C.A., Rickman, C.B., Izatt, J.A., Farsiu, S.: Automatic segmentation of closed-contour features in ophthalmic images using graph theory and dynamic programming. Biomed. Opt. Express **3**(5), 1127–1140 (2012)
6. Coscas, G.: Optical Coherence Tomography in Age-Related Macular Degeneration. Springer Science & Business Media (2009)
7. De Boer, J.F., Cense, B., Park, B.H., Pierce, M.C., Tearney, G.J., Bouma, B.E.: Improved signal-to-noise ratio in spectral-domain compared with time-domain optical coherence tomography. Opt. Lett. **28**(21), 2067–2069 (2003)
8. Fu, T., Liu, X., Liu, D., Yang, Z.: A deep convolutional feature based learning layer-specific edges method for segmenting oct image. In: Ninth International Conference on Digital Image Processing (ICDIP 2017). vol. 10420, p. 1042029. International Society for Optics and Photonics (2017)
9. Girish, G.N., Thakur, B., Chowdhury, S.R., Kothari, A.R., Rajan, J.: Segmentation of intra-retinal cysts from optical coherence tomography images using a fully convolutional neural network model. IEEE J. Biomed. Health Inform. (2018)
10. Huang, D., Swanson, E.A., Lin, C.P., Schuman, et al.: Optical coherence tomography. Science **254**(5035), 1178–1181 (1991)
11. Ioffe, S., Szegedy, C.: Batch normalization: accelerating deep network training by reducing internal covariate shift (2015). arXiv preprint arXiv:1502.03167
12. Ramachandran, P., Zoph, B., Le, Q.V.: Searching for activation functions. CoRR (2017). arXiv preprint arXiv:1710.05941
13. Roy, A.G., Conjeti, S., Carlier, S.G., Dutta, P.K., Kastrati, A., Laine, A.F., Navab, N., Katouzian, A., Sheet, D.: Lumen segmentation in intravascular optical coherence tomography using backscattering tracked and initialized random walks. IEEE J. Biomed. Health Inform. **20**(2), 606–614 (2016)
14. Roy, A.G., Conjeti, S., Carlier, S.G., Houissa, K., König, A., Dutta, P.K., Laine, A.F., Navab, N., Katouzian, A., Sheet, D.: Multiscale distribution preserving autoencoders for plaque detection in intravascular optical coherence tomography. In: 2016 IEEE 13th International Symposium on Biomedical Imaging (ISBI). IEEE, pp. 1359–1362 (2016)
15. Roy, A.G., Conjeti, S., Karri, S.P.K., Sheet, D., Katouzian, A., Wachinger, C., Navab, N.: Relaynet: retinal layer and fluid segmentation of macular optical coherence tomography using fully convolutional networks. Biomed. Opt. Express **8**(8), 3627–3642 (2017)

16. Yeh, A.T., Kao, B., Jung, W.G., Chen, Z., Nelson, J.S., Tromberg, B.J.: Imaging wound healing using optical coherence tomography and multiphoton microscopy in an in vitro skin-equivalent tissue model. J. Biomed. Opt. **9**(2), 248–254 (2004)
17. Yu, F., Koltun, V.: Multi-scale context aggregation by dilated convolutions (2015). arXiv preprint arXiv:1511.07122

Cosaliency Detection in Images Using Structured Matrix Decomposition and Objectness Prior

Sayanti Bardhan and Shibu Jacob

Abstract Cosaliency detection methods typically fail to perform in the situation where the foreground has multiple components. This paper proposes a novel framework, Cosaliency via Structured Matrix Decomposition (CSMD), for efficient detection of cosalient objects. This paper addresses the issue of saliency detection for images with multiple components in the foreground, by proposing a superpixel level, objectness prior. In this paper, we further propose a novel fusion technique to combine objectness prior to the cosaliency object detection framework. The proposed model is evaluated on challenging benchmark cosaliency dataset that has multiple components in the foreground. It outperforms prominent state-of-the-art methods in terms of efficiency.

Keywords Cosaliency detection · Matrix decomposition · Fusion · Objectness

1 Introduction

Human beings have an ability to automatically allocate attention to the relevant object or location in a scene at any given time. This helps humans to focus their attention on the most relevant attribute present in their field of view. Saliency refers to the computational identification of image regions that attract the attention of humans. Many studies have been conducted over the last decade that aims to imitate the human visual system for locating the most conspicuous objects from a scene.

However, in recent years, with rapid growth in the volume of data, the necessity to process multiple relevant images collaboratively has also risen. This challenging issue is addressed as cosaliency detection problem. Cosaliency refers to detecting the common and salient regions in an image group, containing related images, with

S. Bardhan (✉)
Visualization and Perception Lab, Department of Computer Science and Engineering, Indian Institute of Technology Madras, Chennai, India
e-mail: sayantibardhan@gmail.com

S. Bardhan · S. Jacob
Marine Sensor Systems, National Institute of Ocean Technology, Chennai, India

© Springer Nature Singapore Pte Ltd. 2020
B. B. Chaudhuri et al. (eds.), *Proceedings of 3rd International Conference on Computer Vision and Image Processing*, Advances in Intelligent Systems and Computing 1022, https://doi.org/10.1007/978-981-32-9088-4_25

Image RFPR [20] HS [21] SMD[17]   SPL[18] Objectness Prior CSMD   GT

Fig. 1 Challenging cases for iCoSeg cosaliency dataset with multiple components

factors like unknown categories and locations. Problems of visual cosaliency can form a vital preprocessing step to a plethora of computer vision problems like image co-segmentation, object co-location among others.

Researchers in recent years have employed several approaches for the cosaliency detection problem. Methods employing low-level, mid-level and high-level features, top-down and bottom-up approaches have been among the explored methodologies for cosaliency detection [1–18]. The recent works on these methods and more have been discussed in details in Sect. 2. However, the majority of the cosaliency detection models today are based on bottom-up role and do not provide high-level constraint [19]. Such methods have limitation in situations when:

• Foreground is composed of multiple components (Column 1 of Fig. 1 shows such an image). State-of-the-art cosaliency methods, like RFPR [20], HS [21], SPL [18] and SMD [17] fail to provide the entire object-level foreground (shown in column 2–4 in Fig. 1) in such situations and hence fail to generate cosaliency maps close to the ground truth (given in column 8 of Fig. 1).

To solve this challenge, here we propose a novel approach named as Cosaliency via Structured Matrix Decomposition (CSMD). The approach that is adopted by this paper to address the aforementioned issue, i.e., the key contributions of the paper are summarized below as:

• We propose superpixel level objectness prior to cosaliency detection. Objectness value indicates the likelihood of a superpixel to belong to an object. Incorporation of this high-level prior preserves the completeness of the detected cosaliency map which in turn helps in detection of cosalient objects in images where foreground has multiple components.
• The model proposes a novel fusion strategy for fusing the objectness priors in CSMD, the proposed cosaliency framework.

The rest of the paper is organized as follows. Section 2 describes the recent works done on cosaliency problem. Section 3 details CSMD, the proposed cosaliency detection model. Section 4 describes the experimental setup, experimental results and its thorough comparison with other prominent state-of-the-art cosaliency detection models. This section also gives quantitative performance analysis and computation time analysis. Finally, Sect. 5 concludes the paper.

2 Recent Works

Researchers in recent years have employed several approaches towards the cosaliency detection problem. Methods have employed low-level features [1–3] like SIFT descriptors, Gabor filters and colour histograms to extract the cosalient regions assuming that cosalient objects have low-level consistency. Other approaches [4–7] have utilized the output of the previously stated algorithms as mid-level features to exploit detection of the cosalient regions. Methods have also explored high-level semantic features for cosaliency detection. These methods work on the principle that multiple images share stronger consistency in high-level concepts. In addition to this, from the existing literature, it can also be seen that cosaliency approaches have used border connectivity cue [8], objectness cue [9], center cues [2, 10], intra and inter image similarity cues to aid detection of cosalient regions in images.

CoSaliency problems, much like saliency problems, have also been classified as Top-Down and Bottom-up saliency approaches. However, techniques involving combination of both top-down and bottom-up frameworks have been a recent trend. A subset of such studies includes exploiting the low-rank matrix recovery (LR) theory [11–17]. Peng et al. [17] introduced, Structured Matrix Decomposition (SMD), a modification in the low-rank recovery (LR) models by introducing a Laplacian regularization and a tree-structured sparsity inducing regularization. SMD solved the inherent problem of the low-rank recovery models [12–14].

Today, most of the cosaliency methods propose a new model with redesigned algorithms for cosaliency detection. However, in [18], different from existing models, an efficient saliency-model-guided visual cosaliency detection scheme (ESMG) is reported to make existing saliency frameworks work well for cosaliency detection problems. Li et al. proposed a framework [18] where given a saliency model for single image, a cosaliency map is obtained by a detection model based on the query under the ranking framework. The model aims at developing a higher precision cosaliency detection model starting from existing proven saliency map.

Difference with other cosaliency methods: The cosaliency methods today fail to generate efficient cosaliency maps under conditions where the foreground consists of multiple components. A prime reason for the same is that most of the cosaliency methods do not incorporate high-level constraints. The proposed method incorporates a high level prior, objectness, to address such an issue. Objectness prior in cosaliency model ensures that the cosaliency map is complete. In addition to this, the various fusion techniques for the priors are studied in this paper and an exponential technique is observed to give the best results. This technique is used to incorporate the objectness prior to the proposed method. It can be observed that the proposed framework is efficient in detecting objects with multiple components in the foreground.

3   Methodology

The proposed Cosaliency via Structured Matrix Decomposition (CSMD) method utilizes the structured matrix decomposition framework for saliency detection. It incorporates superpixel level objectness prior map and a novel fusion technique. The detailed model of the proposed cosaliency detection framework is illustrated in a flowchart form in Fig. 2. The key steps of the proposed framework are detailed below:

i.   *Image Abstraction*: Low-level features are obtained from the input image, as given in Fig. 2a. Low-level features such as Gabor filter [22], steerable pyramids [23] and RGB colour is extracted from the image [17, 24] and stacked vertically to form a D-dimensional feature vector. Here, in this paper, D is kept as 53. After the low-level feature extraction, the image is segmented by simple linear iterative clustering (SLIC) algorithm [25]. Here the image decomposes into superpixels $\{B_i\}_{i=1....N}$, where each superpixel is represented by a feature vector, f_i. Further stacked feature vectors form a feature matrix. Thus, each image is now represented as a feature matrix, F, of dimension D × N, where D is the dimension of the feature vector for each superpixel and N is no. of superpixels present in an image.

ii. *Structured Index Tree Construction*: Next, as illustrated in Fig. 2b, an index tree is constructed on each superpixel, B_i, of the image. As in [17], the affinity between adjacent superpixels is computed. The affinity matrix is computed by the following equation:

$$C_{ij} = \begin{cases} \exp\left(-\frac{\|f_i - f_j\|^2}{2\sigma^2}\right), & if\ (B_i, B_j) \in V \\ 0, & otherwise \end{cases} \quad (1)$$

where V is the set of adjacent superpixel pairs, which are neighbours of neighbours or neighbours in the image. Superpixels are then merged with their neighbours based on their affinity. Image segmentation [26] algorithm (based on graphs) is utilized to merge the neighbouring superpixels. A hierarchical fine-to-course segmented image is obtained.

iii. *Fusion of Priors*: High-level priors are based on human perception. Priors like location [24], colour [24] and background [27] is computed, in Fig. 2c. Location

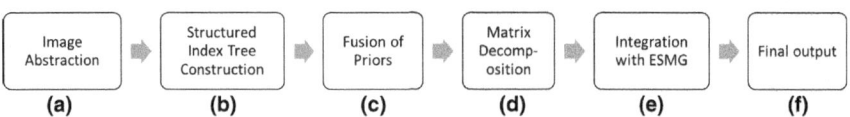

| Image Abstraction | Structured Index Tree Construction | Fusion of Priors | Matrix Decomposition | Integration with ESMG | Final output |
| (a) | (b) | (c) | (d) | (e) | (f) |

Fig. 2 Framework for the proposed model for cosaliency detection method

prior is generated using Gaussian distribution formulated on the distance of the pixels from image center. Background prior computes how well the image region and image boundaries are connected. Colour prior is based on human eye sensitivity to warm colours like yellow and red. The aforementioned three priors are then fused with the proposed objectness prior to a novel fusion strategy. It is the fusion of this objectness prior in this stage that enables preserving the completeness of the detected saliency map in the final stage and helps in the detection of salient objects when the foreground has multiple components. Column 6 of Fig. 1 shows the objectness prior map in the proposed framework and column 7 of Fig. 1 shows the improvement in detection of the co-salient object due to incorporation of the same. It can be seen that the salient object completeness stays intact in the proposed model. The detailed description of the objectness prior and the fusion technique employed is described in Sects. 3.1 and 3.2 of this paper.

iv. *Matrix Decomposition*: Peng et al. [17] proposed Structured Matrix Decomposition (SMD), with l_∞ norm, to form a sparse matrix S and a low-rank matrix L by decomposition of the feature matrix. This assumes that image can be represented by S, a sparse matrix corresponding to the salient object and L, a low-rank matrix corresponding to the redundant background. Utilizing this SMD framework proposed, F, the feature matrix and the index tree is decomposed into the S and L matrix. As in [17], the saliency map is obtained from sparse matrix, S by the equation

$$Sal(B_i) = \|s_i\|_1 \tag{2}$$

where, s_i represents ith column from S. As in [17], the prior map is integrated into the SMD framework, such that each superpixel is associated with a penalty value. A superpixel with larger prior is assigned less penalty and this highlights such a superpixel in the final saliency map.

v. *Integration with ESMG*: A two-stage guided detection pipeline, ESMG, was proposed by Li et al. [18] that exploits a single image saliency map to generate cosaliency maps. Utilizing this work, single image saliency map from Fig. 2d, is taken as the initial query and an efficient manifold ranking (EMR) is chosen as the ranking function [28]. This forms the first stage of the algorithm. EMR has an advantage of discovering the underlying geometry of the given database and significantly reduce computational time of the traditional framework of manifold ranking. The ranking scores obtained by EMR, in this stage, is assigned directly to the pixels as their saliency values in the first stage. This stage brings out the cosalient parts not detected in single image saliency map. Subsequently, in the second stage, queries in every image is assigned as a query turn by turn. As cosalient objects exist in every image group and given that it was successfully detected by the first stage, the second stage pops out the cosalient objects in the other images. These maps are then fused, as in [18], to give the final cosaliency map.

The key novelty of this work lies in the integration of the proposed objectness prior to cosaliency detection and its fusion in the model. This is further described in details in Sects. 3.1 and 3.2.

3.1 Objectness Prior

A major challenge in cosaliency detection is when the foreground consists of multiple components. Cosaliency algorithms, typically under such circumstances fail to detect the entire object-level foreground. Figure 1 exhibits such a case. The improvement in cosaliency detection in CSMD is largely attributed to the fact that the proposed method incorporates the usage of objectness. Objectness helps in preserving completeness of the entire object and can be used to image the entire extent of the foreground. Hence, in the proposed CSMD in order to model the object completeness, Objectness prior is incorporated.

Objectness prior gives a measure of the likelihood of a window to contain an object. Alexe et al. defined objectness measure [29] and argued that an object should have any one of the properties:

- A closed boundary well defined in space.
- Different appearance from its surrounding.
- Sometimes stands out as salient and is unique within the image.

In [29], objectness measure is computed by combining several image cues such as edge density, colour contrast, superpixel straddling and multi-scale saliency. In this paper, we define objectness for a window, $wd = (x, y, w, h)$, with width w, height h and top-left corner at (x, y), as the likelihood $Pr_{obj}(wd)$, that the window wd contains the object.

To compute pixel wise objectness, we sample m_w windows, such that $W = \{wd_k\}_{k=1}^{m_w}$. Then we slide and calculate the Objectness map, O, over each pixel data,

$$O(k, v) = \sum_{(k,v) \in wd \,\wedge\, wd \in W} Pr_{obj}(wd) \tag{3}$$

m_w here, in this paper, is taken as 10,000. A similar objectness measure has been used by Jin et al. [30] for image thumbnailing.

However, in this paper, we extend this idea further and define objectness for superpixels as

$$O(B_i) = \frac{1}{|B_i|} \sum_{(k,v) \in B_i} O(k, v) \tag{4}$$

where, B_i is ith superpixel. We compute the objectness value for all the pixels in a superpixel and obtain the average of the same. The computed value of objectness becomes the new value of objectness for the entire superpixel. This process is repeated for all the superpixels, to obtain the superpixel level objectness map. The objectness prior is then integrated with the other high-level priors, further detailed in Sect. 3.2. It is observed that the superpixel level objectness map gave better final saliency map than pixel level or region level objectness prior.

3.2 Fusion of Priors

As explained above, high-level priors like location, colour and background priors are fused with objectness prior in Fig. 2c, to generate efficient saliency map. The appropriate fusion of these four priors is required to generate the desired high precision saliency map. Studies have utilized various methods of fusing the priors to generate the prior map.

Priors can be multiplied all together to obtain the final prior map. Peng et al. [17] have employed such a strategy. Advantages of both multiplicative and additive effect on priors can be utilized to generate the prior map, by obtaining a threshold (TH) by Otsu's method. Further, a fusion strategy can be defined as

$$PM(i) = \begin{cases} \frac{1}{N}\sum_{k=1}^{N} P_k(i), & \textit{if } P_k(i) > TH \\ \prod_{k=1}^{N} P_k(i), & \textit{otherwise} \end{cases} \tag{5}$$

where P_k are the individual prior maps and N is the number of prior maps to be combined for any ith superpixel. Past works [18, 31] have employed such a combination strategy for saliency maps. Here, we propose a combination of priors in the fashion given by the equation:

$$PM = \exp(LP + OP + BP) \times CP \tag{6}$$

where OP is the proposed objectness prior, LP is location prior, BP is the background prior and CP is the colour prior.

To illustrate the effect of different combination strategy of priors on the saliency map, a comparative study was conducted on the cosaliency dataset. The Precision–Recall (PR) curve for cosaliency map on iCoSeg [7] dataset at the output of Fig. 2d was plotted to draw a comparison between the various combination strategies. Figure 3 gives this PR curve. Two different exponential combination strategies are shown in the figure, as a yellow and blue curve. More exponential combinations were studied but none outperformed the one given by the Eq. 6, hence just two of them, are shown as an example in Fig. 3. The magenta curve gives the output obtained by Eq. 5. The cyan curve gives the output obtained by multiplication of all the priors. It can be seen that the curve obtained by Eq. 6 gives the best result and is hence, chosen as the final prior combination strategy.

Fig. 3 Precision–Recall curve for different combination strategies

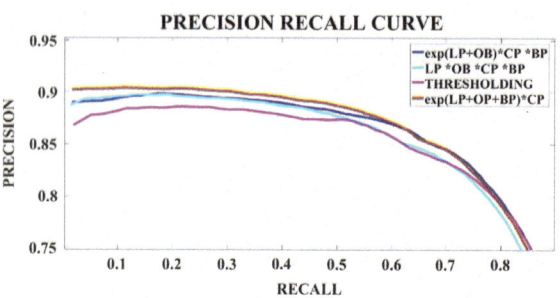

4 Experimental Setup and Results

4.1 Experimental Setup

To evaluate CSMD, the proposed framework, results are observed on two challenging benchmark cosaliency datasets: iCoSeg [7] and Image pair [7]. iCoSeg dataset has 643 images of cosalient images in 38 groups and Image pair dataset has 105 image pairs. iCoSeg dataset has multiple cosalient objects and complex background, whereas, in the image pair dataset, images are uncluttered. The proposed CSMD cosaliency framework is compared with recent state-of-the-art algorithms, HS [21], SMD [17], SPL [18], CP [7] and RFPR [20]. A study between the proposed algorithm and the state-of-the-art methods is drawn using PR and F-measure curve for comparison purpose.

4.2 Results: Visual Comparison

Results with iCoSeg dataset: Fig. 4 shows the performance of the CSMD algorithm in comparison to other state-of-the-art methods on iCoSeg dataset. The proposed algorithm is compared with the other four prominent algorithms, HS [21], SMD [17], SPL [18] and RFPR [20]. Results in Fig. 4 shows that the CSMD algorithm gives cosaliency map closest to the ground truth compared to other prominent algorithms. The prime reason for this improvement in the detection of the cosalient regions is the inclusion of the objectness prior. Objectness prior gives a measure of the superpixels belonging to an object. This helps in the detection of salient objects even in conditions where the foreground consists of multiple components, as can be seen in Fig. 4. It can be seen specifically in row 4, 7, 8 and 9 that the saliency maps detected by other methods are incomplete. However, the saliency map detected by the CSMD method ensures the completeness of the saliency map is kept intact.

Results with Image Pairs dataset: Fig. 5 gives a visual comparison between the performance of CSMD, SMD [17], SPL [18] and CP [7]. It can be observed that for images where multiple components are present is the foreground is best detected by

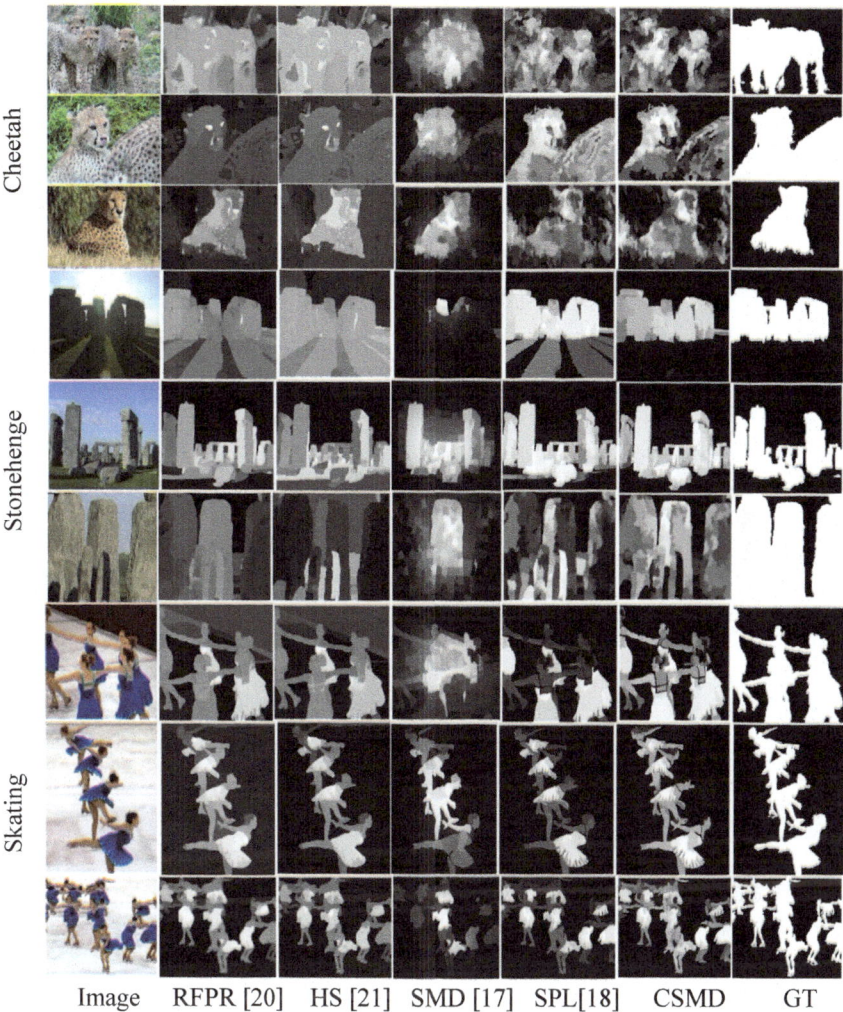

Fig. 4 Results of algorithms on iCoSeg dataset. Our method (CSMD) gives results closest to the Ground Truth (GT)

the proposed CSMD framework. Row 2, 3 and 5 are the specific cases where our framework performs better than the other state-of-the-art models.

Proposed CSMD involves the integration of objectness prior with a strategic fusion methodology in the cosaliency framework. The intuition behind such an integration is that it would prevent the saliency map from being scattered and incomplete. The results in Figs. 4 and 5 confirms this intuition.

Image   CP [7]   SMD [17]   SPL [18]   CSMD   GT

Fig. 5 Results of algorithms on Image Pairs dataset. Our method (CSMD) gives results closest to the Ground Truth (GT)

4.3 Results: Computation Time and Quantitative Performance Analysis

Quantitative evaluation of the algorithms is done using Precision–Recall (PR) and F-measure curve. Figure 6 gives the Precision–Recall curve for iCoSeg dataset. It can be seen that proposed CSMD framework outperforms SPL, SMD, RFPR and HS, the state-of-the-art cosaliency detection methods. Figure 7 shows the F-measure curves for CSMD, SPL, SMD, RFPR and HS methods. It can be observed that CSMD performs better than SPL, RFPR, SMD and HS methods.

Figures 8 and 9 gives Precision–Recall (PR) and F-measure curves for Image Pairs dataset. Figure 8 shows that CSMD framework gives the best performance

Fig. 6 PR curve for iCoSeg
dataset

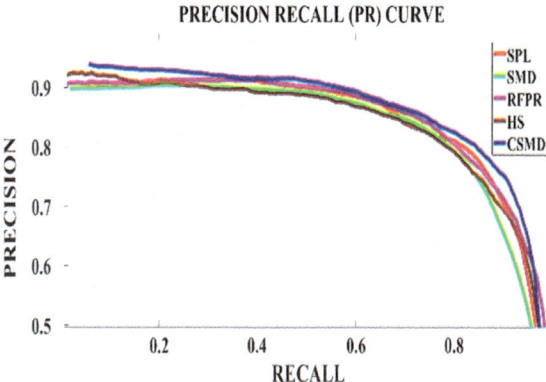

Fig. 7 F-measure curves for
iCoSeg dataset

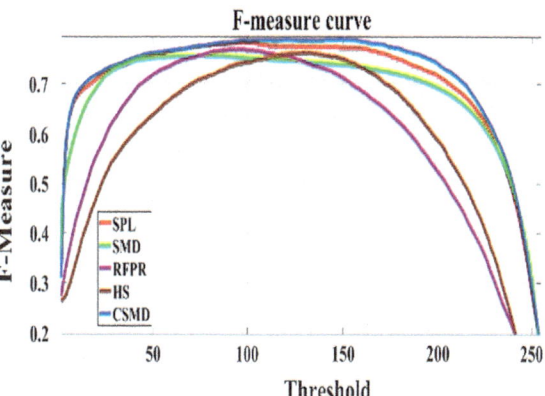

Fig. 8 PR curve for Image
Pairs dataset

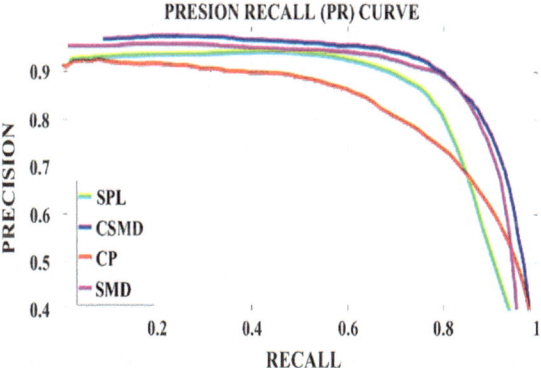

Fig. 9 F-measure curves for
Image Pairs dataset

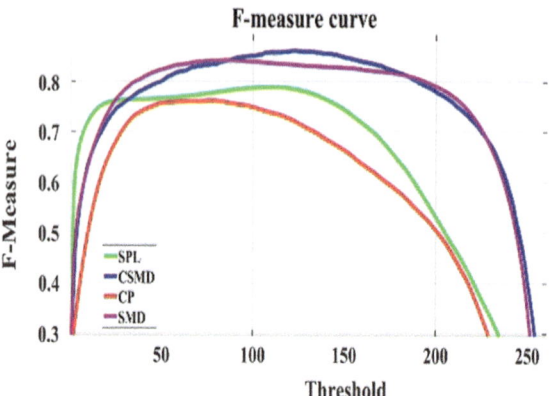

in comparison to SMD, CP and SPL methods. The same can be observed with the
F-measure curve given in Fig. 9.

CSMD method takes 9.15 s per image, whereas methods like HS [21] and RFPR
[20] takes 144.04 s and 162.55 s respectively. However, time taken by CSMD is
comparable to the time taken by SPL [18] and SMD [17] methods. Hence, CSMD
method gives efficient saliency map and is computationally less expensive.

5 Conclusion

The paper attempts to solve the cosaliency detection problem for images with multiple components in the foreground by incorporation of Objectness prior. The paper
also proposes a fusion strategy for integrating this prior in the cosaliency framework. The proposed method is tested with challenging datasets of iCoSeg, which
has images with foreground containing multiple components. The results show that
CSMD method outperforms the state-of-the-art algorithms and is capable of detecting the cosalient objects in such situations.

Acknowledgements The authors thank the faculty and members of Visualization and Perception
Lab, Department of Computer Science and Engineering, IIT-Madras, for the constant guidance and
support for this work. The authors would also like to acknowledge Marine Sensor Systems Lab,
NIOT and Director, NIOT for their support.

References

1. Chang, K.-Y., Liu, T.-L., Lai, S.-H.: From co-saliency to co-segmentation: an efficient and fully unsupervised energy minimization model. In: CVPR, pp. 2129–2136 (2011)
2. Huazhu, F., Cao, X., Zhuowen, T.: Cluster-based co-saliency detection. IEEE Trans. Image Process. **22**(10), 3766–3778 (2013)
3. Tan, Z., Wan, L., Feng, W., Pun, C.-M.: Image co-saliency detection by propagating superpixel affinities. In: ICASSP, pp. 2114–2118 (2013)
4. Cao, X., Tao, Z., Zhang, B., Huazhu, F., Feng, W.: Self-adaptively weighted co-saliency detection via rank constraint. IEEE Trans. Image Process. **23**(9), 4175–4186 (2014)
5. Chen, H.-T.: Preattentive co-saliency detection. In: ICIP, pp. 1117–1120 (2010)
6. Li, H., Meng, F., Ngan, K.: Co-salient object detection from multiple images. IEEE Trans. Multimed. **15**(8), 1896–1909 (2013)
7. Li, H., Ngan, K.: A co-saliency model of image pairs. IEEE Trans. Image Process. **20**(12), 3365–3375 (2011)
8. Ye, L., Liu, Z., Li, J., Zhao, W.-L., Shen, L.: Co-saliency detection via co-salient object discovery and recovery. Signal Process. Lett. **22**(11), 2073–2077 (2015)
9. Li, L., Liu, Z., Zou, W., Zhang, X., and Le Meur, O.: Co-saliency detection based on region-level fusion and pixel-level refinement. In: ICME, pp. 1–6 (2014)
10. Chen, Y.L., Hsu, C.-T.: Implicit rank-sparsity decomposition: applications to saliency/co-saliency detection. In: ICPR, pp. 2305–2310 (2014)
11. Yan, J., Zhu, M., Liu, H., Liu, Y.: Visual saliency detection via sparsity pursuit. IEEE Signal Process. Lett. **17**(8), 739–742 (2010)
12. X. Shen and Y. Wu: A unified approach to salient object detection via low rank matrix recovery. In: CVPR, 2012, 2296–2303
13. Lang, C., Liu, G., Yu, J., Yan, S.: Saliency detection by multitask sparsity pursuit. IEEE Trans. Image Process. **21**(3), 1327–1338 (2012)
14. Zou, W., Kpalma, K., Liu, Z., Ronsin, J.: Segmentation driven low-rank matrix recovery for saliency detection. In: BMVC, pp. 1–13 (2013)
15. Candès, E., Li, X., Ma, Y., Wright, J.: Robust principal component analysis. J. ACM **58**(3), 1–39 (2011)
16. Liu, G., Lin, Z., Yan, S., Sun, J., Yu, Y., Ma, Y.: Robust recovery of subspace structures by low-rank representation. IEEE Trans. Pattern Anal. Mach. Intell. **35**(1), 171–184 (2013)
17. Peng, H., Li, B., Ji, R., Hu, W., Xiong, W., Lang, C.: Salient object detection via low-rank and structured sparse matrix decomposition. IEEE Trans. Pattern Anal. Mach. Intell. **39**(4), 818–832 (2017)
18. Li, Y., Fu, K., Liu, Z., Yang, J.: Efficient saliency-model-guided visual co-saliency detection. IEEE Signal Process. Lett. **22**(5), 588–592 (2014)
19. Zhang, D., Fu, H., Han, J., Wu, F.: A review of co-saliency detection technique: fundamentals, applications, and challenge 1–16 (2016). arXiv:1604.07090
20. Li, L., Liu, Z., Zou, W., Zhang, X., Le Meur, O.: Co-saliency detection based on region-level fusion and pixel-level refinement. In: ICME, pp. 1–6 (2014)
21. Liu, Z., Zou, W., Li, L., Shen, L., LeMeur, O.: Co-saliency detection based on hierarchical segmentation. IEEE Signal Process. Lett. **21**(1), 88–92 (2014)
22. Feichtinger, H.G., Strohmer, T.: Gabor analysis and algorithms: theory and applications. Springer, Berlin (1998)
23. Simoncelli, E.P., Freeman, W.T.: The steerable pyramid: a flexible architecture for multi-scale derivative computation. In: ICIP, pp. 444–447 (1995)
24. Shen, X., Wu, Y.: A unified approach to salient object detection via low rank matrix recovery. In: CVPR, pp. 2296–2303 (2012)
25. Achanta, R., Shaji, A., Smith, K., Lucchi, A., Fua, P., Süsstrunk, S.: Slic superpixels compared to state-of-the-art superpixel methods. IEEE Trans. Pattern Anal. Mach. Intell. **34**(11), 2274–2282 (2012)

26. Felzenszwalb, P., Huttenlocher, D.: Efficient graph-based image segmentation. Int. J. Comput. Vision **59**(2), 167–181 (2004)
27. Zhu, W., Liang, S., Wei, Y., Sun, J.: Saliency optimization from robust background detection. In: CVPR, pp. 2814–2821 (2014)
28. Xu, B., Bu, J., Chen, C., Cai, D., He, X., Liu, W., Luo, J.: Efficient manifold ranking for image retrieval. In: ACM SIGIR, pp. 525–534 (2011)
29. Alexe, B., Deselaers, T., Ferrari, V.: Measuring the objectness of image windows. IEEE Trans. Pattern Anal. Mach. Intell. **34**(11), 2189–2202 (2012)
30. Sun, J., Ling, H.: Scale and object aware image thumbnailing. Int. J. Comput. Vision **104**(2), 135–153 (2013)
31. Li, Y., Fu, K., Zhou, L., Qiao, Y., Yang, J., Li, B.: Saliency detection based on extended boundary prior with foci of attention. In: ICASSP, pp. 2798–2802 (2014)
32. Roy, S., Das, S.: Multi-criteria energy minimization with boundedness, edge-density and rarity, for object saliency in natural images. In: ICVGIP 2014, 55:1–55:8 (2014)
33. Batra, D., Kowdle, A., Parikh, D., Luo, J., Chen, T.: Icoseg: interactive co-segmentation with intelligent scribble guidance. In: CVPR, pp. 3169–3176 (2010)
34. Roy, S., Das, S.: Saliency detection in images using graph-based rarity, spatial compactness and background prior. In: VISAPP, pp. 523–530 (2014)

Agriculture Parcel Boundary Detection from Remotely Sensed Images

Ganesh Khadanga and Kamal Jain

Abstract The object-based image analysis (OBIA) is extensively used nowadays for classification of high-resolution satellite images (HRSI). In OBIA, the analysis is based on a group of pixels known as objects. It differs from the traditional pixels-based methodology, where individual pixels are analyzed. In OBIA, the image analysis consists of image segmentation, object attribution, and classification. The segmentation process thus identifies a group of pixels and are known as objects. These objects are taken for further analysis. Thus segmentation is an important step in OBIA. In order to find out the boundary of agriculture parcels, a two-step process is followed. First, the segmentation of the images is done using the statistical region merging (SRM) technique. Then the boundary information and center of the segmentation are found out using MATLAB. The best fit segment was found out using trial and errors. The extracted boundary information is very encouraging and it matches the parcel boundaries recorded in revenue registers. The completeness and precision analysis of the plots are also quite satisfactory.

Keywords MATLAB · Segmentation · Cadastral parcel · OBIA · HRSI

1 Introduction

The object-based image analysis (OBIA) is extensively used nowadays for classification of high-resolution satellite images (HRSI). In the traditional pixel-based analysis each pixel is taken into account and the classifications are done based on the spectral characteristics of the pixel. But in the object-based analysis, a group of objects is identified based on certain homogeneity condition [1]. This group of homogeneous pixels is known as an object. These objects were taken up for classification of the features in the image. Thus the OBIA analysis is based on this group of pixels. During the classification stage the contextual relationship like neighborhood, spatial, and

G. Khadanga (✉) · K. Jain
Civil Engineering Department, Geomatics Group, Indian Institute of Technology Roorkee,
Roorkee, India
e-mail: ganesh@nic.in

© Springer Nature Singapore Pte Ltd. 2020
B. B. Chaudhuri et al. (eds.), *Proceedings of 3rd International Conference on Computer Vision and Image Processing*, Advances in Intelligent Systems and Computing 1022,
https://doi.org/10.1007/978-981-32-9088-4_26

textural properties are also taken into account. The identification of homogeneous pixels is generally done through a segmentation process. Thus segmentation is an important step in OBIA [3, 9].

Traditionally various segmentation methodologies [6] are evolved for segmentation of the image. The goal of image segmentation is to partition an image into multiple groups of pixels to simplify the representation of an image into something that is meaningful and easier to understand. Segmentation is, however, regarded as an ill-poised problem as the segmentation results obtained by the users may differ from one application to the other [4, 5]. The image may be segmented into groups of pixels via a number of segmentation methods [7, 12, 13]. These include histogram-based methods, edge detection [15, 16, 17], region growing, and clustering-based methods. However, most of these methods are not suitable for noisy environments. Those which are robust to the noisy environment are computationally expensive. Most of the algorithms regard the problem of image segmentation as an optimization problem based upon a homogeneity criterion for groups of pixels or image segments.

Region growing and merging methods start with a set of seed pixels as input, and followed by aggregation of neighboring pixels based on a homogeneity condition defined by the user. The method iteratively adds neighboring pixels that are connected and spectrally similar to the seed point. The region grows until no more pixels can be added. Varied results may be obtained using region growing and merging methods as it highly depends on the choice of seed points, similarity criteria used for merging adjacent regions, and also the merging order. But identification of the seed is a challenge in region growing methods.

In this paper, statistical region merging (SRM) [10, 14] algorithm is used for the segmentation of the image. The details of the SRM segmentation are available in reference [14]. The SRM has been used by researchers in many cases and can be applied to images with multiple channels and noisy images [14].

1.1 Proposed Techniques and Approaches for Image Segmentation

Many researchers have used image processing techniques for extraction of cadastral boundary information from HRSI. Babawuro and Beiji [2] used geo-rectified HRSI and applied morphological operations (dilation and erosion) and Hough transform to the extracted Canny edges to find the field boundaries. The e-cognition software is used [8, 11, 13] for extraction of cadastral parcels using multiresolution segmentation.

The SRM segmentation algorithm is basically an image generation model and it reconstructs the regions from the observed image [14]. The SRM algorithm is similar to region growing and merging segmentation and the merging is based on statistical tests. The SRM segmentation is having a predicate part to decide the merging of two neighboring regions and the other part decides the order followed to test the merging of regions.

A brief description of the SRM algorithm is described below. Let I be the observed image, and |I| be the number of pixels in the image, where |·| denotes the cardinal number. Let Rl be a set of regions with l pixels, the merging predication on the two candidate regions R and R' can be defined as

$$P(R, R') = \begin{cases} \text{true, if} \left\| \bar{R} - \overline{R'} \right\| \leq \sqrt{b^2(R) + b^2(R')} \\ \text{false otherwise} \end{cases} \tag{1}$$

$$b(R) = g\sqrt{(1/(2Q|R|)(\ln|R_{|R|}|/\delta)} \tag{2}$$

where $\delta = 1/(6|I|2)$, |R| is the number of pixels in the image region R, g is the gray level (usually 256), and Q is the spectral scale factor to evaluate the feasibility to merge two regions. It also controls the number of the regions in the final result. As three color channels (R, G, and B) exist in remote sensing images, the decision to merge two regions are done when P(R(p), R(p')) of any color channel returns true.

As per the 4-neighborhood rule, there are $N < 2|I|$ adjacent pixel pairs in the observed image I. Let SI be the set of all these pixel pairs in I and f(p, p') be a real-valued function to measure the similarity of two pixels in the pixel pair (p, p'). In SRM, the sorting of all the pairs in SI is done by increasing order of f(p, p'). The traversing is done only once in this order. The spectral information is thus the key factor based on which the merging is done.

In a region growing approach, homogeneity is the key criteria for region growing segmentation. The selection of seed in a region growing methods is a challenge. The results may vary and depend on the choice of the seed. In order to overcome these issues, experiments were done and the two-stage approach is explored. First, the image is segmented with SRM techniques. The center points of the segments were identified through MATLAB (regionprops) function. The SRM algorithm implementation in the java programming language is taken up for segmentation. The segmented image is then processed in MATLAB for getting the boundary of the segments. The steps are shown in Fig. 1. The initial image and the resultant image are shown in Fig. 2.

1.2 Experiments and Results

The original image of the study region (Agra, India (UP), resolution—0.5 m, GeoEye-1, year-2009) is shown in Fig. 2. The segmented image with SRM segmentation is shown in the middle of Fig. 2. The segmentation is carried out using SRM techniques for various parameters. The multiscale parameter (Q) is adjusted such that the segmentation result matches with the most land parcel in the image. The ground truth is recorded using MATLAB tools. Then the number of correctly detected pixels (TP), wrongly detected pixels (FP), and not correctly identified pixels (FN) are identified

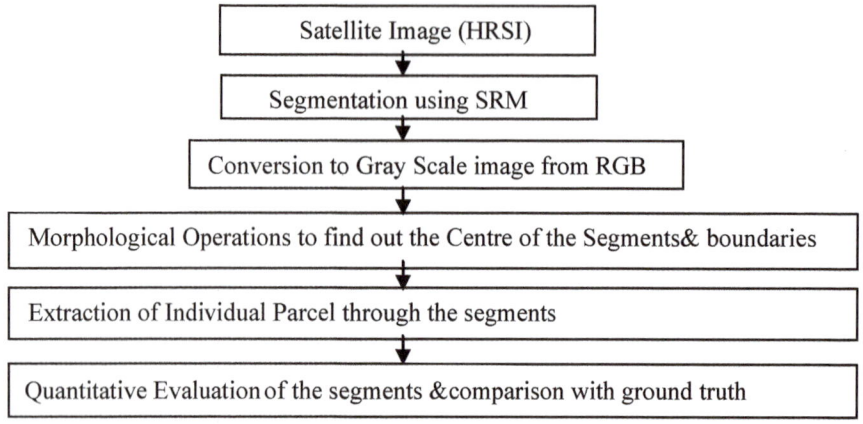

Fig. 1 Structure of the identification of the parcels

Fig. 2 Original image (left), segmented image (center), and parcels (right)

using segmented result and ground truth. The result of completeness and precision is shown in Table 1.

1.3 Discussions and Conclusion

The HRSI image was segmented using SRM segmentation technique. Then the MATLAB programs were developed using regionprops image processing function to extract the centroids of the segments and boundary of the segments. These homogeneous segments match the boundaries of the cadastral parcels as available in the image. The TP, FP, and FN values were processed from the ground truth and the segments. It is observed that the quality of the final result depends on the initial segment. In the top portion of the image, two plots (Fig. 2 right, plot no. 11 and 6) were seen as one in the first segment output. Subsequently, both the segments were identified as one cadastral parcel. This may not create any issue as long as the boundary of the segment is proper for the combined plots. In the later stage, this one

Table 1 Completeteness and precision of the extracted plots from the HRSI

Plots	TP	FP	FN	Completeness	Precision
1	4736	90	836	0.84	0.98
2	1892	60	396	0.82	0.97
3	455	88	273	0.62	0.83
4	1395	46	379	0.78	0.97
5	3862	0	1239	0.76	1
6	13,069	15	2572	0.84	0.99
7	1019	74	958	0.52	0.93
8	7898	10	2875	0.74	0.99
9	477	70	422	0.53	0.87
10	3675	4	679	0.84	0.99
11	8638	90	1494	0.85	0.98
12	7204	0	1019	0.87	1
13	552	52	871	0.38	0.91
14	216	26	295	0.42	0.89

segment can be divided into two parcels using any GIS editing tools. The completeness and correctness of each segment are shown in Table 1. It is also observed that after the identification of segments, the properties (area, perimeter, texture, etc.) of each of these segments can also be generated. Further, the boundaries can be converted to a vector format for land parcel identification and land registration activities. This study can be further enhanced with multiple segment methodologies and other segment quality parameters like Jaccard Index.

References

1. Baatz, M., Schäpe, A.: Multiresolution segmentation—an optimisation approach for high quality multi-scale image segmentation. AGIT Symposium, Salzburg (2000)
2. Babawuro, U., Beiji, Z.: Satellite imagery cadastral features extractions using image processing algorithms: a viable option for cadastral science. Int J Comput Sci Issues (IJCSI) 9(4), 30 (2012)
3. Blaschke, T.: Object based image analysis for remote sensing. ISPRS J Photogramm Remote Sens 65, 2–16 (2010)
4. Blaschke, T., Strobl, J.: What's wrong with pixels? Some recent developments interfacing remote sensing and GIS. Zeitschrift fu'r Geoinformations systeme 6, 12–17 (2001)
5. Castilla, G., Hay, G.J.: Image-objects and Geographic Objects. In: Blaschke, T., Lang, S., Hay, G. (eds.) Object-Based Image Analysis, pp. 91–110. Springer, Heidelberg, Berlin, New York (2008)
6. Dey, V., Zhang, Y., Zhongm, M.: A review of image segmentation techniques with remote sensing perspective. In: ISPRS, Vienna, Austria, vol. XXXVIII, July 2010
7. eCognition User and Reference Manual
8. Fockelmann, R.: Agricultural parcel detection with Definiens eCognition. Earth observation Case Study, GAF, Germany (2001)

9. Hay, G.J., Castilla, G.: Object-based image analysis: strength, weakness, opportunities, and threats (SWOT). In: 1st International Conference on Object-Based Image Analysis (OBIA 2006), Salzburg, Austria, 4–5 July 2006

10. Haitao, L., Haiyan, G., Yanshun, H., Jinghui, Y.: An efficient multiscale SRMMHR (Statistical Region Merging and Minimum Heterogeneity Rule) segmentation method for high-resolution remote sensing imagery. IEEE J. Sel. Top. Appl. Earth Obs. Remote Sens. **2**(2) (2009)

11. Jung, R.W.: Deriving GIS-ready thematic mapping information from remotely sensed maps using Object-Oriented Image Analysis Techniques. In: ASPRS/MAPPS Conference, San Antonia, Texas (2009)

12. Marpu, P.R., Neubert, M., Herold, H., Niemeyer, I.: Enhanced evaluation of image segmentation results. J. Spat. Sci. **55**(1), 55–68 (2010)

13. Navulur, K.: Multispectral Image Analysis Using the Object-Oriented Paradigm. CRC Press (2007)

14. Nock, R., Nielsen, F.: Statistical region merging. IEEE Trans. Pattern Anal. Mach. Intell. **26**(11) (2004)

15. Singh, P.P., Garg, R.D.: A Hybrid approach for information extraction from high resolution satellite imagery. Int. J. Image Graph. **13**(2), 1340007(1–16) (2013)

16. Singh, P.P., Garg, R.D.: Information extraction from high resolution satellite imagery using integration technique. In: Intelligent Interactive Technologies and Multi-media, CCIS, vol. 276, pp. 262–271 (2013)

17. Singh, P.P., Garg, R.D.: Land use and land cover classification using satellite imagery: a hybrid classifier and neural network approach. In: Proceedings of International Conference on Advances in Modeling, Optimization and Computing, IIT Roorkee, pp. 753–762 (2011)

Eigenvector Orientation Corrected LeNet for Digit Recognition

V. C. Swetha, Deepak Mishra and Sai Subrahmanyam Gorthi

Abstract Convolutional Neural Networks (CNNs) are being used popularly for detecting and classifying objects. Rotational invariance is not guaranteed by many of the existing CNN architectures. Many attempts have been made to acquire rotational invariance in CNNs. Our approach '**Eigenvector Orientation Corrected LeNet (EOCL)**' presents a simple method to make ordinary LeNet [1] capable of detecting rotated digits, and also to predict the relative angle of orientation of digits with unknown orientation. The proposed method does not demand any modification in the existing LeNet architecture, and requires training with digits having only single orientation. EOCL incorporates an 'orientation estimation and correction' step prior to the testing phase. Using Principal Component Analysis, we find the maximum spread direction (Principal Component) of each test sample and then align it vertically. We demonstrate the improvement in classification accuracy and reduction in test time achieved by our approach, on rotated-MNIST [2] and MNIST_rot_12k test datasets, compared to other existing methods.

Keywords Convolutional neural networks · Rotation invariance · Eigenvector · Digit recognition · Principal component

V. C. Swetha · D. Mishra (✉)
Department of Avionics, Indian Institute of Space Science and Technology,
Trivandrum 695547, Kerala, India
e-mail: deepak.mishra@iist.ac.in

V. C. Swetha
e-mail: swethavc481@gmail.com

S. S. Gorthi
Department of Electrical Engineering, Indian Institute of Technology,
Tirupati 517506, AP, India
e-mail: rkg@iittp.ac.in

© Springer Nature Singapore Pte Ltd. 2020
B. B. Chaudhuri et al. (eds.), *Proceedings of 3rd International Conference on Computer Vision and Image Processing*, Advances in Intelligent Systems and Computing 1022,
https://doi.org/10.1007/978-981-32-9088-4_27

1   Introduction

Character recognition is a basic, at the same time, an important aspect of Image Processing and Machine Learning. It helps to translate handwritten characters to machine-encoded texts. For better maintenance and storage, always we prefer digital form of data. Another important aspect of character recognition is to help visually impaired people. A handwritten document can be converted to digital format and further to voice or any other format, which blind or visually impaired people can understand (e.g., text to speech conversion), through character recognition. Convolutional Neural Networks have proved their efficiency in detecting and predicting correct labels of characters. Hand-crafted features can be used for the same. One such approach, character recognition using DTW-Radon feature is explained in [3]. In this approach, computational complexity and time spent in feature extraction is high because of the DTW matching. CNN itself learns the 'best' features from samples provided during training phase. Once properly trained, CNN gives the best accuracy along with least test time for image or character recognition compared to most of the conventional methods, which uses handcrafted features [4].

The practical scenario demands CNN to be invariant to many transformations, one of it being rotation of the character or the object. But as such, convolutional neural networks do not guarantee any rotational invariance or equivariance property.

The proposed method, 'Eigenvector Orientation Corrected LeNet (EOCL)', classifies rotated digits correctly, even though the network is not trained with rotated digits. With a simple preprocessing step, EOCL incorporates a **'pre-CNN orientation estimation and correction'** to the test digits and then passes the orientation corrected digits to pretrained LeNet-5 network. EOCL relies on the observation that, the maximum spread of handwritten digits happens to be in the vertical direction when the digits are not rotated. This observation is exploited to find out the orientation of each test sample in our approach. Experiments justify the validity of this approach through better test accuracy on rotated MNIST and MNIST_rot_12k datasets (Fig. 7, Tables 2 and 3) along with lesser test time compared to many other existing networks. At the same time, EOCL does not demand any kind of training data augmentation or increase in number of network parameters. Another advantage of the proposed method is that pretrained LeNet networks can be used without any architectural modification for classifying rotated digits.

2   Related Works

Many methods have been proposed to incorporate rotational invariance in CNNs. Data augmentation, selection of rotation invariant features, incorporating rotational invariance or rotational equivariance in CNN architectures are some of them. The following subsections discuss some of the existing approaches, which deal with rotational invariance in CNNs.

2.1   Data Augmentation

Data augmentation refers to training the network with original and transformed images. CNN can be trained explicitly on rotated versions of images along with the original (with no rotation) images. One of the drawbacks of this approach is that it results in an enlarged training dataset, which leads to a higher training time.

2.2   Rotational Invariant Features

Some other methods extract rotational invariant features from the training and test-ing data samples. Scale Invariant Feature Transform (SIFT) [5], Speeded-Up Robust Features (SURF) [6], Fourier Mellin Moments [7] are some examples of rotational invariant features of images. These methods may not require any architectural dif-ference for the existing CNN architectures.

2.3   Rotational Invariant CNN Architectures

Many other methods put forward the concept of modifying existing CNN architec-tures to make them invariant to rotation. Harmonic Networks (HNet) [8] replaces regular CNN filters with circular harmonics to achieve patch-wise translational and rotational invariance. In Rotation Equivariant Vector Field Networks (RotEqNet) [9], convolutional filters are applied at multiple orientations and results in vector fields representing magnitude and angle of the highest scoring orientation at every spatial location. RICNN [10] defines another rotation invariant architecture in terms of differential excitation and distance map.

2.4   Spatial Transformer Network

Spatial Transformer Network (STN) [11] puts forward the idea of a differentiable module (Spatial Transformer) which learns almost all transformations during train-ing. During testing, with the help of learned transformation Θ, it actively transforms the input feature map or image, conditioned on the feature map or image itself. But, for learning the transformations from input images, STN should be trained with transformed images. It can be considered as a type of training data augmentation.

2.5 *Multiple Instance Testing*

RIMCNN [12] uses pretrained LeNet network without any architectural change, for digit recognition. Multiple instances of same digit are used during testing. It compares the prediction scores for the same test image, rotated by different angles. The label with maximum prediction score is chosen as the correct label for that test sample, and the corresponding input rotation gives the hint for its orientation. RIMCNN compares the prediction scores for arbitrary test input rotations while our method EOCL predetermines and corrects the rotation of each test sample.

3 Proposed Method: Eigenvector Orientation Correction for Rotated Digits

Proposed method (Fig. 1) uses the concept of eigenvectors to correct the orientation of rotated digits. The orientation correction is based on the observation that, maximum spread or elongation of the handwritten digits will be mainly in the vertical direction, when they are not rotated (Fig. 2). In our approach, the maximum spread direction of the digit is found using Principal Component Analysis, which will be used further to de-rotate and align the rotated characters back to a unique orientation. Principal component of an image can be found by either eigenvalue decomposition of the data covariance (or correlation) matrix or singular value decomposition (SVD) of the data matrix.

The covariance matrix of the image can be found out as

$$\Sigma = \begin{bmatrix} \sigma(x, x) & \sigma(x, y) \\ \sigma(y, x) & \sigma(y, y) \end{bmatrix} \tag{1}$$

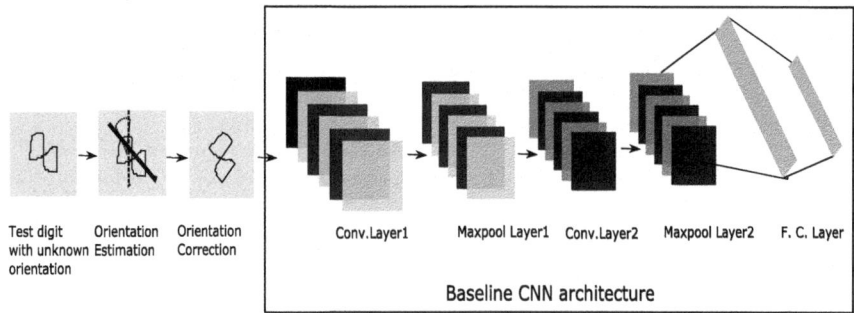

Fig. 1 Proposed method. Principal component based orientation estimation and correction are used as preprocessing steps. The de-rotated test digits are given to the baseline LeNet-5 network

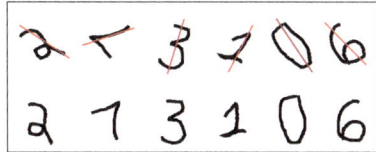

Fig. 2 Example of orientation estimation and correction using principal component analysis. First row contains test digits having arbitrary orientations, line segment showing their principal component. Second row shows corresponding de-rotated or orientation corrected digits

---

Algorithm 1 Orientation Estimation & Correction

Input :
52×52 test digit image with unknown orientation
Orientation Extraction :
(a) Extract the coordinates X of image of each test digit $X = \begin{bmatrix} x \\ y \end{bmatrix}$, where x and y denotes the x and
y coordinates of the test digit, having non zero intensity values.
(b) Compute covariance matrix Σ of X.
(c) Compute eignvalues and the eignvectors of the Σ.
(d) The eigenvector corresponding to maximum eigen value is called Principal Component of the data, and gives the direction of maximum spread of the digit.
Orientation Correction :
(a) Measure the angle θ between vertical axis and the principal component of the test digit image.
(b)This angle θ decides how much we should rotate the digit image back to get the aligned one.
Output :
52×52, orientation corrected test digit and its angle of orientation ρ.

$\sigma(x, x)$ gives the variance of the image in x direction and $\sigma(y, y)$ is the variance in y direction. $\sigma(x, y)$ or $\sigma(y, x)$ gives the covariance of the image. Eigenvectors of Σ can be found from the characteristic equation

$$\Sigma \mathbf{v} = \lambda \mathbf{v}, \qquad (2)$$

where $\mathbf{v}$ is the eigenvector matrix corresponding to eigen values λ of the covariance matrix Σ. For the coordinates of test images, eigen value decomposition gives two eigen values and their corresponding eigenvectors. We choose the eigenvector corresponding to the larger eigen value of the covariance matrix (Principal Component). By measuring the orientation of this principal component with respect to the vertical axis (or any reference direction), we can find the orientation of each test digit. We can get the aligned image by rotating the test image back by the same angle, which is given to the LeNet-5 network trained with digits of single orientation. The orientation estimation and correction steps are explained in Algorithm 1.

For images or digits of same category, it is observed that the maximum spread direction (principal component of the sample) happens to be in the same relative direction of the image. In our approach, this principal component direction is chosen as the reference direction for aligning the images or digits. The images are de-rotated such that their principal component coincides with the vertical axis ('pre-CNN orientation correction'). So, if we are given with digit image of any orientation, after the orientation correction preprocessing step, it will be mapped to the same target digit image.

4 Experimental Analysis—Rotated Digit Classification

The proposed method uses LeNet-5 [1] network trained on binarized MNIST training dataset. We validate our approach using MNIST [2] and MNIST_rot_12k test datasets. For testing, MNIST test dataset is rotated with angles spanning from 0° to 350° with a step of 10°. We analyze the performance of EOCL on MNIST-rot-12k test dataset also. MNIST_rot_12k consists of handwritten digits having arbitrary orientations. The training and testing methods are explained in following sections. Whole experiments are carried out in Intel(R) Xeon(R) X5675 supported with 24 GB RAM and GEFORCE GTX 1080 Ti, 11 GB GPU.

4.1 Training a LeNet Network

We use LeNet-5 architecture as our baseline as shown in Fig. 1. The architecture consists of two convolutional layers with kernel size 5×5, two subsampling layers with window size 2×2 and a fully connected layer. The training is done on MNIST train dataset having single orientation. The proposed method does not require any augmentation in training dataset. This ensures a lesser training time (6 s/epoch)compared to other methods which demand an augmented training dataset for the same objective. The training images are first binarized and then the network is trained using normal backpropagation. Experiments were conducted without binarization also. Better results were obtained when binarized images were used for training. Considering the ambiguity in rotated 6 and 9 digits, images corresponding to digit 9 are excluded from the train and test data sets. Out of 60,000 MNIST training images, we use 54,000 training images corresponding to 9 classes (0–8).

4.2 Orientation Correction for Rotated Digits

The testing phase is preceded by a simple orientation estimation and correction step as shown in Fig. 1. Prior to testing, in the Orientation Correction step, we correct the orientation of each test digit and align it to a reference direction. For estimating the orientation, we find principal component of each test digit. For the specific case of digits, in most of the cases, it is observed that the maximum spread direction (Principal Component) is same as the vertical direction for a perfectly oriented digit without any rotation (Fig. 2). By measuring the angle between principal component

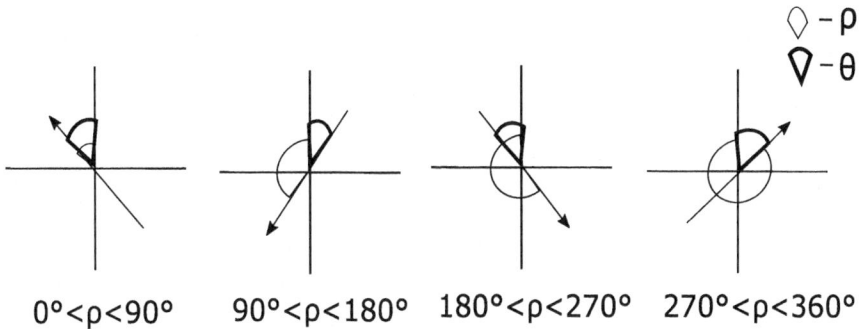

Fig. 3 Four different cases (quadrants) of digit rotation, spanning $0°-360°$. ρ denotes the angle of rotation in anti clockwise from vertical axis. θ is the angle obtained from principal component analysis of the test image

of data and vertical axis (or any other reference axis), we can estimate the relative orientation of the test digit. The reverse rotation (de-rotation) by the same angle can be done to align the test image. In this way, EOCL not only classifies the rotated digits, but also predicts the approximate orientation of each test digit.

As shown in Fig. 3, there can be four different cases for angle measurement using Principal Component Analysis. We take care of these cases in our approach. For angle of rotation specifically from $-90°$ ($270°$) to $90°$, we consider two de-rotation angles $-\theta$ and θ degrees, where θ is the absolute value of the angle between principal component of the data and reference vertical axis. We de-rotate the test image by these two angles and passes the de-rotated test image individually to the trained network in both cases. In this way, corresponding to each test image, two de-rotated images should be passed through the pre trained network. To get the correct de-rotated image back, we compare the prediction scores for each de-rotated image. Out of the two de-rotated image, the one which gives maximum prediction score is chosen as the correct de-rotated image and the corresponding de-rotation angle gives the correct angle of orientation of the test image. Similarly, for angle of orientations from $90°$ to $270°$, as shown in Fig. 3, we consider two de-rotation angles $180 - \theta$ ad $\theta - 180$ and compare the prediction scores for test image de-rotated by these angles. We call this approach as Case 1. As a generalization for angle of rotation from $0°$ to $360°$, we consider these four de-rotation angles at a time. Here, corresponding to each test image, we generate 4 de-rotated images and compare the prediction scores to choose the one corresponding to maximum prediction score. This approach is called as Case 2. Case 2 leads to an increased test time and a slight decrease in test accuracy, but gives a generalized approach which can be used to classify digits, rotated at any angle. Figure 4 shows the test accuracy profile for EOCL, when Case 2 approach is considered.

The whole experiment is based on the observation that in most of the cases, maximum spread of handwritten digits is in the vertical direction when it is aligned without any rotation. We can also use the eigenvector corresponding to the smaller

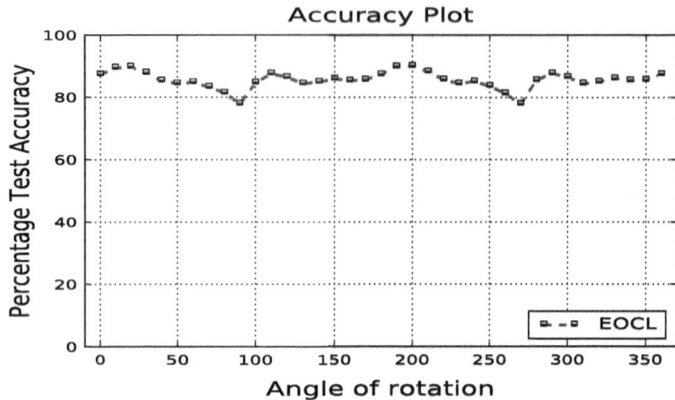

Fig. 4 EOCL test accuracy profile for angle of rotations from 0° to 350°, on MNIST test dataset (test images corresponding to digit 9 are excluded from the test dataset). Orientation angle 90° gives a maximum test error of 22%. The accuracy plot is symmetric about 180°

eigenvalue and get the angle with vertical axis for more accurate results. In that case, we have to compare the prediction score for each test image rotated by this new angle along with that of principal component angle and should choose the label with maximum score. This will lead to an increased test time but will give better classification accuracy.

4.3 EOCL on Rotated MNIST and MNIST_rot_12k Test Datasets

From the MNIST test dataset, samples corresponding to the digit 9 are excluded. So, the testing is done on remaining 9,000 images corresponding to digits 0–8. For MNIST dataset, rotated by different angles, the average test time is 25 s when EOCL is used for testing. We conduct our experiments on MNIST_rot_12k dataset also. 45,136 test images corresponding to digits 0–8 are used in this case and takes 510 s for testing. Table 1 shows the average test accuracy percentage over different ranges of digit rotation, when two de-rotation angles are considered for each range (Case 1, Sect. 4.2). Table 2 compares the performance of EOCL with different number of angles chosen for de-rotation.

Table 3 compares the performance of different architectures on MNIST_rot_12k test dataset. EOCL does not require any training with rotated digits. It can be trained with digits having single orientation and can be used to classify digits having arbitrary orientations. So, for the comparison, we train STN [11], RICNN [10] and EOCL with MNIST training dataset, excluding the digit 9. We use two approaches here. In MNIST_rot_12k dataset, horizontal flipped images and their rotated versions are also included. To take this condition also into consideration, a LeNet-5 network is trained

Table 1 Percentage test accuracy for EOCL over different ranges of orientation of MNIST test dataset. Two de-rotation angles are considered per range (Case 1, Sect. 4.2)

Angle of orientation	Average test accuracy (%)
$0°-90°$	85.58
$90°-180°$	86.22
$180°-270°$	85.49
$270°-350°$	85.84

Table 2 Average test accuracy (%) of EOCL over the range of rotation $0°-350°$, of MNIST test dataset. Considering 2 and 4 de-rotation angles for each test digit (Case 1, Case2, Sect. 4.2). N denotes the number of de-rotation angles used at a time

Method	N	Test accuracy (%)	Average test time (s)
Case 1	**2**	**85.94**	**25**
Case 2	4	81.98	53

Table 3 Percentage test accuracy for different methods, trained on MNIST training dataset (single orientation of digits) and tested on MNIST_rot_12k test dataset

Method	Test accuracy (%)
STN [11]	44.41
RICNN [10]	52
EOCL-H	62
EOCL-HT	**75**

with digits from MNIST train dataset and their horizontally flipped images. In testing, we consider four de-rotation angles as explained in Sect. 4.2. This approach is referred to as **EOCL-H** in this paper. In the second approach, we use only MNIST training dataset for training a LeNet-5 network, not their horizontally flipped versions. During testing, we consider eight de-rotation angles, four corresponding to the test image and the other four corresponding to horizontally flipped test image. This approach is referred as **EOCL-HT**. In EOCL-HT, horizontal flipping is used during testing. RICNN [10] introduces a rotational invariant feature which is same for an image and its horizontally flipped version. Localization module in STN [11] is also claimed to be learning almost all transformation. In this way, using horizontal flipping in EOCL and comparing the performance of EOCL with those of others do not introduce any discrimination. From Table 3, it is obvious that the performance of EOCL is the best among the three approaches.

Digit Wise Classification Error Analysis: We analyze the performance of EOCL for each digit separately. Also we compare the performance of EOCL with that of the baseline LeNet-5 for each digit. Figure 5 shows the test error profile for digits

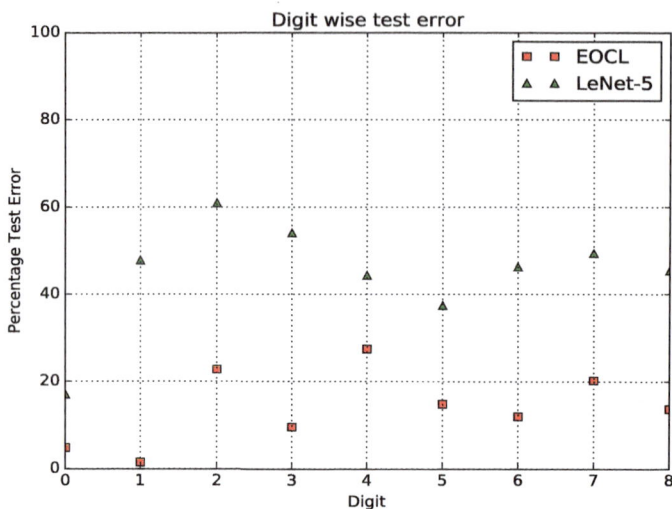

Fig. 5 Comparison of test error of baseline LeNet-5 and EOCL, for the range of rotation angles 0°–90° on MNIST test dataset. For all the digits (0–8), EOCL gives lesser classification error than that of the LeNet-5 architecture

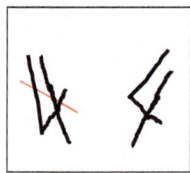

Fig. 6 Example of wrong orientation estimation and correction for the digit 4. Line segment shows the direction of maximum spread (principal component)

0, 1, ..., 8, for angle of rotation ranging from 0°–90°. Obviously, EOCL performance is better than that of LeNet. The digit 4 gives a higher test error compared to all other digits when EOCL is used. As shown in Fig. 6, the maximum spread of the digit 4 can be in an orientation other than the vertical direction, which is deviating from our assumption that the principal component direction of digits which are not rotated (normal MNIST test dataset) will be same as the vertical axis direction. Since we are validating our approach on handwritten characters, sometimes this assumption may go wrong as the characters are highly subjective. Even then, our experiments (Tables 2 and 3, Fig. 7) show that the proposed method gives a simple but efficient approach to classify rotated digits.

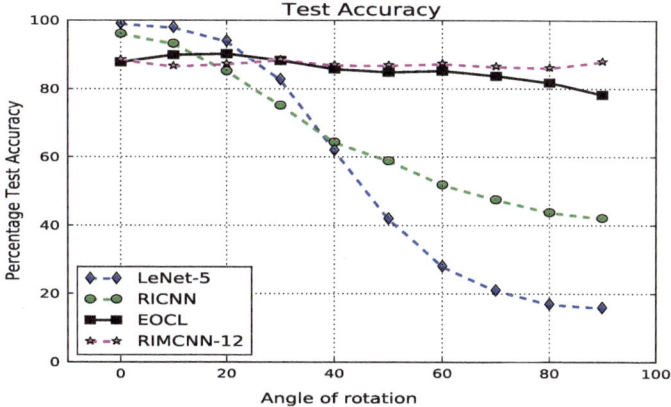

Fig. 7 Test accuracy on rotated MNIST dataset for different methods. The comparison is done over the range of orientations of test digits from 0° to 90°

4.4 Comparing Performance of EOCL with Other Approaches

The proposed method does not require any rotation in training. It can be trained with digits having single orientation. So, the number of training images is lesser in EOCL compared to other methods which are using training data augmentation [11]. It results in lesser training time compared to such data augmented methods. Many of the approaches can not estimate the orientation of test digits (eg: RICNN [10]) while our approach, EOCL, can be used to extract the relative orientation of each test digit. We are not modifying the existing CNN architecture for rotation invariance, which ensures the number of parameters of the network not more than that of the baseline architecture. Suitable pretrained models can be used without any modification in our approach. Figure 7 gives the comparison of the test accuracy on MNIST test dataset with rotations ranging from 0° to 90° for normal LeNet-5 architecture [1], RICNN [10], RIMCNN [12] and the proposed method EOCL. EOCL is computationally efficient and guarantees higher classification accuracy on rotated digits with least test time.

RIMCNN [12] test time is on an average 155 s when test is done with 12 different instances of test image, on the MNIST test dataset consisting of 9,000 test samples. For the same test dataset, EOCL takes an average test time of 25 s, if we are considering only two de-rotation angles as discussed in Case 1 (Sect. 4.2). It is 53 s, if 4 angles are considered at a time as in Case 2 (Sect. 4.2). Also, our approach has a comparable accuracy as that of RIMCNN [12] with 12 instances of test image and with same number of network parameters.

5 Conclusions

We propose an effective approach, EOCL (Eigenvector Orientation Corrected LeNet), for classifying rotated digits and estimating their angle of orientation. Our approach uses pretrained LeNet-5 architecture, trained using handwritten digits having only single orientation. By using an Principal Component based de-rotation step prior to testing with CNN, EOCL estimates and corrects the orientation of digits having arbitrary rotation. Our method ensures higher accuracy for rotated digit classification with lesser test time and does not require any increase in the number of network parameters.

The concept of 'EOCL' can be extended to other datasets like characters of different languages, Fashion MNIST, CIFAR etc. choosing any appropriate baseline CNN, by incorporating the eigenvector based orientation correction step prior to both training and testing. The eigenvector based 'Orientation estimation and correction' step can be combined with any baseline CNN, without demanding any architectural change and can be used for rotation invariant classification of characters and objects.

References

1. Lecun, Y., Bottou, L., Bengio, Y., Haffner, P.: Gradient-based learning applied to document recognition. Proc. IEEE **86**(11), 2278–2324 (1998). https://doi.org/10.1109/5.726791
2. Deng, L.: The MNIST database of handwritten digit images for machine learning research. IEEE Signal Process. Mag. **29**(6), 141–142 (2012). https://www.microsoft.com/en-us/research/publication/the-mnist-database-of-handwritten-digit-images-for-machine-learning-research/
3. Santosh K.C.: Character recognition based on DTWRadon. In: International Conference on Document Analysis and Recognition (2011). https://doi.org/10.1109/ICDAR.2011.61
4. Bhandare, A., Bhide, M., Gokhale, P., Chandavarka, R.: Applications of convolutional neural networks. Int. J. Comput. Sci. Inf. Technol. **7**(5), 2206–2215 (2016)
5. Lowe, D.G.: Distinctive image features from scale-invariant keypoints. Int. J. Comput. Vis. **14**(1), 234–778 (2004). https://doi.org/10.1023/B:VISI.0000029664.99615.94
6. Bay, H., Tuytelaars, T., Van Gool, L.: SURF: speeded up robust features. Lecture Notes in Computer Science, vol. 3951(1), pp. 404–417. Springer, Berlin (2006). https://doi.org/10.1007/11744023_32
7. Zhang, H., Li, Z., Liu, Y.: Fractional orthogonal Fourier-Mellin moments for pattern recognition. In: CCPR, vol. 662(1), 766–778. Springer, Singapore (2016). https://doi.org/10.1007/978-981-10-3002-4_62
8. Worrall, D.E., Garbin, S.J., Turmukhambetov, D., Brostow, G.J.: Harmonic networks: deep translation and rotation equivariance. arXiv:1612.04642
9. Marcos, D., Volpi, M., Komodakis, N., Tuia, D.: Rotation equivariant vector field networks. arXiv:1612.09346
10. Kandi, H., Jain, A., Velluva Chathoth, S. et al.: Incorporating rotational invariance in convolutional neural network architecture. Pattern Anal. Appl. (2018). https://doi.org/10.1007/s10044-018-0689-0
11. Jaderberg, M., Kimonyan, K., Zisserman, A., Kavukcuoglu, K.: Spatial transformer networks. arXiv:1506.02025
12. Jain, A., Subrahmanyam, G.S., Mishra, D.: Stacked features based CNN for rotation invariant digit classification. In: Pattern Recognition and Machine Intelligence, 7th International Conference, PReMI Proceedings (2017). https://doi.org/10.1007/978-3-319-69900-4_67

A Reference Based Secure and Robust Zero Watermarking System

Satendra Pal Singh and Gaurav Bhatnagar

Abstract In this paper, a new approach for zero-watermarking has been proposed based on log-polar mapping, all phase biorthogonal sine transform and singular value decomposition. The core idea is to produce a zero watermark to protect the copyright of the image. For this purpose, the host image is transformed into the log-polar domain followed by the all phase biorthogonal sine transform (APBST). The transformed coefficients are then divided into nonoverlapping blocks and some blocks are selected based on secret key. These blocks are finally used to formulate a reference matrix which is utilized to generate a zero watermark for the host image. A detailed experimental analysis is conducted to demonstrate the feasibility of the proposed algorithm against various image/signal processing distortions.

Keywords Zero watermarking · Log-polar mapping · Singular value decomposition (SVD)

1 Introduction

In the digital era, modern society increasingly accesses the digitized information due to wide availability of Internet and digital device such as mobile, digital camera, etc. As a result, the distribution of multimedia data such as image, video, and audio, over the social networks is exponentially increased. At the same time, widespread availability of cheap and powerful image processing software has made illegal image reproduction, copying, tempering, and manipulation an easy task. Therefore, copyright protection and authentication become an important issue for the security of digital data. These problems can be effectively addressed by digital watermarking technology [1, 2].

In traditional watermarking system [3], an identification mark (watermark) is embedded into the host image which can be extracted at the later stage to verify

S. P. Singh (✉) · G. Bhatnagar
Department of Mathematics, Indian Institute of Technology Jodhpur, Jodhpur, India
e-mail: pg201383504@iitj.ac.in

© Springer Nature Singapore Pte Ltd. 2020
B. B. Chaudhuri et al. (eds.), *Proceedings of 3rd International Conference on Computer Vision and Image Processing*, Advances in Intelligent Systems and Computing 1022, https://doi.org/10.1007/978-981-32-9088-4_28

the ownership of the digital image. However, insertion of watermark degrades the quality of the image and hence a tradeoff can be realized between robustness and imperceptibility for these type of embedding algorithms and therefore less suitable for medical and remote sensing applications. More precisely, loss-less watermarking schemes are suitable for these type of images to protect the copyright without altering the image data and therefore zero-watermarking techniques become very popular in comparison to the traditional watermarking techniques. Generally, zero watermarking techniques extract the inherent features of the host image to produce a watermark for image authentication and verification. A number of zero-watermarking techniques have been proposed in the literature [4–6], but most of them are mainly based on local invariant features in frequency domain. In [5], authors have represented a zero-watermarking scheme based on higher order cumulants. This scheme preserves a good robustness against common signal processing operations but less effective against geometric operations, specially large-scale rotations. In [6], authors have constructed a zero watermark using the discrete wavelet transform and chaotic sequence. The detailed coefficients at three-level wavelet decomposition are used in the generation of zero-watermark using chaotic sequences. In [7], authors have generated a zero watermark by producing a feature vector map based on lifting wavelet transform and Harris corner detector. In [8], authors have proposed a zero watermarking scheme based on DWT and principal component analysis (PCA). The PCA is applied on details wavelet coefficients to produce a feature vector of required size and the resultant image is registered with the help of a binary watermark in intellectual property right (IPR) database. In [9], authors have reported a watermarking scheme where quaternion exponent moments (QEM) is used in the generation of feature image followed by the construction of verification image with the help of a scrambled binary watermark.

In this paper, a robust and secure zero watermarking algorithm based on log-polar mapping (LPM) and all phase biorthogonal sine transform (APBST) is presented to resist the various signal and image processing attacks. The input image is transformed using LPM and APBST simultaneously. Both the coefficient matrices are then divided into nonoverlapping blocks. Some blocks of each coefficient matrices are selected using the same secret key and subjected to zig-zag scan for extraction of the significant coefficients. The selected coefficients of each block are stacked into a vector, then collection of all these vectors construct a reference matrix followed by its factorization using singular value decomposition. Finally, a feature vector is generated using the left and right singular vectors. The feature vector along with binary watermark produce a verification image or zero watermark to protect the image.

The rest of the paper is arranged as follows: Sect. 2 provides the brief description of log-polar mapping and all phase biorthogonal sine transform. Section 3 presents the proposed zero watermarking system followed by the experimental results in Sect. 4. Finally, Sect. 5 summarizes the concluding remarks.

2 Preliminaries

2.1 Log-Polar Mapping

Log-polar mapping is a nonuniform sampling method and a space-variant transformation which maps the Cartesian coordinates of an image into the log-polar coordinates. The main characteristic of this map is defined by the property of rotation and scale invariance, i.e., if an image is rotated or scaled with respect to center in cartesian domain then output turns to be transitional shift in log-polar domain. Mathematically, the log-polar coordinates [10] can be defined as follows:

$$(\xi, \eta) \equiv \left(\log \left(\frac{\rho}{\rho_0} \right), q \cdot \theta \right) \tag{1}$$

where (ρ, θ) denote the polar coordinates and (x, y) are the usual Cartesian coordinates. These polar coordinates can be defined as

$$(\rho, \theta) = \left(\log \sqrt{(u - \tilde{u})^2 + (v - \tilde{v})^2}, \arctan \frac{(v - \tilde{v})}{(u - \tilde{u})} \right) \tag{2}$$

where $(\tilde{u}, \tilde{v})$ and (u, v) denotes the center and sampling pixel in Cartesian coordinate system. Here ρ and θ are called radial distance and angle measured from the center. The logarithm of radial distance and angle represent the log-polar coordinate system. Due to discretization, $(x, y) = (\xi, \eta), 0 \le x < N_r, 0 \le y < N_s$ where N_r and N_s is number of ring and sectors in log-polar image, and $q = N_s/2\pi$ is the respective angular resolution [10]. Let $I(u, v)$ be an input image which is rotated by an angle α and then scaled by the factor s to produce a geometrically transformed image. Let I' be an output image and corresponding coordinates can be obtained as follows:

$$\begin{aligned} I'(u, v) = I[s(u - \tilde{u}) \cos \alpha + s(v - \tilde{v}) \sin \alpha + \tilde{u}, \\ s(u - \tilde{u}) \sin \alpha + s(v - \tilde{v}) \cos \alpha + \tilde{v}] \end{aligned} \tag{3}$$

In log-polar coordinate system

$$\begin{aligned} u = e^\rho \cos \theta \\ v = e^\rho \sin \theta \end{aligned} \tag{4}$$

Then Eq. (2) can be described as follows:

$$I'(\log \rho, \theta) = I[(\log \rho, \log s), (\theta + \alpha)] \tag{5}$$

The rotation applied in spatial domain results the linear shift along the η-axis whereas scaling operation in the spatial domain reflect a linear shift along the ξ-axis in log-polar image. Figure 1 illustrate the mechanism of log-polar mapping.

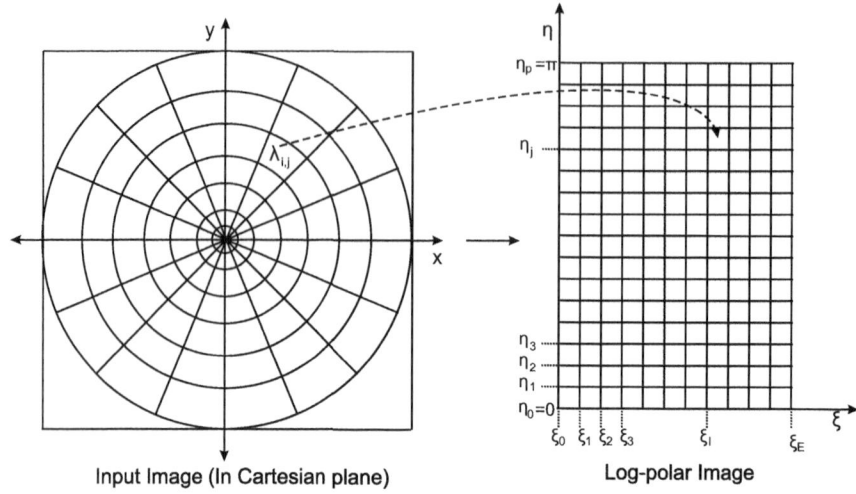

Input Image (In Cartesian plane)   Log-polar Image

Fig. 1 Basic mechanism of Log-polar mapping

2.2 All Phase Biorthogonal Sine Transform

All Phase Biorthogonal Sine Transform (APBST) [11] is a mathematical transform and widely used in the image compression, image denoising, and other image processing application. The basic idea is inspired by the sequence filtering in the frequency domain. It has more advantage than the conventional transform due to better energy compaction in the low frequency component. Discrete sine transform (DST) [11] matrix can be defined as follows:

$$S_{i,j} = \frac{2}{\sqrt{(2K+1)}} \sin\left[\frac{(2i+1)(j+1)\pi}{(2K+1)}\right] \quad i,j = 1 \ldots K-1 \qquad (6)$$

For DST, all phase digital filtering can be designed using a digital sequence. For this purpose, consider a digital sequence $z(t)$, where each member of sequence correspond to K different values. The average of these values can be assigned as the all phase filtering. The output response can be defined as

$$y(t) = \sum_{i=0}^{K-1} \sum_{j=0}^{K-1} [H_{i,j} \, z(t - i + j)] \qquad (7)$$

where

$$H_{i,j} = \frac{1}{K} \sum_{i=0}^{K-1} F_K(m) S(i,m) S(j,m) \qquad (8)$$

Substituting Eq. (8) into Eq. (7) and after solving, the output

$$y(t) = \sum_{\tau=-(K-1)}^{K-1} h(\tau)\, z(t-\tau) = h(t) * z(t) \tag{9}$$

where $h(\tau)$ represents the unit impulse response and can be express as given below:

$$h(\tau) = \begin{cases} \displaystyle\sum_{i=\tau}^{K-1} H_{(i,i-\tau)} & \tau = 0, 1\ldots, K-1 \\[2em] \displaystyle\sum_{i=0}^{\tau+K-1} H_{(i,i-\tau)} & \tau = -1, -2\ldots, -K+1 \end{cases} \tag{10}$$

Also, $H_{i,j} = H_{j,i}$. From Eqs. (8) and (10), we have

$$h(\tau) = \sum_{m=0}^{K-1} V(\tau, m) F_K(m) \qquad \tau = 0, 1\ldots, K-1 \tag{11}$$

Matrix representation of Eq. (11) can be expressed as $h = VF$, where the transformation corresponding to matrix V is used to describe the relationship between unit-pulse time response in time domain and sequence response in transform domain. This matrix V is know as APBST matrix and elements of V can be computed as

$$V_{i,j} = \frac{1}{K} \sum_{i=0}^{K-1-i} S(j, \ell) S(j, \ell + i) \tag{12}$$

From Eq. (6) and (12), APBST matrix can be estimated as follows:

$$V_{i,j} = \begin{cases} \frac{1}{K} & i = 0,\ j = 0, 1\ldots, K-1 \\[1em] \frac{4}{K(2K+1)} * \beta & i = 1, \ldots, K-1 \\ & j = 0, 1\ldots, K-1 \end{cases} \tag{13}$$

where

$$\beta = \sum_{i=0}^{K-1-i} \sin\left[\frac{(2j+1)(\ell+1)\pi}{(2K+1)}\right] \sin\left[\frac{(2j+1)(\ell+i+1)\pi}{(2K+1)}\right] \tag{14}$$

3 Proposed Zero Watermarking System

In this section, a new approach to zero watermarking scheme has been discussed. The main objective is feature extraction based on the invariance transformation. The invariant features are achieved by the log-polar mapping and their frequency component through APBST transformation. These extracted features and a logo image is then used to construct a verification image for authentication purposes. Without loss of generality, let $\mathcal{I}$ and w be grayscale host and binary watermark images respectively of sizes $M \times M$ and $n \times n$.

3.1 Watermark Generation Process

The watermark generation process can be summarized as follows:

1. Obtain a log-polar image $\mathcal{I}_{LP}$ form the input image $\mathcal{I}$.

$$\mathcal{I}_{LP} = \text{LPM}(\mathcal{I}) \tag{15}$$

2. Obtain a coefficient matrix $\mathcal{I}_{LP}^{f}$ by applying the APBST on each nonoverlapping block of the log-polar image $\mathcal{I}_{LP}$.

$$\mathcal{I}_{LP}^{f} = \text{APBST}(\mathcal{I}_{LP}) \tag{16}$$

3. Each block $(B_i | i = 1, \dots M \times N/s^2)$ of $\mathcal{I}_{LP}$ and $\mathcal{I}_{LP}^{f}$ are accessed using same secret keys A_{Key}. The parameter s denotes the size of the block.
4. Selected blocks B_s is subjected to zig-zag scan and ℓ significant coefficients with their corresponding frequency components are stacked to generate a column vector.
5. Construct a Reference matrix R_M by combining these vectors and perform the singular value decomposition.

$$R_M = U_M S_M V_M^T \tag{17}$$

6. Obtain a feature vector f_v using the k columns of U_M and V_M as follows.

$$f_v = [U_M^{(k)}, V_M^{T(k)}] \tag{18}$$

7. Arrange f_v into two-dimensional matrix $P \times Q$ to construct a feature image F_M and obtain a binary feature image F_M^B through using threshold T.

$$F_M^B(i, j) = \begin{cases} 1, & F_M(i, j) \geq T \\ 0, & F_M(i, j) < T \end{cases} \tag{19}$$

8. Obtain a scrambled watermark $\bar{w}$ based on Arnold cat map [12] from the input watermark w.
9. Produce a zero watermark/verification image by performing XOR operation between the F_M^B and $\bar{w}$.

$$V_{img} = XOR(F_M^B, \bar{w}) \qquad (20)$$

10. Finally, verification image along with all secret keys are signed by image owner using a digital signature technique [13] and register into intellectual property right (IPR) database for copyright protection. Mathematically,

$$D_{sig} = Sig_{opk}(V_{img}, P_{Key}) \qquad (21)$$

where Sig_{opk} denote a digital signature function based on owner's private key P_{Key}.

3.2 Watermark Verification Process

Verification steps can be summarized as follows:

1. Firstly, validate the digital signature D_{sig} using the private key P_{Key} to confirm the security. After successful verification, authentication process may proceed further otherwise the process will be terminated.
2. Obtain a feature vector f_v^* and feature image F_M^{B*} from the possibly attacked image I_A by using the secret key A_{Key} as described in Steps 1–6 of Sect. 3.1.
3. Obtain a scrambled logo by using XOR operation between the binary feature image F_M^{B*}

$$\bar{w}* = XOR(F_M^{B*}, V_{img}) \qquad (22)$$

4. Obtain the original logo watermark using inverse scrambling process and evaluate the authenticity of retrieved logo to authenticate the image.

4   Experimental Results

The efficiency of proposed zero watermarking technique is investigated through a detailed experimental analysis using MATLAB platform. The standard grayscale images such as Lena, Pepper, Cameraman and Barbra of size 512×512 are considered as the host image in the experiments to evaluate the performance of the proposed technique. However, the visual results are shown only for Lena and Cameraman image, due to limited space availability. These images are depicted in Fig. 2a,

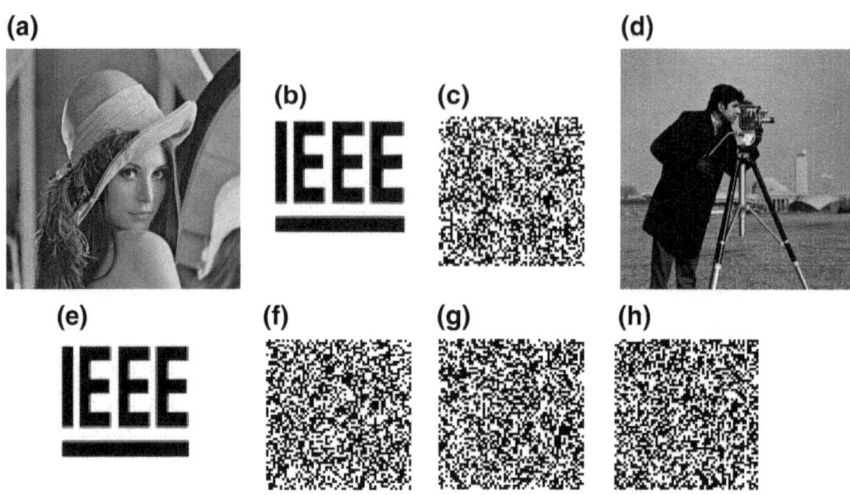

Fig. 2 **a** Lena image, **b** Watermark logo image, **c** Verification image, **d** Cameraman image, **e** Extracted watermark image, **f** Extracted watermark with Cameraman image, **g**, **h** Extracted with wrong keys

d. A binary watermark logo namely 'IEEE' of size 64×64 is used in the construction of zero watermark/Verification image. The binary watermark and verification image are depicted Fig. 2b, c respectively.

The authenticity of test image 'Lena' is examined by extracting the watermark logo image for copyright protection. The watermark image is extracted from the Lena image using the verification image and compared with the original watermark. The extracted watermark is same as the original one and depicted in Fig. 2e. The process extended to check the validity of proposed scheme by considering the same verification logo and different image other than Lena namely 'Cameraman' have been taken for watermark extraction and it does not extract the logo similar to original 'IEEE' logo. The extracted logo is shown in Fig. 2f. This essentially implies that image fails in the verification process. Further, effectiveness of the proposed scheme is analyzed by considering the verification logo and test image with wrong secret keys. This also fails to extract an estimate of original watermark. The retrieved logos are shown in Fig. 2g, h. Hence proposed framework is secure as none is able to extract the original watermark without the knowledge of true secret keys.

The performance of the proposed algorithm is investigated using different kind of image distortions such as salt and pepper noise, Gaussian noise addition, JPEG compression, median filter, Gaussian blur, resizing, rotation, cropping, sharpening, UnZign and contrast adjustment. The degree of robustness of the proposed technique is evaluated through watermark extraction from distorted image by considering the above attacks using bit error rate (BER). Mathematically, BER can be defined as follows:

$$BER = \frac{C_B}{m \times n} \times 100\% \tag{23}$$

Table 1 Comparison of BER results of proposed method and other zero watermarking method for extracted watermark logo

Attacks	Bit error rate			
	Proposed	Reference [9]	Reference [13]	Reference [14]
Salt and pepper (0.01)	0.0090	0.0088	0.0465	0.0698
Gaussian noise (0.01)	0.0349	0.0131	0.0285	0.0850
JPEG compression (60%)	0.0076	0.0066	0.0092	0.0066
Median filter (3 × 3)	0.0046	0.0049	0.0164	0.0178
Gaussian blur (20%)	0.0073	0.0125	0.0075	0.0092
Resizing (512 → 256 → 512)	0.0054	0.0062	0.0075	0.0096
Rotation (2°)	0.0828	0.0000	0.1776	0.3245
Cropping (1/4)	0.0349	0.1242	0.1978	0.1601
Sharpen (50%)	0.0120	0.1120	0.0251	0.0285
Contrast adjustment (70%)	0.0417	0.0185	0.0344	0.0621
UnZign (6, 6)	0.0417	0.0185	0.0344	0.0621

where $m \times n$ denotes the size of the watermark image and C_B represents the number of error bits. The lower BER defines the closeness between the original and retrieved logo. The BER for the retrieved watermark from Lena image is found to be zero. For detailed attack analysis, firstly, Lena image is subjected to salt and pepper noise (density $= 0.01$) and then watermark image is retrieved from the noisy distorted images. The BER is computed between the retrieved logo and the original one, which is listed in Table 1, whereas retrieved logo and distorted image are shown in Fig. 3a, i. Similarly, the efficiency of proposed image is tested against the additive Gaussian noise (mean $= 0$, variance $= 0.01$). The respective BER is listed in Table 1, and corresponding distorted image with extracted logo is shown in Fig. 3(b, ii).

Robustness of proposed scheme is also tested against JPEG compression (60%) and median filtering with (3 × 3) window. The distorted images and their extracted logos are depicted in Figs. 3c, d, iii, iv respectively. The watermark image is also reconstructed after blurring (20%) and resizing operation. For image resizing, firstly image size is scaled down to 256 and scale up to the original size. These distorted images and corresponding retrieved logo are listed in Figs. 3 e, f, v, vi. The efficiency of the proposed scheme also measured image rotation (2%) and cropping (1/4). The rotated and cropped images and respective extracted logos are depicted in Fig. 3 g, h, vii, viii respectively. In addition, the extracted logos from the distorted image after

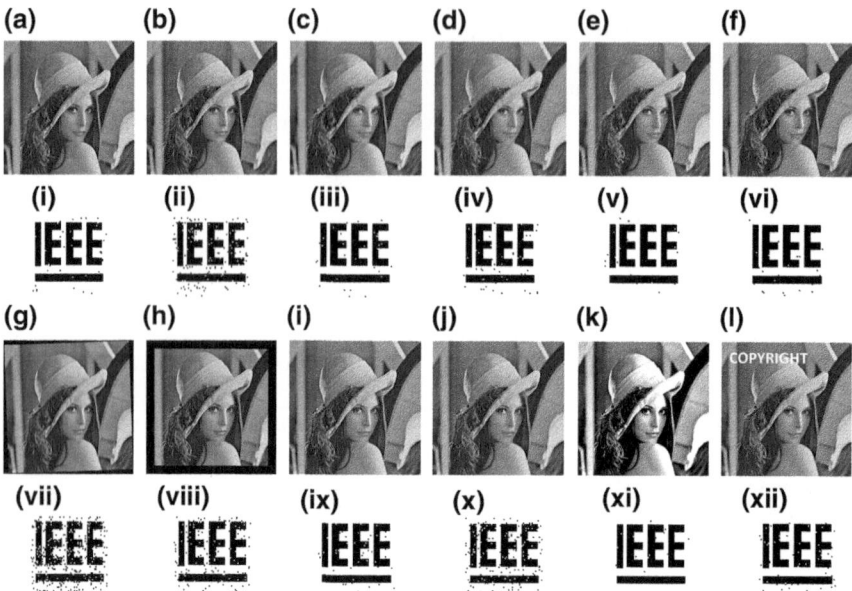

Fig. 3 Demonstration of attacked images and retrieved logos: **a** Salt and pepper noise (noise density = 0.01), **b** Additive gaussian noise (mean=0, variance=0.01), **c** JPEG compression (60%), **d** Median filer (3 × 3), **e** Blurring (20%), **f** Resizing (512→ 256→ 512), **g** Rotation (2%), **h** Cropping (1/4), **i** UnZign (6, 6) **j** Sharpening (50%), **k** Contrast adjustment (70%), **l** Image tempering; Second and fourth row shows the corresponding retrieved logos

image sharpening (50%) and contrast adjustment (70%) are depicted in Figs. 3x, xi and corresponding distorted images are shown in Figs. 3j–k. Also, the host image is subjected to UnZign attack in which same number of row and column are deleted randomly and then scaled up to original size of the image. In the next experiment, the performance of the scheme has been measured against image tampering. Both of attacked images and corresponding extracted watermarks are depicted in Figs. 3ix, xii, i, l. The BER for all the attacks are listed in Table 1.

For comparative analysis, the BER obtained from the proposed scheme are compared with the existing schemes [9, 13, 14]. From Table 1, it can be observed that proposed zero watermarking scheme gives lower BER in the comparison of the other schemes and therefore, proposed scheme is better than existing schemes in terms of robustness and optimization. The BER against JPEG compression have been computed and depicted in Fig. 4. From the figure, it can be seen that our scheme produce lower bit error rate among all.

The consistency of the performance of the proposed scheme also examined by determining the accuracy rate (AR) for the retrieved image. The accuracy rate can be defined mathematically as follows:

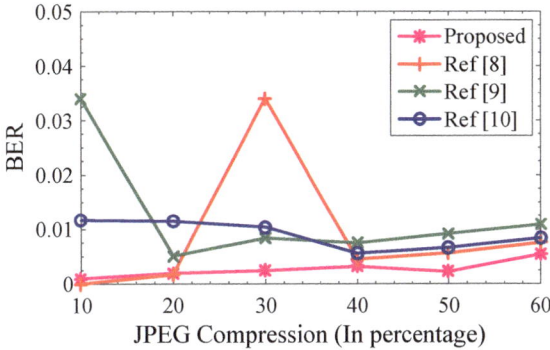

Fig. 4 Comparison of BER results obtained after JPEG compression with other schemes

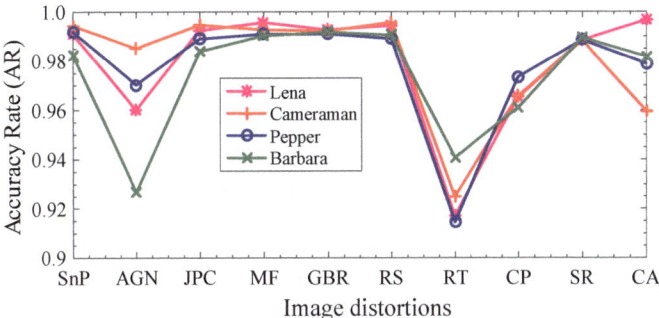

Fig. 5 Accuracy rate for different standard images after image distortions (SnP = Salt and pepper noise, AGN = Additive Gaussian noise, JPC = JPEG compression, MF = Median filtering, GBR = Gaussian blurring, RS = Resizing, RT = Rotation, CP = Cropping, SR = Sharpening, CA = Contrast adjustment.)

$$AR = \frac{C_C}{m \times n} \times 100\% \tag{24}$$

where $m \times n$ denote the size of the watermark image and C_c represent the number correctly classified bits. The estimated AR of Lena and Cameraman image is one, which essentially proves that the proposed technique is capable of perfect verification. Further, AR for all the experimental images against underlying distortions are also estimated and shown in Fig. 5.

Finally, the consistency of proposed scheme is also measured by estimating AR against the JPEG compression and image cropping for different images namely Lena, Cameraman, pepper and Barbra. The accuracy rate have been shown in Figs. 6 and 7. From the figure, it can be observe minimum AR is 97 and 94.80% for jpeg compression and cropping. From the detailed statistical analysis, it can be summarized that proposed scheme gives equivalent performance to [9], however better than [13, 14].

Fig. 6 Accuracy rate for standard images after JPEG compression

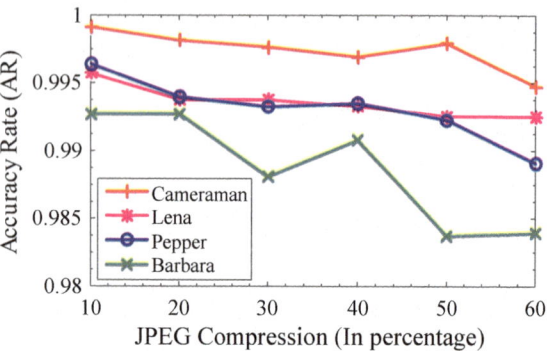

Fig. 7 Accuracy rate for different standard images after cropping

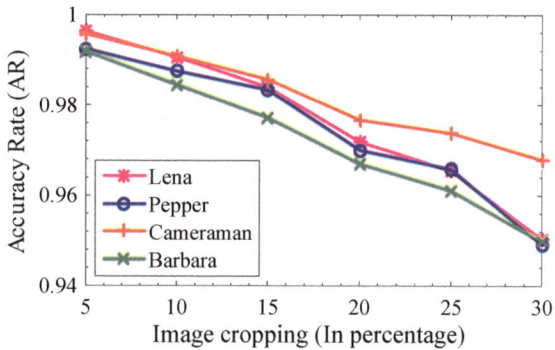

5 Conclusion

In this paper, a novel zero watermarking technique based on log-polar mapping and APBST have been presented. The proposed method extracts the internal features of host media by utilizing the invariance property of log-polar mapping and correspond-ing descriptive details using the APBST. These features generate a digital signature using a binary watermark for the authentication purpose. A detailed experimental analysis has been conducted for validation of the scheme and simulated results indi-cate that the proposed scheme have lower BER and good accuracy rate against several kinds of image distortions. The proposed scheme demonstrates the better robustness in compare to the existing schemes.

Acknowledgements The authors gratefully acknowledges the support of Science and Engineering Research Board, DST, India for this research work.

References

1. Katzenbeisser, S., Petitcolas, F.A.P.: Information Hiding Techniques For Steganography and Digital Watermarking. Artech House, Boston (2002)
2. Singh, S.P., Bhatnagar, G.: A new robust watermarking system in integer DCT domain. J. Vis. Commun. Image Represent. **53**, 86–101 (2018)
3. Singh S.P., Bhatnagar G.: A novel chaos based robust watermarking framework. In: International Conference on Computer Vision and Image Processing, CVIP, vol. 2, pp. 439–447 (2017)
4. Wjtowicz, W., Ogiela, M.R.: Digital images authentication scheme based on bimodal biometric watermarking in an independent domain. J. Vis. Commun. Image Represent. **38**, 1–10 (2016)
5. Wen, Q., Sun, T.F., Wang, S.X.: Concept and application of zero-watermark. Acta Electron. Sin. **31**(2), 214–216 (2003)
6. Kou, J.K., Wei, L.X.: Zero-watermarking algorithm based on piecewise logistic chaotic map. Comput. Eng. Des. **34**(2), 464–468 (2013)
7. Li, F., Gao, T., Yang, Q.: A novel zero-watermark copyright authentication scheme based on lifting wavelet and Harris corner detection. Wuhan Univ. J. Nat. Sci. **5**, 408–414 (2010)
8. Leng X., Xiao J., Wang Y.: A robust image zero-watermarking algorithm based on DWT and PCA. In: Communications and Information Processing, pp. 484–492 (2012)
9. Wang, C.P., Wang, X.Y., Xia, Z.Q., Zhang, C., Chen, X.J.: Geometrically resilient color image zero-watermarking algorithm based on quaternion exponent moments. J. Vis. Commun. Image Represent. **41**, 247–259 (2016)
10. Young, D.: Straight lines and circles in the log-polar image. In: British Machine Vision Conference, pp. 426–435 (2000)
11. Saxena, A., Fernandes, F.C.: DCT/DST based transform coding for intra-prediction in image/video coding. IEEE Trans. Image Process. **22**(10), 3974–3981 (2013)
12. Sui, L., Gao, B.: Color image encryption based on gyrator transform and Arnold transform. Opt. Laser Technol. **48**, 530–538 (2016)
13. Chen, T.H., Horng, G., Lee, W.B.: A publicly verifiable copyright-proving scheme resistant to malicious attacks. IEEE Trans. Ind. Electron. **52**(1), 327–334 (2005)
14. Chang, C.C., Lin, P.Y.: Adaptive watermark mechanism for rightful ownership protection. J. Syst. Softw. **81**(7), 1118–1129 (2008)

Two-View Triangulation: A Novel Approach Using Sampson's Distance

Gaurav Verma, Shashi Poddar, Vipan Kumar and Amitava Das

Abstract With the increase in the need for video-based navigation, the estimation of 3D coordinates of a point in space, using images, is one of the most challenging tasks in the field of computer vision. In this work, we propose a novel approach to formulate the triangulation problem using Sampson's distance, and have shown that the approach theoretically converges toward an existing state-of-the-art algorithm. The theoretical formulation required for achieving optimal solution is presented along with its comparison with the existing algorithm. Based on the presented solution, it has been shown that the proposed approach converges closely to Kanatani–Sugaya–Niitsuma algorithm. The purpose of this research is to open a new frontier to view the problem in a novel way and further work on this approach may lead to some new findings to the triangulation problem.

Keywords Triangulation · Stereovision · Monocular vision · Fundamental matrix · Epipolar geometry

1 Introduction

Over the past few decades, the development of computer vision techniques for estimating the position of a point in space has been the goal of many research programs for applications ranging from guidance and control to 3D structure reconstruction. This process of computing the position of a point in space, from its projection on two views, is known as triangulation. The basic method of triangulation is carried out by

---

The work was carried out during the first author's affiliation with CSIR-Central Scientific Instruments Organisation, Chandigarh, India.

G. Verma · S. Poddar · V. Kumar · A. Das (✉)
CSIR-Central Scientific Instruments Organisation (CSIO), Chandigarh 160030, India
e-mail: amtds06@gmail.com

Academy of Scientific Innovation and Research (AcSIR), CSIR-CSIO Campus,
Chandigarh 160030, India

B. B. Chaudhuri et al. (eds.), *Proceedings of 3rd International Conference on Computer Vision and Image Processing*, Advances in Intelligent Systems and Computing 1022,
https://doi.org/10.1007/978-981-32-9088-4_29

339

intersecting the back-projected rays that are corresponding to the 2D image points from two views and is popularly known as direct linear transform (DLT), which minimizes the algebraic distance [1]. However, due to the presence of noise in the corresponding 2D image points, such as quantization noise, geometric error [2], and feature matching uncertainty [3], the problem of triangulation becomes nontrivial. The feature matching uncertainty refers to the presence of outliers, which arises due to the matching of features which are similar but do not correspond to each other in the two views. On the contrary, the quantization noise, during imaging of the point in space, imparts uncertainty to the location of the image points because of which the location of the 3D point can only be ascertained to lie within a specific region (the shaded region, as shown in Fig. 1a). Although several feature detectors return image coordinates with sub-pixel resolution, these coordinates still remain affected by a certain amount of noise [4]. Matthies and Shafer [2] have proposed a noise model to reduce the error, which was reiterated by Maimone et al. [5]. This was followed by the geometric error analysis method, for modeling the quantization error as projected pyramids and the uncertainty region as an ellipsoid, around the polyhedron intersection of the pyramids [6]. Recently, Fooladgar et al. [7] addressed the problem of quantization error and proposed a model to analyze the error in the camera CCD (charge-coupled device).

The geometric error and the feature matching uncertainty affect the position of the points in the image plane. This leads to the deviation of the image points from their respective epipolar lines and the back-projected rays may not intersect, as shown in Fig. 1b, which results in triangulation error. This research is a step toward reducing the effect of geometric error and feature matching uncertainty, with computationally less expensive method.

In order to obtain the best estimate of 3D coordinates of the point in space, several methods have been proposed in the past. The midpoint method proposed in [8], aims at finding the midpoint of the shortest distance between the two skewed rays. Although few improvements have been explored with this method, they do not

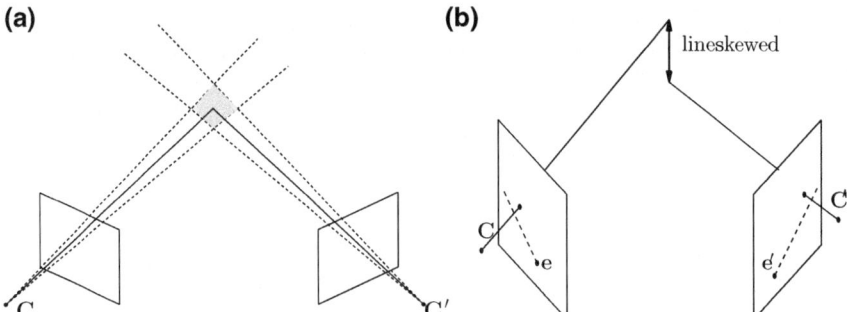

Fig. 1 Triangulation error. **a** Uncertainty region due to quantization error. **b** Skewed back-projected lines due to deviation of image points from the epipolar lines. **C** and **C′** represent the camera centers and **e** and **e′** are the epipoles for the two views respectively

provide an optimal output due to various approximations [9]. An optimal method to minimize the error [1] is by using the L_2 re-projection error in (1)

$$L = d(\mathbf{x}, \hat{\mathbf{x}})^2 + d(\mathbf{x}', \hat{\mathbf{x}}')^2 \tag{1}$$

as the cost function. Here, $d(\hat{\mathbf{x}}, \hat{\mathbf{x}}')$ is the Euclidean distance between the vectors $\hat{\mathbf{x}}$, $\hat{\mathbf{x}}'$ where the vectors denote the re-projection of the 3D point corresponding to the homogeneous coordinates $\mathbf{x}$ and $\mathbf{x}'$ in the two views respectively. Usually, nonlinear optimization techniques (e.g., Levenberg–Marquardt) are used to obtain a correct solution, but evolutionary algorithms have also been used in the minimization of triangulation error. Vite-Silva et al. [10] have used multi-objective evolutionary algorithm, whereas [11] implemented particle swarm optimization (PSO) for finding the optimal solution. However, these algorithms are quite time consuming and hence can be used if there are only few noisy corresponding points.

One of the landmark works in this field was that of Hartley and Sturm [9], who reformulated the cost function by computing the Euclidean distance between the image point and corresponding epipolar line. Henceforth, we will refer to this method as *polynomial*. The highlight of the method was that the dimensionality of the cost function was reduced to a single variable, with the solution satisfying the epipolar constraint is given by

$$\mathbf{x}'^{\top} \mathbf{F} \mathbf{x} = 0, \tag{2}$$

where $\mathbf{F}$ is the fundamental matrix [1]. Though the method is theoretically optimal, it suffers from the disadvantage of finding the roots of a 6 degree polynomial, which needs a significant amount of computation for high precision, making the task computationally expensive.

Kanazawa and Kanatani [12] implemented a concept similar to that of Hartley and Sturm and tried to find the optimality with reduced computational requirement. Further to this, [13] improved upon the previous method by introducing Lagrange's multiplier along with the absolute error in the cost function to obtain the solution iteratively, referred to here as *Kanatani−Sugaya−Niitsuma algorithm* or *KSN*. This method has lesser computational requirements as compared to the *polynomial* and computes the minima in relatively very short time. Recently, [14] proposed a more efficient approach (`niter2` as mentioned in [14]), which converges very quickly and finds an optimal solution with error relative to that of *polynomial*, of the order of $\pm 10^{-10}$. This method is referred to as *niter* in this paper. The major advantage of this method is its convergence in just two iterations which are further combined to form non-iterative method. Kanatani and Niitsuma [15] have extended the *KSN* method by introducing a constraint that the points lie on a planar surface. This is an extension of the *KSN* method by Kanatani et al. [13], for planar surfaces.

In this work, we propose a different approach toward the formulation of the triangulation problem from two views, which employs Sampson's distance [1, 16], as the cost function. It is considered because the Sampson's distance takes us to the first-order approximation of the point (of which position is to be estimated) [1]. We need to find the 2D points which minimize the error as well as satisfy the epipolar constraint.

It will further be shown that our approach with the Sampson's distance as the cost function converges to *Kanatani–Sugaya–Niitsuma algorithm* or *KSN* [13].

2 Theoretical Background

In this work, the uncalibrated case is being considered, where the camera calibration matrix is not known. The coordinates of the matched image points are given by $\mathbf{x}$ and $\mathbf{x}'$, in the homogeneous coordinate system, for first and second view, respectively. If the camera calibration matrix is known, i.e., for the calibrated case, the fundamental matrix $\mathbf{F}$ can be replaced with essential matrix $\mathbf{E}$, and the coordinates $\mathbf{x}$ and $\mathbf{x}'$ can be expressed in normalized coordinates. These points may not exactly satisfy (2) due to the error $\Delta\mathbf{x}$ and $\Delta\mathbf{x}'$ in the points $\mathbf{x}$ and $\mathbf{x}'$, respectively. Therefore, we need to find the correct estimate of the coordinates which are denoted here as $\hat{\mathbf{x}}$ and $\hat{\mathbf{x}}'$ and satisfy the epipolar constraint as

$$\hat{\mathbf{x}}'^{\top} \mathbf{F} \hat{\mathbf{x}} = 0. \tag{3}$$

In order to keep the third component of $\Delta\mathbf{x}$ and $\Delta\mathbf{x}'$ in homogeneous coordinates to zero, we describe a matrix

$$\mathbf{S} = \begin{pmatrix} 1 & 0 & 0 \\ 0 & 1 & 0 \end{pmatrix} \tag{4}$$

similar to [14]. Hence, $S^{\top}\Delta\mathbf{x}$ will bring the third component of $\Delta\mathbf{x}$ to zero and the corrected points can be represented as

$$\hat{\mathbf{x}} = \mathbf{x} - \mathbf{S}^{\top}\Delta\mathbf{x}, \quad \hat{\mathbf{x}}' = \mathbf{x}' - \mathbf{S}^{\top}\Delta\mathbf{x}'. \tag{5}$$

Therefore, the epipolar constraint (3) can be written as

$$(\mathbf{x}' - \mathbf{S}^{\top}\Delta\mathbf{x}')^{\top} \mathbf{F}(\mathbf{x} - \mathbf{S}^{\top}\Delta\mathbf{x}) = 0 \tag{6}$$

Further, we will use

$$\mathbf{F}_{in} = (\mathbf{SFS}^{\top})^{-1} \tag{7}$$

which represents the inverse of the upper left 2×2 sub-matrix of $\mathbf{F}$.

3 Optimal Triangulation Method Using Sampson's Distance

The re-projection error (or the geometric error) in (1) is complex in nature, and its minimization requires the estimation of both the homography matrix and the points $\hat{\mathbf{x}}$ and $\hat{\mathbf{x}}'$. The geometric interpretation of the error leads to Sampson error that lies between the algebraic and geometric cost functions in terms of complexity, but gives a close approximation to geometric error [1]. In this work, we reformulate methodology by employing Sampson's distance for the error formulation, as compared to the Euclidean distance used previously in [13] and [14]. The mathematical formulation required for the proposed method, referred hereafter as Sampson's Distance-based Triangulation Method (*SDTM*), is described further.

3.1 Reformulation of the Problem

We begin by reformulating the cost function using Sampson's distance, which is given according to [1] as

$$
\begin{aligned}
d_s &= \frac{(\mathbf{x}'^{\top}\mathbf{F}\,\mathbf{x})^2}{(\mathbf{Fx})_1^{\ 2} + (\mathbf{Fx})_2^{\ 2} + (\mathbf{F}^{\top}\mathbf{x}')_1^{\ 2} + (\mathbf{F}^{\top}\mathbf{x}')_2^{\ 2}} \\
&= \frac{a^2}{b}
\end{aligned}
\tag{8}
$$

where $(\mathbf{Fx})_j$ represents the jth entry of the vector $\mathbf{Fx}$, a and b are denoted as

$$
a = \mathbf{x}'^{\top}\mathbf{F}\,\mathbf{x} \ \text{ and,}
$$

$$
b = (\mathbf{Fx})_1^{\ 2} + (\mathbf{Fx})_2^{\ 2} + (\mathbf{F}^{\top}\mathbf{x}')_1^{\ 2} + (\mathbf{F}^{\top}\mathbf{x}')_2^{\ 2}
$$

This gives a first-order approximation to geometric error, which may provide good results if higher order terms are small as compared to the first-order [1]. The equation (8) is of the form similar to the cost function,

$$
d(\mathbf{x}', \mathbf{Fx})^2 + d(\mathbf{x}, \mathbf{F}^{\top}\mathbf{x}')^2
$$

used by [9], aiming to minimize the distance between the feature point and its corresponding epipolar line. However, this cost function gives slightly inferior results as compared to the Sampson's distance [1, 17]. Therefore, Sampson's distance is relatively a better option. In the proposed framework, (5) and (8) have been combined to yield the minimization function as formulated in (9)

$$f_s = \frac{[(\hat{\mathbf{x}}' + \mathbf{S}^\top \Delta \mathbf{x}')^\top \mathbf{F}\,(\hat{\mathbf{x}} + \mathbf{S}^\top \Delta \mathbf{x})]^2}{(\mathbf{b}_1)_1{}^2 + (\mathbf{b}_1)_2{}^2 + (\mathbf{b}_2)_1{}^2 + (\mathbf{b}_2)_2{}^2} \tag{9}$$

where,

$$\mathbf{b}_1 = \mathbf{F}\,(\hat{\mathbf{x}} + \mathbf{S}^\top \Delta \mathbf{x}) \text{ and, } \mathbf{b}_2 = \mathbf{F}^\top\,(\hat{\mathbf{x}}' + \mathbf{S}^\top \Delta \mathbf{x}')$$

However, to impose the constraint strictly, we introduce the Lagrange's multiplier λ, and hence the minimization problem can be reformulated as

$$f = f_s + \lambda (\mathbf{x}' - \mathbf{S}^\top \Delta \mathbf{x}')^\top \mathbf{F}(\mathbf{x} - \mathbf{S}^\top \Delta \mathbf{x}) \tag{10}$$

3.2 Optimal Solution

In order to find an optimal solution, it is required to solve (10) and obtain the value of $\Delta \mathbf{x}$ and $\Delta \mathbf{x}'$ for which a minima is attained. Therefore, computing the partial derivatives of f with respect to $\Delta \mathbf{x}$ and $\Delta \mathbf{x}'$ and equating both the derivatives to zero will yield the following equations:

$$\Delta \mathbf{x} = \frac{a}{b}\,\mathbf{F}_{in}\,[(\mathbf{F}^\top \mathbf{x}')_1\,\mathbf{s}_1(:,1) + (\mathbf{F}^\top \mathbf{x}')_2\,\mathbf{s}_1(:,2)]$$
$$-\mathbf{F}_{in}\left[(\mathbf{SF})\hat{\mathbf{x}} - \frac{\lambda b}{2a}(\mathbf{SF})\hat{\mathbf{x}}\right] \tag{11}$$

$$\Delta \mathbf{x}' = \frac{a}{b}\,\mathbf{F}_{in}^\top\,[(\mathbf{Fx})_1\,\mathbf{s}_2(:,1) + (\mathbf{Fx})_2\,\mathbf{s}_2(:,2)]$$
$$-\mathbf{F}_{in}^\top\left[(\mathbf{SF}^\top)\hat{\mathbf{x}}' - \frac{\lambda b}{2a}(\mathbf{SF}^\top)\hat{\mathbf{x}}'\right] \tag{12}$$

where

$$\mathbf{s}_1 = \mathbf{SF}, \quad \mathbf{s}_2 = \mathbf{SF}^\top$$

and $\mathbf{s}_i(:,j)$ represents the jth column of the matrix $\mathbf{s}_i$.

It is now imperative to find the value of λ which satisfies the constraint given in (10), and has been found experimentally to be

$$\lambda = \frac{2\,(\mathbf{x}'^\top \mathbf{F}\,\mathbf{x})}{(\mathbf{Fx})_1{}^2 + (\mathbf{Fx})_2{}^2 + (\mathbf{F}^\top \mathbf{x}')_1{}^2 + (\mathbf{F}^\top \mathbf{x}')_2{}^2} = \frac{2a}{b}. \tag{13}$$

This results to

$$\Delta \mathbf{x} = \frac{a}{b}\,\mathbf{F}_{in}\,[(\mathbf{F}^\top \mathbf{x}')_1\,\mathbf{s}_1(:,1) + (\mathbf{F}^\top \mathbf{x}')_2\,\mathbf{s}_1(:,2)] \tag{14}$$

$$\Delta \mathbf{x}' = \frac{a}{b} \, \mathbf{F}_{in}^{\top} \, [(\mathbf{Fx})_1 \, \mathbf{s}_2(:, 1) + (\mathbf{Fx})_2 \, \mathbf{s}_2(:, 2)] \tag{15}$$

which eventually reduces to,

$$\Delta \mathbf{x} = \frac{a}{b} \, \mathbf{S} \mathbf{F}^{\top} \mathbf{x}' \tag{16}$$

$$\Delta \mathbf{x}' = \frac{a}{b} \, \mathbf{S} \mathbf{F} \mathbf{x} \tag{17}$$

which is similar to the first iteration of *KSN* method. A detailed derivation for the optimal solution is provided in Appendix 1 and the mathematical tools related to matrix operations can be found in [18]. Equations (16) and (17) represent the estimated error in the measurement of the coordinates of the points in two views.

4   Experiments and Results

A number of experiments were carried out on the data set obtained from [14] for evaluation and analysis of the proposed method in comparison to the *niter*. The results are obtained by comparing the mean re-projection error, i.e., mean of the square of Euclidean distances, between the given 2D points and the optimal points found by the two methods for the three different configurations.

The data sets consist of 10×9 random point clouds of 10000 points each, projected onto images of 1024×1024 pixels, having focal length of 512 pixels, for all the three configurations namely lateral, orbital and the forward. The point clouds are Gaussian distributed with standard deviation of $\frac{1}{4}$ relative to the baseline. Each set of points of the 10×9 group corresponds to a certain image-space Gaussian noise with standard deviation of $\sigma = 2^n$ pixels, where $n = -5, -4, -3, \ldots + 4$ and distance from the center of the baseline $\delta = 2^d$, with $d = 0, 1, 2, \ldots, 8$.

The fundamental matrix is computed using the image points, by employing the 8-point algorithm [19]. For cases where essential matrix is used, computation can be done using the 5-point algorithm [20]. Stewénius et al. [21] proposed a novel version of the 5-point algorithm and the *MATLAB* code for the same has been made available by the author at http://vis.uky.edu/~stewe/FIVEPOINT/.

The advantage of using the 5-point algorithm is that they are less sensitive to outliers since they use fewer points to compute the (essential) matrix [22], but since the image points are already given in the data set, we assume that there are no outliers present. As the absolute re-projection error is quite close for both the methods, the difference between the mean re-projection error of the methods is plotted against the varying noise levels n in Figs. 2, 3, and 4, for different values of d. The difference in the re-projection error is calculated here as $\Delta E = E_{SDTM} - E_{niter}$ where, E_{SDTM} and E_{niter} are the mean re-projection error found by the *SDTM* method and *niter*,

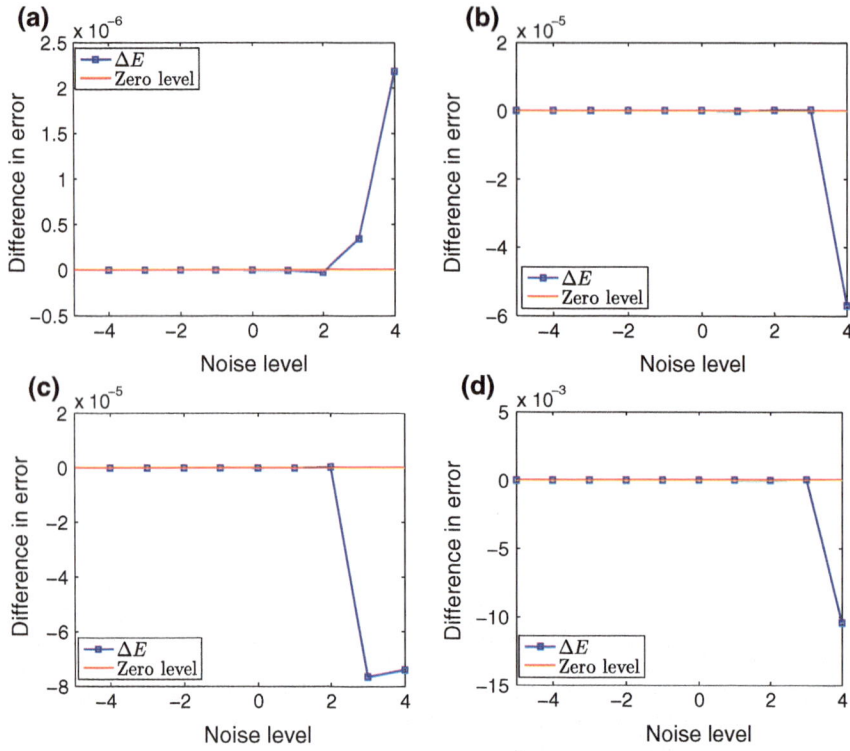

Fig. 2 Difference in mean re-projection error for lateral configuration (synthetic data) **a** $d = 0$, **b** $d = 1$, **c** $d = 2$ and **d** $d = 3$

respectively, averaged over 10000 points at each noise level at a given value of d. Hence, a negative value signifies an improved performance of *SDTM* over *niter*. Table 1 shows the epipolar error for varying values of d and n for the forward configuration. Here, the value in each cell is for varying values of n and d. From the results, it is observed that *SDTM* has performed better than *niter*. As observed from the analysis, *SDTM* method performs comparatively better in quantitative terms. For cases where $d \geq 6$ and $n \geq 2$, we have found out the error value ΔE to be positive, i.e., *niter* has performed better and the difference in the error is of the order of 10^{-1}, this has been the limitation of the *SDTM* method.

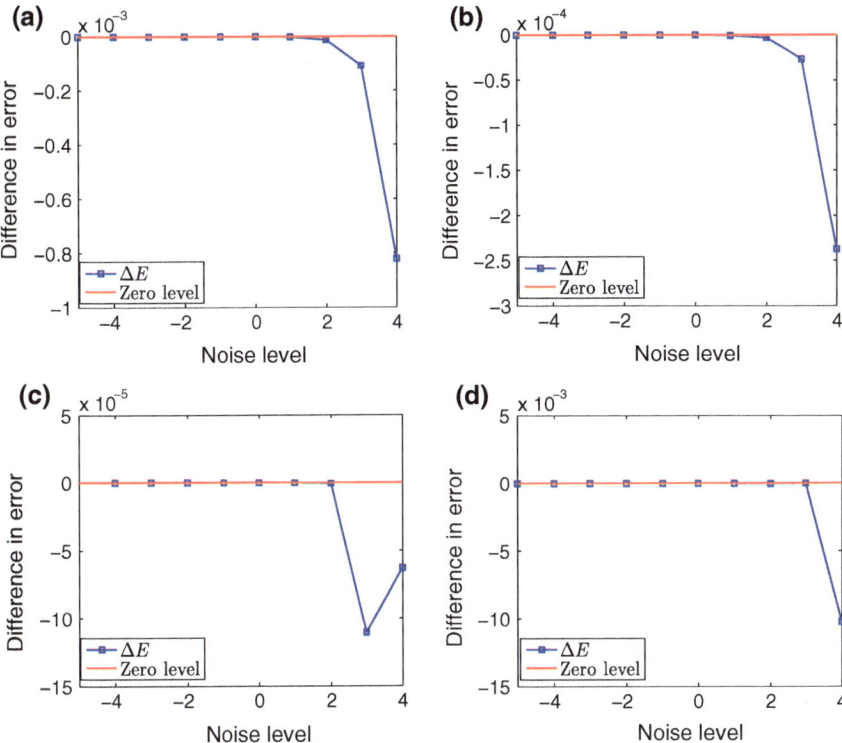

Fig. 3 Difference in mean re-projection error for orbital configuration (synthetic data) **a** $d = 0$, **b** $d = 1$, **c** $d = 2$ and **d** $d = 3$

5 Discussions

The similarity of the proposed approach with *KSN* can be found by studying the first iteration of the *KSN* method which is given in Algorithm 1 and represented in a simplified form in [14]

1. $n_k = \mathbf{SFx'}$
2. $n_k{'} = \mathbf{SF}^\top \mathbf{x}$
3. $c = \dfrac{a}{b}$
   (where a and b are as defined for *SDTM*)
4. $\Delta \mathbf{x} = c n_k$
5. $\Delta \mathbf{x'} = c n_k{'}$

Algorithm 1: First iteration of *KSN* algorithm

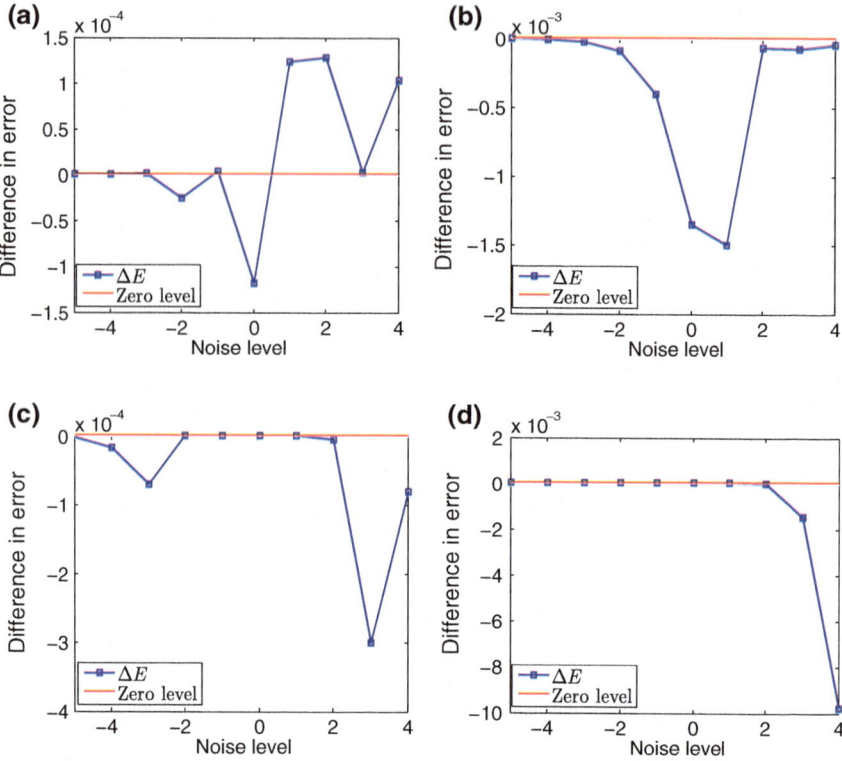

Fig. 4 Difference in mean re-projection error for forward configuration (synthetic data) **a** $d = 0$, **b** $d = 1$, **c** $d = 2$ and **d** $d = 3$

Here, it can be observed that the value of $\Delta\mathbf{x}$ and $\Delta\mathbf{x}'$ are same as in (16) and (17), i.e., the proposed approach converges to the first iteration of *KSN*.

The proposed *SDTM* approach contains novelty in the form that it has never been tried earlier for the purpose of resolving the triangulation problem. Hence, the authors find it an opportunity to explore the area with a new tool for formulation of the problem. Though, the method converges to the existing *KSN* method but these two methods differ significantly in the way they have been formulated. However, geometrical interpretation of the *SDTM* is required in order to further explore the method for novel and more accurate solution to the triangulation problem.

Table 1 Epipolar error for data in forward configuration

	$d=0$	$d=1$	$d=2$	$d=3$
$n=-5$	$-8.94e-11$	$-1.19e-10$	$-3.79e-11$	$-4.29e-11$
$n=-4$	$-6.76e-11$	$-3.37e-09$	$-3.39e-11$	$2.80e-10$
$n=-3$	$6.29e-10$	$5.24e-09$	$-3.98e-09$	$-9.39e-10$
$n=-2$	$-6.10e-08$	$-1.42e-10$	$-3.29e-09$	$5.68e-10$
$n=-1$	$-6.86e-09$	$-2.47e-07$	$1.30e-09$	$2.85e-09$
$n=0$	$-8.49e-07$	$2.51e-06$	$-3.98e-08$	$-1.56e-07$
$n=1$	$8.40e-06$	$-2.68e-07$	$2.35e-08$	$-5.93e-08$
$n=2$	$6.78e-06$	$-7.87e-07$	$-1.22e-06$	$-8.93e-06$
$n=3$	$9.24e-05$	$2.92e-06$	$4.99e-06$	$-2.86e-05$
$n=4$	$-4.39e-07$	$3.72e-06$	$-3.32e-05$	-0.00014

6 Conclusion

The approach using the Sampson's distance as a cost function is a novel way to formulate the triangulation problem and it can readily be observed that it converges to the first iteration of the *KSN* method. The work presented here is an effort to revisit the triangulation problem in a different manner and explore the grey areas. This attempt can be said to find a different path to formulate the problem and can be taken forward to look for further possibilities in this area. The future work may involve the geometrical analysis of the solution and may lead to a new solution to the triangulation problem.

Acknowledgements The authors would like to thank Dr. Peter Lindstrom, Lawrence Livermore National Laboratory, for sharing with us the data set on which he has tested his method in [14].

Appendix 1: Derivation of the Lagrangian

To start with, we consider the Lagrangian in (10)

$$f = f_s + \lambda(\mathbf{x}' - \mathbf{S}^\top \Delta \mathbf{x}')^\top \mathbf{F}(\mathbf{x} - \mathbf{S}^\top \Delta \mathbf{x})$$

where, f_s is from (9).

Computing the partial derivatives of f with respect to $\Delta \mathbf{x}$ and $\Delta \mathbf{x}'$ yields,

$$\frac{\partial f}{\partial \Delta \mathbf{x}} = -\lambda \mathbf{s}_2 \hat{\mathbf{x}}' + \frac{2a(\mathbf{s}_2 \hat{\mathbf{x}}' + \mathbf{s}_2 \mathbf{S}^\top \Delta \mathbf{x}')b}{b^2} - \frac{2[(\mathbf{Fx})_1 \mathbf{s}_2(:, 1) + (\mathbf{Fx})_2 \mathbf{s}_2(:, 2)]a^2}{b^2}, \tag{18}$$

$$\frac{\partial f}{\partial \Delta \mathbf{x}'} = -\lambda \mathbf{s}_1 \hat{\mathbf{x}} + \frac{2a(\mathbf{s}_1 \hat{\mathbf{x}} + \mathbf{s}_1 \mathbf{S}^\top \Delta \mathbf{x})b}{b^2} - \frac{2[(\mathbf{F}^\top \mathbf{x}')_1 \, \mathbf{s}_1(:, 1) + (\mathbf{F}^\top \mathbf{x}')_2 \, \mathbf{s}_1(:, 2)]a^2}{b^2} \tag{19}$$

respectively. Here, we have considered

$$(\mathbf{F} \, (\hat{\mathbf{x}} + \mathbf{S}^\top \Delta \mathbf{x}))_1 = (1, 0, 0) \, \mathbf{F} \, (\hat{\mathbf{x}} + \mathbf{S}^\top \Delta \mathbf{x}) \tag{20}$$

and,

$$(\mathbf{F} \, (\hat{\mathbf{x}} + \mathbf{S}^\top \Delta \mathbf{x}))_2 = (0, 1, 0) \, \mathbf{F} \, (\hat{\mathbf{x}} + \mathbf{S}^\top \Delta \mathbf{x}) \tag{21}$$

Therefore, their derivatives will become

$$\mathbf{SF}(1, 0, 0)^\top = \mathbf{s}_1(:, 1),$$

$$\mathbf{SF}(0, 1, 0)^\top = \mathbf{s}_1(:, 2)$$

respectively. Similarly for $\mathbf{F}^\top \, (\hat{\mathbf{x}}' + \mathbf{S}^\top \Delta \mathbf{x}')_1$ and $\mathbf{F}^\top \, (\hat{\mathbf{x}}' + \mathbf{S}^\top \Delta \mathbf{x}')_2$. This will lead to (18) and (19).

Now, equating $\dfrac{\partial f}{\partial \Delta \mathbf{x}}$ and $\dfrac{\partial f}{\partial \Delta \mathbf{x}'}$ to zero, we get (11) and (12). From here, the solution given in (16) and (17) can be obtained, by keeping the value of λ, as in (13).

References

1. Hartley, R., Zisserman, A.: Multiple View Geometry in Computer Vision, vol. 2. Cambridge University Press, Cambridge (2000)
2. Matthies, L., Shafer, S.A.: Error modeling in stereo navigation. IEEE J. Robot. Autom. **3**(3), 239–248 (1987)
3. Scaramuzza, D., Fraundorfer, F.: Visual odometry [tutorial]. IEEE Robot. Autom. Mag. **18**(4), 80–92 (2011)
4. Sünderhauf, N., Protzel, P.: Stereo odometry–a review of approaches. Technical report, Chemnitz University of Technology (2007)
5. Maimone, M., Cheng, Y., Matthies, L.: Two years of visual odometry on the mars exploration rovers. J. Field Robot. **24**(3), 169–186 (2007)
6. Wu, J.J., Sharma, R., Huang, T.S.: Analysis of uncertainty bounds due to quantization for three-dimensional position estimation using multiple cameras. Opt. Eng. **37**(1), 280–292 (1998)
7. Fooladgar, F., Samavi, S., Soroushmehr, S., Shirani, S.: Geometrical analysis of localization error in stereo vision systems (2013)
8. Beardsley, P.A., Zisserman, A., Murray, D.W.: Navigation using affine structure from motion. In: Computer Vision ECCV'94, pp. 85–96. Springer, Berlin (1994)
9. Hartley, R.I., Sturm, P.: Triangulation. Comput. Vis. Image Underst. **68**(2), 146–157 (1997)

10. Vite-Silva, I., Cruz-Cortés, N., Toscano-Pulido, G., de la Fraga, L.G.: Optimal triangulation in 3d computer vision using a multi-objective evolutionary algorithm. In: Applications of Evolutionary Computing, pp. 330–339, Springer, Berlin (2007)
11. Wong, Y.-P., Ng, B.-Y.: 3d reconstruction from multiple views using particle swarm optimization. In: 2010 IEEE Congress on Evolutionary Computation (CEC), pp. 1–8. IEEE (2010)
12. Kanazawa, Y., Kanatani, K.: Reliability of 3-d reconstruction by stereo vision. IEICE Trans. Inf. Syst. **78**(10), 1301–1306 (1995)
13. Kanatani, K., Sugaya, Y., Niitsuma, H.: Triangulation from two views revisited: Hartley-sturm vs. optimal correction. In: Practice, vol. 4, p. 5 (2008)
14. Lindstrom, P.: Triangulation made easy. In: 2010 IEEE Conference on Computer Vision and Pattern Recognition (CVPR), pp. 1554–1561. IEEE (2010)
15. Kanatani, K., Niitsuma, H.: Optimal two-view planar scene triangulation. In: Computer Vision–ACCV 2010, pp. 242–253. Springer, Berlin (2011)
16. Sampson, P.D.: Fitting conic sections to very scattered data: an iterative refinement of the bookstein algorithm. Comput. Graph. Image Process. **18**(1), 97–108 (1982)
17. Zhang, Z.: Determining the epipolar geometry and its uncertainty: a review. Int. J. Comput. Vis. **27**(2), 161–195 (1998)
18. Petersen, K.B., Pedersen, M.S.: The matrix cookbook. Technical University of Denmark, pp. 7–15 (2008)
19. Longuet-Higgins, H.: A computer algorithm for reconstructing a scene from two projections. In: Fischler M.A., Firschein, O. (eds.) Readings in Computer Vision: Issues, Problems, Principles, and Paradigms, pp. 61–62 (1987)
20. Nistér, D.: An efficient solution to the five-point relative pose problem. IEEE Trans. Pattern Anal. Mach. Intell. **26**(6), 756–770 (2004)
21. Stewénius, H., Engels, C., Nistér, D.: Recent developments on direct relative orientation. ISPRS J. Photogramm. Remote. Sens. **60**(4), 284–294 (2006)
22. Szeliski, R.: Computer Vision: Algorithms and Applications. Springer, Berlin (2010)

Entry–Exit Video Surveillance: A Benchmark Dataset

V. Vinay Kumar, P. Nagabhushan and S. N. Roopa

Abstract Techniques to automate video surveillance around places where cameras are forbidden due to privacy concerns are yet under-addressed. This can be achieved by building conceptual models and algorithms to investigate the credibility of monitoring of events using the video frames captured by mounting the cameras so as to have the view of the entrances of such camera-forbidden areas. Evaluation of these models and algorithms require standard datasets. The proposal here is to introduce a new benchmark dataset—"EnEx dataset" as no traces specific to the problem were found in the literature. The dataset comprises of video frames captured in 5 different locations accounting 90 entry–exit event pairs based on 9 different sequences involving 36 participants. Ground statistics of the dataset is reported. This work ventures a new sub-domain for research in the area of automated video surveillance.

Keywords Automated video surveillance · Dataset · Entry–exit surveillance · Appearance transformation · Private areas

1 Introduction

The theory of video surveillance, particularly in the indoor environments is often reflected as privacy breach [1] and hence, is limited to situations where public safety is at jeopardy. Detailed discussion on "privacy concerns" as well as "surveillance for crime prevention" and "public safety" can be studied from [1–7]. However, monitoring of "private areas" where there is higher expectation of privacy by individu-

V. Vinay Kumar (✉)
Department of Studies in Computer Science, University of Mysore, Mysuru, India
e-mail: vkumar.vinay@ymail.com

P. Nagabhushan
Indian Institute of Information Technology Allahabad, Allahabad, India
e-mail: pnagabhushan@hotmail.com

S. N. Roopa
Mysuru, India

© Springer Nature Singapore Pte Ltd. 2020
B. B. Chaudhuri et al. (eds.), *Proceedings of 3rd International Conference on Computer Vision and Image Processing*, Advances in Intelligent Systems and Computing 1022, https://doi.org/10.1007/978-981-32-9088-4_30

als (addressed as "subjects" hereafter), for example, washrooms, changing rooms, breastfeeding rooms, is considered illegal. Even in manual surveillance of the private areas, monitoring is confined to entrances and anything beyond that is unlawful [1] and the same is applicable for automated video surveillance.

Contemporary video surveillance systems, facilitated by multiple cameras, have scope for analysis of every event that occurs in its ambit. Intelligent multi-camera video surveillance notions are studied in detail in the classical papers [8, 9]. Person re-identification techniques in particular, are studied widely in the recent years and a significant number of datasets have been introduced and are publicly available [10–29]. These datasets are captured in unconstrained environments such as educational institutions, shopping areas, transportation gateways such as railway and bus stations, airports. They play a significant role in validating person re-identification algorithms. Detailed survey on the conventional person re-identification techniques can be found in [29–31]. However, it is challenging to design the surveillance model where there is a temporal gap between the entry and exit which envisage the possibility of change in appearance of the subject. The objective of the research is to automate the surveillance of people entering and exiting the private areas by replicating the procedure followed in manual surveillance. Design and evaluation of algorithms for the analysis require standard datasets. The aforementioned datasets available in the literature are limited to conventional person re-identification and hence, there is a need for a standard dataset that incorporates the problem with the following dimensions.

1. Reidentifying subjects that have endured rational transformation when they move out of camera catchment area.
2. Prediction of exit of the subjects with reference to their order of entry.
3. Unsupervised classification of objects the subjects possess.

The problem can be addressed by mounting the camera(s) so as to have the view of the entrances to capture the possible amount of knowledge. The video frames are to be extracted and effective algorithms are to be developed to analyze every event occurring. The proposal is to introduce a benchmark dataset that accomplish various dimensions of the problem thus overcoming the limitations in the current state of the art. Conventionally, it is named "EnEx dataset". The ground truth provided can be utilized for effective evaluation of performances of algorithms and comparative studies to be proposed in the aforementioned context. Variety in data with respect to subjects' attire, height, age, and gender are reported in detail.

The dataset is unique to support performance evaluation of low-level as well as high-level vision algorithms that intend to solve the following problems:

1. High variations between cognition and recognition of the subject with respect to appearance, behavior, and timing which are natural as there is temporal gap between entry and exit.
2. Unusual behavior of subject; such as blocking of entrance.
3. Delay in exiting with reference to the order of entry.
4. Partially masked heads of subjects during entry and exit.
5. Collision among subjects occluding each other while entering and exiting simultaneously.

6. Different patterns in front and back views of the subjects' attires.
7. Suspicious behavior when a subject carries an unsupervised object while entering and exits empty-handed.
8. Additionally, subject face recognition in videos by approximation using the mugshots provided as ground truth.

The paper is organized as follows. Camera placement and dataset captured locations are discussed in Sect. 2. Various possible video sequences at the entrances of private areas are discussed in Sect. 3. Section 4 emphasize the ground truth and ground statistics of the dataset followed by details about the availability of the dataset for research in Sect. 5.

2 Dataset Collection

The dataset is collected by capturing videos at the entrances of five different private areas. The video sequences include private areas in an office and an educational institution. The videos contain subjects entering and exiting private areas in different orders. In addition, three unusual scenarios that involve blocking of entrance talking in a mobile phone, dropping of items carried while entering and occluding the faces to exhibit variations in the appearances. Illustrations of video sequences along with the ground truth is provided in the next sections.

Camera Deployment:

Algorithms for optimal placement of camera are studied from [32–34]. However, based on the knowledge gain from literature, the camera intrinsic attributes were handset to capture maximum possible information. The camera was placed at a distance approximately 4 m from the entrances. With this range, it was legitimate to capture individuals of height ranging between 4.5 and 6.5 ft. Camera was placed such that its axis was non-orthogonal to the possible trajectory of the subjects in order to provide more scope to obtain ample tracking attributes. Illumination was from natural lighting at all the places as the videos were captured in the daytime. The camera deployment is summarized with attributes such as location, camera description, subjects, and sequences details and is tabulated in Table 1. Figure 1 shows sample background images captured at the entrances of private areas.

3 Video Sequences

In order to address various dimensions of Entry–Exit Video surveillance problem, it is important to consider all possible sequences in Entry–Exit scenario. Following are the set of sequences that represents the variety in the dataset.

Table 1 Camera deployment—location—sequences summary

Location	Camera	Subjects count	Sequence count
Office private area	Sony cybershot DSC-H10	11	16
Changing room area	Sony cybershot DSC-H10	1	2
Institute private area 1	Sony cybershot DSC-H10	17	4
Institute private area 2	Sony cybershot DSC-H10	4	4
Institute private area 3	Sony cybershot DSC-H10	9	2

Fig. 1 Sample background images of the entrances

3.1 Sequence 1

Subjects enter and exit the private areas one at a time in the order $i = 1, 2, 3, \ldots n$ where n is the number of subjects participated in a location l. Subject $(i+1)$ enter only after subject i exits. The video sequences are captured at locations $l = 1, 2, 3, 4$, and 5. The assumptions are—the entrance corridors and the private areas are narrow and hence only one subject enter and exit at a time and maximum one subject can be inside the private area.

3.2 Sequence 2

Subjects enter the private areas one at a time in the order $i = 1, 2, 3, \ldots n$ and exit in the same order where n is the number of subjects participated in a location l. Subjects

1–n enter one by one After subject n enters, subjects 1–n exits n the same order. The video sequences are captured at locations $l = 1, 3, 4$, and 5. The assumptions are—the entrance corridors are narrow and hence only one subject enter and exit at a time and the private area is big enough to accommodate more than n subjects at any instance t.

3.3 Sequence 3

Subjects enter the private area one at a time such that subject $(i+1)$ enters following subject i but subject $(i+2)$ enters only after subject i exits. The video sequence is captured at locations $l = 1$. The assumptions are—the entrance corridors are narrow and hence only one subject enter and exit at a time. The private areas can accommodate maximum of two subjects at any instance t.

Note: Totally seven subjects enter and exit but subject 7 exits before subject 6.

3.4 Sequence 4

Similar to sequence 3 but with an assumption that maximum of three subjects can be inside the private area at any instance t.

Unusual: Subject 1 enters first but exits at the end exhibiting delay in exit with reference to the order of entry.

3.5 Sequence 5

Captured at locations 1 and 4. A subject blocks the entrance corridor for a finite interval without entering but talking on phone instead. In the sequence captured at location 1, the subject blocks the entry and exit of other subjects by partially occluding the subjects. But in the sequence captured at location 4, the subject enters the private area after blocking the entrance for a finite interval.

3.6 Sequence 6

Two sets of data:

Normal: Subjects enter and exit the private area with their heads occluded with masks.

Unusual: Subjects enter the private area with their heads occluded with masks but exits unmasked.

3.7 Sequence 7

Exhibition of suspicious behavior where the subject carry an object while entering and exits empty- handed. The data is presented in two sets similar to sequence 6 to provide the expected behavior (subjects exiting carrying the same object) in one set and the actual behavior (subjects dropping the objects in the private area and exiting empty-handed) in the other.

3.8 Sequence 8

Multiple subjects are allowed to move randomly to enter and exit the private area. Occlusion among subjects entering and exiting simultaneously are captured. Every event is recorded labeled with frame numbers range. The assumptions are—the entrance corridors are broad enough to allow more than one subject to enter and exit freely and the private area is big enough to accommodate more than n subjects at any instance t.

3.9 Sequence 9

Location 2 is assumed to be a dress trial room. A subject enters the private area with finite number of outfits in hand and exits in different attire.
  Two sets of data:

Expected:

  a. Exit in the initial attire with the carried outfits in hand.
  b. Exit in different attire with the initial outfits in hand.

Actual:

  a. Exit in the initial attire empty-handed.
  b. Exit in different attire empty-handed.

4   Ground Truth and Organization of Dataset

Video footages are stored as series of frame images encoded in PNG format. Every video clip is complemented with ground truth attributed with frame number pairs $[\underline{n}, \overline{n}]$ that represents begin and end of events, segmented images of subjects along with the segmented coordinates extracted from frames ranging between $[\underline{n}, \overline{n}]$. The data is classified and stored in directories based on the sequences discussed in the previous section. Each sequence directory consists of data extracted from different locations in the corresponding directories. The data in each location directory is organized as follows. a. Raw data, b. Sequences, and c. Labeled events data. Labeled events data supports researchers to perform high-level analysis on the frames that range between $[\underline{n}, \overline{n}]$.

4.1 Ground Truth

The ground truth for EnEx dataset is organized as follows:

1. Background images of each location are stored in the directory location data/backgrounds in PNG encoded format as shown in Fig. 1.
2. Every sequence video clip is attributed with a XML document that stores information of every event in the sequence in the following structure.
   $< SequenceID > < Event > < EventID > < locationID > < Subject >$
   $< SubjectID > < BeginFrame > < EndFrame > < EventType >$
   $< /Event > < /SequenceID >$
3. Subjects segmented in every frame they are captured.
4. Collection of all the segmented images of a subject is stored in the corresponding directories subject data/segmented images/SubId. Sample segmented images for subjects during entry and exit are as shown in Fig. 2.
5. Additionally, mugshots for each subject are captured from approximately one meter and are stored in the directory subject data/mugshots in PNG encoded format.

4.2 Ground Statistics of the Dataset

The dataset comprises of participation of 36 subjects in total (16 male and 20 female participants of age ranging between 22 and 31) captured in 5 different locations. Maximum sequences are captured in location 1. The dataset varies with subjects possessing different objects such as doll, bags, trolley, jackets, mobile phones and head masks such as jacket hoods, stoles, shower caps, and helmets. Location 3 is of two opposite, symmetric entrances of boys and girls waiting rooms in an educational institution which gave scope to involve highest number of subjects. Subjects were attired in different range of colors in every location. Few of these statistics are presented in Fig. 3.

Fig. 2 Variety in EnEx dataset

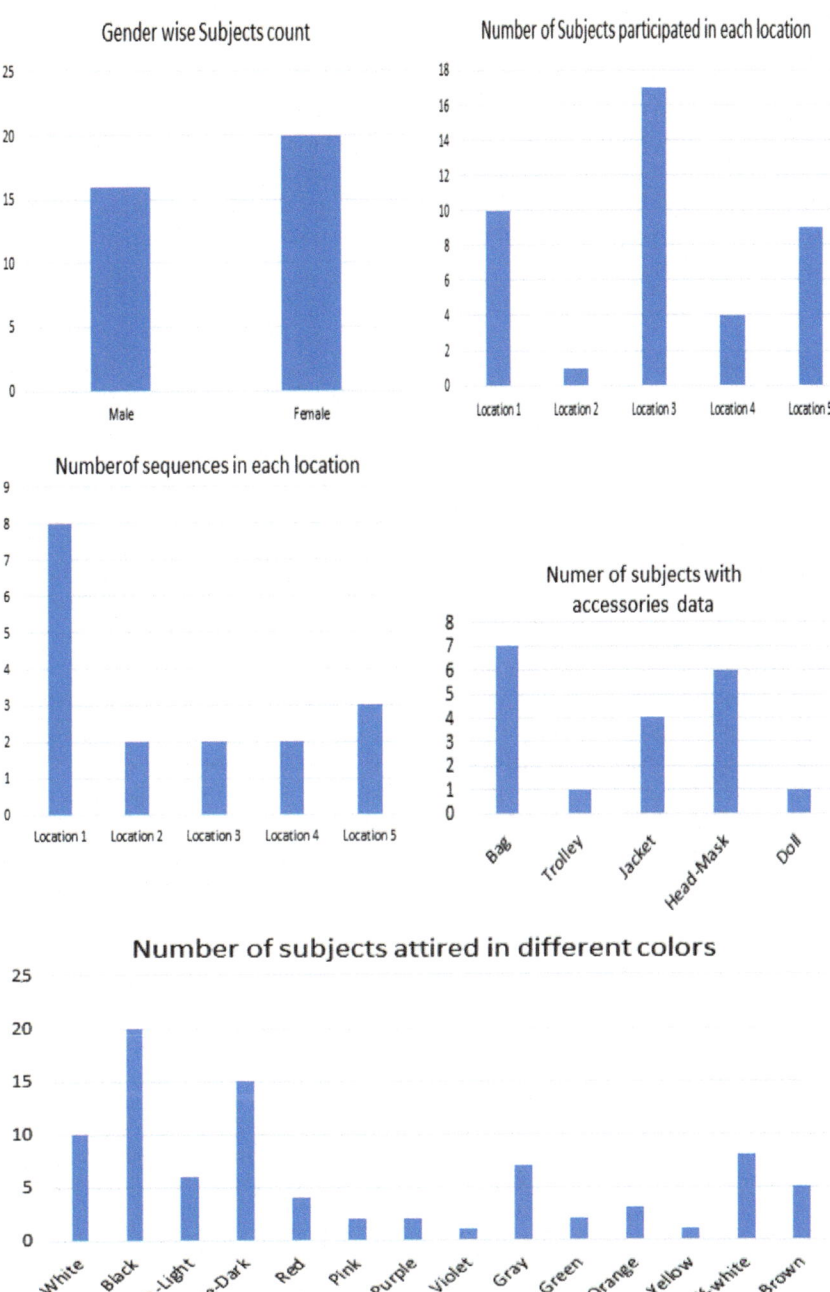

Fig. 3 Ground statistics of EnEx dataset

5 Availing the Dataset

The EnEx data is available as open source for academics and research purposes only. Details about availing the dataset will be made available in the institution website corresponding to the first author or the first author can be directly contacted. As mentioned in the previous section, all the video clips are stored as series of frames. Researchers using EnEx dataset in support of their research are requested to cite this paper appropriately.

6 Conclusion

The EnEx dataset is made available to support active research on monitoring of events at the entrances of camera-forbidden areas. Additionally, the dataset supports researches on independent problems such as person re-identification, head-occlusion detection and unmasking, gait-matching, face recognition by approximation.

The ground truth creation required over 100 human-hours effort. The ground truth can be further extended to provide sequence of gait structures of subjects for every event and work is continued for the accomplishment. Once accomplished, it will be made available as add-on for the dataset.

The dataset can be further extended to collection of video sequences involving kids to extend the range of height of the subjects, collection of video footages during night time that involves artificial illumination as all the sequences except for the ones captured in location 2 are captured in daytime. One of the objectives of our research is to explore the possibility of monitoring of events at the entrances of camera-forbidden areas with a single camera and hence the dataset collection was confined to single camera. This work can further be expanded for collection of video sequences using more than one camera.

The dataset is expected to continue to evolve over time urging for active research in the context discussed in this paper. This novel idea ventures new sub-domain in the area of automated video surveillance.

Acknowledgements The first author acknowledges Union Grants Commission, Government of India for providing financial aid to carry out this research work. The research scholars and students of the Department of studies in Computer Science, University of Mysore, Mysuru and the students of NIE Institute of Technology, Mysuru are acknowledged for their active participation in the efforts of collection of dataset and willingness to reveal their identities for publication with the consent of their respective educational institutions for research purposes only.

References

1. Westin, Alan F.: Privacy and Freedom, 5th edn. Atheneum, New York (1968)
2. von Silva-Tarouca Larsen, B.: Setting the Watch: Privacy and the Ethics of CCTV Surveillance, p. 226. Hart Publishing, Oxford (2011)

3. Macnish, K.: The Ethics of Surveillance: An Introduction. Taylor and Francis Group, London (2017)
4. Rajpoot, Q.M., Jensen, C.D.: Video surveillance: privacy issues and legal compliance, pp. 69–92 (2015). https://doi.org/10.4018/978-1-4666-8502-4.ch004
5. Chen, A., Biglari-Abhari, M., Wang, K.: Trusting the computer in computer vision: a privacy-affirming framework (2017). https://doi.org/10.1109/CVPRW.2017.178
6. Alexandrie, G.: Surveillance cameras and crime: a review of randomized and nat-ural experiments. J. Scand. Stud. Criminol. Crime Prev. **18**(2), 210–222 (2017). https://doi.org/10.1080/14043858.2017.1387410
7. Mathur, G., Bundele, M.: Research on intelligent video surveillance techniques for suspicious activity detection critical review. In: 2016 International Conference on Recent Advances and Innovations in Engineering (ICRAIE), Jaipur, pp. 1–8 (2016)
8. Wang, X.: Intelligent multi-camera video surveillance: a review. Pattern Recognit. Lett. **34**, 3–19 (2013). https://doi.org/10.1016/j.patrec.2012.07.005
9. Ibrahim, S.W.: A comprehensive review on intelligent surveillance systems. Commun. Sci. Technol. **1**, 7–14 (2016)
10. Gray, D., Brennan, S., Tao, H.: Evaluating appearance models for recognition, reacquisition, and tracking. In: Proceedings of IEEE International Workshop on Performance Evaluation for Tracking and Surveillance (PETS) (2007)
11. Schwartz, W.R., Davis, L.S.: Learning discriminative appearance-based models using partial least squares. In: Proceedings of the XXII Brazilian Symposium on Computer Graphics and Image Processing (SIBGRAPI'2009), Rio de Janeiro, Brazil, 11–14 October 2009 (2009)
12. Cheng, D.S., Cristani, M., Stoppa, M., Bazzani, L., Murino, V.: Custom pictorial structures for re-identification. In: BMVC (2011)
13. Zheng, L., Wang, S., Shen, L., Tian, L., Bu, J., Tian, Q.: Person re-identification meets image search. Technical report (2015)
14. Ma, X., Zhu, X., Gong, S., Xie, X., Hu, J., Lam, K.-M., Zhong, Y.: Person re-identification by unsupervised video matching. Pattern Recognit. 65, 197–210 (2017). (PR)
15. Wang, T., Gong, S., Zhu, X., Wang, S.: Person re-identification by discriminative selection in video ranking. IEEE Trans. Pattern Anal. Mach. Intell. 38(12), 2501–2514 (2016). (TPAMI)
16. Wang, T., Gong, S., Zhu, X., Wang, S.: Person re-identification by video ranking. In: Proceedings of European Conference on Computer Vision (ECCV), Zurich, Switzerland, September 2014 (2014)
17. Liu, C., Gong, S., Loy, C.C.: On-the-fly feature importance mining for person re-identification. Pattern Recognit. 47(4), 1602–1615 (2014). (PR)
18. Gou, M., Karanam, S., Liu, W., Camps, O., Radke, R.J.: DukeMTMC4ReID: a large-scale multi-camera person re-identification dataset. In: 2017 IEEE Conference on Computer Vision and Pattern Recognition Workshops (CVPRW), Honolulu, HI, pp. 1425–1434 (2017)
19. Baltieri, D., Vezzani, R., Cucchiara, R.: 3DPes: 3D people dataset for surveillance and forensics. In: Proceedings of the 1st International ACM Workshop on Multimedia Access to 3D Human Objects, Scottsdale, Arizona, USA, pp. 59–64, 28 November–1 December 2011 (2011)
20. Hirzer, M., et al.: Person re-identification by descriptive and discriminative classification. In: Image Analysis (2011)
21. Wang, S.M., Lewandowski, M., Annesley, J., Orwell, J.: Re-identification of pedestrians with variable occlusion and scale. In: IEEE International Conference on Computer Vision (ICCV), Barcelona, Spain, 6–13 November 2011 (2011). (2011 IEEE International Conference on Computer Vision Workshops (ICCV workshops))
22. Martinel, N., Micheloni, C., Piciarelli, C.: Distributed signature fusion for person re-identification. In: ICDSC (2012)
23. Bialkowski, A., et al.: A database for person re-identification in multi-camera surveillance networks. In: DICTA (2012)
24. Li, W., Zhao, R., Wang, X.: Human Re-identification with Transferred Metric Learning, pp. 31–44. Springer, Berlin (2012)

25. Li, W., Wang, X.: Locally aligned feature transforms across views. In: CVPR, pp. 3594–3601 (2013)
26. Li, W., et al.: DeepReId: deep filter pairing neural network for person re-identification. In: CVPR (2014)
27. Das, A., Chakraborty, A., Roy-Chowdhury, A.K.: Consistent re-identification in a cam-era network. In: ECCV (2014)
28. Figueira, D., et al.: The HDA+ data set for research on fully automated re-identification systems. In: ECCV Workshops (2014)
29. Karanam, S., Gou, M., Wu, Z., Rates-Borras, A., Camps, O., Radke, R.J.: A systematic evaluation and benchmark for person re-identification: features, metrics, and datasets. IEEE Trans. Pattern Anal. Mach. Intell. 99, 1–14 (2018)
30. Loy, C.C., Liu, C., Gong, S.: Person re-identification by manifold ranking. In: 2013 IEEE International Conference on Image Processing, Melbourne, VIC, pp. 3567–3571 (2013). https://doi.org/10.1109/ICIP.2013.6738736; Author, F.: Article title. Journal 2(5), 99–110 (2016)
31. Zhao, R., Ouyang, W., Wang, X.: Unsupervised salience learning for person re-identification. In: 2013 IEEE Conference on Computer Vision and Pattern Recognition, Port-land, OR, pp. 3586–3593 (2013). https://doi.org/10.1109/CVPR.2013.460
32. Al-Hmouz, R., Challa, S.: Optimal placement for opportunistic cameras using genetic algorithm. In: ICISSNIP, pp. 337–341 (2005)
33. Indu, S., Chaudhury, S.: Optimal sensor placement for surveillance of large spaces. In: Proceedings of ACM/IEEE International Conference on Distributed Smart Cameras, September 2009 (2009)
34. Pito, R.: A solution to the next best view problem for automated surface acquisition. IEEE Trans. Pattern Anal. Machl Intell. 21, 1016–1030 (1999)

Recent Advancements in Image-Based Prediction Models for Diagnosis of Plant Diseases

Shradha Verma⊙, Anuradha Chug⊙ and Amit Prakash Singh⊙

Abstract India plays a significant role in the world as a major contributor to the overall food industry. Farming being a major occupation, crop protection from plant diseases has become a serious concern. The occurrence of plant diseases is hugely dependent on environmental factors which are uncontrollable. Ongoing agricultural research and the advanced computational technologies can be coalesced to determine an effective solution, which can improve the yield and result in better harvests. The objective is to minimize the economic and production losses. Researchers have developed several modeling techniques, viz. Artificial Neural Networks (ANN), Support Vector Machines (SVM), and Deep Learning for prediction and preferably early detection of diseases in plants. Mostly, images of the diseased plant are given as input to these prediction models. Early detection leads to minimizing the pesticide usage, hence, resulting in lowering the expense along with the ecological impact. Also, it is essential that the prediction techniques are nondestructive in nature. This paper delves into the Machine Learning based plant disease prediction models, along with a brief study of diverse imaging techniques, viz., RGB, multi- and hyperspectral, thermal, fluorescence spectroscopy, etc., catering to different features/parameters of a plant, useful in the prediction of diseases. Around 35 research papers from reputed peer-reviewed journals were studied and analyzed for this systematic review. It highlights the latest trends in plant disease diagnosis as well as identifies the future scope with the application of modern era technologies.

Keywords Imaging · Machine Learning · Plant diseases · Neural Networks · Deep Learning · Agriculture

S. Verma (✉) · A. Chug · A. P. Singh
USICT, GGSIP University, New Delhi, Delhi, India
e-mail: verma.shradha@gmail.com

A. Chug
e-mail: anuradha@ipu.ac.in

A. P. Singh
e-mail: amit@ipu.ac.in

© Springer Nature Singapore Pte Ltd. 2020
B. B. Chaudhuri et al. (eds.), *Proceedings of 3rd International Conference on Computer Vision and Image Processing*, Advances in Intelligent Systems and Computing 1022, https://doi.org/10.1007/978-981-32-9088-4_31

1 Introduction

Pests and diseases in plants lead to post-harvest crop losses, accounting for major economic deprivations in the agricultural domain. With increasing population and heavy crop production losses, for sustainability purposes, one must think of safety measures for the crop. Traditional methods encompass visual inspections by experienced individuals, be it the farmer or a botanist. Detecting plant diseases in a lab is another option. It comprises bringing back samples from the field and conducting a microscopic evaluation and diagnostic experiments. While this is an accurate procedure, it is time-consuming, not to mention costly, requiring laboratory equipment setup and is highly labor-intensive. Precision Agriculture necessitates the use of automated and innovative technologies for data collection and analysis in plant pathology. Research work spanning for more than a decade has led to several algorithms and techniques that apply the advancements in the computing technologies to the field of agriculture, the latest being the Machine Learning models. A weather-based prediction model for rice blast was suggested in Kaundal et al. [1]. They developed two different models, cross-location and cross-year, and validated them separately with Regression (REG), Artificial Neural Networks (namely General Regression Neural Networks (GRNN) and Backpropagation Neural Networks (BPNN)), and SVM techniques. Rice blast/weather data for five consecutive years (2000–2004) was accumulated from five different locations and was used to compute, r: the correlation coefficient; r^2: coefficient of determination, and %MAE: percent mean absolute error. The mentioned weather data included six predictor variables namely maximum and minimum relative humidity, maximum and minimum temperature, rainfall, and finally, rainy days per week. With the empirical results, it was revealed that SVM outperformed the other three approaches with maximum r^2 and minimum %MAE. An SVM-based approach has been studied and developed by Pal et al. [2], which identifies the R (resistance) genes in plants that counter the disease-causing pathogens. They suggest that the identification of these proteins can be applied for the improvement of crops and corresponding yield. Their work utilized an assembly of 112 R proteins from 25 different plant species as the positive dataset, and 119 non-R proteins dataset as the negative set, selected out of all known protein sequences of these 25 plants. The dataset was divided into 80–20 train-test pairs, Protr package of R tool was used for extraction of eight features such as amino acid composition, amino acid properties also known as correlation, conjoint triad descriptors, etc. Data from these features was scaled between 0 and 1. Out of these, only those features which would perform as discriminately as possible were selected for further processing. Freely available online tool libSVM was utilized to implement the supervised learning model. The model was validated and optimized using the 10-fold cross-validation technique. The parameters computed for performance evaluation were specificity, sensitivity, accuracy, and Matthews Correlation Coefficient (MCC). DRPPP identified 91.11% (accuracy) of the protein sequences as R genes.

A monitoring system for plant disease detection can prove to be a costly endeavor, hence, Petrellis [3] suggests a low-cost image processing approach with reduced complexity that can be implemented on a smart phone. Instead of computing and extracting several features, it focuses mainly on the visible lesions, i.e., number of spots/lesions, their coverage, and gray level, which can fairly indicate the status of the plant. The technique was applied to images of tangerine tree leaves and achieved higher than 90% accuracy in the measurement of spots and their area. Another low-cost weatherproof Thermal-RGB imaging technique has been proposed by Osroosh et al. [4]. The thermal (in binary) and RGB (jpeg format) images are overlaid and aligned in real time, utilized to create two separate masks, one for average temperature calculation, the other for canopy coverage. This system can also be set up as a File Transfer Protocol (FTP) server for sending files to a remote computer via FTP protocol.

166 hyperspectral images of tomato leaves, both healthy and infected, that were collected by Lu et al. [5], aimed to detect the yellow leaf curl disease, affecting the tomato leaves. Spectral dimension analysis was performed after manually identifying three Regions of Interests (ROIs), namely background, healthy and diseased, followed by band selection. Camargo and Smith [6] have described an image-based identification model for visual symptoms of a disease, simply by preprocessing and analyzing colored images. The algorithm works in four stages: preprocessing, to change the RGB image into a different and suitable color transformation model (Hue, Saturation, and Value (HSV), I1I2I3); enhancement, to highlight the affected regions; segmentation, to separate the affected regions; and finally, post-processing to remove unwanted sections of the image. It successfully identified the correct target area in the image with different intensities. Such methods can be automated and used to determine critical factors based on visual symptoms. Images of diseased plants showing clear visible symptoms (such as rots, wilts, tumors, and lesions), taken in the field or laboratory, can be used for training a neural network in order to predict and identify the disease. The most prominent part of the plant mainly used for prediction of diseases in almost all studies is the plant leaf, which can be used to classify and predict the disease. This paper also briefly describes the various imaging techniques, viz., RGB, hyperspectral, thermal spectroscopy [7, 8], etc., being employed for monitoring plant health. Finally, this paper outlines a review of the current research trends in the application of Machine Learning and Neural Network approaches in the prediction of pests in crops, when trained on these images. The study mainly focuses on automated, nondestructive procedures.

2 Remote Sensing and Imaging Techniques

Recent research in the sensor-based techniques has led to the identification and development of imaging technologies that are extensively being utilized for detection as well as identification of plant diseases. Optical properties of a plant are exploited to detect plant stress levels and disease severity. These noninvasive remote sensing approaches, viz., RGB, multi- and hyperspectral, thermal, fluorescence spectroscopy [7, 8], etc., have the ability to predict the onset of diseases in plants, resulting in early treatment and prevention of spread to other potential hosts. However, RGB requires the symptoms to be visible, that necessitate the manifestation of the disease, causing changes in the appearance of the plant. These sensors can be deployed in the field on mobile platforms or up high on a tower monitoring the maximum area.

RGB are three broad bands in the electromagnetic spectrum (visible range) having wavelengths ranging from 475 nm (blue) to 650 nm (red). RGB images can be captured via a digital camera or a smart phone, where a color image is the combination of these three wavelengths. These images can be preprocessed, converted into other color representation models (YCrCb, HSI), and further used for feature extraction followed by the quantification of the disease. This is the most widely used and easily available tool for disease diagnosis in plant pathology. Figure 1 shows three Apple rot leaf images, colored, gray scale, and segmented, taken from the PlantVillage repository.

One of the most promising and emerging imaging techniques is Hyperspectral imaging that has demonstrated a high potential in crop protection, as it can capture subtle modifications in the growth of a plant. Spectral Vegetation indices, viz., Normalized Difference Vegetation Index (NDVI), Simple Ratio Index (SRI), Photochemical Reflectance Index (PRI), Plant Senescence Reflectance Index (PSRI) [7, 8, 20], etc., are derived from these images that pertain to specific biochemical as well as biophysical characteristics of a plant. These indices can be employed to figure out either the disease outbreak or determine the stress level [10, 11] in a plant, as increased stress levels make the plant susceptible to pathogens. Hyperspectral sensors have evolved from multispectral sensors, both operating in the visible as well as non-visible bands of the electromagnetic spectrum. The major difference between both is the number of bands and width of these bands. Multispectral images refer to 3–10 broader bands while hyperspectral images capture high level of spectral data pertaining to thousands of narrow bands. RGB images can be thought of as 3-band multispectral images. Similarly, hyperspectral images capture the spectral data for each pixel in the image, having x- and y- spatial axes and z- spectral axis comprising the reflectance data for every band at each pixel. Thus, this imaging requires highly sophisticated hardware for capturing images and huge disk storage.

Fig. 1 Apple rot images of a leaf (colored, grayscale, and segmented) from PlantVillage dataset [9]

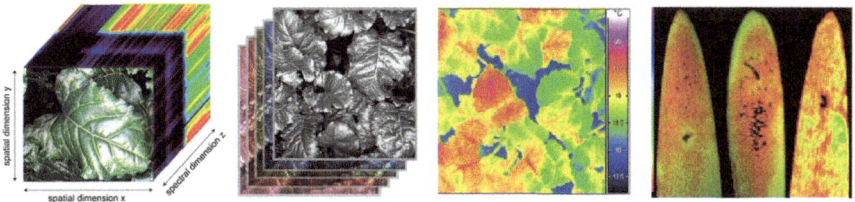

Fig. 2 (left to right) Hyperspectral, multispectral, thermal, and fluorescence spectroscopy images [7]

Thermal Imaging or infrared thermography [7] detects infrared radiation from plant leaves. Certain pathogens infect the plants leading to an increase in temperature of the leaf surface. Different levels of infection may also have varied heat signatures. Any abnormality may be an indication of a disease outbreak. Hence, these temperature changes are captured, maybe even prior to any actual physical change in the appearance of a plant, and utilized to control the invasion of disease. It can also be used for the preparation of irrigation schedules. Fluorescence Spectroscopy [7] is most commonly used to determine the quality of food products. It is also a useful tool to sense pests at an early stage. The molecules of the object under consideration are excited by a beam of light. The excited object emits light, also known as fluorescence, which is observed under the UV light, that captures the distinct colors being emitted. This technique can be used to determine stress level in plants, presence of pests, chlorophyll content as well as nutrient deficiency in the plant. Figure 2 depicts a sample of hyperspectral, multispectral, thermal and fluorescence spectroscopy images.

3 Prediction Models for Plant Diseases

3.1 RGB Imaging

Mohanty et al. [12] proposed in their paper a smart phone-assisted image-based plant disease diagnosis system. Their work acquired 54,306 images (from PlantVillage dataset) of plant leaves, having 38 class labels of 14 crops and 26 diseases. Each label depicted a pair of the crop and its corresponding disease. The images of plant leaves were downsized to 256×256 and were then used for predicting the crop–disease pair. Three different versions of images, namely colored, gray scale, and leaf segmented were used in the experiment. To restrict overfitting, all experiments were conducted across a wide span of train-test splits such as 80–20, where 80% of all datasets were used for training and 20% for testing. Similarly, other splits were 60–40, 50–50, 40–60, and 20–80. The performances of two popular deep convolutional neural network architectures were analyzed: AlexNet and GoogLeNet. Values of mean precision, mean recall, and mean F1 score were computed for each experiment and mean F1 score was used for performance evaluation of various configurations. After the completion of 30 epochs, the overall performance of the model was found better for colored images as compared to segmented or gray scale with the highest mean F1 score. Their work clearly states that while training is a time-consuming process, classification takes no time, and hence can be implemented on a smartphone, paving way for the smartphone-assisted plant disease diagnosis systems. Plant image data may be augmented with location and time for improved precision.

Singh and Misra [13] have devised an algorithm for the segmentation of colored images based on Genetic Algorithms (GA) that can be further utilized for automated detection as well as classification of diseases in plants. Plant leaf images captured via a digital camera go through the preprocessing stage whereby they are clipped, smoothed, enhanced, and finally, the green-colored pixels are masked (i.e., a zero intensity is assigned to the RGB components of that pixel where the green component has a value below the precomputed threshold). The resultant image is then segmented by the use of GA. Color co-occurrence technique is employed for co-occurrence feature computation (for the H image) including local homogeneity, contrast, cluster shade, cluster prominence, and energy, which are later compared to the conforming feature values saved in the feature library. The SVM classifier and Minimum Distance Criterion (MDC) are used for the classification. The proposed algorithm was trained and tested on 10 plant species. Both the classifiers when implemented with the proposed algorithm, give improved results with a detection accuracy of 95.71% and 93.63%, respectively, over 86.54% with k-means.

Camargo and Smith [14] have studied the use of texture-related or pattern-related features on plant leaves for a robust plant disease identification system based on visual symptoms. Overall a total of 117 cotton crop images were segmented and their features such as shape, lacunarity, dispersion, etc., were extracted. A subset of features that provided trivial or no information whatsoever were discarded. Co-occurrence matrix was used for texture calculation. SVM was employed as a Machine Learning tool for the classification. To identify the most appropriate classification model, three different methodologies were used: (1) single feature as input, (2) group of features as input, and lastly (3) all but one feature as input to the classifier. The first approach resulted in poor classification ($\leq 50\%$). The second approach, when texture group was used as input, resulted in the highest classification rate with 83%. The third approach was utilized to distinguish and remove unnecessary features. One by one, features which increased the classification rate were retained and the ones which decreased the value of the classification rate were discarded. A total to 45 out of 53 remained with a classification accuracy of 93.1%. For evaluation, 7-fold cross validation was used.

Bashish et al. [15] have proposed an image processing based solution to automatically detect and classify plant diseases. The entire algorithm is divided into four phases: color space transformation of images which is device independent, i.e., the RGB image is converted into HSI image; segmentation via k-means clustering; extraction of texture-related features via color co-occurrence method; and lastly these features are given as input to a pretrained neural network. Eleven textural features such as angular moment, mean intensity level, variance texture feature, correlation, entropy, etc., were calculated. Testing was carried out on five diseases, namely early scotch, ashen mold, cottony mold, tiny whiteness, and late scorch. Feed Forward Backpropagation Neural Network was utilized with 10 hidden layers, no. of inputs same as the number of textural features and the number of outputs was six (no. of diseases plus one for a normal leaf). MSE was used for the performance measurement and iterations were set to 10,000. On the basis of the experiment, the NN classifier gave 93% classification accuracy.

Liu et al. [16] have worked on image processing and recognition to identify wheat crop diseases, viz., stripe rust, leaf powdery mildew, leaf rust, and leaf blight, by means of multi-class Radial Basis Function (RBF) SVM classifier. Daubechies wavelet transformation was used for enhancing the image and noise filtering purposes. In the segmentation process, via the mathematical morphology, the healthy area is blackened and diseased area of the leaf is left unaltered. Three schemes are used for feature extraction and compared for classification accuracy later: (a) mean R, G, B values, (b) normalized mean of R, G, B values, and (c) green ratios R/G, B/G. The second scheme resulted in the best recognition rate with 96%. Wang et al. [17] have employed the Apple black rot leaf images from the PlantVillage dataset to explore the deep learning based automated identification of plant disease severity. Each image belonged to one of the severity classes/stages: healthy stage, early stage, middle stage, and end stage. Finally, images of 1644 healthy, 137 early, 180 middle, and 125 end stage were utilized further with a train-test split of 80–20. Image preprocessing included scaling, sampling, sample-wise normalization, and several other modifications to generalize and avoid overfitting. In their work, they have compared two distinct architectures, i.e., developing a shallow network from the beginning and fine-tuned deep models via transfer learning. The training parameters for both architectures were training algorithm, learning rate, batch size, and early stopping. Shallow networks of 2, 4, 6, 8, 10 convolutional layers were trained, whereby the best performance of 79.3% accuracy was given by eight convolutional layers while in the fine-tuned network models, VGG16, VGG19, Inception-v3, and ResNet50, VGG16 model proved to be superior with 90.4% test accuracy.

Sabrol and Kumar [18] suggested a method of identifying five types of tomato plant diseases by extracting 10 intensity-based features from their images and giving them as input to the decision tree, for classification. Images were categorized into six classes: healthy, bacterial leaf spot, septoria leaf spot, fungal late blight, bacterial canker, and leaf curl. These images were scaled to 256×256 and converted into color segmented and binary images by employing Otsu's segmentation. After the color space transformation, 10 statistical features for each extracted intensity component were computed, i.e., $mean_x$, $mean_y$, $mean_z$, std_x, std_g, std_b, $skewness_r$, $skewness_g$, $skewness_b$, and $corr(x, y)$. Gini index was used as the splitting criterion for creating a decision tree, finally having 80 nodes. This algorithm resulted in 78% overall recognition accuracy. Rice is a widely consumed serial grain which is affected by a seed-borne disease called Bakanae disease. Chung et al. [19] have examined the seeds at early stages, for their color and morphological characteristics (elongated seedling or stunted growth, seedling height/length, leaf angle) to distinguish between healthy and diseases grains, employing the SVM classifier giving 87.9% accuracy, with 10-fold cross validation. Genetic Algorithms were used to identify the differentiating parameters to train a cascade of two SVMs.

3.2 Hyperspectral Imaging

Rumpf et al. [20] proposed a methodology with experimental evidences, for early prediction and determination of Sugar beet diseases, solely based on hyperspectral reflectance data. Hyperspectral data was collected from healthy leaves, along with leaves inoculated beforehand with pathogens namely Uromyces betae, Cercospora beticola, or Erysiphe betae. These pathogens caused leaf rust, cercospora leaf spot, and powdery mildrew respectively, in the plant. After the completion of inoculation, the spectral data was registered every day for 21 days. For automatic classification purposes, nine spectral vegetation indices such as NDVI, Simple Ratio, Structure Insensitive Vegetation Index, Pigments Specific Simple Ratio, Anthocyanin Reflectance Index, Red Edge Position, and Modified Chlorophyll Absorption Integral were utilized as features for classification. This paper focused on three supervised learning techniques, viz., ANN, Decision Trees, and SVM. The inoculated leaves were identified even before any visible symptoms appeared, indicated by the changes in their spectral reflectance properties. These changes were due to the modifications in the content of water at the infected sites, cell death, resistance reactions by the plant tissue, etc., SVM proved to be a better approach with minimum classification error percentage.

Singh et al. [21] utilized three statistical classifiers (namely linear, mahalanobis, and quadratic) and BPNN, to distinguish between healthy wheat kernels and damaged wheat kernels (due to insects feeding on them). A few selected features from a total of 230, extracted from color images along with near infrared (NIR) hyperspectral data, were given as input to the classifiers, where Quadratic Discriminant Analysis (QDA) achieved the highest classification accuracy of 96.4% for healthy and 91–100% for insect-ruined wheat kernels. The statistical features extracted from the NIR hyperspectal imaging contributed most to the classification accuracy.

Xie et al. [22] collected over 120 hyperspectral tomato leaf images (in the range 380–1023 nm) of healthy as well as early blight infected each, and 70 infected with late blight. They have tried to utilize both the spectral data as well as textural features for detecting these fungal diseases in tomatoes. They have analyzed the spectral curves in the attempt to establish a method for selecting specific wavelengths for effective disease identification. They also acquired textural features in these wavelengths and finally replaced the full wavelengths with the identified ones in the successive projections algorithm-extreme learning machine (SPA-ELM) model for classification. The paper demonstrates the potential of textural features in these specific wavelengths for classification accuracy. Zhao et al. [23] also have investigated the potential of hyperspectral images for finding the pigment content (chlorophyll, carotenoid) in cucumber leaves due to the angular leaf spot (ALS) disease. The bacterial disease causes a reduction in the leaf pigmentation as well as changes in their spatial spread. Hyperspectral images of 196 cucumber leaves, in various stages of severity, were taken in the range 380–1030 nm. They have also identified nine significant wavelengths by analyzing the spectral variation of the leaves, then developing PLSR to establish the regression technique. Chlorophyll content and caroteroid content were computed at every pixel of the hyperspectral images and displayed the dispersion map with the aid of MATLAB.

3.3 Other Imaging Techniques

Heckmann et al. [24] employed the gas exchange data and leaf optical reflectance data for the evaluation of photosynthetic capacity in 1000 crops of Brassica (C_3) or Zea (C_4). Their work indicates that crop photosynthetic capacity can be greatly improved by exploiting the leaf reflectance phenotyping and subsequently would result in significant yield increase. Initially, they studied and compared the reflectance spectra of 36 different plant species using their developed leaves. For modeling, partial least square regression along with recursive feature elimination was used. This process required robust feature selection as most of the plant species exhibit similar spectral features. For validation, k-fold cross validation, with varying values of k, was employed. To determine the photosynthetic capacity of plant species, parameter like maximal rate of carboxylation (V_{cmax}) and the maximal rate of electron transport (J_{max}) were computed by utilizing the A-C_i curves, where A_{max} is the maximal photosynthetic rate and C_i represents crops C_3 and C_4.

Raza et al. [25] in their paper have extracted and combined visible data, thermal data, and depth data from various kinds of images for remotely detecting and identifying diseased regions in tomato plants. They inoculated 54 out of 71 tomato plants with powdery mildew causing fungus namely Oidium Neolycopersici. After the inoculation, their stereo visible light and thermal images were captured for 14 straight days. The symptoms of the disease became visible after 7 days. The depth information is derived from the disparity map. Post the pixel-wise alignment process of color, thermal, and depth data, the SVM classifier is used for their separation into the background or plant pixels. Alignment is done to ensure that each pixel in these images is consistent with the identical physical location. Two-fold feature extraction methods, i.e., local as well as global, are used. The local method utilizes information from depth, color, temperature, etc., at each pixel for feature extraction and global exploits data directly from all pixels such as mean, standard deviation, variance, etc. Plants were classified as healthy and diseased based on both techniques via the SVM classifier. For performance evaluation, parameters like average sensitivity, accuracy, specificity, and positive predictive value (PPV) were computed. Sankaran et al. [26] have investigated the potential of visible and near infrared (VNIR) spectroscopy for detecting HLB disease in citrus orchards. 93 infected along with 100 healthy citrus trees were utilized to collect 989 spectral characteristics, which were further processed to reduce the count of the features, and used for input to the prediction algorithm, with QDA yielding the highest accuracy.

Table 1 summarizes a few of the approaches investigated and proposed by researchers. Out of the techniques studied, 56% utilized RGB Imaging, 25% hyperspectral, and for 19%, other imaging techniques were employed, such as thermal, visible, and infrared spectroscopy, etc. Use of RGB imaging indicates the manifestation of the disease to a higher severity level that leads to visible symptoms. Hence, early detection is highly dependent on capturing spectral data and its interpretation. Such data reflects even the subtle modifications in biological as well as chemical properties of a plant, due to the attack of pathogens. The implementation of IoT sensors can be utilized to detect patterns in the values collected from the environmental variables, a smart system in which various environmental parameters such as temperature, humidity, pressure, CO_2, water level, etc., along with images, can be

Table 1 Summary of the imaging techniques and corresponding statistical approaches for plant disease identification and classification

Publication and year	Imaging technique	Statistical approach	Results/highest classification accuracy	Crop	Disease
Wang et al. (2017) [17]	RGB imaging	Deep learning (VGG16, VGG19, Inception-v3, ResNet50)	80.0–90.4%	Apple	Apple black rot
Mohanty et al. (2016) [12]	RGB imaging (PlantVillage dataset)	Deep learning models (GoogleNet and AlexNet)	99.35%	14 crops	26 diseases
Sabrol and Kumar (2016) [18]	RGB imaging	Decision trees	78%	Tomato	Late blight, bacterial leaf spot, septoria leaf, leaf curl, bacterial canker
Zhao et al. (2016) [23]	Hyperspectral reflectance	Partial Least Square Regression (PLSR)	Correlation coefficient: 0.871 and 0.876 for Chl and Car resp.	Cucumber	Angular Leaf Spot (ALS) disease
Singh and Misra (2016) [13]	RGB imaging	SVM classifier + Minimum distance criterion + k-means	MDC with k-means: 86.54% MDC with proposed technique: 93.63% SVM with proposed technique: 95.71%	Rose, beans leaf, lemon leaf, banana leaf	Banana leaf: Early scorch disease; Lemon leaf: Sun burn disease; Beans and rose: Bacterial disease; Beans leaf: Fungal disease

(continued)

Table 1 (continued)

Publication and year	Imaging technique	Statistical approach	Results/highest classification accuracy	Crop	Disease
Chung et al. (2016) [19]	RGB imaging	SVM	Accuracy: 87.9% and Positive predictive value: 91.8%	Wheat	Bakanae disease
Raza et al. (2015) [25]	Thermal and stereo visible light imaging	SVM classifier with linear kernel	90%	Tomato	Powdery mildew
Xie et al. (2015) [22]	Hyperspectral imaging	Extreme Learning Machine (ELM) classifier model	97.1–100% in testing sets, while features dissimilarity, second moment and entropy, had the highest classification accuracy of 71.8, 70.9, and 69.9%	Tomato	Early blight, late blight
Liu et al. (2013) [16]	RGB imaging	RBF SVM	96%	Wheat	Stripe rust, powdery mildew, leaf rust, leaf blight
Sankaran et al. (2011) [8]	Visible near infrared spectroscopy	LCA, QDA, K-nearest neighbor, SIMCA	98% (QDA)	Citrus orchards	Huanglongbing (HLB)
Al-Bashish et al. (2011) [15]	RGB imaging	k-means clustering and Feedforward Backpropagation Neural Network	89.50%	General leaf images	Late scorch, cottony mold, ashen mold, tiny whiteness, early scorch
Rumpf et al. (2010) [20]	Hyperspectral reflectance	Decision trees, ANN, SVM + RBF	97%	Sugar beet	Powdery mildew, sugar beet rust, cercospora leaf spot
Singh et al. (2010) [21]	SWIR hyperspectral + RGB imaging	Linear, quadratic, Mahalanobis, and a BPNN classifier	96.4% healthy, 91–100% damaged (QDA)	Wheat kernels	Grain borer, rice weevil, red flour beetle, rusty grain beetle
Camargo and Smith (2009) [14]	RGB imaging	SVM	90%	Cotton	Southern green stink bug, bacterial angular, Ascochyta blight
Huang (2007) [27]	RGB imaging	BPNN	89.60%	Phalaenopsis	Phalaenopsis seedling disease

observed and timely action can be taken to prevent any hazards. By the use of image processing and IoT, data can be collected and analyzed, leading to the timely prediction of most probable diseases in crops, using data analytics. Use of smart phones and the development of mobile applications employing deep learning algorithms for disease classification is already under implementation. Deep learning combined with hyperspectral data is another approach that can prove to be monumental in early disease detection. But certain limitations are posed on this approach due to the unacceptable spatial resolution of the imaging equipment, choice and identification of accurate wavelength, not to mention the expensive hardware required and computationally intensive algorithms. Multiple sources of data combined with weather forecasting models (due to the dependency of pathogens requiring suitable conditions to thrive) can be taken into account for developing products and services for farmers to exercise.

4 Conclusion and Future Direction

As listed in Table 1, different forms of non-destructive, noninvasive imaging techniques used for remote sensing in agronomy can be exploited to strengthen the technology-driven approach in Precision Agriculture. Pathogens are present in the soil, water, air, only requiring suitable environmental conditions to infect the plant. With the ever-increasing population, the world cannot afford food gain losses to plant diseases. Hence, early prediction, detection, and quantification of plant diseases is highly essential. In future, decreased hardware costs may allow these sensors to be deployed largely in the fields, collecting both spectral data as well as spatial data for further interpretation. Also, there are several factors that can be worked upon such as different parts of a plant being used for disease prediction, Iot-based sensors being supplemented and deployed along with imaging sensors for data collection and subsequent analysis. Work needs to be done to make the existing methods reach the farmers effectively, to be user-friendly, secure, and accurate.

References

1. Kaundal, R., Kapoor, A.S., Raghava, G.P.S.: Machine learning techniques in disease forecasting: a case study on rice blast prediction. BMC Bioinform. 7(485), 1–16 (2006)
2. Pal, T., Jaiswal, V., Chauhan, R.S.: DRPPP: a machine learning based tool for prediction of disease resistance proteins in plants. Comput. Biol. Med. (2016). http://dx.doi.org/10.1016/j.compbiomed.2016.09.008
3. Petrellis, N.: Plant disease diagnosis based on image processing, appropriate for mobile phone implementation. In: Proceedings of 7th International Conference on Information and Communication Technologies in Agriculture, Food and Environment (HAICTA-2015), Greece (2015)
4. Osroosh, Y., Khot, L.R., Peters, R.T.: Economical thermal-RGB imaging system for monitoring agricultural crops. Comput. Electron. Agric. 147, 34–43 (2018)
5. Lu, J., Zhou, M., Gao, Y., Jiang, H.: Using hyperspectral imaging to discriminate yellow leaf curl disease in tomato leaves. Springer, Precision Agriculture (online) (2017)

6. Camargo, A., Smith, J.S.: Image pattern classification for the identification of disease causing agents in plants. Comput. Electron. Agric. **66**, 118–125 (2009)
7. Mahlein, A.K.: Plant disease detection by imaging sensors—parallels and specific demands for precision agriculture and plant phenotyping, plant disease. APS J. **100**(2), 241–251 (2016)
8. Sankaran, S., Mishra, A., Ehsani, R., Davis, C.: A review of advanced techniques for detecting plant diseases. Comput. Electron. Agric. **72**, 1–13 (2010)
9. PlantVillage Homepage. https://plantvillage.org/. Last accessed 25 Apr 2018
10. Lowe, A., Harrison, N., French, A.P.: Hyperspectral image analysis techniques for the detection and classification of the early onset of plant disease and stress. Plant Methods **13**(80), 1–12 (2017)
11. Ye, X., Sakai, K., Okamoto, H., Garciano, L.O.: A ground-based hyperspectral imaging system for characterizing vegetation spectral features. Comput. Electron. Agric. **63**, 13–21 (2008)
12. Mohanty, S.P., Hughes, D.P., Salathé, M.: Using deep learning for image-based plant disease detection. Front. Plant Sci. **7**, Article 1419, 1010 (2016)
13. Singh, V., Misra, A.K.: Detection of plant leaf diseases using image segmentation and soft computing techniques. Inf. Process. Agric. **4**, 41–49 (2017)
14. Camargo, A., Smith, J.S.: An image-processing based algorithm to automatically identify plant disease visual symptoms. Biosys. Eng. **102**, 9–12 (2009)
15. Al-Bashish, D., Braik, M., Bani-Ahmad, S.: Detection and classification of leaf diseases using K means-based segmentation and neural-networks-based classification. Inform. Technol. J. **10**, 267–275 (2011)
16. Liu, L., Zhang, W., Shu, S., Jin, X.: Image recognition of wheat disease based on RBF support vector machine. In: International Conference on Advanced Computer Science and Electronics Information (ICACSEI 2013), pp. 307–310, Published by Atlantis Press, China (2013)
17. Wang, G., Sun, Y., Wang, J.: Automatic image-based plant disease severity estimation using deep learning. Hindawi Comput. Intell. Neurosci., Article ID 2917536, 1–8 (2017)
18. Sabrol, H., Kumar, S.: Intensity based feature extraction for tomato plant disease recognition by classification using decision tree. Int. J. Comput. Sci. Inf. Secur. (IJCSIS) **14**(9), 622–626 (2016)
19. Chung, C.L., Huang, K.J., Chen, S.Y., Lai, M.H., Chen, Y.C., Kuo, Y.F.: Detecting Bakanae disease in rice seedlings by machine vision. Comput. Electron. Agric. **121**, 404–411 (2016)
20. Rumpf, T., Mahlein, A.K., Steiner, U., Oerke, E.C., Dehne, H.W., Plümer, L.: Early detection and classification of plant diseases with support vector machines based on hyperspectral reflectance. Comput. Electron. Agric. **74**, 91–99 (2010)
21. Singh, C.B., Jayasa, D.S., Paliwala, J., White, N.D.G.: Identification of insect-damaged wheat kernels using short-wave near-infrared hyperspectral and digital colour imaging. Comput. Electron. Agric. **73**, 118–125 (2010)
22. Xie, C., Shao, Y., Li, X., He, Y.: Detection of early blight and late blight diseases on tomato leaves using hyperspectral imaging. Sci. Rep. **5**(16564), 1–11 (2015). https://doi.org/10.1038/srep16564
23. Zhao, Y.R., Li, X., Yu, K.Q., Cheng, F., He, Y.: Hyperspectral imaging for determining pigment contents in cucumber leaves in response to angular leaf spot disease. Sci. Rep. **6**(27790), 1–9 (2016). https://doi.org/10.1038/srep27790
24. Heckmann, D., Schluter, U., Weber, A.P.M.: Machine learning techniques for predicting crop photosynthetic capacity from leaf reflectance spectra. Mol. Plant, Cell Press. **10**, 878–890 (2017)
25. Raza, S.A., Prince, G., Clarkson, J.P., Rajpoot, N.M.: Automatic detection of diseased tomato plants using thermal and stereo visible light images. PLoSONE **10**, e0123262 (2015). https://doi.org/10.1371/journal.pone.0123262
26. Sankaran, S., Mishra, A., Maja, J.M., Ehsani, R.: Visible-near infrared spectroscopy for detection of Huanglongbing in citrus orchards. Comput. Electron. Agric. **77**, 127–134 (2011)
27. Huang, K.Y.: Application of artificial neural network for detecting Phalaenopsis seedling diseases using color and texture features. Comput. Electron. Agric. **57**, 3–11 (2007)

Linear Regression Correlation Filter: An Application to Face Recognition

Tiash Ghosh and Pradipta K. Banerjee

Abstract This paper proposes a novel method of designing a correlation filter for frequency domain pattern recognition. The proposed correlation filter is designed with linear regression technique and termed as linear regression correlation filter. The design methodology of linear regression correlation filter is completely different from standard correlation filter design techniques. The proposed linear regression correlation filter is estimated or predicted from a linear subspace of weak classifiers. The proposed filter is evaluated on standard benchmark database and promising results are reported.

Keywords Linear regression · Correlation filter · Face recognition

1 Introduction

Application of correlation filters in pattern recognition is reported in several research papers with promising results [6, 9, 11, 13, 17, 19]. In [1], a generalized regression network based correlation filtering technique is proposed for face recognition under poor illumination condition. A preferential digital optical correlator is proposed in [2] where the trade-off parameters of the correlation filter are optimized for robust pattern recognition. A class-specific subspace based nonlinear correlation filtering technique is introduced in [3]. Basic frequency domain pattern (face) recognition techniques are carried out by correlation filter by cross-correlating the Fourier transform of test face image with a synthesized template or filter, generated from Fourier transform of training face images and processing the resulting correlation output via inverse Fourier transform. The process of correlation pattern recognition is pictorially given in Fig. 1. The process given in Fig. 1 can be mathematically summarized in the following way. Let $\mathbf{F}_x$ and $\mathbf{F}_h$ denote the 2D discrete Fourier transforms (DFTs)

T. Ghosh (✉) · P. K. Banerjee
Department of Electrical Engineering, Future Institute of Engineering and Management,
Kolkata, India
e-mail: tiazghosh@gmail.com

© Springer Nature Singapore Pte Ltd. 2020
B. B. Chaudhuri et al. (eds.), *Proceedings of 3rd International Conference on Computer Vision and Image Processing*, Advances in Intelligent Systems and Computing 1022,
https://doi.org/10.1007/978-981-32-9088-4_32

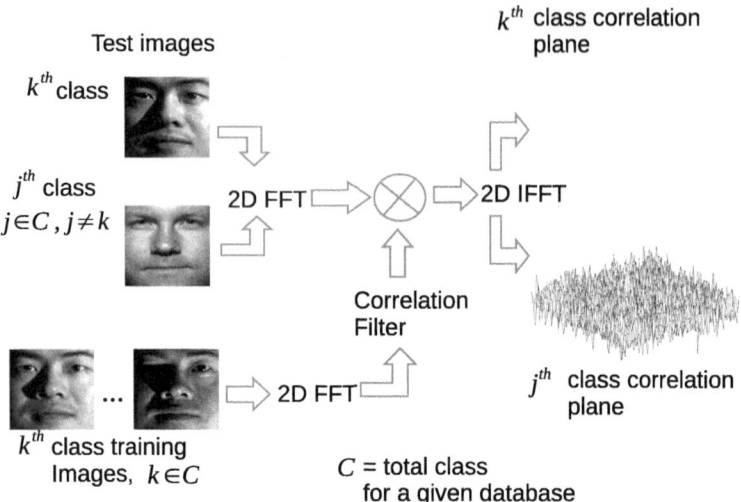

Fig. 1 Basic frequency domain correlation technique for face recognition

of 2D image x and 2D filter h in spatial domain, respectively, and let $\mathbf{F}_y$ is the Fourier-transformed probe image. The correlation output g in space domain in response to y for the filter $\mathbf{F}_h$ can then be expressed as the inverse 2D DFT of the frequency domain conjugate product as

$$g = FFT^{-1}[\mathbf{F}_y \circ \mathbf{F}_h^*] \tag{1}$$

where $\circ$ represents the element-wise array multiplication, $*$ stands for complex conjugate operation, and FFT is an efficient algorithm to perform DFT. As shown in Fig. 1, the information of N number of training images from kth face class ($k \in C$), out of total C number of face classes for a given database, are Fourier transformed to form the design input for kth correlation filter. In ideal case, a correlation peak with high value of peak-to-sidelobe ratio (PSR) [8] is obtained, when any Fourier-transformed test face image of kth class is correlated with kth correlation filter, indicating authentication. However, in response to impostor faces no such peak will be found as shown in Fig. 1 corresponding to jth correlation plane.

Traditional correlation filters used for pattern recognition purpose are commonly categorized in constrained and unconstrained filters. Due to several advantages of unconstrained correlation filters [12] over constrained filters, pattern recognition tasks are generally carried out with the former one. Several correlation filters are developed as unconstrained minimum average correlation energy (UMACE) [12], maximum average correlation height (MACH) [12], optimal trade-off synthetic discriminant (OTSDF) [16], and optimal trade-off MACH (OTMACH) [7] for pattern recognition. Basic fundamental blocks needed for designing these correlation filters are minimization of average correlation energy (ACE) of the correlation plane, minimization of the average similarity measure (ASM), maximization of correlation peak height of the correlation plane. Detailed design technique of traditional correlation filters can be found in [14].

Apart from traditional design method of correlation filter, this paper proposes a linear regression based correlation filter estimation from a span of subspace of similar objects. Basic concept used in designiing the linear regression correlation filter (LRCF) is that images from a same class lie on a linear subspace. The proposed method does not include the characteristics of correlation planes like ASM and/or ACE and peak height. Without specifying the correlation plane characteristics this design technique solely depends on some weak classifiers, known as matched filters, obtained from Fourier-transformed training images of a specific class.

Proposed design technique of LRCF: In this paper, a simple and efficient linear regression based correlation filter for face recognition is proposed. Sample images from a specific class are known to lie on a subspace [4]. Using this concept class, specific model of correlation filters is developed with the help of sample gallery images, thereby defining the task of correlation filter design as a problem of linear regression [5]. Gallery images are Fourier transformed which leads to matched filters. These matched filters are representative of weak classifiers which correspond to each gallery image. Design of proposed filter assumes that the matched filters for a specific class lie in a linear subspace. Probe image is also Fourier transformed, which results in a probe-matched filter. Least-squares estimation is used to estimate the vectors of parameters for a given probe-matched filter against all class models. The decision rule is based on frequency domain correlation technique. The probe image is classified in favor of the class with most precise estimation.

Proposed LRCF is tested on a standard database and promising results are given. Rest of the paper is organized as follows: In Sect. 2, the mathematical formulation and design algorithm of LRCF is discussed. Extensive experimental results are reported in Sect. 3. The paper concludes in Sect. 4.

Nomenclature used in this paper:

C : total number of classes,

N : total number of images in each class,

$x^{(k)(i)} \in \Re^{d_1 \times d_2}$: spatial domain ith image of kth class,

$\mathbf{F}_x \in \Re^{d_1 \times d_2}$: DFT of x,

$y \in \Re^{d_1 \times d_2}$: spatial domain probe image,

$\mathbf{F}_y \in \Re^{d_1 \times d_2}$: DFT of y,

$h^{(k)}, \in \Re^{d_1 \times d_2}$: spatial domain kth class correlation filter,

$\mathbf{F}_h^{(k)}, \in \Re^{d_1 \times d_2}$: DFT of $h^{(k)}$,

$g^{(i)}, \in \Re^{d_1 \times d_2}$: spatial domain correlation plane in response to ith image,

$\mathbf{F}_g^{(i)}, \in \Re^{d_1 \times d_2}$: DFT version of $g^{(i)}$,

$m = d_1 \times d_2$,

$\mathbf{X}^{(k)} \in \Re^{m \times N}$: kth class design matrix,

$\mathbf{f}_x, \mathbf{f}_y, \in \Re^{m \times 1}$: vector representation of $\mathbf{F}_x$ and $\mathbf{F}_y$, respectively.

$\theta^{(k)}, \in \Re^{N \times 1}$: parameter vector of kth class,

DFT : Discrete Fourier Transform.

2   Mathematical Formulation of LRCF

Let C number of distinguished classes with $N^{(k)}$ number of training images from
kth class, $k = 1, 2, \ldots, C$. Each grayscale image $x^{(k)(i)}$ is then Fourier transformed.
This $\mathbf{F}_x^{(k)(i)}$ is a simple matched filter generated from $x^{(k)(i)}$. These $N^{(k)}$ number of
matched filters can be termed as weak classifier for the kth class. Using the concept
that patterns from same class lie in a linear subspace, a class-specific model $\mathbf{X}^{(k)}$,
known as design matrix, is developed with the weak classifiers from kth class. The
class-specific model is designed in such a way that each column of $\mathbf{X}^{(k)}$ contains
$\mathbf{f}_x^{(k)(i)}$, which is lexicographic version of $\mathbf{F}_x^{(k)(i)}$. The design matrix mathematically
expressed as

$$\mathbf{X}^{(k)} = [\mathbf{f}_x^{(k)(1)}, \mathbf{f}_x^{(k)(2)}, \ldots, \mathbf{f}_x^{(k)(N)}] \in \Re^{m \times N}, k = 1, 2, \ldots, C \tag{2}$$

Each Fourier-transformed vector $\mathbf{f}_x^{(k)(i)}$ spans a subspace of $\Re^m$ also called column
space of $\mathbf{X}^{(k)}$. At the training stage, each class k is represented by a vector subspace
$\mathbf{X}^{(k)}$, which can be called as predictor or regressor for class k.

Let y be the probe image to be classified in any of the classes $k = 1, 2, \ldots, C$.
Fourier transform of y is evaluated as $\mathbf{F}_y$ and lexicographically ordered to get $\mathbf{f}_y$. If
$\mathbf{f}_y$ belongs to the k th class, it should be represented as a linear combination of weak
classifiers (matched filters) from the same class, lying in the same subspace, i.e.,

$$\mathbf{f}_y = \mathbf{X}^{(k)}\theta^{(k)}, k = 1, 2, \ldots, C \tag{3}$$

where $\theta^{(k)} \in \Re^{N \times 1}$ is the vector of parameters. For each class, the parameter vector
can be estimated using least-squares estimation method [18]. The objective function
can be set for quadratic minimization problem as

$$\begin{aligned} J(\theta^{(k)}) &= \|\mathbf{f}_y - \mathbf{X}^{(k)}\theta^{(k)}\|^2 \\ &= (\mathbf{f}_y - \mathbf{X}^{(k)}\theta^{(k)})^T (\mathbf{f}_y - \mathbf{X}^{(k)}\theta^{(k)}) \\ &= \mathbf{f}_y^T \mathbf{f}_y - 2\theta^{(k)T}\mathbf{X}^{(k)T}\mathbf{f}_y + \theta^{(k)T}\mathbf{X}^{(k)T}\mathbf{X}^{(k)}\theta^{(k)} \end{aligned} \tag{4}$$

Setting

$$\frac{\partial J(\theta^{(k)})}{\partial \theta^{(k)}} = 0$$

gives

$$\theta^{(k)} = (\mathbf{X}^{(k)T}\mathbf{X}^{(k)})^{-1}\mathbf{X}^{(k)T}\mathbf{f}_y \tag{5}$$

The estimated vector of parameters $\theta^{(k)}$ along with the predictors $\mathbf{X}^{(k)}$ is used to
predict the correlation filter in response to the probe-matched filter $\mathbf{f}_y$ for each class
k. The predicted correlation filter obtained from the linear regression technique is
termed as linear regression correlation filter whihc is having the following mathematical expression:

$$\hat{\mathbf{f}}_y^{(k)} = \mathbf{X}^{(k)} \theta^{(k)}, k = 1, 2, \ldots, C$$
$$= (\mathbf{X}^{(k)T} \mathbf{X}^{(k)})^{-1} \mathbf{X}^{(k)T} \mathbf{f}_y$$
$$= \mathbf{H}^{(k)} \mathbf{f}_y \tag{6}$$

Class-specific $\mathbf{H}^{(k)}$, known as *hat matrix* [15], is used to map the probe-matched filter $\mathbf{f}_y$ into a class of estimated LRCFs $\hat{\mathbf{f}}_y^{(k)}$, $k = 1, 2, \ldots, C$.

2.1 Design Algorithm of LRCF

Algorithm: Linear Regression Correlation Filter (LRCF).
Inputs: Class model matched filters design matrix $\mathbf{X}^{(k)} \in \mathfrak{R}^{m \times N}$, $k = 1, 2, \ldots, C$, and probe-matched filter $\mathbf{f}_y \in \mathfrak{R}^{m \times 1}$.
Output: kth class of y

1. $\theta^{(k)} \in \mathfrak{R}^{N \times 1}$, $k = 1, 2, \ldots, C$, is evaluated against each class model

$$\theta^{(k)} = (\mathbf{X}^{(k)T} \mathbf{X}^{(k)})^{-1} \mathbf{X}^{(k)T} \mathbf{f}_y, k = 1, 2, \ldots, C$$

2. Estimated probe-matched filter is computed for each parameter vector $\theta^{(k)}$ in response to probe-matched filter $\mathbf{f}_y$ as

$$\hat{\mathbf{f}}_y^{(k)} = \mathbf{X}^{(k)} \theta^{(k)}, k = 1, 2, \ldots, C$$

3. Frequency domain correlation:

$$\hat{\mathbf{F}}_y^{(k)} \otimes \mathbf{F}_y, k = 1, 2, \ldots, C$$

4. PSR metric is evaluated for each class, C number of PSRs obtained.
5. y is assigned to kth class if[1]

$$PSR^{(k)} = \max\{PSRs\} \geq thr, k = 1, 2, \ldots, C$$

Decision-making: Probe-matched filter $\mathbf{f}_y$ is correlated with C numbers of estimated LRCFs $\hat{\mathbf{f}}_y^{(k)}$, $k = 1, 2, \ldots, C$ and highest PSR value is searched from C correlation planes. The maximum PSR obtained is further tested with a preset threshold and the probe image is classified in class k for which the above conditions are satisfied.

---

[1] thr: hard threshold selected empirically.

Table 1 Randomly chosen five training sets

set 1	23	36	27	32	5	51	46	15	38	37	6	17	21	54	9	49	42	56	44	47
set 2	60	2	48	56	10	16	1	40	22	25	59	42	31	43	29	37	39	24	27	6
set 3	57	19	13	38	58	23	30	36	59	4	61	6	45	56	28	50	9	42	14	18
set 4	11	40	21	48	30	23	36	60	32	13	34	7	64	15	9	39	59	20	18	54
set 5	63	56	16	7	17	15	47	41	6	44	8	49	48	26	12	27	64	39	28	33

3 Experimental Results

3.1 Database and Preparation

The Extended Yale Face Database B [10] contains 38 individuals under 64 different illumination conditions with 9 poses. Only frontal face images are taken for experiments. Frontal cropped face images are readily available from website.[2] All grayscale images are downsampled to a size of 64 × 64 for experimental purpose.

Each class of yale B database contains 64 images. In training stage, out of 64 images 20 images are randomly chosen 5 times and five sets of $\mathbf{X}^{(k)}$ are designed. For each set, the size of design matrix $\mathbf{X}^{(k)}$ is therefore 4096 × 20. Hence, for kth class, five such predictors $\mathbf{X}$ are formed, each of which contains different sets of 20 weak classifiers or matched filters (Tables 1 and 2).

In testing stage, rest of the images are taken, i.e., no overlapping is done. Probe image of size 64 × 64 is taken and Fourier transformed. The lexicographic version $\mathbf{f}_y$ from $\mathbf{F}_y$ is then used to estimate the LRCFs $\mathbf{f}_y^{(k)}$, $k = 1, 2, \ldots, C$ using the parameter vector $\theta^{(k)}$ and predictor class $\mathbf{X}^{(k)}$.

3.2 Performance of LRCF

Correlation plane: In frequency domain correlation pattern recognition, the authentic image should give a distinct peak in correlation plane whereas the imposter will not. To establish the above statement, in this case one probe image from class 1 is taken and tested with both class 1 and class 2 LRCF. Figure 2 shows the proposed method works well for authentication purpose as the probe image from class 1 when correlated with estimated LRCF originated from $\mathbf{X}^{(1)}$ subspace and $\theta^{(1)}$ parameters give a correlation plane with distinct peak having high PSR value of 66.65. The same

---

[2]http://vision.ucsd.edu/~leekc/ExtYaleDatabase/ExtYaleB.html.

Table 2 Randomly chosen five test sets. It can be observed that no overlapping is done in testing stage, i.e., image index of training set1 is completely different from test set1 and that is true for all other sets

set 1	1	2	3	4	7	8	10	11	12	13	14	16	18	19	20	22	24	25	26	28	29	30
	31	33	34	35	39	40	41	43	45	48	50	52	53	55	57	58	59	60	61	62	63	64
set 2	3	4	5	7	8	9	11	12	13	14	15	17	18	19	20	21	23	26	28	30	32	33
	34	35	36	38	41	44	45	46	47	49	50	51	52	53	54	55	57	58	61	62	63	64
set 3	1	2	3	5	7	8	10	11	12	15	16	17	20	21	22	24	25	26	27	29	31	32
	33	34	35	37	39	40	41	43	44	46	47	48	49	51	52	53	54	55	60	62	63	64
set 4	1	2	3	4	5	6	8	10	12	14	16	17	19	22	24	25	26	27	28	29	31	33
	35	37	38	41	42	43	44	45	46	47	49	50	51	52	53	55	56	57	58	61	62	63
set 5	1	2	3	4	5	9	10	11	13	14	18	19	20	21	22	23	24	25	29	30	31	32
	34	35	36	37	38	40	42	43	45	46	50	51	52	53	54	55	57	58	59	60	61	62

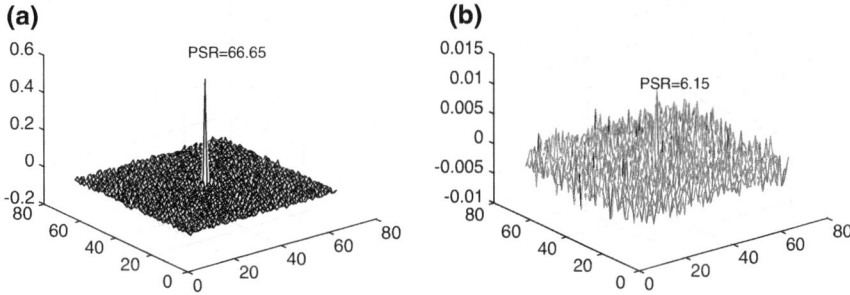

(a) PSR=66.65

(b) PSR=6.15

Fig. 2 **a** authentic correlation plane, **b** imposter correlation plane

probe image when tested with LRCF originated from $\mathbf{X}^{(2)}$ and $\theta^{(2)}$ provides no such peak with a PSR value of 6.15 in the correlation plane.

PSR distribution: This phase of experiment is categorized into two ways: (1)single probe image is taken from a specific class and tested on all class LRCFs and (2) all probe images from all classes are taken and tested with a specific class LRCF. As five sets of predictor class models are available for each class, for each model one estimated LRCF $\hat{\mathbf{f}}_y^{(k)}$ is evaluated. Then single probe image (matched filter) from a specific class is correlated with five estimated LRCFs of each class. Five such PSR values are then averaged out for each class and 38 PSR values are obtained. Figure 3 shows the PSR distribution of randomly selected single probe image taken from some specific classes 10, 24, 15, 20. It is evident from Fig. 3 that the probe image from a specific class when tested with LRCF of the same class gives high PSR value and in other cases low PSRs are obtained.

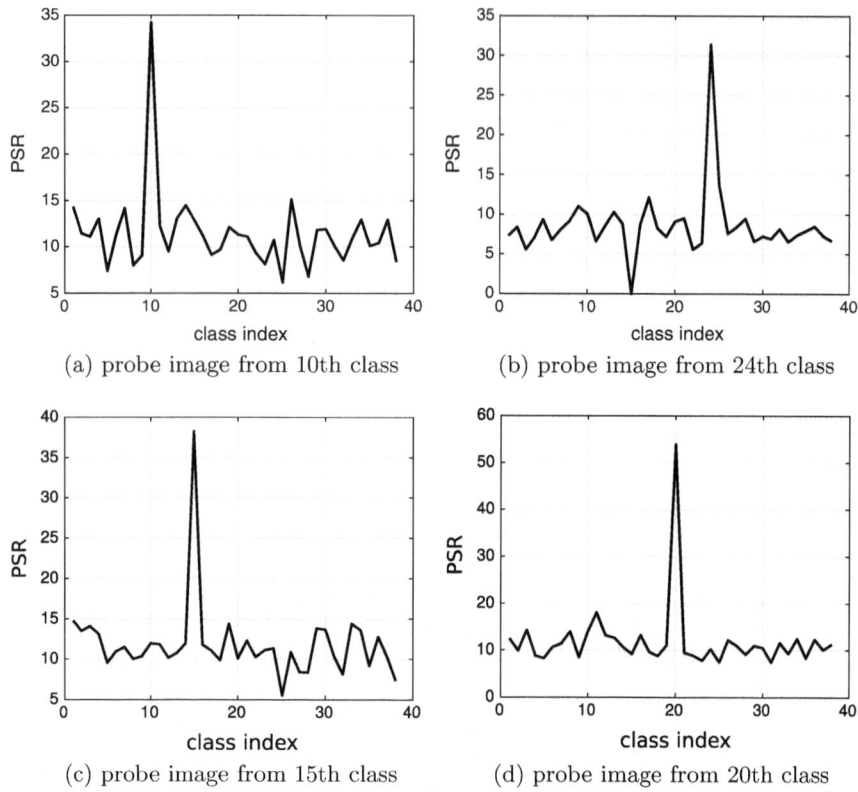

(a) probe image from 10th class   (b) probe image from 24th class

(c) probe image from 15th class   (d) probe image from 20th class

Fig. 3 PSR distribution: single probe image response for all classes

Another experiment has been performed where all 44 probe images are taken from each of 38 classes and response to a specific class is observed. Class numbers 5, 10, 25, and 35 are taken for testing purpose. PSR values are averaged out for each class as five sets of training and test sets are available. Figure 4 shows high PSR values when probe images from class 5, 10, 25, and 35 are taken. Clear demarcation is observed between the authentic and imposter PSR values in all four cases.

F1-score and Confusion Matrix: Another experiment is performed to measure the performance of the proposed LRCF by evaluating precision, recall, and F1-score. Precision and recall are mathematically expressed as

$$precision = \frac{tp}{tp + fp} \qquad recall = \frac{tp}{tp + fn} \tag{7}$$

where tp : true positive, fp: false positive, and fn: false negative. F1-score is further measured as a harmonic mean of precision and recall as

$$F1 = 2\frac{precision \times recall}{precision + recall} \tag{8}$$

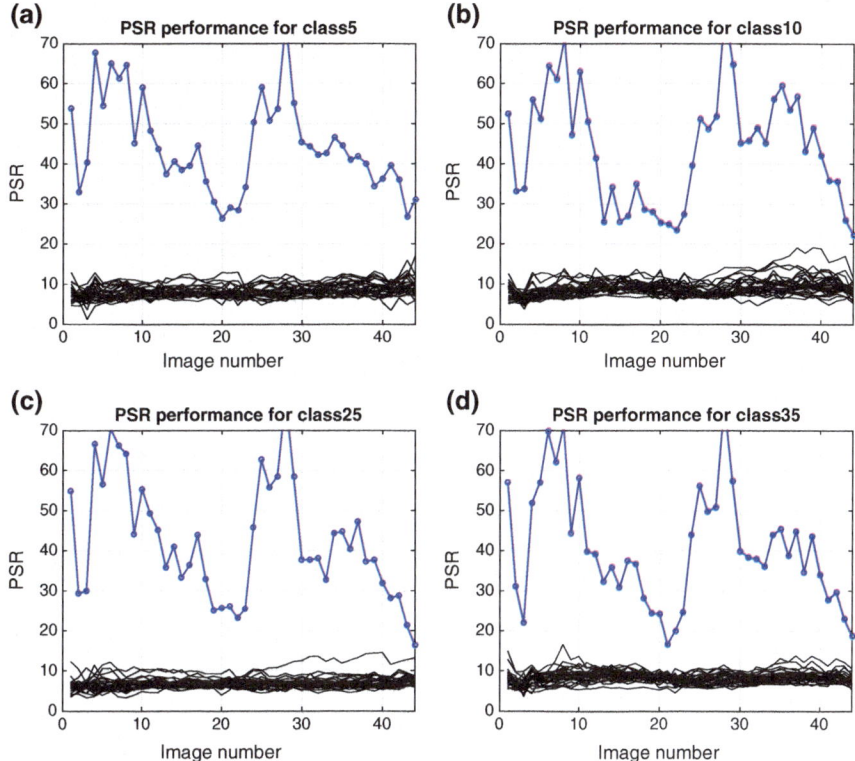

Fig. 4 PSR distribution: All probe image responses for a specific class. All 38×44 probe images are tested with class 5, 10, 25, 35. Clear demarcation of PSR values is observed

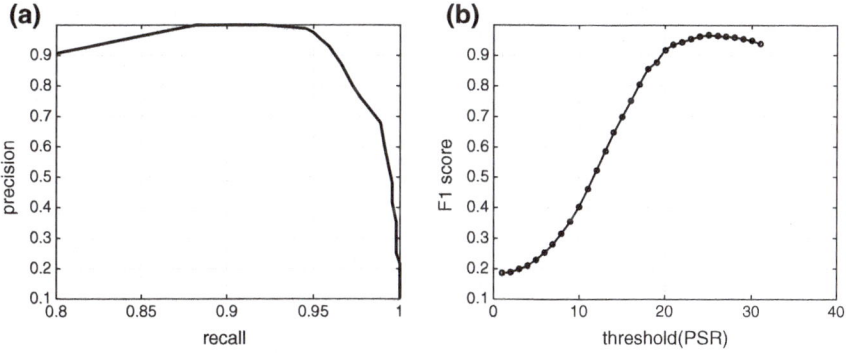

Fig. 5 **a** Precision versus recall curve, **b** F1-score versus threshold (PSR) curve

It is observed from Fig. 5a that high precision high recall is obtained with the proposed method. This ensures that the LRCF classifier performs well in classification. Precision can be seen as a measure of exactness, whereas recall is a measure

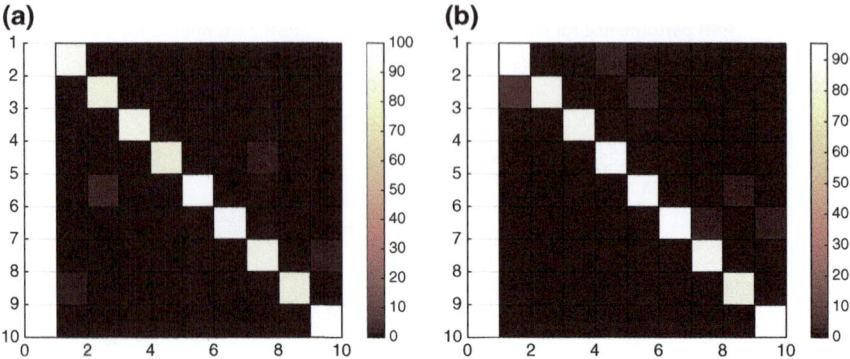

Fig. 6 Confusion matrix: **a** for set2, **b** for set3

of completeness. High precision obtained from the experimental results ensures that the proposed filter (LRCF) returns substantially more relevant results than irrelevant ones, while high recall ensures that LRCF returns most of the relevant results. Highest F1-score of value 0.9663 is obtained at PSR value 17 as observed from Fig. 5b. Two confusion matrices are developed for randomly chosen 10 classes out of 38 with hard threshold of PSR value 17 as obtained from F1-score curve (Fig. 6).

Comparative study: A comparative performance analysis of proposed filter is performed with traditional state-of-the-art correlation filter like UMACE and OTMACH. Each filter is trained with four sets of training images and tested with four test sets. Each case grayscale images are downsampled to 64×64. Both UMACE and OTMACH along with proposed LRCF are evaluated on 38 classes for varying threshold values of PSR from 5 to 20 with 0.5 increment. From the set of PSR values, confusion matrix, precision, and recall are calculated. To analyze the performance of the filters, F1-score is determined. Figure 7 shows comparative F1-curves for four sets of training–testing images. It is observed from Fig. 7 that comparatively higher F1-score is achieved with proposed LRCF filters. Also, the high F1-score is obtained at high PSR values. This result establishes the fact obtained in PSR distribution results, refer Fig. 4.

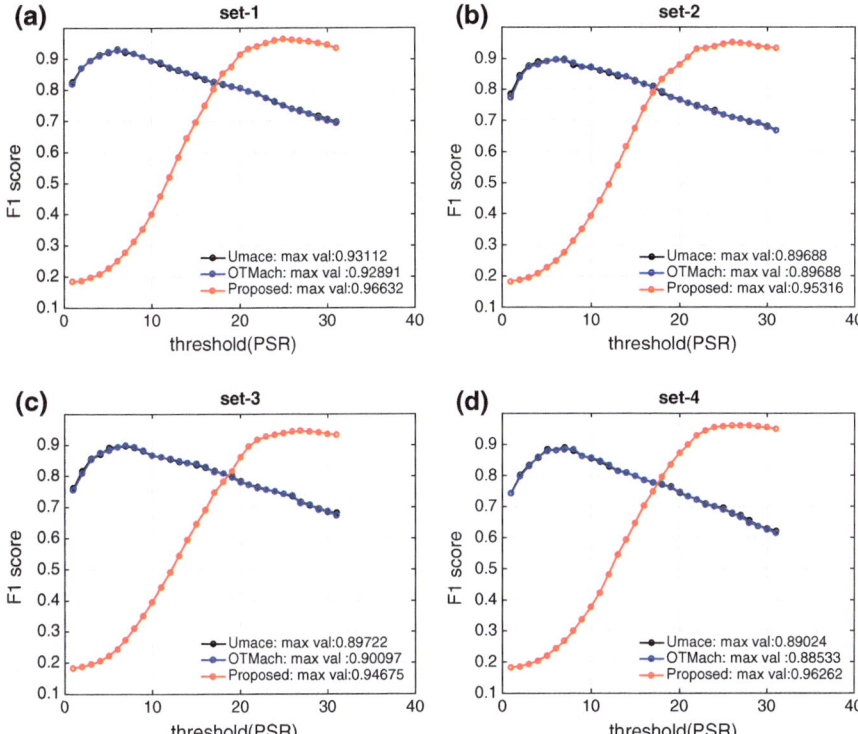

Fig. 7 Comparative F1-score is shown. Better F1-score achieved at high PSR values for proposed filter compared to others for four sets of train–test pairs

4 Conclusion

A new design method of correlation filter is proposed. Linear regression method is used to estimate the correlation filter for a given probe image and training-matched filter subspace. Estimated LRCF is correlated with probe filter for classification purpose. The proposed method is tested on a standard database and promising results with high precision high recall are reported. High F1-score, obtained in case of proposed LRCF, ensures the robustness of the classifier in face recognition application. This paper handles only with frontal faces under different illumination conditions. How to extend the application of proposed filter in pose variation, noisy conditions, and other object recognition needs further research.

References

1. Banerjee, P.K., Datta, A.K.: Generalized regression neural network trained preprocessing of frequency domain correlation filter for improved face recognition and its optical implementation. Opt. Laser Technol. **45**, 217–227 (2013)
2. Banerjee, P.K., Datta, A.K.: A preferential digital-optical correlator optimized by particle swarm technique for multi-class face recognition. Opt. Laser Technol. **50**, 33–42 (2013)
3. Banerjee, P.K., Datta, A.K.: Class specific subspace dependent nonlinear correlation filtering for illumination tolerant face recognition. Pattern Recognition Letters **36**, 177–185 (2014)
4. Belhumeur, P.N., Hespanha, J.P., Kriegman, D.J.: Eigenfaces versus fisherfaces: recognition using class specific linear projection. IEEE Trans. Pattern Anal. Mach. Intell. **19**(7), 711–720 (1997)
5. Trevor, H., Robert, T., Jerome, F.: The Elements of Statistical Learning. Springer, Berlin (2009)
6. Jeong, K., Liu, W., Han, S., Hasanbelliu, E., Principe, J.: The correntropy mace filter. Pattern Recognit. **42**(9), 871–885 (2009)
7. Johnson, O.C., Edens, W., Lu, T.T., Chao, T.H.: Optimization of OT-MACH filter generation for target recognition. In: Proceedings of the SPIE 7340, Optical Pattern Recognition, vol. 7340, pp. 734008–734009 (2009)
8. Kumar, B., Savvides, M., Xie, C., Venkataramani, K., Thornton, J., Mahalanobis, A.: Biometric verification with correlation filters. Appl. Opt. **43**(2), 391–402 (2004)
9. Lai, H., Ramanathan, V., Wechsler, H.: Reliable face recognition using adaptive and robust correlation filters. Comput. Vis. Image Underst. **111**, 329–350 (2008)
10. Lee, K., Ho, J., Kriegman, D.: Acquiring linear subspaces for face recognition under variable lighting. IEEE Trans. Pattern Anal. Mach. Intell. **27**(5), 684–698 (2005)
11. Maddah, M., Mozaffari, S.: Face verification using local binary pattern-unconstrained minimum average correlation energy correlation filters. J. Opt. Soc. Am. A **29**(8), 1717–1721 (2012)
12. Mahalanobis, A., Kumar, B., Song, S., Sims, S., Epperson, J.: Unconstrained correlation filter. Appl. Opt. **33**, 3751–3759 (1994)
13. Levine, M., Yu, Y.: Face recognition subject to variations in facial expression, illumination and pose using correlation filters. Comput. Vis. Image Underst. **104**(1), 1–15 (2006)
14. Alam, M.S., Bhuiyan, S.: Trends in correlation-based pattern recognition and tracking in forward-looking infrared imagery. Sensors (Basel, Switzerland) **14**(8), 13437–13475 (2014)
15. Imran, N., Roberto, T., Mohammed, B.: Linear regression for face recognition. IEEE Trans. Pattern Anal. Mach. Intell. **32**(11), 2106–2112 (2010)
16. Refregier, Ph: Filter design for optical pattern recognition: multi-criteria optimization approach. Opt. Lett. **15**, 854–856 (1990)
17. Rodriguez, A., Boddeti, V., Kumar, B., Mahalanobis, A.: Maximum margin correlation filter: a new approach for localization and classification. IEEE Trans. Image Process. **22**(2), 631–643 (2013)
18. Seber, G.: Linear Regression Analysis. Wiley-Interscience (2003)
19. Yan, Y., Zhang, Y.: 1D correlation filter based class-dependence feature analysis for face recognition. Pattern Recognit. **41**, 3834–3841 (2008)

Storm Tracking Using Geostationary Lightning Observation Videos

**Nora Elizabeth Joby, Nimisha Susan George, M. N. Geethasree,
B. NimmiKrishna, Noora Rahman Thayyil and Praveen Sankaran**

Abstract It was recently observed and proved by geoscientists that lightning observations from space peaked as a precursor to severe weather occurrences like flash
floods, cloudbursts, tornadoes, etc. Thus, total lightning observations from space may
be used to track such disasters well in advance. Satellite-based tracking is especially
important in data-sparse regions (like the oceans) where the deployment of ground-
based sensors is unfeasible. The Geostationary Lightning Mapper (GLM) launched
in NASA's GOES-R satellite which maps lightning by near-infrared optical transient
detection is the first lightning mapper launched in a geostationary orbit. Sample
time-lapse videos of these total lightning observations have been published by the
GOES-R team. This work describes the challenges, optimizations and algorithms
used in the application of tracking filters like the Kalman filter and particle filter for
tracking lightning cells and hence storms using these videos.

Keywords Storm track · Geostationary lightning mapper · Satellite video
tracking · Disaster prediction · Lightning · Remote sensing · Computer vision

N. E. Joby · N. S. George · M. N. Geethasree · B. NimmiKrishna · N. R. Thayyil · P. Sankaran (✉)
National Institute of Technology, Calicut, India
e-mail: psankaran@nitc.ac.in

N. E. Joby
e-mail: norajoby@gmail.com

N. S. George
e-mail: nimisharsn196@gmail.com

M. N. Geethasree
e-mail: mngeethasree97@gmail.com

B. NimmiKrishna
e-mail: nimmikrishnab@gmail.com

N. R. Thayyil
e-mail: noora.thayyil@gmail.com

© Springer Nature Singapore Pte Ltd. 2020   391
B. B. Chaudhuri et al. (eds.), *Proceedings of 3rd International Conference on Computer
Vision and Image Processing*, Advances in Intelligent Systems and Computing 1022,
https://doi.org/10.1007/978-981-32-9088-4_33

1 Introduction

Conventionally satellite weather prediction technologies and storm tracking are performed using wind velocity measurements and observation of reflectivity of clouds using optical, infrared or microwave imaging of the surface of earth using radiometers. Recently, it was proven that total lightning (TL), combined number of cloud-to-ground (CG) and Intra-cloud (IC) lightning can be used as an indicator and precursor of severe weather like storms, hurricanes, flash floods, cloudbursts, etc [1].

The Geostationary Lightning Mapper (GLM) aboard GOES-R satellite is the first operational lightning mapper flown in a geostationary orbit. It is a single-channel, near-infrared optical transient detector that can detect the transient changes in an optical scene, denoting the presence of lightning. The GLM has a near-uniform spatial resolution of approximately 10 km, and captures images in the visible spectrum in the near-infrared region. It collects information such as the frequency, location and extent of lightning discharges to identify intensification of storms, which are often accompanied by an increased lightning activity [2]. Observation from geostationary orbit has several advantages: (1) ability to view lightning against a constant scene, (2) continuous observation of weather over the specific field of view, (3) coverage over a large area, (4) ability to detect both CG and IC strikes and (5) range-independent sensitivity and detection efficiency [3]. It observes lightning at storm-scale resolutions across most of the western hemisphere with low latency and a near-uniform spatial resolution of approximately 10 kms [2].

This work performs satellite video processing to first cluster lightning cells and then implement tracking filters (Kalman filter and particle filter) to predict storm motion. Conventional methods of clustering and tracking cannot be used directly for tracking lightning cells due to the nature of their occurrence and motion. This work discusses the optimizations and variations in algorithms that were used to tune these conventional techniques for this application.

2 Literature Survey

There has been several works in the direction of using high altitude winds and precipitation data for storm tracking [4, 5]. Precipitation-based techniques, however, are ground based, and require the deployment of a network of sensors.

Cloud-to-Ground (CG) lightning has been used as one of the parameters used in Consolidated Storm Prediction for Aviation (CoSPA) [5]. Studies have shown that Intra-Cloud lightning (IC) is more correlated to storm severity than CG Lightning [6, 7]. Systematic total lightning and abrupt increase in it is a precursor to severe weather of all kinds—wind, supercell storms, hail and tornado [1, 6]. Hence, using total lightning along with the above-mentioned parameters can improve lead time of severe weather prediction and alerting. Liu et al. [6] use total lightning information available through The Earth Networks WeatherBug Total Lightning Network (WTLN), a ground-based lightning detector network, to track tornadoes.

In the work by Liu et al. [6], a polygon is used to describe a lightning cell and it is calculated every minute with a 6 min data window. The path and direction of storm movement were determined by correlating cell polygons over a time. Lighting flash rate, cell speed and area of storm were thus found using movement speed and movement direction. Practically a lead time advantage of about 30 min before tornado touch down was obtained using this method. Lakshmanan et al. [8] have put forward a review work that describes how to parameterize the efficiency of storm-tracking methods. According to this, longer, linear tracks are better, and an important concern is how the algorithm handles the splitting and merging of storms.

3 Data and Methods

3.1 Video Dataset

First phase of the clustering and tracking experiments was performed on non-operational video data that has been released at the website of GOES-R satellite. Since GOES-R is a geostationary satellite over the Americas, the video is the imagery of lightning against the scene of a part of the western hemisphere. Three videos were chosen for the study here. Video 1 is dated 5 May 2017, and is a time-lapse video of time-lapse factor $2000 \times realtime$. Video 2 is dated 19 June 2017 with time-lapse factor $6000 \times realtime$ and video 3 is dated 29 April 2017 with time-lapse factor $6000 \times realtime$. The exact resolution is not provided and needs to be approximated using empirical calculations.

3.2 Proposed Clustering Algorithm

A brute-force clustering approach using Euclidean distance is developed which is similar to a combined first and second passes of the connected components algorithm. This gives similar results as a flat kernel approach of mean shift algorithm [9], and has lesser computational demand compared to DBScan. K-means clustering cannot be used here, as lightning cells dynamically evolve in each frame, and thus cannot fix K in advance.

Inputs: Matrix c, the data of the lightning detections made in a particular frame. Each row represents one detection—the x- and y-coordinates.

Initialize: Initialize *label* as 0. Define threshold distance $d0$ for clustering. Append a column of row-labels initialized to all 0s at the end of the object c. Sweep the rows of detections i from top to bottom.

1. If *label*==0 for a row i, row-label ← *label* +1 and increment *label* as *label* ← *label* +1.

2. for j from i+1 to end (rows):

a. Calculate L2 norm (Euclidean distance) d of row j with row i.

b. If $d < d0$ and row-label==0, row-label (j) ← row-label(i).

c. Else if $d < d0$ and row-label $\sim= 0$, row-label (j) ← min(row-label(i), row-label(j)).

If min(row-label(i), row-label(j)) == row-label(j))

row-label(i) ← row-label(j)

i ← i-1;

label ← label-1

3. Repeat steps 1 and 2 until end of rows is reached.

Output: The row-labels of each row correspond to the cluster they belong to.

The part of algorithm shown in bold italics corresponds to the second pass of clustering. It was observed that performing second pass made clustering worse, as it caused a 'bridging-effect' of collapsing seemingly separate clusters as a supercluster because of close-lying outliers. It does well to exclude this part of code and a one-pass algorithm satisfies our requirement.

3.3   Preprocessing of Videos and Tracking Without Prediction

1. Each frame of the video is read one by one and cropped for the relevant parts required for processing, to remove labels and logos that came as part of the video.

2. Video 1 and Video 3 were converted to greyscale image and thresholded to make the lightning detections. However, Video 2 is a coloured video with a contoured map of countries; hence, thresholding was done on the RGB channels separately to get lightning detections.

3. From this thresholded image, the connected components were identified and their centroids and bounding boxes stored into variables. These are the centroids and bounding boxes of each lightning event/detection.

4. A clustering algorithm is implemented which groups each identified lightning event into clusters.

5. Based on the clustering/labelling above, the bounding box for the entire cluster is found out. This is done by finding the coordinates of the top-left event and the bottom-right event.

6. The centroids of each cluster are found out by simply averaging the x- and y-coordinates of the events grouped into a particular cluster. Along with this, the number of strikes occurring inside each cluster is calculated, which is defined as the 'frequency' of the cluster.

7. The objects are annotated on the video frames and saved as image files. This is later compiled into a video which lets us observe the results visually.

3.4 Tracking with Prediction

(a) Kalman Filtering

Prediction methods were incorporated into the existing detection-based algorithm (3.3). Kalman filtering was first attempted because of its simplicity and fundamental nature, assuming the system to be linear and following Gaussian distribution. Steps 1–6 of the detection algorithm (3.3) are employed as the initial steps for finding the centroids and bounding boxes of each lightning cluster/track. Kalman filtering is performed on each of these tracks as explained below:

1. Initializing tracks: Data structures called 'tracks' are initialized for each new cluster detected.

2. Predict New Locations of Tracks: New positions of each cluster predicted using Bayesian inference of Kalman filter.

3. Assign detection to tracks: Detections made in present frame are assigned to existing tracks. Thus, there will be assignments, unassigned tracks, and unassigned detections.

4. Update assigned tracks: Wherever assignments are made, the predicted centroid locations are replaced by new ones and their Bayesian model updated.

5. Update unassigned tracks: The parameters, age and consecutive invisible count of unassigned tracks, are increased by 1.

6. Delete lost tracks: If a track is invisible for too long or does not satisfy a minimum visibility criterion, it is deleted.

7. Update unassigned detections: For every unassigned detection, a new track object is created and initialized.

8. Annotation and display: The predictions (or detections, if present in the frame) of reliable tracks are annotated and displayed as the tracking video.

(b) Particle filtering

Due to limitations of Kalman filter in correctly modelling nonlinear and non-Gaussian system, particle filter is employed for lightning-based storm tracking. It does not need any initial state or transition model and even works for multimodal distributions with non-Gaussian noise. Particle filter uses a set of entities called particles, which are sample points of probability distribution function of the position of centroid of each cluster. Steps 1–6 of algorithm (3.3) are used for detecting lightning events from a frame and to group them into clusters. Particle filtering is implemented using the following steps:

1. Initializing tracks: Data structure 'track' initialized for each freshly detected cluster, along with the uniformly distributed particles associated with it. Here it was observed that the best tracking results (accuracy) were obtained when the initial number of particles was around 400.

2. Updating particles: From the resampled particles available from previous iteration/frame, new location of particles (and thus tracks) are predicted based on a transition matrix.

3. Assigning to tracks: Detections in current frame are assigned to existing tracks as in Kalman filter (3.4a).

4. Calculation of log likelihood: Based on the lightning detections in the present frame, the likelihood of each particle in a particular track to be the next position of centroid is calculated as follows:

$$L(k) = -log(\sqrt{2}\pi(Xstd\_rgb)) - \frac{0.5}{(Xstd\_rgb.\hat{2})}(x - Xrgb\_trgt)^2 \quad (1)$$

Here $Xstd\_rgb$ is the standard deviation from a threshold colour/intensity value $Xrgb\_trgt$ used to detect a lightning event from a frame and x is the RGB value of each particle.

5. Resampling particle: Particles are resampled according to their likelihood values calculated above. It is the process of deleting irrelevant particles with very less weight and increasing the weight of more probable particles. Updation of unassigned tracks, deletion of lost tracks and updation of unassigned detections are implemented in the same manner as in the case of Kalman filter.

6. Annotation and display: Both particles and their mean positions are displayed.

3.5 Disaster Predictors from Video Data

The validation of the tracking process and its efficiency to track and predict storm is performed by defining parameters called 'disaster predictors' that are indicative of the size and effect of storm. These are to be compared with ground truth data prepared from weather reports.

Kalman and particle filters predict the next location of the storm centre and its region of risk (bounding box). Distance between the present and next (predicted) location of track gives the immediate displacement '$dis1$' which when divided by time lapse of one frame gives the instantaneous velocity of our track. Average velocity is calculated by computing the distance between the next and first position of track '$dis2$', divided by the time lapse corresponding to the age of track.

$$Inst\_Vel = dis1 * mperpixel * Frate/timeLapseFactor \quad (2)$$

$$Avg\_Vel = dis2 * mperpixel * Frate/(age\_of\_track * TimeLapseFactor) \quad (3)$$

$mperpixel$ is the ratio mapped between one pixel on the video to distance on the ground. $Frate$ is the frame rate of the video and $TimeLapseFactor$ the time duration corresponding to each frame transition in video. The other disaster predictors are as follows:

- **The region of risk** is mapped by the bounding boxes that a track encloses. For practical purposes, this may be extended to a few more miles to the right, left, top or bottom of a bounding box.
- **Intensity of strikes** in a storm (number of strikes that has been associated with the track) can be directly related to the intensity parameter of the tracks.
- **Expected frequency of strikes** (in the next instant) is calculated by dividing the track's intensity of strikes (total number of strikes so far) by its age.

For particle filters, the velocity of each track is computed using the mean position of particles associated with each track. The difference is in how 'dis' is computed, and the remaining procedure used in Kalman can be adopted here also.

4 Results

The results of the algorithms described above are verified in two steps:

1. Visual perception of videos
The video files are annotated with the tracking labels and verified visually. The clustering of storms is shown in Fig. 1. Kalman and particle filters implemented on the three video files gave good tracking results visually, with Kalman characterized by more 'jerkiness' in the path of tracks. Particle filter was able to smooth the trajectory better.

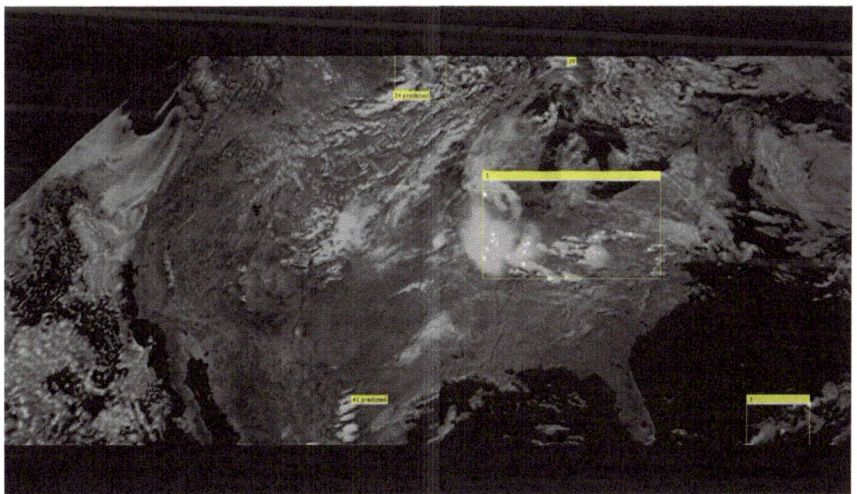

Fig. 1 Tracking of Vid1 using Kalman filter

It was observed that the proposed approach could deal with merging and breaking up of storms fairly well. One of the branches of a split storm will carry a new ID and be regarded as a new track, and the weaker track in merged storms will die out after a while.

2. Comparison of disaster predictors with ground truth data.

The disaster predictors developed above are compared with the ground truth data which are the weather reports obtained from NOAA [10, 11]. For evaluating the efficiency of a tracking method, four approaches are used as given below:
(1) Confusion matrix,
(2) Tracked storm's velocity should fall in the actual velocity range of storms,
(3) Positive correlation between simulated parameters and reported ground truth values and
(4) Root Mean Square Error (RMSE) estimation.

For confusion matrix (Table 1), a positive occurs when our simulation predicts storm at a particular location in the time corresponding to the next frame, and a negative occurs in the absence of such a prediction. Truth/falsehood of occurrence is the presence/absence of an actual storm according to the NOAA weather reports (ground truth).

For disaster prediction, a false negative (inability to predict a storm which actually occurred) is the most costly, and it can be seen that our methods perform excellent in this respect. Particularly, Kalman filter performs really good by giving maximum true positives. Various other detection efficiency parameters are also calculated in Table 2:

Sensitivity: TP/(TP+FN),
Specificity: TN/(TN+FP),
Positive likelihood ratio = sensitivity/(1-specificity),
Negative Likelihood ratio = (1-sensitivity)/specificity,
Positive Predictive Value = TP/(TP+FP) and
Negative Predictive Value = TN/(TN+FN).

For the next two approaches of evaluation, the calculation of tracking velocity is required as explained in Sect. 3.5, which is to be compared with high-speed wind data

Table 1 Confusion matrix for prediction by Kalman and particle filters for video dataset

		Reported results	
		Present	Absent
Simulated results	Present	True Positive (TP)	False Positive (FP)
		Kalman:32	Kalman:1
		Particle:33	Particle:5
	Absent	False Negative (FN)	True Negative (TN)
		Kalman:0	Kalman:0
		Particle:0	Particle:1

Table 2 Detection efficiency parameters of tracking techniques

	Kalman	Particle
Sensitivity	1	1
Specificity	0	0.1667
Positive likelihood ratio	1	1.2
Negative likelihood ratio	0	0
Positive predictive value	0.9696	0.86842
Negative predictive value	inifinity	1

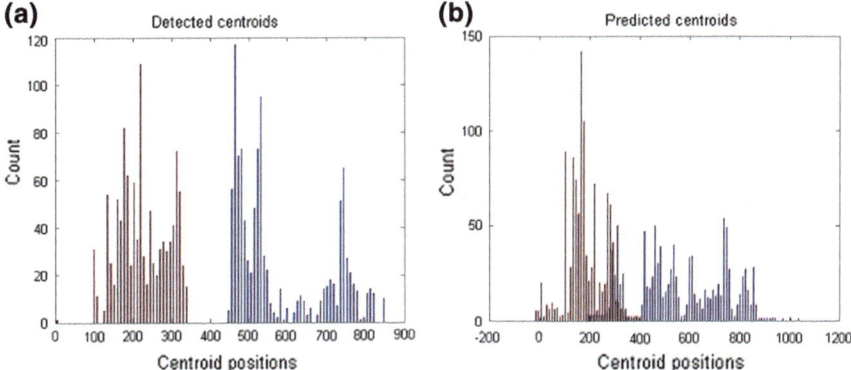

Fig. 2 a Detected locations. **b** Predicted locations: Red histogram represents x-coordinates and blue histogram represents y-coordinates for video 3, Kalman filter

from NOAA. The ground truth validation is, however, restricted by the number of wind reports available at that period of time, especially since some time-lapse videos cover a very short duration of time during which sufficient number of ground truth data points are not available. In general, the instantaneous velocity values are higher than the average velocity values, as averaging essentially smooths out the trajectory. Also, there is a chance that the simulated velocities which are representative of cloud motion are slightly higher than the reported velocities which are ground-based wind velocity measurements. This effect is due to shear effect of high altitude winds.

For the video files, Kalman filter gave instantaneous velocities significantly higher than the reported range, because of the sudden jumps as visible in the tracking video. Usage of Kalman filter is based on the assumption that the states (centroid positions) follow normal distribution, which is unimodal in nature. However, it was found that this is not a valid assumption to make in this case. The effects of the state variable not following a normal distribution can be clearly seen in the simulated results Fig. 2a, b.

In Fig. 2, the x-coordinates of the detected positions (states) show a nearly uni-modal nature whereas the y-coordinates follow a bimodal distribution. Hence, in the predicted centroid distribution, the x-coordinates showed a clear concentration around a value whereas the y-coordinates were spread out, giving lots of jerkiness

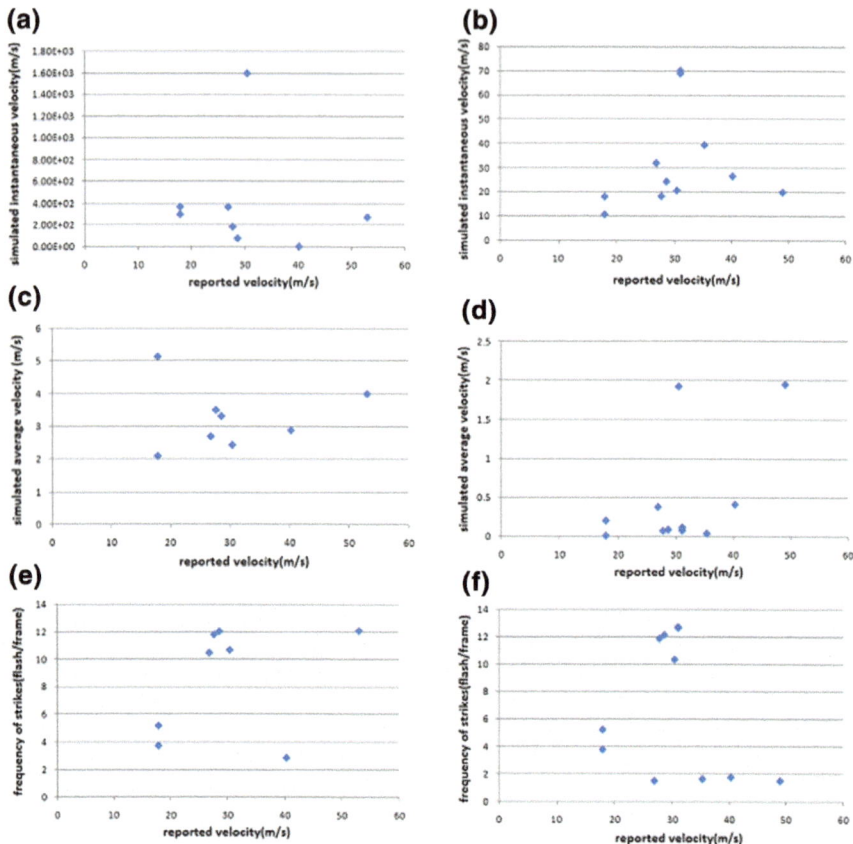

Fig. 3 **a–f** Plots of all simulated parameters (y-axis) versus reported velocity (x-axis) for video 3: Kalman on column 1 and particle on column 2

or large jumps between states. This affirms the inference that our scenario cannot be modelled using a Kalman filter which requires the states to be normally distributed.

Based on these inferences, the results may be improved by using particle filter that does not assume a unimodal distribution for the state space. Thus, by implementing particle filter, we obtained smoother tracks and corresponding velocity values fall in the reported range. It was also visually verified that the best tracking results were obtained (for all three videos) when the initial number of particles was around 400.

For correlation studies, the simulated average velocity, simulated instantaneous velocity and simulated intensity were plotted against reported velocity from weather reports (Fig. 3). A good correlation between these values and ground truth should be characterized by a high positive correlation value.

The correlation results while applying tracking filters to the video files are shown in Table 3. The linear regression plots for Kalman or particle filters results did not show significant correlation between the simulated and reported values (ground truth).

Table 3 Correlation results of using Kalman and particle filters on video data

Filter	File	No. of test points	Sim. avg vel v/s reported vel	Sim. inst vel v/s reported vel	Simulated v/s reported vel
Kalman	Vid 1	3 (not enough)	0.71513	0.54876	–0.2531
	Vid 2	21	–0.0627	0.27631	0.03114
	Vid 3	8	0.06324	–0.0969	0.32353
Particle	Vid 1	3 (not enough)	0.54639	0.34959	0.99998
	Vid 2	19	–0.1497	–0.1905	–0.3544
	Vid 3	11	0.54235	0.36848	–0.2497

Table 4 RMSE results of using Kalman and particle filters on GLM video

Video datasets	Kalman			Particle		
	vid 1	vid 2	vid 3	vid 1	vid 2	vid 3
Avg vel	16.9198	135.1563	29.1516	28.1662	118.5697	31.1402
Inst vel	2.05E+03	2.26E+03	598.9222	7.1878	7.8825	19.7101

This shows scope for further improvement of methods and approaches. Correlation may be improved by filtering out absurd values. Root Mean Square Error (RMSE) in estimating the storm velocity (unit: m/s) was also considered for the videos (Table 4). The relative low values of RMSE for particle filter are due to the fact that the simulated velocities are in the same range as reported velocities unlike in Kalman filter.

5   Conclusions

This work exploits the newly released total lightning optical data available from the world's first Geostationary Lightning Mapper (GLM) in order to track and predict storms with considerable lead time. There is a need to develop new methods for clustering and tracking due to the dynamically evolving nature of lightning-cluster data. Kalman filter often fails to smoothly track lightning clusters and to predict storm velocities in the reported range, as the states here cannot always be assumed to be as normally distributed. Particle filter was able to model the motion very well (smooth tracking and velocities in the correct range) as it does not operate under this assumption. Both filters, especially Kalman, showed excellent results in the confusion matrix with nearly all true positives. This provided ground truth validation for our assumption that lightning activity can be used as a reliable indicator of severe weather.

The challenge of the work was that neither the data nor the methods are standardized. So a proper scheme for quantification of results and validation had to be developed. In few cases, the lack of availability of enough weather data as ground truth limited the process of testing and validating the results. The statistical validation has been performed here using confusion matrix, admissible range criteria, RMSE, linear regression and correlation factor. One major advantage of our algorithm is that they could deal with merging and breaking off of storms fairly well. All these have been achieved without employing machine learning techniques, thus significantly reducing complexity and storage-cum-computational demands on the processor.

The above experiments show that total lightning alone may not be a reliable parameter to calculate storm velocities, but the detection efficiency of predicting a storm using it is near-perfect. Thus, tracking storms by using total lightning in conjunction with other information like terrain, climate history, wind patterns, etc. can increase the lead time of severe weather prediction significantly.

Acknowledgements We would like to acknowledge CLASS, NOAA and GOES-R Series Program team for providing access to the GLM data and for the extended support regarding usage and documentation of satellite video imagery.

References

1. Lynn, B.H., Yair, Y., Price, C., Kelman, G., Clark, A.J.: Predicting cloud-to-ground and intra-cloud lightning in weather forecast models. Weather. Forecast. **27**, 1470–1488 (2012). https://doi.org/10.1175/WAF-D-11-00144.1
2. Daniels, J., Goodman, S.: Towards GOES-R launch: an update on GOES-R algorithm and proving ground activities (2015)
3. Finke, U., Betz, H.D., Schumann, U., Laroche, P.: Lightning: Principles Instruments and Applications. Springer, Dordrecht (1999)
4. Niemczynowicz, J.: Storm tracking using rain gauge data. J. Hydrol. **93**(1–2), 135–152 (1987). https://doi.org/10.1016/0022-1694(87)90199-5
5. Wolfson, M.M., Dupree, W.J., Rasmussen, R.M., Steiner, M., Benjamin, S.G., Weygandt, S.S.: Consolidated storm prediction for aviation (CoSPA), In: Integrated Communications, Navigation and Surveillance Conference, Bethesda, pp. 1–19. Bethesda, MD (2008)
6. Liu, C., Heckman, S.: Using total lightning data in severe storm prediction: global case study analysis from north America, Brazil and Australia. In: International Symposium on Lightning Protection, Fortaleza, pp. 20–24 (2011). https://doi.org/10.1109/SIPDA.2011.6088433
7. Liu, C., Heckman, S.: The application of total lightning detection and cell tracking for severe weather prediction. In: 91st American Meteorological Society Annual Meeting, pp. 1–10. Seattle (2011)
8. Lakshmanan, V., Smith, T.: An objective method of evaluating and devising storm-tracking algorithms. Weather. Forecast. **25**, 701–709 (2010). https://doi.org/10.1175/2009WAF2222330.1
9. Comaniciu, D., Meer, P.: Mean shift, a robust approach toward feature space analysis. IEEE. Trans. Pattern Anal. Machine Intell. 603–619 (2002)
10. http://www.spc.noaa.gov/exper/archive/event.php?date=YYMMDD/. Accessed 4 Jan 2018
11. https://www.accuweather.com/. Accessed 7 Feb 2018
12. Johnson, J.T., MacKeen, P.L., Witt, A., Mitchell, E.D., Stumpf, G.J., Eilts, M.D., Thomas, K.W.: The storm cell identification and tracking algorithm: an enhanced WSR-88D algorithm. Weather. Forecast. **13**, 263–276 (1998). https://doi.org/10.1175/1520-0434(1998)013<0263:TSCIAT>2.0.CO;2

Human Head Pose and Eye State Based Driver Distraction Monitoring System

Astha Modak, Samruddhi Paradkar, Shruti Manwatkar, Amol R. Madane
and Ashwini M. Deshpande

Abstract One of the major causes of road accidents is driver distraction. Driver distraction is diversion of attention away from activities critical for safe driving. Driver distraction can be categorized into drowsiness and inattentiveness. Drowsiness is a condition in which the driver feels sleepy, therefore cannot pay attention toward road. Inattentiveness is diversion of driver's attention away from the road. Our system provides facility for monitoring driver's activities continuously. The in-car camera is mounted to capture live video of driver. Viola–Jones algorithm is used to identify the driver's non-front-facing frames from video. Inattentiveness is detected if the system identifies consecutive frames having non-frontal face. Drowsiness is identified by continuous monitoring of the eye status, which is either "open" or "closed" using horizontal mean intensity plot of eye region. Once the system detects the distraction, alert is generated in the form of audio. This will reduce the risk of falling asleep in long distance traveling during day and night time.

Keywords Driver distraction · Drowsiness · Eye state inattentiveness · In-car camera

A. Modak (✉) · S. Paradkar · S. Manwatkar · A. M. Deshpande
EXTC Department, CCEW, Pune, India
e-mail: modakastha@gmail.com

S. Paradkar
e-mail: samrudhi.paradkar@gmail.com

S. Manwatkar
e-mail: mshruti28@gmail.com

A. M. Deshpande
e-mail: ashwini.deshpande@cumminscollege.in

A. R. Madane
TCS, Pune, India
e-mail: amol.madane@tcs.com

© Springer Nature Singapore Pte Ltd. 2020
B. B. Chaudhuri et al. (eds.), *Proceedings of 3rd International Conference on Computer Vision and Image Processing*, Advances in Intelligent Systems and Computing 1022,
https://doi.org/10.1007/978-981-32-9088-4_34

1   Introduction

Advanced driver assistance system (ADAS) is a vehicle-based intelligent safety system. ADAS offers the technologies and real-time testing strategies that alert the driver about the potential problem well in advance. It assists the driver in recognizing and reacting to risky driving situations. ADAS features offer technologies that warn driver about potential collision. These systems are designed with human–machine interface (HMI). They lead to less number of road accidents.

ADAS relies on inputs from multiple data sources, including LiDaR, radar, ultrasonic sensors and night vision sensors, which allow vehicle to monitor the surrounding areas. ADAS features provide automatic braking, adaptive cruise control, incorporate GPS-based traffic warnings, keep the driver in the correct lane and also show what is in blind spots. Other ADAS features include parking assistance, lane departure warning, drowsiness monitoring, tire pressure monitoring and night vision. These features provide drivers with important information about their surroundings to increase safety.

ADAS plays an important role in providing assistance to distracted drivers. Driver distraction detection is one of the significant features of ADAS. There are accumulating evidences that driver distraction is the leading cause of considerable percentage of all road crashes. Driver distraction means driver's attention is drawn away from the primary task of driving. The activities other than driving take the driver's attention away from the road. It compromises the safety of the driver.

Driver distraction includes driver drowsiness and driver inattention. Drowsiness is an inability to perform the driving activity due to untreated sleep disorder, fatigue, medications or shift work. The sensation of sleep reduces the driver's level of attention, resulting in dangerous situations. In order to prevent accidents, the state of drowsiness of the driver should be monitored. The duration of driving task also plays a major role in influencing drowsiness. Drowsiness detection system assists in long-distance driving.

Inattention arises while distributing attention between various activities which are ongoing along with driving. Inattentiveness is classified into three categories as visual, manual and cognitive. Visual distraction involves taking driver's eyes off the road, while manual distraction involves taking driver's hands off the wheel. Cognitive distraction occurs when driver take his/her mind off the task of driving, such as texting or thinking deeply.

As driver distraction has become a major factor in fatalities on roadways, three different approaches exist for driver distraction detection. First method is based on bioelectric and nervous signal, the second is monitoring vehicle's steering motion and lane changing pattern and third is driver facial expression monitoring.

The first approach, that is, bioelectric and nervous signal feature includes heart rate variability (HRV), galvanic skin response (GSR) or body temperature to estimate driver distraction. The physiological signals such as electrocardiogram (ECG), electromyogram (EMG) and electrooculogram (EoG) are used to detect driver drowsiness. These techniques are very intrusive because number of different electrodes is attached to driver's body for measuring such signals.

The second approach is on driver steering wheel movement, how frequently the driver is changing lane, acceleration pedal movement or steering wheel grip pressure. These techniques are subject to several limitations, including the driver's expertise in driving, external environment or weather conditions outside. For example, the roads need to be well maintained with proper lane markings if distraction detection system is based on lane changing pattern. Therefore, this adds up to the maintenance cost along with detection system development cost. The third approach is the driver's face monitoring system which investigates the driver's physical conditions, such as head pose, eye movements or eye blinking frequency. The limitations of first and second techniques are overcome by the third approach, since the face monitoring is not intrusive as only camera needs to be installed inside the vehicle. Also, vehicle type or external environment does not play any role in detection as this technique is based on directly processing the driver's images captured by the camera.

2 Literature Survey

Many efforts have been reported in the literature on developing real-time distraction monitoring systems [1–7]. Distraction detection is usually based on eye blink rate [2, 7], gaze movement [2] and head rotation [5, 7].

Panicker and Nair [1] have proposed a novel method for detecting open eyes based on iris–sclera pattern analysis (ISPA) for an in-vehicle driver drowsiness monitoring. Elliptical approximation and template matching techniques are used for face detection. Face is detected using face templates of direction vectors. Distinctive anatomical elements such as eyebrows, nose and lips in the face region are used for creating the templates. Level of drowsiness is determined using percentage of eye closure (PERCLOS) measure. The proposed method for open eye detection uses the basic concepts of image processing, such as morphological and Laplacian operations. The algorithm mentioned gets affected with variance in camera position. In real-time situation, the camera position may change due to a sudden jerk on road or due to a sharp turn which affects localization and analyzing of the iris–sclera region.

Maralappanavar et al. [2] have proposed driver's gaze estimation technique in which driver's gaze direction is estimated as an indicator of driver's attentiveness. The proposed algorithm detects the gaze with the help of face, eye, pupil, eye corners detection followed by gaze categorization on whether the driver is distracted or not. Eye region extraction is done by cropping of two-thirds of upper region from a detected face followed by horizontal and vertical projection of cropped region, and at last, template matching is done to get the extracted left and right eye region. Then the pupil detection is done by finding left and right pupil boundaries using morphological and zero-crossing detection. Also, eye corners are found. The distance between left eye corner and pupil boundary is found to determine the direction of gaze. Similar approach is applied for right side eye as well. Depending upon these distances, gaze direction is given as centre gazing, left gazing and right gazing. Moreover, left and right gazing detection is determined as inattentiveness. This technique only deals with small degrees of deviation of eye direction from the centre. The developed system only deals with front-facing drivers. The system could not give any gaze estimation for driver showing head movements.

Yan et al. [3] proposed a system for real-time drowsiness detection based on gray-scale simulation system and PERCLOS to establish a fatigue model. Calculation of approximate position of the driver's face is based on edge detection using the Sobel operator. Small template matching is used to analyze the position of eye. The system uses the quick sort method to confirm the distribution range of black pixels in eye region to determine the state of eye. The weak point of the system is the processing of gray-scale images to analyze the skin region rather than color images. Also, the accuracy of Sobel operator suffers in noisy condition and hence cannot detect face properly. In spite of mentioning that the system works on driver wearing glasses, no proof of concept for the same has been described in the paper.

Kong et al. [4] provide a driving fatigue detection system based on machine vision and AdaBoost algorithm. A practical machine vision approach based on improved AdaBoost algorithm is built to detect driving fatigue by checking the eye state in real time. The system proposes a strategy to detect eye state instead of detecting eye directly. It includes face detection using classifiers of front face and deflected face. Then the region of eye is determined by geometric distribution of facial organs. Finally, trained classifiers of open and closed eyes are used to detect eyes in the candidate region. The indexes which consist of PERCLOS and duration of closed state are extracted in video frames in real time. It determines the rate of driver's fatigue. The system is transplanted into smart phone or tablet which could be widely used for driving fatigue detection in daily life. Conditions under poor illumination affect the performance of the system.

Moreno et al. [5] proposed an algorithm that considers three decisive factors, which are eye movement, head movement and mouth movement, to determine distraction state. For the determination of eye state, canny eye edge detection algorithm followed by Hough transform is used to find the location of iris. However, partial opening of eye may lead to false detection of eye state. Since Hough transform may not find the circle in such case and presence of iris could not be detected, it leads to false classification of eye state.

Eren et al. [6] proposed a system using the inertial (smart phone) sensors, namely the accelerometer, gyroscope and magnetometer. Using these sensors, deflection angle information, speed, acceleration and deceleration are measured, and estimate commuting safety by statistically analyzing driver's behavior. The system relates the measured data to analyze driver's behavior. Driver behavior is estimated as safe or unsafe using Bayesian classification, and optimal path detection algorithm. This work uses no external sensors, like video cameras, microphones, LiDaR and radar. Instead, it uses in-built smart phone sensors.

Batista [7] presents a framework that combines a robust facial features location with an elliptical face modeling to measure driver's vigilance level. The proposed system deals with the computation of eyelid movement parameters and attention of head (face) point. The main focus of the proposed system is to recover and track the three degrees (yaw, pitch and roll) of rotational freedom of a human head. This work is achieved without any prior knowledge of the exact shape of the head and face being observed. So, the proposed system is independent of the scale of the face image.

3 Problem Definition

We present a method for driver's continuous behavior monitoring. We define two broad, meaningful categories as safe and unsafe. Safe driving activities are those which are required for driving. Unsafe driving activities are unaccepted and need to be avoided.

- Safe = {"road", "left mirror", "right mirror", "top mirror", "eye open", "eye blinking"}.
- Unsafe = {"talking to co-passengers", "looking away from road", "eye closed", "Drowsy"}.

Our proposed method for distraction detection is used to provide monitoring of safe and unsafe driver behavior.

4 Proposed Methodology

4.1 Introduction

For the real-time monitoring system, it is necessary that the system monitors driver's behavior continuously frame-by-frame. Figure 1 shows the flowchart of the system and it is evident that the system is a continuous monitoring system. We have proposed two different approaches for inattention and drowsiness detection. Input parameters to the algorithms are the frames taken by the live video captured by in-car camera. The driver's images captured from camera are processed frame-by-frame for detection of inattention, followed by drowsiness. Direction of head is the indicator for driver's attention. Therefore, inattention algorithm is based on detection of direction of head coarse. Head pose direction detection technique is used for inattention detection. Once the system detects inattention, an audio alert is raised. However, if the system detects the frame to be attentive frame, it is then processed for checking drowsiness. For drowsiness detection, eye state of the driver is considered. The closed eye state corresponds to two conditions i.e. eyes blinking and drowsiness. Therefore, time-based thresholding approach is used to distinguish between the two cases. Once drowsiness is detected, audio alert is generated and the next frame is input for processing which makes the system a continuous monitoring system.

Fig. 1 Flowchart of the
proposed system

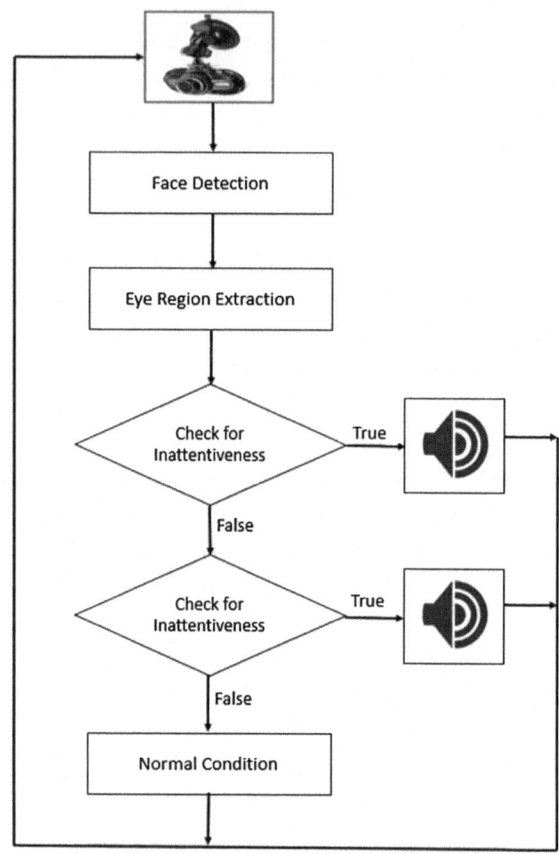

Fig. 1 Flowchart of the proposed system

4.2 Face Detection

The face detection algorithm is based on Viola–Jones face detection technique. There are three main approaches working together for implementation of Viola–Jones algorithm. The first step involves a new image representation called as the "integral image". It allows the features used by detector to be computed very quickly. The second step involves implementation of learning algorithm called AdaBoost.

The third step deals with the construction of "strong" classifier by cascading different "weak" classifier. The system detects frontal faces by moving a sub-window over the image. In each stage of the classifier, specific region is defined by the current location of the sub-window as a face or not a face.

4.3 Eye Region Extraction

The face detection algorithm detects the face region from the input image. The detected face is enclosed with rectangle, tight bounding box. It is specified as a four-element vector [x y width height], which is given by the left top corner coordinates along with right bottom corner coordinates. The next step is eye region extraction. Many different approaches, such as template matching, skin color-based methods, vertical and horizontal projections and neural network, have been developed for the same, but the effectiveness of these approaches has not been satisfactory enough. Since such approaches increase the complexity in detection of eye, we proposed a simple technique, which is a symmetry-based technique considering the following facts.

1. Eyes are equidistant from the major axis of detected face;
2. Eyes are located above the minor axis of detected face.

Therefore, no explicit eye detection is performed and a simple symmetry-based approach is applied. The first step in eye region extraction is obtaining the top half-segment of facial image which is above the minor axis, as shown in Fig. 2. The top half-segment is partitioned at the major axis of the face, which is calculated from the face width. The height of the detected face is considered as h, width as w and eye region is extracted at a distance of (0.2 * h) from the top. The size of rectangular eye window is (0.48 * h) in height and (w) in width. The separation results in two individual regions, that is, left and right eye regions, as shown in Fig. 2. These obtained regions are processed independently to extract the symptoms

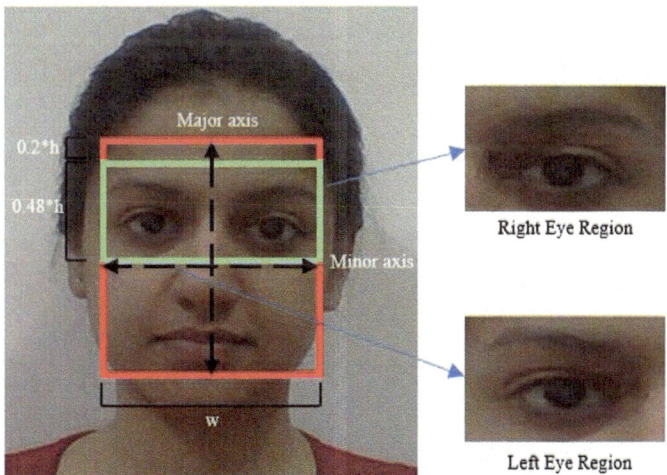

Fig. 2 Eye extraction using symmetry-based approach

of drowsiness. Since no skin segmentation or horizontal projections approach is used for eye region, the proposed method for eye extraction is not sensitive to illumination, skin color or ethnicity.

4.4 Drowsiness

Introduction

In the proposed system, state of the eye region (open or closed) of the detected face is considered as decision parameter to determine the drowsiness of the driver. We calculate the distance between eyebrows and upper edge of eyes to determine the state of the eyes, as shown in Fig. 3. From these Figs. 4 and 5, it is evident that this distance is much smaller for open eye than closed eye. In our approach, the concept of intensity change is used, since eyebrows and eye are the darkest regions of the face.

Horizontal Average Plotting

For plotting, we calculate the horizontal average intensity for each y-coordinate of the extracted right eye and extracted left eye regions. This is called the horizontal average, since the averages are taken for all the horizontal values at the particular

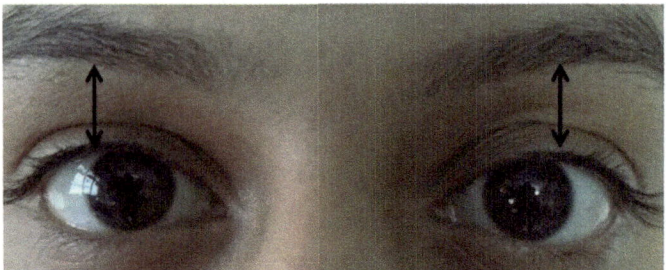

(a). Distance between eyebrow and upper edge of open eye

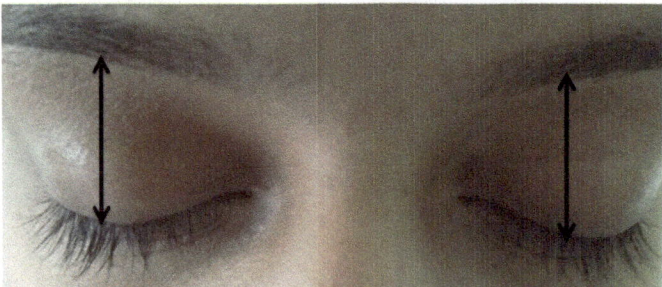

(b). Distance between eyebrow and upper edge of close eye

Fig. 3 Distance between eyebrow and upper edge

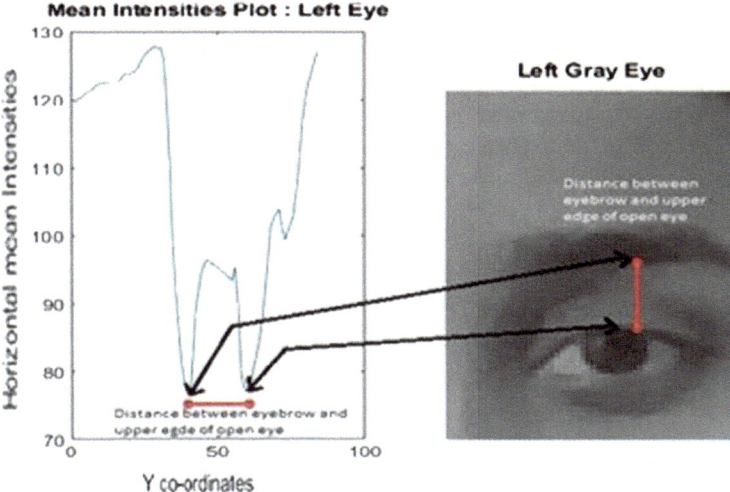

Fig. 4 Horizontal mean intensity plot for open eyes

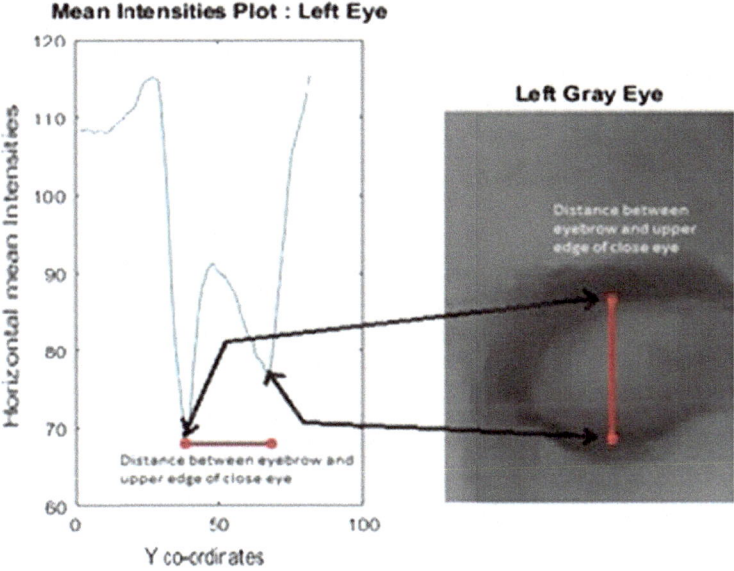

Fig. 5 Horizontal mean intensity plot for closed eyes

y-coordinate. The valley points in the plot of the horizontal values indicate intensity changes. The valleys are found by finding the change in slope from negative to positive. From the plot shown in the Figs. 4 and 5, it is evident that there are many small valleys. These valleys do not represent the intensity changes but result from small differences in the averages. After obtaining the plot for horizontal averages, the next step is to find the most significant valleys, that is, the two minimum valley points corresponding to eyebrows and upper edge of the eye. This is based on the notion that from the top of the face, gradually moving down pixel by pixel, the first intensity change is the eyebrow, and the next change is the upper edge of the eye. Therefore, we obtain first valley at eyebrows and second valley at upper edge of the eye, as shown in Figs. 4 and 5.

Determining the Eye State

In this stage, the state of the eyes (open or closed) is determined by the distance between the first and second minimum valley points. These valley points are obtained from the horizontal mean intensity plot shown in Figs. 4 and 5. The distance between two minima is the factor for classification of eye state. This distance is larger for closed eye as compared to open eye state. Therefore, this distance gives us the classification of state of the eye. It is invariant to any changes in gender, age and color.

Obtaining Reference Frame

Because of different eyelid behaviors in different individuals, estimating eye state based on absolute values of eyebrows and upper edge distances is not suitable for robust driver drowsiness systems. For example, distances between eyebrows and eyes in Japanese or Chinese men/women are lower than Mid-East, European and American men/women. Therefore, for developing a robust and adaptive system, the reference frame of that particular driver under test is obtained. The reference frame is the first frame obtained after the start of the system. This is considered as a reference under the notion that at the start of the vehicle, the driver is awake and attentive. History of the reference frame data is maintained as absolute vector. Figure 4 is the reference frame. The distance between two valleys in this figure is stored as reference valley distance for open eye. These vector values are compared with the next obtained frames to determine the eye state. If the valley distance in the upcoming frame is larger than the reference values, then such frame is classified as closed eye. If the distances are less than or equal to the reference values, such frames are classified as open eye.

Drowsiness Detection and Alert Generation System

The frames classified as containing closed eye are processed. A series of consecutive closed eye frames satisfies the drowsiness criteria. We need to avoid blinking of eye state, to avoid misclassification of eye blinking as drowsiness. We go with a time-based approach. So, after encountering the first closed eye frame, the consecutive frame is checked. When consecutive frames are not classified as closed eye for more than 3 s, it is marked as eye blink. However, obtaining closed eye frames continuously for more than 3 s confirms drowsiness. Audio alert is generated on drowsiness detection.

4.5 Inattentiveness

Coarse Head Pose Direction Detection

Input video frames are first processed through the frontal profile face detector to determine coarse head pose direction. This face detector comprises fast AdaBoost cascade classifiers operating on Haar feature descriptors proposed by Viola and Jones [8]. Frontal profile detector detects the face after locating the left eye, right eye, mouth and nose. These are located again using AdaBoost cascade classifiers trained specifically to detect the respective facial regions. A 14-dimension feature vector comprising normalized face part locations and sizes is defined as follows:

$$x = (x_{le}, y_{le}, s_{le}, x_{re}, y_{re}, s_{re}, x_n, y_n, w_n, h_n, x_m, y_m, w_m, h_m) \tag{1}$$

In Eq. 1, x and y are spatial locations, s denotes the side length, and w and h denote the width and height, respectively. Subscripts le, re, n, m denote left eye, right eye, nose and mouth, respectively. Unavailability of any of these features marks the encounter of non-frontal pose. Driver's attention is determined by the head pose which is expected to be in front direction. The frames classified as containing frontal poses satisfy the criteria of attentiveness (Fig. 6).

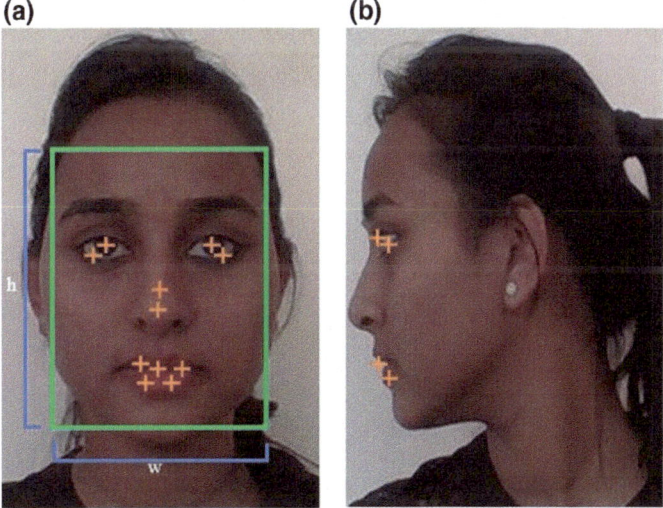

Fig. 6 Features points for frontal profile detector

Non-frontal Poses Frame Processing and Alert Generation System
In this stage we obtain frames classified as containing non-frontal poses from the previous step. Frames of driver looking at the side mirrors of the vehicle are also non-frontal poses. Encounter of first non-frontal pose frame indicates the needs to check head pose in consecutive frames. This is because the level of inattentiveness is determined by head pose in a series of consecutive frames. The distinguishable factor between looking at the mirror and inattentive case is the time for which consecutive non-frontal pose frames are obtained. The time-based approach is used as the decision parameter. After encountering first non-frontal frame, if consecutive non-frontal frames are not obtained for more than 3 s, then this case is detected as driver looking at the mirror. If consecutive non-frontal frames are classified for more than 3 s, then the system detects inattentiveness of driver. Hence audio alert is generated, which sounds "STAY ALERT" and potential risk is avoided.

5   Experimental Result

Our system has been run on PC with Intel core i5 2.40 GHz processor and the execution platform is Windows 10. MATLAB software has been used for algorithm development. The execution platform is Windows 10 powered by Intel core i5 2.40 GHz processor with 8 GB RAM.

We tested our eye state classification algorithm on publicly available database [9] "The Database of Faces", an archive of AT&T Laboratories Cambridge. The database consists of 40 distinct individuals having 10 images each, with different facial expressions (open/closed eyes, smiling/not smiling). We have tested our eye state detection on 290 images from the database. The system performs efficiently with 96.5% accuracy calculated using Eq. 2 with 280 correctly classified individuals eye state. The results obtained after eye state classification in MATLAB has been shown in Figs. 7 and 8.

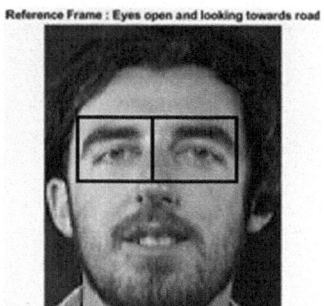 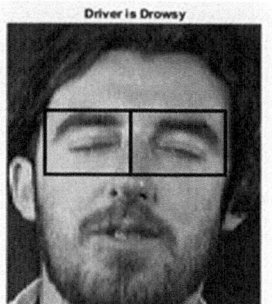

Fig. 7   Result of eye state detection on AT & T dataset (male)

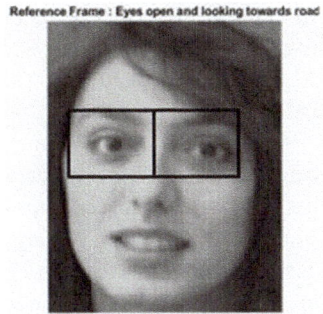
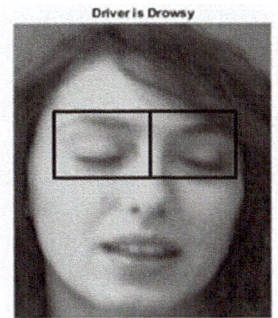

Fig. 8 Result of eye state detection on AT & T dataset (female)

$$Accuracy = \frac{Correctly\ Classified\ Individuals}{Total\ Testing\ Class\ of\ Individuals} \tag{2}$$

Also, 50 students and staff members from Cummins Engineering College served as testing subjects in controlled environment having white background pixels. These subjects were asked to simulate typical driving tasks such as looking in front with eyes open for 10 s followed by looking in right mirror for 30 ms and left mirror for 45 ms. They were also asked to simulate typical distracted driving behavior which were looking back to talk to co-passenger for more than 5 s, looking outside window continuously for 5 s and eye closed for more than 3 s. The resolution of these videos was 573 × 814 pixels and the frame rate was 30 fps. We recorded the videos at different times of whole 24 h period to test the system at varied illumination. The inattention detection algorithm was run on all 50 videos. The results of inattention detection directly taken after detection using MATLAB software are shown in Fig. 9.

In order to distinguish between the attentive and non-attentive frames, system marks green rectangle when the front profile face detector gives detection output. In Fig. 9a, as the driver is front facing, the front profile face detector gives the result marking green rectangle. In Fig. 9b, the front profile face detector could not find a face. It again checks for next frame for 3 s. Before 3 s, a front facing has arrived. Therefore, detector detects that the driver was looking at the mirror and displays the

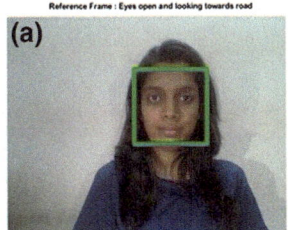
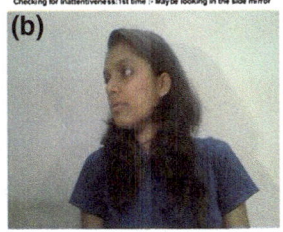

Fig. 9 Results of inattention detection algorithm on own dataset

same at the top of the frame. In Fig. 9c, the front profile detector was unable to find a face for more than 3 s and therefore inattentiveness is detected, displays at the top to be invalid condition and audio alert "STAY ALERT" is generated for the same frame. Figure 9c image corresponds to that frame at which audio alarm was produced by the system.

The videos were also tested to detect symptoms of drowsiness. Black rectangle shows the extracted left and right eye region by symmetry-based approach. After processing the images by plotting horizontal mean intensity and getting valley distances, the system classifies the frame's eye state as open or close. In Fig. 10a, the system classifies the frame as open eye frame and displays "Reference frame: Eyes open and looking today's road" at the top of the image. Whereas, Fig. 10b is classified into eye close category of driver. However, there is a need to check for consecutive frame to avoid false alarming for eye blinking. The system checks for eye state of consecutive frame for 3 s. If its eye state is closed for 3 s, then only audio alarm is raised "Time to take a break". Figure 10b image corresponds to that frame at which audio alarm was produced by the system and "Driver is drowsy" is displayed at the top of the frame.

System performance based on accuracy using Eq. 2 was measured for detection of inattention and drowsiness. To determine the system performance in varying lightening condition, different time slots were considered. Time gap between each time slot is 4 h. They are considered as time slot between 8 a.m. and 12 p.m., 12 p.m. and 4 p.m., 4 p.m. and 8 p.m. and 8 p.m. and 8 a.m. Figure 11 shows the accuracy percentage at different time slots of the day.

In the morning slot, both the detection was carried with 100% efficiency. The drowsiness detection and inattention rate got affected by small degree mainly at 4 p.m. to 8 p.m., because at this time of the day, the driver's face appears to be dark and bleached out in lightness. But from system's face detection point of view, the visible shade that the system gets to see in the driver's image is outside its programmed accepted range of facial hues. So, the system could not detect faces from the images in such cases resulting in false detection of distraction.

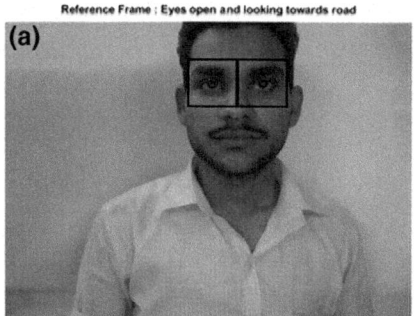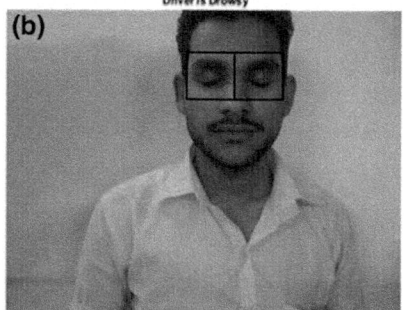

Fig. 10 Results of drowsiness detection algorithm on own dataset

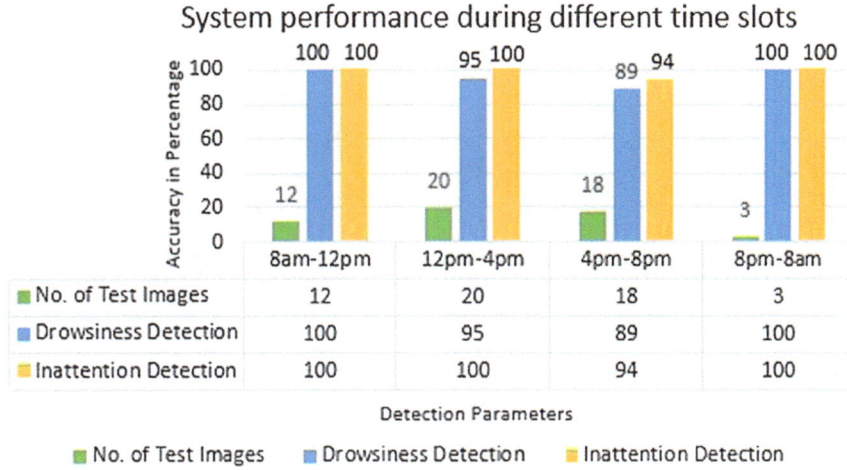

System performance during different time slots

Detection Parameters	8am-12pm	12pm-4pm	4pm-8pm	8pm-8am
■ No. of Test Images	12	20	18	3
■ Drowsiness Detection	100	95	89	100
■ Inattention Detection	100	100	94	100

■ No. of Test Images   ■ Drowsiness Detection   ■ Inattention Detection

Fig. 11 System performance at different times of 24 h period

The individuals whose distractions are correctly classified are referred as true positive (TP). The cases where the individuals distracted are not correctly detected are referred as false negative (FN). System detects distraction parameter with TP of 49 and FN of 4. Overall accuracy provided by the system is 92%.

For testing in real-time environment, three video datasets in real-time environment were considered for detection of inattention and drowsiness to check the effect of real-time scenarios. Figure 12 shows the results of inattention detection and Fig. 13 shows the results of drowsiness. A total of 243 frames taken from videos were tested, out of which 238 gave correct classification of driver state. One video dataset with

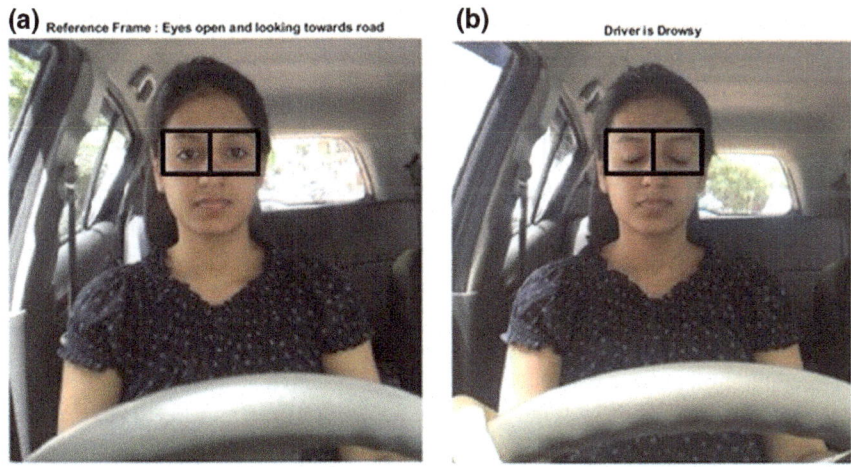

(a) Reference Frame : Eyes open and looking towards road   **(b)** Driver is Drowsy

Fig. 12 Results of drowsiness in real-time environment

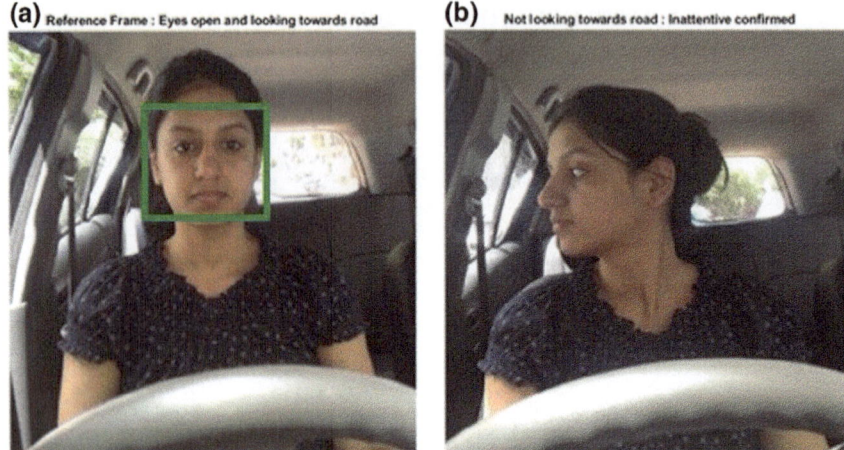

Fig. 13 Results of inattentiveness in real-time environment

co-passenger seating with driver was also tested, as shown in Fig. 14. For such cases, there is a need to distinguish between driver and co-passengers to avoid processing of co-passenger's faces and false alarming. To handle such cases, we use the concept of largest face detector using Viola–Jones method. This is based on the fact that the driver is nearest to the camera; therefore its face appears to be the largest. The largest face is classified by the detector which scales the detection resolution using scale factor. It scales the search region at increments between size of smallest detectable face and size of largest detectable face. Largest face location is targeted by clustering points of interest and hierarchically forming candidate regions. Once the driver face is classified using this concept, then that detected face region is marked as ROI for processing in the upcoming frames, as shown in Fig. 14. The system does not get affected by the presence of any number of co-passengers.

Special Cases

Case 1: Driver with Pet
Figure 15 shows a person holding the pet animal. There can be a situation of driver sitting with the pet animal in the driver seat. Since the face detection algorithm Viola and Jones [8] is trained with human faces only, therefore, the system detects and processes only the human faces in the frame. The system does not raise any kind of false alarming by processing pet's face instead of human.

Case 2: Driver with Covered Head
Figure 16a shows driver with covered head. The system performance on such driver is tested. The results of inattentiveness detection are shown in Fig. 16b. For inattentiveness detection 14-dimension feature vector is used, which includes left eye, right eye, nose and mouth, respectively. As detection of all these features do not get affected because of covered head, the system detects all cases of distraction without any false alarming.

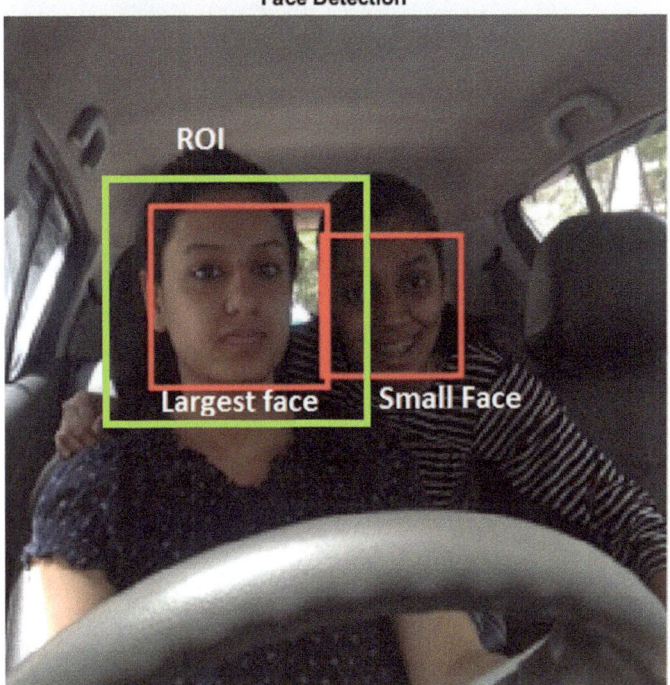

Fig. 14 ROI detection for driver seating with co-passenger

6 Conclusion

In this paper, a driver distraction monitoring method is proposed using vision-based approach. The novelty of algorithm is in adaptive feature extraction using symmetry-based approach without explicit eye detection stage and eye state classification techniques using horizontal plot projection. Fast feature descriptors and intensity change plot-based classifier are chosen to increase the computational speed. Warning system in the audio alert form is incorporated.

7 Future Scope

Immediate future efforts include additional experimental validation across a wider range of drivers and incorporating advanced warning mechanisms, such as in-car ECU synchronized sensor-based hardware alerts, for example, vibrating seat belt for drowsiness and vibrating seat for inattentiveness.

Fig. 15 Face detection on human with pet

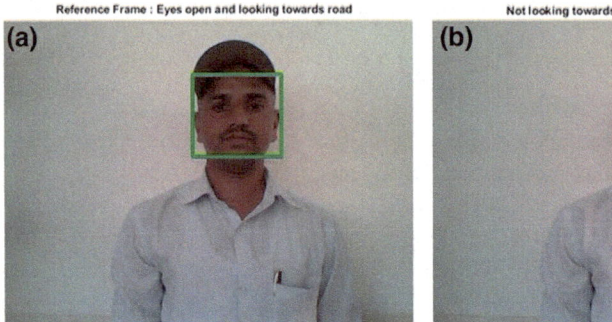

Fig. 16 Results of driver with covered head

Acknowledgements The authors express their deep sense of gratitude and indebtedness to Dr. Amol R. Madane (TCS) who was very kind to provide us with an opportunity to work under his immense expertise. His prompt inspiration, suggestions with kindness and dynamism enabled us to shape the present work as it shows. We would like to express our sincere gratitude toward our project guide Dr. A. M. Deshpande for her constant encouragement and valuable guidance.

References

1. Panicker, A.D., Nair, M.S.: Open-eye detection using iris–sclera pattern analysis for driver drowsiness detection. Sadhana **42**(11), 1835–1849 (2017)
2. Maralappanavar, S., Behera, R.K., Mudenagudi, U.: Driver's distraction detection based on gaze estimation. In: 2016 International Conference on Advances in Computing, Communications and Informatics (ICACCI), Jaipur, India, pp. 2489–2494 (2016)
3. Yan, J.-J., Kuo, H.-H., Lin, Y.-F., Liao, T.-L.: Real-time driver drowsiness detection system based on PERCLOS and grayscale image processing. In: International Symposium on Computer, Consumer and Control, Xi'an, China, pp. 243–246 (2016)
4. Kong, W., Zhou, L., Wang, Y., Zhang, J., Liu, J., Ga, S.: A system of driving fatigue detection based on machine vision and its application on smart device. J. Sens. (2015)
5. Moreno, R.J., Sánchez, O.A., Hurtado, D.A.: Driver distraction detection using machine vision techniques. Ingeniería y Competitividad **16**(2), 55–63 (2014)
6. Eren, H., Makinist, S., Akin, E., Yilmaz, A.: Estimating driver behavior by a smartphone. In: IEEE Intelligent Vehicles Symposium Alcalá de Henares, Spain, pp. 234–239 (2012)
7. Batista, J.: A drowsiness and point of attention monitoring system for driver vigilance. In: IEEE Intelligent Transportation Systems Conference, USA, pp. 702–708 (2007)
8. Viola, P., Jones, M.: Robust real-time face detection. Kluwer Int. J. Comput. Vis. **57**(2), 137–154 (2004)
9. AT & T Laboratories Cambridge. http://www.cl.cam.ac.uk/research/dtg/attarchive/facedatabase.html

Discriminative BULBPH Descriptor with KDA for Palmprint Recognition

Deepti Tamrakar and Pritee Khanna

Abstract This work proposes Block-wise uniform local binary pattern histogram (BULBPH) followed by kernel discrimination analysis (KDA) as descriptor for palmprint recognition. BULBPH provides distribution of uniform patterns (such as line and wrinkles) in local region and can be better used as palmprint features. KDA is applied on BULBPH to reduce dimension and enhance discriminative capability using chi-RBF kernel. The experiments are conducted on four palmprint databases and performance is compared with related descriptors. It is observed that KDA on BULBPH descriptor achieves more than 99% accuracy with 4.04 decidability index on four palmprint databases.

Keywords Palmprint · Blockwise uniform local binary pattern · Kernel discriminant analysis

1 Introduction

Many palmprint recognition systems are proposed in literature for palmprint images captured in the natural light [7, 30]. Still recognition accuracy is required to be improved by maximizing the separability between different classes of palmprints. In a few cases texture features are not adequate to differentiate palm lines. Also, sometimes textures are not perceptible due to distorted image.

A large variety of palmprint feature extraction approaches are based on principal line [5], texture coding [6, 8, 19], statistical [2, 10], and subspace learning [1]. Recently, histogram of pattern information on image such as gradient, orientation, and phase are proposed as prominent palmprint descriptors. For various palmprint

D. Tamrakar
Jabalpur Engineering College, Jabalpur, India

P. Khanna (✉)
PDPM Indian Institute of Information Technology, Design and Manufacturing, Jabalpur, Jabalpur, India
e-mail: pkhanna@iiitdmj.ac.in

© Springer Nature Singapore Pte Ltd. 2020
B. B. Chaudhuri et al. (eds.), *Proceedings of 3rd International Conference on Computer Vision and Image Processing*, Advances in Intelligent Systems and Computing 1022,
https://doi.org/10.1007/978-981-32-9088-4_35

recognition applications, higher classification rates are obtained with Block-wise histogram as compared to the global histogram [2, 3, 12, 13, 23]. Hierarchical multi-scale LBP histograms computed from sub-windows are also considered as prominent features for various palmprint classification techniques [2, 10, 11, 23, 24].

Local binary pattern (LBP) is a pixel coding scheme that denotes the relation between the pixel in the center and its neighbour pixels. LBP histogram has been widely utilized for prominent local palmprint texture feature extraction [2, 12, 23]. Earlier, LBP histograms are extracted from scalable sub-windows on palmprint for different radius value to extract non-uniform and uniform features [2, 23]. Its combination with other image transformation methods like Gabor, Wavelet, complex directional filter bank is widely used for feature extraction [28, 29]. Palm lines and wrinkles are uniform patterns, and their LBP is known as uniform LBP (ULBP). Uniform LBP histograms of sub-blocks are computed from either gradient response of ROI given by sobel operators [12] or transformed ROI given by Gabor and complex directional filter bank (CDFB) [12]. Thus, the coefficients extracted from multi-scale and multi-orientation filtering using a CDFB are shift and gray scale palmprint features [10]. In these methods, the final palmprint descriptor is high dimensional because it is computed by concatenating ULBP histograms of all sub-blocks into a single vector.

In most of the works, features with more discriminative power are extracted from large sized descriptor by applying linear dimension reduction techniques such as principal component analysis (PCA) and linear discriminant analysis (LDA) [10] on them. Recently, kernel based dimension reduction techniques have been found more suitable for low dimension discriminative information extraction from non-linear features. Features can be mapped on a linear feature space using kernel function without explicit knowledge of non-linear mapping. Kernel based dimension reduction techniques such as kernel PCA and kernel discriminant analysis (KDA) are proposed for non-linearly separable features. These methods are sensitive to the type of kernel function. KDA calculates projection vectors by maximizing the ratio of between-class covariance and with-in class covariance [26].

This work aims to develop an accurate, fast, and more discriminative palmprint identification and verification system. A histogram-based BULBPH (blockwise uniform local binary patterns histogram) feature [21] earlier proposed by us is a dominant texture descriptor that represents the spatial relation between textural regions of an image. It is observed that BULBPH is a high dimensional descriptor with correlated and redundant elements because it is computed by the concatenation of ULBP histograms of all sub-blocks of ROI (Region of Interest). Therefore, the dimension of BULBPH descriptor must be reduced to design real-time palmprint recognition system. Due to non-linear nature of histogram based descriptor, linear subspace dimension reduction techniques do not enhance classification performance of the system [10]. Non-linear dimension reduction techniques which maximize the discriminant between classes would be more suitable in this case. The proposed system maps non-linear BULBPH features onto linear feature space with a suitable kernel function and kernel based discriminant analysis technique is applied on mapped features to find uncorrelated reduced size feature vector. The work is focused to find a suitable kernel to map BULBPH descriptor onto linear space such that discriminant analysis can be applied to find better accuracy. Further, discriminant analysis of the feature is used to achieve low dimension feature with higher separability between dissimilar classes.

The manuscript is organized as follows. Section 2 explains BULBPH and KDA based palmprint recognition system. Experimental setup and results are discussed in Sect. 3. Section 4 compares performance of the proposed approach with related works. Finally, Sect. 5 concludes the work.

2 Proposed Palmprint Recognition System

Flow diagram of the proposed palmprint recognition system is shown in Fig. 1. It consists of five basic steps; palmprint acquisition, ROI extraction, BULBPH computation, dimension reduction using KDA, and nearest neighbor classification for verification or identification purposes. ROI of each palmprint is extracted by referencing the gap between two fingers [20]. ULBP operator is applied on the ROI and resulting image is divided into square blocks of same size. BULBPH descriptor is a single vector that is obtain by the concatenating the histograms of all blocks. These features are mapped onto low dimensional feature space with the help of a projection matrix computed by KDA. This increases the ratio of between-class scatter matrix and with-in class scatter matrix. A suitable kernel function is identified through experimentation. The test palmprint features are also mapped from non-linear feature space to linear feature space using resultant projection matrix. Nearest neighbor (NN) classifier based on Euclidean distance verifies and identifies test palmprint. Experiments are performed on four gray-scale palmprint databases collected through scanner as well as camera.

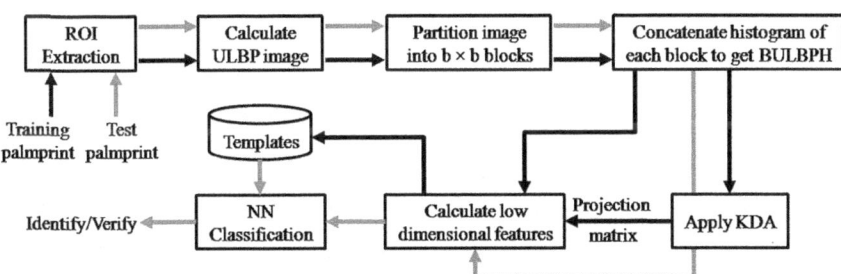

Fig. 1 Framework of the proposed palmprint recognition system

2.1 Block-Wise ULBP Histogram (BULBPH) as Descriptor

LBP represents the properties of local spatial texture configuration of an image and it is adopted for many pattern recognition applications due to its superior properties such as computational competency, simplicity, and discriminative power [11]. LBP (P, R) operator is computed on the basis of the variation between gray value of a central pixel $g_c(x_c, y_c)$ and gray values of its P circular neighbor pixels in radius R.

$$\text{LBP (P, R)} = \sum_{p=0}^{p=P-1} s(g_c - g_p)2^p$$

$$\text{where } s(x) = \begin{cases} 1 & x \geq 0; \\ 0 & x < 0. \end{cases} \tag{1}$$

The neighbors coordinates g_p are given by $(x_c + Rcos(2\pi p/P), y_c + Rsin(2\pi p/P))$. Bi-linear interpolation technique is used to interpolate sample points of valid neighbor pixels. LBP (P, R) operator produces 2^P binary patterns formed by P neighbor pixels. Each pixel is labeled by a integer value corresponding to the obtained binary number.

ULBP refers to uniform emergence patterns which have restricted alteration or discontinuities, i.e., it contains at most two bitwise change from 0 to 1 or vice versa, when the binary string is considered circular ($U \leq 2$). The value between 0 to $P(P - 1) + 1$ is mapped to all possible uniform patterns (for $U \leq 2$) using a look-up table. $P(P - 1) + 2$ is mapped to all non-uniform binary patterns (for $U > 2$). ULBP image is obtained by mapping ULBP value of each pixel in the original image. ULBP contains about 90% of local micro-patterns, such as edges, spots, and at areas of a natural image for $U \leq 2$. Uniform patterns are statistically more stable and less prone to noise than LBP [15, 31].

Palmprint ROI is converted into ULBP coded image. It contains micro patterns of the image that represent certain textural properties based on the neighborhood pixel. However, the histogram is unable to find the spatial uniformity of the coded image. Hence, ULBP coded image is divided into non-overlapped blocks of size $b \times b$, say $(B_{11}, B_{12}, \ldots, B_{bb})$ and a histogram is computed for each block. Histogram of each sub-block is concatenated into a single feature vector to get the final descriptor as

$$\mathcal{H} = \underset{i=1}{\overset{m}{\|}} \underset{j=1}{\overset{m}{\|}} h_{i,j} \tag{2}$$

where $\mathcal{H}$ is BULBPH descriptor, h_{ij} is ULBPH descriptor of block B_{ij} and $\|$ is the concatenation operation. Figure 2 shows an example of ULBP image that is divided into 4×4 non-overlapped sub-blocks and resultant BLUBPH descriptor.

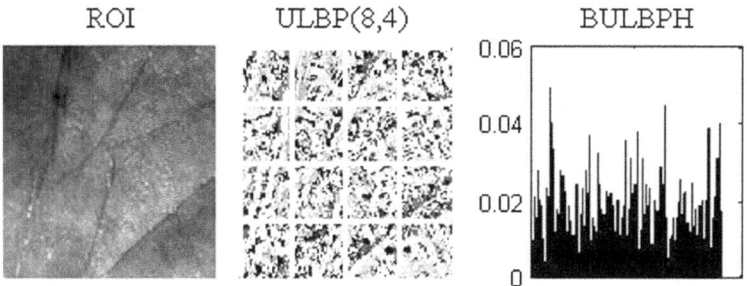

Fig. 2 BULBPH descriptor extracted from a palmprint ROI

2.2 *KDA for Dimension Reduction and Enhanced Discrimination*

In order to enhance its discriminative power and computational efficiency, uncorrelated and prominent features can be extracted from BULBPH descriptor by applying a suitable dimension reduction technique on it. KDA is found more suitable to learn a compact non-linear feature space $\mathcal{R}$ in this case as BULBPH is a non-linearly separable descriptor.

Let l number of BULBPH descriptors of dimension r for palmprints enrolled in the training phase are denoted as $\mathcal{H} = \mathcal{H}_1, \mathcal{H}_2, \ldots, \mathcal{H}_l$. Kernel mapping function is given as $\phi : \mathcal{H} \in \mathcal{R}^{l \times r} \rightarrow \phi(\mathcal{H}) \in \mathcal{F}^{l \times l}$. Let discriminant vector $\boldsymbol{w}$ is computed by linear discriminant that maximizes the ratio of between-class scatter matrix S_B^ϕ and with-in class scatter matrix S_W^ϕ in linear feature space $\mathcal{F}$ [17]. Optimization function $J(\boldsymbol{w})$ is defined as

$$J(\boldsymbol{w}) = \frac{\boldsymbol{w}^T S_B^\phi \boldsymbol{w}}{\boldsymbol{w}^T S_W^\phi \boldsymbol{w}} \tag{3}$$

Eigenvector $\boldsymbol{w}$ corresponding to the largest eigenvalue λ of $S_W^{\phi^{-1}} S_B^\phi$ matrix denotes the KDA projection matrix.

$$S_B^\phi = \sum_{i=1}^{C} l_i \left(\boldsymbol{m}_i^\phi - \boldsymbol{m}_o^\phi \right) \left(\boldsymbol{m}_i^\phi - \boldsymbol{m}_o^\phi \right)^T \tag{4}$$

$$S_W^\phi = \sum_{i=1}^{C} \sum_{k=1}^{l_i} \left(\phi \left(\mathcal{H}_k^i \right) - \boldsymbol{m}_i^\phi \right) \left(\phi \left(\mathcal{H}_k^i \right) - \boldsymbol{m}_i^\phi \right)^T \tag{5}$$

Here l_i palmprints belong to ith class/palm, i.e, $\sum_{i=1}^{C} l_i = l$ and C is the number of classes/palms. Histogram template of kth palmprint of ith palm class is denoted by $\mathcal{H}_k^i$.

$$m_i^\phi = \frac{1}{l_i} \sum_{k=1}^{l_i} \phi\left(\mathcal{H}_k^i\right) \tag{6}$$

$$m_o^\phi = \frac{1}{C} \sum_{i=1}^{C} \left(m_i^\phi\right) \tag{7}$$

Further theory of reproducing kernels says that any solution $w \in \mathcal{F}$ must lie in the span of all training samples in $\mathcal{F}$. Expansion of w in terms of existing coefficients α_j is given as

$$w = \sum_{j=1}^{l} \alpha_j \phi(\mathcal{H}_j) \tag{8}$$

By means of the above equation and the explanation of m_i^ϕ, Eqs. 6 and 8 are combined by matrix multiplication.

$$w^T m_i^\phi = \frac{1}{l_i} \sum_{j=1}^{l} \sum_{k=1}^{l_i} \alpha_j \kappa(\mathcal{H}_j, \mathcal{H}_k^i) = \alpha^T M_i \tag{9}$$

where $M_i = \frac{1}{l_i} \sum_{k=1}^{l_i} \kappa(\mathcal{H}, \mathcal{H}_k^i)$ and $M_o = \sum_{i=1}^{C} M_i$. $\kappa(\mathcal{H}, \mathcal{H}')$ is kernel function calculated by inner product of $(\phi(\mathcal{H}).\phi(\mathcal{H}'))$ [16]. By using Eq. (9), numerator of Eq. (3) is given as

$$w^T S_B^\phi w = \alpha^T M \alpha \tag{10}$$

where $M = \sum_{i=1}^{c} (M_i - M_o)(M_i - M_o)^T$. Now Eqs. (4) and (8) are combined to get m_i^ϕ, which is used to write denominator of Eq. (3) given as

$$w^T S_W^\phi w = \alpha^T N \alpha \tag{11}$$

where $N = \sum_{i=1}^{C} K_i(I - 1_{l_i})K_i^T$, K_i is defined as $\kappa(\mathcal{H}_n, \mathcal{H}_m^i)$ that denotes the kernel matrix for class i of dimension $l \times l_i$, I refers to the identity matrix, and 1_{l_i} is the matrix with all entries as $\frac{1}{l_i}$. Equations (10) and (11) are combined to rewrite Eq. (3), which is maximized to get linear discriminant coefficient α in feature space $\mathcal{F}$.

$$J(\alpha) = \frac{\alpha^T M \alpha}{\alpha^T N \alpha} \tag{12}$$

The leading eigen vector of $N^{-1}M$ is represented as the coefficient matrix α of size $l \times (C - 1)$ [17]. Features of query palmprint, $\mathcal{H}_q$, are projected onto $\mathcal{F}$ as

$$< \alpha, \phi(\mathcal{H}_q) > = \sum_{j=1}^{l} \alpha_j \kappa(\mathcal{H}_q, \mathcal{H}_j) \tag{13}$$

$$= \alpha^T K(\mathcal{H}_q, \mathcal{H}) \tag{14}$$

where $K(\mathcal{H}_q, \mathcal{H}) = [\kappa(\mathcal{H}_q, \mathcal{H}_1), \kappa(\mathcal{H}_q, \mathcal{H}_2), \ldots, \kappa(\mathcal{H}_q, \mathcal{H}_l)]$. The mapping of features can be linear or non-linear that depends on kernel function. In most of the applications, either RBF or polynomial kernel is used to map the dataset into linear feature space. RBF kernel is more suitable for circular linear separable descriptors and Euclidean distance metric is more suitable for similarity matching. As BULBPH is a histogram based feature, chi-square metric is more suitable for matching. RBF kernel function can be modified with chi-square distance (referred as chi-RBF in [22]).

3 Experimental Results and Discussions

Experiments are carried out on Intel Quad Core processor @2.83GHz with 8GB RAM using MATLAB®. Experiments are conducted on four gray-scale palmprint databases—PolyU 2D [25], CASIA [18], IITD [9]. The performance is evaluated on equal error rate (EER), decidability index (DI) and correct identification rate (CIR). EER evaluates the verification accuracy, while DI measures the separability of genuine and imposter users. Table 1 summarizes characteristics of each database with distribution of palmprints. Two sample palmprints from each database are shown in Fig. 3 [20]. ROIs from these palmprint databases are extracted with a scheme earlier proposed by us [20].

BULBPH descriptor of size 912 is calculated with 4×4 partitions of ULBP image obtained by applying ULBP(8,4) operator on ROI as it has been found more suitable for palmprint feature [21]. It is observed in literature that the application of KDA reduces features upto $C - 1$ dimension for C number of classes [16]. Here, use of KDA reduces the dimension of BULBPH descriptor for four databases as 399 (PolyU 2D), 623 (CASIA), 429 (IITD), and 149 (IIITDMJ). EER, DI, and CIR for all four palmprint databases are calculated using BULBPH descriptor without and with KDA learning.

Figure 4 compares the performance parameters of RBF, chi-RBF, and polynomial kernel. Polynomial function is calculated for degree one with standard deviation of

Table 1 The description of palmprint databases

Database	Total palmprints	Palms	Acquisition device
PolyU 2D [25]	8,000	400	Scanner
CASIA [18]	5,335	624	Camera
IITD [9]	2,400	430	Camera
IIITDMJ	900	150	Scanner

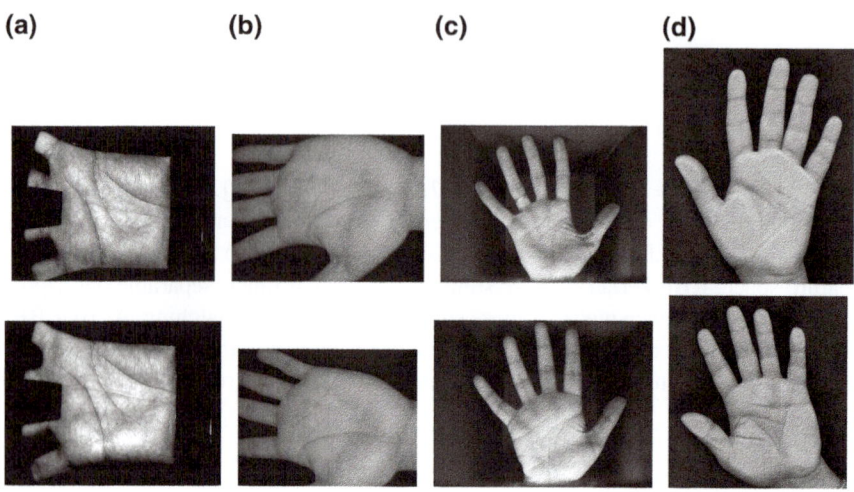

Fig. 3 Sample palmprints **a** PolyU 2D, **b** CASIA, **c** IITD, and **d** IIITDMJ

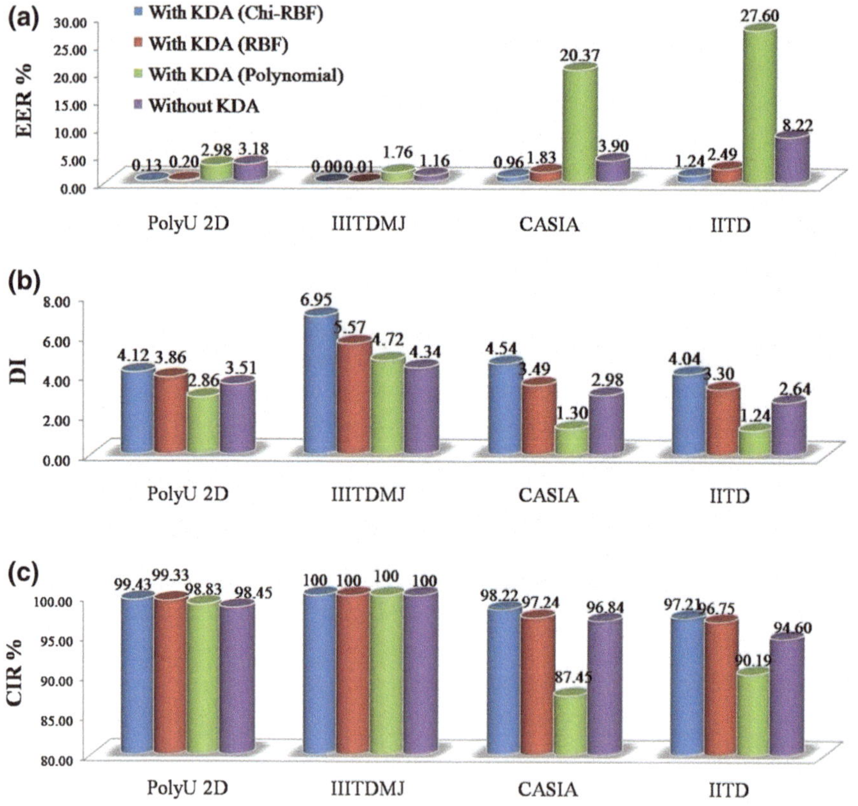

Fig. 4 **a** EER, **b** DI, and **c** CIR values comparison for BULBPH +KDA with three kernel functions

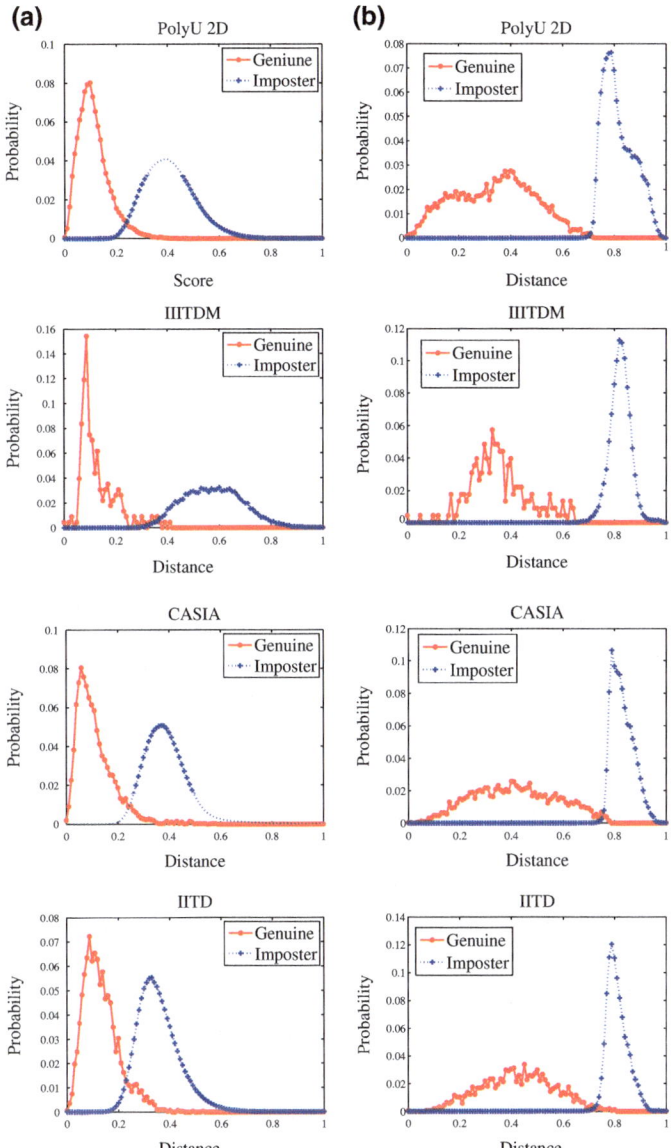

Fig. 5 Genuine and imposter probability distribution curves using BULBPH descriptor **a** without KDA and **b** with KDA (chi-RBF)

RBF and chi-RBF kernel function are taken as 1. The significance of using KDA with BULBPH is clearly visible through these graphs. The value of EER is nearly 0% with increasing value of DI for touch-based palmprints (PolyU and IIITDMJ). In case of touch-less palmprints, EER is reduced by approximately 3% for CASIA databases and 5.73% for IITD using chi-RBF kernel as compared to BULBPH without KDA. Comparison of the performance in terms of EER and DI helps to find usefulness of kernel function used in KDA. The recognition performance is better with chi-RBF kernel among all three kernels, while polynomial kernel is not suitable for BULBPH feature mapping.

Figure 5 compares genuine and imposter probability distribution curves against different distance values. Separability between genuine and imposter users is shown for all four databases without and with KDA using chi-RBF kernel. These curves ensure that KDA using chi-RBF kernel increases separability between genuine and imposter users that helps in improvement of all performance parameters.

3.1 Comparison of Dimension Reduction Techniques on BULBPH

Performance of BULBPH descriptor with dimension reduction techniques like PCA, KPCA, LDA, and KDA are compared to show that KDA is a better choice for dimension reduction in this case. PCA utilizes a linear transformation matrix to obtain low-dimensional linear subspace by minimizing the squared reconstruction error. LDA attempts to extract features that maximize the separability among classes by using their labels. KPCA and KDA are extension of PCA and LDA with kernel methods. Use of PCA and KPCA reduces the dimension of descriptor based on the percentage of variance that is the ratio of sum of the selected eigenvalues to sum of eigenvalues [4]. Table 2 compares the performance parameters obtained by applying

Table 2 Comparison of dimension reduction techniques on BULBPH descriptor

Database→ Feature↓	PolyU 2D			IIITDMJ		
	EER	DI	CIR	EER	DI	CIR
BULBPH	3.18	3.51	98.45	1.16	4.34	100
BULBPH+PCA	7.35	2.81	96.12	5.73	3.40	95.15
BULBPH+KPCA	4.04	3.68	94.42	1.71	5.00	96.92
BULBPH+LDA	2.83	3.86	99.33	1.76	2.98	99.42
BULBPH+KDA	0.13	4.12	99.42	0	6.95	100
Database→	CASIA			IITD		
Feature↓	EER	DI	CIR	EER	DI	CIR
BULBPH	3.90	2.98	96.84	8.23	2.64	94.60
BULBPH+PCA	6.08	2.88	94.27	12.08	2.31	90.16
BULBPH+KPCA	3.57	3.70	96.40	5.57	3.30	94.49
BULBPH+LDA	4.40	3.38	95.60	5.99	3.17	94.30
BULBPH+KDA	0.96	4.54	98.21	1.24	4.04	97.21

linear dimension reduction techniques (PCA and LDA) and nonlinear dimension reduction techniques (KPCA and KDA) on BULBPH descriptor for four palmprint databases.

4 Comparison with the Related Works

Performance of the proposed BULBPH descriptor is compared with some related features for gray scale palmprint databases. This includes palm code [27], competitive code [8], multi-scale LBP [2], complex-directional filter bank with uniform LBP histogram [10], and block-wise histogram of orientated gradient (BHOG) [3]. These techniques are also based on coding features and block-wise histogram features. Table 3 performs the comparison on feature length, EER, DI, and CIR parameters for gray scale palmprint databases collected from scanner and camera. Palm code, competitive code, and BHOG features are more suitable for scanner based recognition system as compared to camera based system. MLBP, CDFB-LBP and BHOG are based on block-wise feature extraction and these methods achieved more accuracy for palmprints captured through camera. BULBPH with KDA descriptor achieves minimum EER, DI value more than 4, and more than 97% accuracy for palmprint recognition.

Table 3 Comparison of BULBPH with KDA descriptor with related descriptors

Database→ Feature↓	PolyU 2D				IIITDMJ			
	Feature size	EER	DI	CIR	Feature size	EER	DI	CIR
Palm code [27]	4096*2 bits	6.65	2.99	93.34	4096*2 bits	4.42	3.7	95.59
Comp code [8]	4096*3 bits	3.97	3.86	96.70	4096*3 bits	2.64	4.63	96.48
MLBP [2]	5024	1.45	4.30	99.15	5024	0.98	3.44	100
CDFB-LBP [10]	399	2.56	3.80	98.68	149	0.44	4.41	100
BHOG [14]	2048	3.16	4.18	97.77	2048	3.08	4.05	96.92
BULBPH [21]	944	3.56	2.87	98.40	944	1.16	4.34	100
BULBPH+KDA	*399*	*0.13*	*4.12*	*99.42*	*149*	*0*	*6.95*	*100*
Database→ Feature↓	CASIA				IITD			
	Feature size	EER	DI	CIR	Feature size	EER	DI	CIR
Palm code [27]	4096*2 bits	23.51	1.33	66.98	4096*2 bits	20.79	1.56	70.01
Comp code [8]	4096*3 bits	10.37	2.41	86.24	4096*3 bits	13.93	2.18	82.64
MLBP [2]	5024	4.63	2.56	96.94	5024	10.71	2.29	93.65
CDFB-LBP [10]	623	3.65	3.22	96.27	623	6.35	3.18	94.30
BHOG [14]	2048	15.27	2.02	46.28	2048	10.89	2.52	85.30
BULBPH [21]	2052	3.95	3.40	96.94	2052	8.23	2.64	94.60
BULBPH+KDA	*623*	*0.96*	*4.54*	*98.21*	*429*	*1.24*	*4.04*	*97.21*

5 Conclusion

This work presents a reliable, fast, and discriminative palmprint verification and identification system. From the experimental results it is found that BULBPH descriptor with 8 neighbors and 4 radius performs well. KDA helps in decreasing the dimension of feature vector by mapping BULBPH descriptor using chi-RBF kernel function. This increases discrimination ability that is visible through lower EER and increased DI achieved with BULBPH+KDA descriptor.

References

1. Cui, J., Wen, J., Fan, Z.: Appearance-based bidirectional representation for palmprint recognition. Multimed. Tools Appl. **1–13**, (2014)
2. Guo, Z., Zhang, L., Zhang, D., Mou, X.: Hierarchical multiscale LBP for face and palmprint recognition. In: 17th IEEE International Conference on Image Processing (ICIP), pp. 4521–4524. IEEE (2010)
3. Hong, D., Liu, W., Su, J., Pan, Z., Wang, G.: A novel hierarchical approach for multispectral palmprint recognition. Neurocomputing **151, Part 1**(0), 511–521 (2015)
4. Hoyle, D.C.: Automatic PCA dimension selection for high dimensional data and small sample sizes. J. Mach. Learn. Res. **9**(12), 2733–2759 (2008)
5. Jia, W., Huang, D.S., Zhang, D.: Palmprint verification based on robust line orientation code. Pattern Recognit. **41**(5), 1504–1513 (2008)
6. Kong, A., Zhang, D., Kamel, M.: Palmprint identification using feature-level fusion. Pattern Recognit. **39**(3), 478–487 (2006)
7. Kong, A., Zhang, D., Kamel, M.: A survey of palmprint recognition. Pattern Recognit. **42**(7), 1408–1418 (2009)
8. Kong, A.K., Zhang, D.: Competitive coding scheme for palmprint verification. In: Proceedings of the 17th International Conference on Pattern Recognition. vol. 1, pp. 520–523. IEEE (2004)
9. Kumar, A.: IIT delhi touchless palmprint database version 1.0. http://www4.comp.polyu.edu.hk/~csajaykr/IITD/Database_Palm.htm (2009)
10. Mu, M., Ruan, Q., Guo, S.: Shift and gray scale invariant features for palmprint identification using complex directional wavelet and local binary pattern. Neurocomputing **74**(17), 3351–3360 (2011)
11. Ojala, T., Pietikainen, M., Maenpaa, T.: Multiresolution gray-scale and rotation invariant texture classification with local binary patterns. IEEE Trans. Pattern Anal. Mach. Intell. **24**(7), 971–987 (2002)
12. Ong Michael, G.K., Connie, T., Jin Teoh, A.B.: Touch-less palm print biometrics: novel design and implementation. Image Vis. Comput. **26**(12), 1551–1560 (2008)
13. Qian, J., Yang, J., Gao, G.: Discriminative histograms of local dominant orientation (D-HLDO) for biometric image feature extraction. Pattern Recognit. **46**(10), 2724–2739 (2013)
14. Raghavendra, R., Busch, C.: Novel image fusion scheme based on dependency measure for robust multispectral palmprint recognition. Pattern Recognit. **47**(6), 2205–2221 (2014)
15. Ren, J., Jiang, X., Yuan, J.: Noise-resistant local binary pattern with an embedded error-correction mechanism. IEEE Trans. Image Process. **22**(10), 4049–4060 (2013)
16. Schölkopf, B., Smola, A., Müller, K.R.: Nonlinear component analysis as a kernel eigenvalue problem. Neural Comput. **10**(5), 1299–1319 (1998)
17. Scholkopft, B., Mullert, K.R.: Fisher discriminant analysis with kernels. In: IEEE Signal Processing Society Workshop Neural Networks for Signal Processing, pp. 23–25 (1999)
18. Sun, Z.: Casia palmprint database (2005)

19. Tamrakar, D., Khanna, P.: Palmprint verification with XOR-SUM Code. Signal Image Video Process. **9**(3), 535–542 (2013)
20. Tamrakar, D., Khanna, P.: Blur and occlusion invariant palmprint recognition with block-wise local phase quantization histogram. J. Electron. Imaging **24**(4), 043006 (2015)
21. Tamrakar, D., Khanna, P.: Occlusion invariant palmprint recognition with ULBP histograms. In: Eleventh International Multi Conference on Information Processing (IMCIP 2015) (Procedia Computer Science), vol. 54, pp. 491–500 (2015)
22. Tamrakar, D., Khanna, P.: Kernel discriminant analysis of Block-wise Gaussian Derivative Phase Pattern Histogram for palmprint recognition. J. Vis. Commun. Image Represent. **40**, 432–448 (2016)
23. Wang, X., Gong, H., Zhang, H., Li, B., Zhuang, Z.: Palmprint identification using boosting local binary pattern. In: 18th International Conference on Pattern Recognition, vol. 3, pp. 503–506. IEEE (2006)
24. Wang, Y., Ruan, Q., Pan, X.: Palmprint recognition method using dual-tree complex wavelet transform and local binary pattern histogram. In: International Symposium on Intelligent Signal Processing and Communication Systems, pp. 646–649. IEEE (2007)
25. Zhang, D.: Polyu palmprint database. Biometric Research Centre, Hong Kong Polytechnic University. http://www.comp.polyu.edu.hk/~biometrics (2012)
26. Zhang, S., Gu, X.: Palmprint recognition based on the representation in the feature space. Opt.-Int. J. Light. Electron Opt. **124**(22), 5434–5439 (2013)
27. Zhang, D., Kong, W.K., You, J., Wong, M.: Online palmprint identification. IEEE Trans. Pattern Anal. Mach. Intell. **25**(9), 1041–1050 (2003)
28. Zhang, W., Shan, S., Gao, W., Chen, X., Zhang, H.: Local gabor binary pattern histogram sequence (lgbphs): a novel non-statistical model for face representation and recognition. In: Tenth IEEE International Conference on Computer Vision (ICCV). vol. 1, pp. 786–791. IEEE (2005)
29. Zhang, B., Shan, S., Chen, X., Gao, W.: Histogram of gabor phase patterns (hgpp): a novel object representation approach for face recognition. IEEE Trans. Image Process. **16**(1), 57–68 (2007)
30. Zhang, D., Zuo, W., Yue, F.: A comparative study of palmprint recognition algorithms. ACM Comput. Surv. (CSUR) **44**(1), 2 (2012)
31. Zhao, Y., Jia, W., Hu, R., Gui, J.: Palmprint identification using LBP and different representations. In: International Conference on Hand-Based Biometrics (ICHB), pp. 1–5. IEEE (2011)

Things at Your Desk: A Portable Object Dataset

Saptakatha Adak

Abstract Object Recognition has been a field in Computer Vision research, which is far from being solved when it comes to localizing the object of interest in an unconstrained environment, captured from different viewing angles. Lack of benchmark datasets clogs the progress in this field since the last decade, barring the subset of a single dataset, *alias* the *Office* dataset, which attempted to boost research in the field of pose-invariant detection and recognition of portable object in unconstrained environment. A new challenging object dataset with 30 categories has been proposed with a vision to boost the performances of the task of object recognition for portable objects, thus enhancing the study of cross domain adaptation, in conjunction to the *Office* dataset. Images of various hand-held objects are captured by the primary camera of a smartphone, where they are photographed under unconstrained environment with varied illumination conditions at different viewing angles. The monte-carlo object detection and recognition has been performed for the proposed dataset, facilitated by existing state-of-the-art transfer learning techniques for cross-domain recognition of objects. The baseline accuracies for existing Domain Adaptation methods, published recently, are also presented in this paper, for the kind perusal of the researchers. A new technique has also been proposed based on the activation maps of the AlexNet to detect objects, alongwith a Generative Adversarial Network (GAN) based Domain Adaptation technique for Object Recognition.

Keywords Domain adaptation · Generative adversarial network (GAN) · Object detection · Object recognition

1 Introduction

Datasets have always played a vital role in research, specially in Computer Vision which forms an integral part in Machine learning. The datasets not only adjudges the strength of the evaluation algorithms but also opens a scope for new challenging

S. Adak (✉)
VP Lab, Department of Computer Science and Engineering, IIT Madras, Chennai, India
e-mail: sapta@cse.iitm.ac.in

© Springer Nature Singapore Pte Ltd. 2020
B. B. Chaudhuri et al. (eds.), *Proceedings of 3rd International Conference on Computer Vision and Image Processing*, Advances in Intelligent Systems and Computing 1022,
https://doi.org/10.1007/978-981-32-9088-4_36

research. The existing datasets are mainly formed by capturing images for both training and testing, either placed on similar background or under same ambient conditions. Moreover most of the large datasets [6, 20] are populated with images obtained from Internet databases, which can be used for various Computer Vision tasks. Recognition of an object not only states the belongingness of an object to a specific class but also localizes the object in an image. There has always been a lack of images of objects that we captured from different angles in these datasets which facilitates the process of effective and pose invariant object detection, using bounding box to represent its location in an image. Often, precise detection of object largely depends on the contextual information, making it essential for objects to belong to their natural environment. For example, it is easy for any machine learning model to perceive an aeroplane flying in the sky than to apprehend it while it is in hangar of an airport. Accuracy provided by the detection algorithms are obtained by calculating the Intersection over Union (IoU) scores and comparing with the ground truth bounding boxes. There has been a deficiency of suitable object detection datasets containing images in their natural unconstrained environment with intact contextual information. Also, the images of a particular object captured from multiple viewing angles are largely absent in the existing datasets. In this paper, a new dataset has been proposed consisting of images of objects captured not only in an unconstrained natural ambiance but also from multiple viewing angles. The proposed dataset consists of readily available portable objects at our daily workplace. The object detection dataset is a first of its kind to the best of my knowledge, consisting of images captured through primary mobile camera.

In the rest of the paper, Sect. 2 details the existing object datasets, recent advances in the field of the object detection and recognition, and Domain Adaptation (DA). Section 3 briefly outlines the proposed dataset, while the Sect. 4 introduces the real world datasets used along with the proposed dataset for the evaluation of Domain Adaptation techniques and discusses about GAN based Domain Adaptation. Section 5 gives the quantitative results of performance analysis for the Domain Adaptation techniques, on the datasets, using recognition rate. Finally, we conclude the paper discussing the major points and future scope in Sect. 6.

2 Related Work

In the recent past several works have been published in the domain of object detection and recognition. A number of datasets have been proposed in last few years of which COIL [24], MIT-CSAIL [36], PASCAL VOC [7], Caltech-6 and Caltech-101 [8], Caltech-256 [14], Graz [25], CIFAR-100 [18], MS-COCO [20], ImageNet [6], etc. has gained visibility and importance. With saturation of performance of the existing algorithms, these datasets have become progressively challenging. COIL [24] consists of objects placed on black background with no clutter. Caltech-6 consists of 6 classes viz. airplanes, cars, faces, leaves, motorcycles, etc., while the Caltech-101 [8] is built in a similar fashion, differing in many objects being clicked in front of a bright background. The MIT-CSAIL [36] dataset consists of $\sim$77, 000 images taken in varied environment, where majority of the images are labeled using the LabelMe [27] annotation tool. PASCAL VOC [7] image dataset includes $\sim$5, 000 fully annotated images spread over 10 object classes, which are fully annotated. Object categories with complex viewing conditions is present in the Graz dataset [25]. CIFAR-100 [18] dataset is consists of 6000 32×32 pixels low resolution images belonging to 100 classes which is again grouped into 20 super-classes. Each image in CIFAR-100 is annotated with a "coarse" label and a "fine" label depending on the super-class and class it belongs. Caltech-256 [14] is built upon popular image classes representing a wide range of natural and artificial objects in diverse settings of lighting conditions, poses, backgrounds, image sizes and camera settings. Compared to Caltech-101, this has more number of images per class, as high as 80. Also the right-left alignment of the objects in Caltech-101 have been taken care off in Caltech-256. ImageNet [6], built upon the hierarchical structure of WordNet [23], consists of $\sim$50 million labeled high resolution samples consisting of 12 subtrees.

In the last few decades, Object class detection has become a major research area in Computer Vision. A number of approaches exist [5, 9], but sliding window approach and part based models are followed by most of the state-of-the-art detectors. In the "sliding window" approach, dense scanning of each image takes place from the top-left corner to the bottom-right corner with rectangular sliding windows at varying scales. For every sliding window, features like image patches, edges, etc., are extracted and fed to classifier alongwith labeled training data to perform offline training. The sliding windows are categorized as a positive or negative samples by the classifier. In order to solve this problem, many state-of-the-art techniques have been proposed recently. Yet, there still lacks a general and comprehensive solution which works under many situation: occlusion, complex background textures, illumination variation and pose, having similar structure over various object classes, large number of object classes etc. Researchers have provided a wide spectrum of local descriptors, dictionary based model and efficient algorithms for classification of objects. The recent state-of-the-art techniques uses the concepts of spatial pyramid matching and Bag-of-Features (BoF) for image classification [10]. In BoF method an image is represented as a histogram of its local features.

A wide range of research has been done in different applications in the fields of Text processing, Speech, Natural Language Processing and Computer Vision [2, 26]. Minimizing the disparity in distributions of source and target domain is the main focus for Domain Adaptation. 'Instance Weighting' is an approach to this problem, where the expected loss is minimized by training on the weighted instances of the source domain [4, 32]. Instances from source and target domains can be projected onto some intermediate domains (common subspace), which reduces the disparity of distribution between the two domains, in the projected subspace [13, 17, 28]. Gopalan et al. [13] suggested a geodesic path between the principal components of source and target domain data, which is considered in the Grassmann manifold to solve the problem. An estimate for the continuous change of the properties in subspaces of source and target domain is given by the intermediate points sampled along the path. Later, Gong et al. [11] proposed that on the geodesic path infinite number of intermediate points can be interpolated which is estimated by the geodesic flow kernel. Some recent computer vision tasks includes application of DA which are discussed in [28–30]. Now-a-days Domain Adaptation using Convolutional Neural Networks (CNN) [21, 33] and Generative Adversarial Networks (GAN) [16, 37, 39] has gained an enormous success in the field of Computer Vision.

3 The Portable Object (PO) Dataset

The proposed dataset consists of portable objects which are part of day-to-day life and are frequently available at workplaces. The high resolution images in the dataset are captured using primary mobile camera of a smartphone (the camera parameters are described later) at different ambient conditions mimicking a real-world office-cabinet scenario.

The dataset includes 40, 607 images distributed among 30 categories (refer Table 1) of portable objects viz. back-pack, bottle, keyboard, mouse, mug, wrist watch, etc. Mobile having primary camera of 16 Mega-pixel (OmniVision—OV16860—PureCel) having Phase Detection Auto Focus (PDAF) technology [22] and $f/2.0$ aperture was used to capture the images. Each class consists of $\sim$1, 350 images (refer Table 1) each of which is captured along multiple angles. The dataset consists of images of dimension 1920×1080 pixels (refer Table 1), thereby enabling a better recognition rate. Due to the objective being recognition and detection of the objects under unconstrained ambient conditions, the background and illumination conditions were chosen arbitrarily. Figure 1 shows an example each from 5 exemplar classes, viz. Scissor, Tape dispenser, Puncher, Backpack and Laptop, of various objects present in the proposed dataset.

We further demonstrate the different viewing angles (θ) at which the images are captured in the dataset in Fig. 2, where the sub-figures (a)–(g) shows the object at focus following an uniform rotation with an interval of $\Delta\theta = \pi/3, \forall\theta \in [0, \pi]$. The proposed dataset virtually opens the door for the Computer Vision researchers to examine their methods on the novel objects (present in our dataset) that are barely available, to the best of our knowledge, in other object recognition or detection datasets. In addition, this dataset acts in conjunction to the Office dataset [28] that has been a benchmark for Domain Adaptation techniques in the field of Transfer learning [4]. The unconstrained nature of proposed dataset is mainly accounted for

Table 1 Characteristic information of the proposed dataset

Characteristics	Values
Number of object classes	30
Total images in the dataset	40607
Images per class	$\sim$1, 350
Dimension of each image	1920×1080 pixels

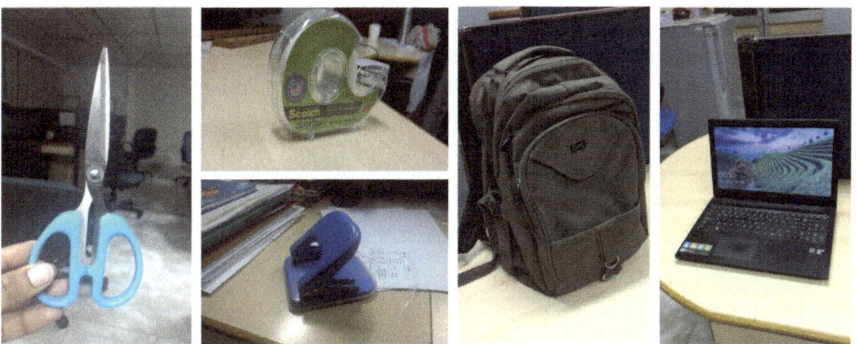

Fig. 1 Figures representing 5 different object categories (chosen at random) present in the proposed portable object dataset

| (a) | (b) | (c) | (d) | (e) | (f) | (g) |

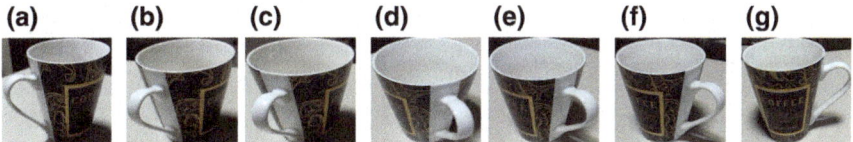

Fig. 2 Figure showing a particular object captured from multiple viewing angles (manually cropped for display purpose); **a–g** represent rotating viewing angles in clockwise direction

the variable lighting conditions, camera shake, motion blur and environmental noise, which were prevalent during the capture of the dataset.

4 Experimental Setup

Experiments on the proposed dataset which includes object detection and recognition on objects across 2 domains: source and target. The detection methodology helps to precisely localize the object of interest. The object recognition task has been attacked evidently to provide a baseline measure to the research community along with extensive study of similar categories over different domains from the Office Dataset [28].

4.1 The Office Dataset

3DA Dataset (Office Data): The dataset reflects the challenging task of object recognition and evaluation of visual domain transformation approaches in real-world environment. The 4106-image dataset consists of 31 categories from three domains, viz. Amazon, DSLR and Webcam. The amazon domain consists of 600 × 600 pixels images web-crawled from the Amazon website, capturing wide intra-class variations. The DSLR images are captured using Digital Single-Lens Reflex camera in realistic environment with higher resolution (1000×1000 pixels) and lower noise. On the other hand, 423×423 pixel images were captured in Webcam are of low resolution containing significant amount of noise.

4.2 Data Preprocessing and Feature Extraction

The primary objective of the proposed dataset lie in the online object detection and recognition using effective Computer Vision techniques to cope with the task at hand. A pretrained model of AlexNet [19] has been used, followed by fine-tuning of the network using our proposed dataset. The AlexNet architecture consists of 9 layers and has been used extensively for large scale object recognition on ImageNet [6] dataset. The filter response at the CONV-3 layer has been recorded as a salient part of the image that facilitates the high classification accuracies of AlexNet. A global max-pooling [35] has been used to localize the objects present in the image based on the filter responses from the CONV-3 layer of the AlexNet. The pre-trained AlexNet model has been adopted from *KerasModelZoo*, while *Keras-vis* helps to localize the image. Figure 3 illustrates our proposed framework for detection (see Fig. 3a), along with the results for an intermediate processing on a *mouse* image (see Fig. 3b). Additional results showing the detection of the objects using yellow rectangular bounding boxes are given in Fig. 4.

Furthermore, the experiments are done on 19 categories of the proposed Portable Object dataset that are common with Amazon, DSLR and Webcam domains of the Office dataset [28]. The benchmark for object recognition in Domain Adaptation has been mostly based on the results of the Office dataset. This results demonstrates the lack of data for testing characterized by moderately low accuracies provided by the state-of-the-art techniques proposed in the recent past. This motivates us to

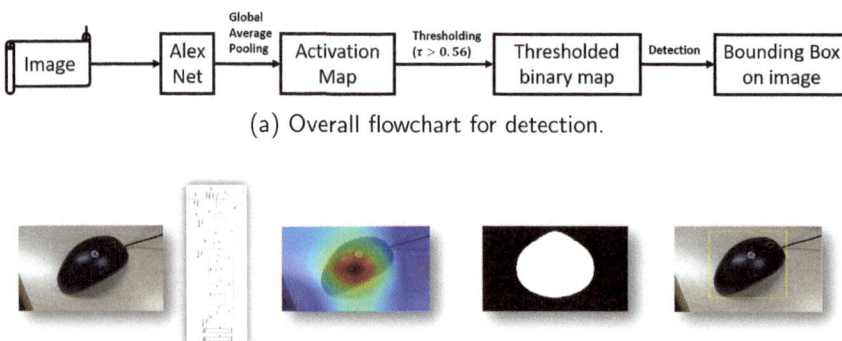

(a) Overall flowchart for detection.

(b) An example showing the intermediate results for our proposed technique.

Fig. 3 The proposed framework for object detection (best viewed in color)

Input Image	Activation map	Bounding box	Input Image	Activation map	Bounding box

Fig. 4 Detected objects localized within rectangular yellow bounding boxes (best viewed in color) alongwith corresponding activation map for each

avail the opportunity in order to complement our dataset with 19 matching object categories with the Amazon, DSLR and Webcam domains of the Office Dataset, such that researchers can bench-test their proposed Domain Adaptation techniques on our dataset. For providing baselines in Domain Adaptation, features like Histogram of Gradients (HOG) [5] and Bag of Words (BoW) [10] using Speed-up Robust Features (SURF) [1] are extracted from the images. Apart from these, the images fed to a pretrained VGG [31] model to extract 4096 dimensional features from the *fc7* (layer before the softmax classifier layer) layer. The images of both the proposed and the Office dataset are resized to 224×224 pixels before feeding them to the pretrained VGG model to match the input layer configuration of VGGNet. The other parameters of this deep model is fine tuned using our proposed dataset optimized by Stochastic Gradient Descent (SGD) [3] algorithm. For extraction of BoW features, the images were also resized to 224×224 pixels and the bag was formed with the Speeded Up Robust Feature (SURF) [1] descriptors. BoW features, with a dictionary size of 512, is obtained by performing K-means clustering on randomly initialized 512 centres. Histogram of these SURF features were then encoded to obtain 512-dimensional BoW features for the images. All the different types of features extracted are then made mean-centralized and scale normalized before carrying out the experiment.

4.3 Domain Adaptation

Images from 2 different domains (the source and target domains) can be matched by projecting the features of both the domains into a common subspace and classifying using the same classifier, which forms the basic motivation for Domain Transformation. The two different domains involved in domain adaptation are souce and target domain. Recently, several transfer learning [26] techniques have been proposed for this purpose. In this paper, we have used Geodesic Flow Kernel (GFK) [11], Extreme Learning Machine-Based Domain Adaptation (EDA) [38] and Max-margin Domain

Transformation (MMDT) [15] to map the cross domain features of the images. The goal of these methods is to jointly learn affine hyperplanes that seperate the source domain classes, and the transformation from the target domain points to the source domain. This enforces the transformed target data to lie on the correct side of the learned source hyperplane.

Data preparation for domain adaptation involves dividing the target domain— 20% for training and 80% for testing. The training subset of the target domain is augmented with source domain and was used for training the model. Various data augmentation techniques were introduced to induce variation in the dataset. The images were randomly flipped in left-right and rotated in both clockwise and anti-clockwise direction upto 5°. The remaining features of the target are being used for testing purpose. For using SVM [34], the augmented training set is used to train the model keeping the rest for testing. All these techniques were made to undergo 20 fold validation where each split of training and testing set is initialized randomly. Mean recognition rates of the techniques are reported in Table 2. Random initializations were incorporated during split formation, by making sure images from all the categories are present in each split.

4.4 GAN Based Domain Adaptation

GAN [12] is built upon two models: a generative model (G) and a discriminative model (D). The generator G learns to mimic the true data distribution p_{data} whereas the discriminator D distinguishes whether a sample belongs to the true distribution or coming from G, by predicting a binary label. The process of alternate training of the models in this network is similar to the two player min-max games. The overall objective function for simultaneous minimizing the loss at G and maximizing the distinguisher D is as follows:

$$\min_{G} \max_{D} v(D, G) = \mathbb{E}_{x \sim p_{data}}[log(D(x))] + \mathbb{E}_{z \sim p_z}[log(1 - D(G(z)))] \quad (1)$$

where, x be an example from true distribution p_{data} and z is the example sampled from the distribution p_z, assumed to be uniform or Gaussian.

In this paper, a complex GAN architecture has been proposed (refer Fig. 5) where two GANs has been implemented for source and target domain images. Respective generators of the GANs has been used to generate the source and corresponding target images and a domain invariant feature space is achieved by sharing the resulting high-level layer features. Thus, a corresponding pair of images from two different distributions is generated by virtue of same noise input. In addition to these, weights in the last few layers of the two discriminators of different domains has been tied to ensure the domain invariableness of feature space. The number of parameters in the network are also reduced due to the weight sharing between the discriminative models. During training phase, apart from adversarial loss ($\mathcal{L}_{adv}$), moment loss ($\mathcal{L}_{moment}$)

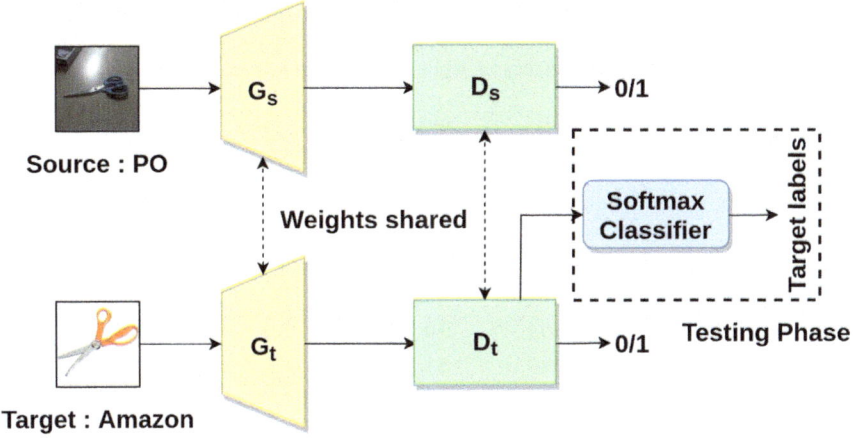

Fig. 5 Proposed model for Domain Adaptation using GAN

is used to train the generators G_s and G_t which matches the higher order moments of the two distributions. The overall objective function for the proposed framework is as follows

$$\min_{G_s, G_t} \max_{D_s, D_t} v(D_s, D_t, G_s, G_t) = \alpha \mathcal{L}_{adv}(D_s, D_t, G_s, G_t) + \beta \mathcal{L}_{moment}(G_s, G_t) \quad (2)$$

where $\mathcal{L}_{adv}$, $\mathcal{L}_{moment}$ are the adversarial and moment losses respectively with α and β being the trade-off controlling parameters between these losses. In Eq. (2),

$$\mathcal{L}_{adv}(D_s, D_t, G_s, G_t) = \mathbb{E}_{x_s \sim p_{src}}[log(D_s(x_s))] + \mathbb{E}_{z \sim p_z}[log(1 - D_s(G_s(z)))]$$
$$+ \mathbb{E}_{x_t \sim p_{tgt}}[log(D_t(x_t))] + \mathbb{E}_{z \sim p_z}[log(1 - D_t(G_t(z)))] \quad (3)$$

where x_s and x_t are the images from the source (p_{src}) and target (p_{tgt}) distributions respectively, $z \sim p_z$. G_s, G_t are the generators while D_s, D_t are the discriminators for source and target data respectively.

$$\mathcal{L}_{moment}(G_s, G_t) = \frac{1}{(p-q)} ||E(G_s(z)) - E(G_t(z))||_2$$
$$+ \sum_{n=2}^{N} \frac{1}{|p-q|^n} ||C_n(G_s(z)) - C_n(G_t(z))||_2 \quad (4)$$

where $C_n(r) = E(r - E(R))^n$ is the vector of all nth order sample central moments and $E(R) = \frac{1}{|R|} \sum_{r \in R} r$ is the expectation.

The corresponding pair of source and target images were passed into two different generators. The deep architectures requires a lot of data during training. Hence, various data augmentation techniques were introduced to induce variation in the

Table 2 Baseline measures (in percentage) for different state-of-the-art techniques of Domain Adaptation, to show a transfer of features from the proposed to the *Office* dataset and vice-versa. *PO* represents the Portable Object dataset, while *Amazon, DSLR, Webcam* are *abbrv.* as *A, D* and *W* respectively

Techniques	PO → A	PO → D	PO → W	A → PO	D → PO	W → PO
SVM + HOG	30.54	29.68	28.05	24.90	23.99	20.09
SVM + BOW	35.34	31.80	33.94	23.45	25.87	26.89
SVM + VGG	25.69	28.62	24.59	37.39	33.94	38.33
GFK + HOG	20.48	25.63	22.29	18.88	33.61	29.77
GFK + BOW	23.82	27.51	26.97	27.69	23.73	23.53
GFK + VGG	49.67	60.12	50.57	41.33	41.64	38.31
MMDT + HOG	39.71	39.93	51.21	55.95	56.62	46.26
MMDT + BOW	51.09	39.56	52.07	47.68	61.89	49.58
MMDT + VGG	59.92	66.07	57.84	57.02	62.73	55.36
EDA + HOG	23.21	34.69	24.41	37.86	42.96	42.35
EDA + BOW	21.88	34.58	23.87	47.29	58.29	48.36
EDA + VGG	52.64	76.45	68.62	56.34	48.61	52.32
GAN based	60.81	78.42	72.34	59.23	65.93	56.07

dataset. The images were randomly flipped in left-right and rotated in both clockwise and anti-clockwise direction upto $5°$. During testing, the images are fed into G_t and the output of the last convolutional layer of D_t is passed through a softmax classifier to predict the target labels. The mean recognition rates obtained using *Office* and *PO* object dataset as source and target and vice-versa are reported in Table 2.

5   Performance Analysis

Domain Adaptation (DA) techniques like GFK [11], EDA [38] and MMDT [15] has been implemented on the extracted HOG [5], BoW [10] and VGG [31] features and recognition accuracies are provided in Table 2. The results show that MMDT on BoW features gives better recognition rate outperforming all other methods using similar feature at test, as depicted in the table, while the VGG features involving GFK outperforms all other features with same DA technique studying the cross domain image recognition of Portable Object (PO) → Amazon (A), Portable Object (PO) → DSLR (D) and Portable Object (PO) → Webcam (W) and vice-versa. The GAN based DA technique mentioned in this paper works better than all other techniques.

Table 2 provides a benchmark for our proposed dataset with several state-of-the-art techniques. The results represent a viable mixture of shallow and deep features with renowned classification algorithms. The result in bold shows the best performance among these well-known DA algorithms. Owing to the low accuracies shown in the table, the proposed dataset provides a scope of improvement among the researchers to develop novel algorithms to tackle this problem. We also encourage the researchers to quote results from Table 2 in order to evaluate the performance of their algorithms, with the baseline measures.

6 Conclusion

This paper proposes a new dataset in the field of portable object detection, available at desk in everyday life. The dataset opens a new avenue for the researchers to benchmark novel detection and recognition algorithms. A new pipeline has been proposed for object detection using AlexNet based CNN architecture, to automate the groundtruth of the image with considerable success. A GAN based DA framework is also provided which will act as a baseline for further experiments in the field of Transfer learning on the proposed *Portable Object* dataset from other popular ones like *Office*. Apart from Domain Adaptation, this dataset can also be used for Object Recognition and Detection purposes. With the increase in availability of subjects, further expansion of this dataset is inevitable.

Acknowledgements We gratefully thank the faculty and researchers of Visualization and Perception Lab, IIT Madras, for their valuable insight into this research.

References

1. Bay, H., Ess, A., Tuytelaars, T., Van Gool, L.: Speeded-up robust features (surf). CVIU **110**(3), 346–359 (2008)
2. Beijbom, O.: Domain adaptations for computer vision applications. arXiv:1211.4860 (2012)
3. Bottou, L.: Large-scale machine learning with stochastic gradient descent. In: COMPSTAT, pp. 177–186. Springer (2010)
4. Dai, W., Yang, Q., Xue, G.R., Yu, Y.: Boosting for transfer learning. In: ICML, pp. 193–200. ACM (2007)
5. Dalal, N., Triggs, B.: Histograms of oriented gradients for human detection. IEEE CVPR **1**, 886–893 (2005)
6. Deng, J., Dong, W., Socher, R., Li, L.J., Li, K., Fei-Fei, L.: Imagenet: a large-scale hierarchical image database. In: IEEE CVPR, pp. 248–255 (2009)
7. Everingham, M., Van Gool, L., Williams, C.K., Winn, J., Zisserman, A.: The pascal visual object classes (voc) challenge. IJCV **88**(2), 303–338 (2010)
8. Fei-Fei, L., Fergus, R., Perona, P.: Learning generative visual models from few training examples: an incremental bayesian approach tested on 101 object categories. CVIU **106**(1), 59–70 (2007)
9. Felzenszwalb, P.F., Girshick, R.B., McAllester, D., Ramanan, D.: Object detection with discriminatively trained part-based models. IEEE TPAMI **32**(9), 1627–1645 (2010)
10. Filliat, D.: A visual bag of words method for interactive qualitative localization and mapping. In: IEEE ICRA, pp. 3921–3926 (2007)
11. Gong, B., Shi, Y., Sha, F., Grauman, K.: Geodesic flow kernel for unsupervised domain adaptation. In: IEEE CVPR, pp. 2066–2073 (2012)
12. Goodfellow, I., Pouget-Abadie, J., Mirza, M., Xu, B., Warde-Farley, D., Ozair, S., Courville, A., Bengio, Y.: Generative adversarial nets. In: NIPS, pp. 2672–2680 (2014)

13. Gopalan, R., Li, R., Chellappa, R.: Domain adaptation for object recognition: an unsupervised approach. In: IEEE ICCV, pp. 999–1006 (2011)
14. Griffin, G., Holub, A., Perona, P.: Caltech-256 object category dataset (2007)
15. Hoffman, J., Rodner, E., Donahue, J., Darrell, T., Saenko, K.: Efficient learning of domain-invariant image representations. arXiv:1301.3224 (2013)
16. Isola, P., Zhu, J.Y., Zhou, T., Efros, A.A.: Image-to-image translation with conditional adversarial networks. In: IEEE CVPR, pp. 1125–1134 (2017)
17. Jhuo, I.H., Liu, D., Lee, D., Chang, S.F.: Robust visual domain adaptation with low-rank reconstruction. In: IEEE CVPR, pp. 2168–2175 (2012)
18. Krizhevsky, A.: Learning multiple layers of features from tiny images (2009)
19. Krizhevsky, A., Sutskever, I., Hinton, G.E.: Imagenet classification with deep convolutional neural networks. In: NIPS, pp. 1097–1105 (2012)
20. Lin, T.Y., Maire, M., Belongie, S., Hays, J., Perona, P., Ramanan, D., Dollár, P., Zitnick, C.L.: Microsoft coco: Common objects in context. In: ECCV, pp. 740–755. Springer (2014)
21. Long, M., Cao, Y., Wang, J., Jordan, M.I.: Learning transferable features with deep adaptation networks. In: ICML, vol. 37, pp. 97–105. JMLR.org (2015)
22. Mandelli, E., Chow, G., Kolli, N.: Phase-detect autofocus (Jan 14 2016), uS Patent App. 14/995,784
23. Miller, G.A., Beckwith, R., Fellbaum, C., Gross, D., Miller, K.J.: Introduction to wordnet: an on-line lexical database. Int. J. Lexicogr. 3(4), 235–244 (1990)
24. Nene, S.A., Nayar, S.K., Murase, H., et al.: Columbia object image library (coil-20) (1996)
25. Opelt, A., Pinz, A.: Object localization with boosting and weak supervision for generic object recognition. In: Image Analysis, pp. 431–438 (2005)
26. Pan, S.J., Yang, Q.: A survey on transfer learning. IEEE TKDE 22(10), 1345–1359 (2010)
27. Russell, B.C., Torralba, A., Murphy, K.P., Freeman, W.T.: Labelme: a database and web-based tool for image annotation. IJCV 77(1), 157–173 (2008)
28. Saenko, K., Kulis, B., Fritz, M., Darrell, T.: Adapting visual category models to new domains. In: ECCV, pp. 213–226 (2010)
29. Samanta, S., Banerjee, S., Das, S.: Unsupervised method of domain adaptation on representation of discriminatory regions of the face image for surveillance face datasets. In: Proceedings of the 2nd International Conference on Perception and Machine Intelligence, pp. 123–132. ACM (2015)
30. Selvan, A.T., Samanta, S., Das, S.: Domain adaptation using weighted sub-space sampling for object categorization. In: ICAPR, pp. 1–5. IEEE (2015)
31. Simonyan, K., Zisserman, A.: Very deep convolutional networks for large-scale image recognition. arXiv:1409.1556 (2014)
32. Sugiyama, M., Nakajima, S., Kashima, H., Buenau, P.V., Kawanabe, M.: Direct importance estimation with model selection and its application to covariate shift adaptation. In: NIPS, pp. 1433–1440 (2008)
33. Sun, B., Feng, J., Saenko, K.: Return of frustratingly easy domain adaptation. In: AAAI, vol. 6, p. 8 (2016)
34. Suykens, J.A., Vandewalle, J.: Least squares support vector machine classifiers. Neural Process. Lett. 9(3), 293–300 (1999)
35. Szegedy, C., Liu, W., Jia, Y., Sermanet, P., Reed, S., Anguelov, D., Erhan, D., Vanhoucke, V., Rabinovich, A.: Going deeper with convolutions. In: IEEE CVPR, pp. 1–9 (2015)
36. Torralba, A., Murphy, K.P., Freeman, W.T.: Sharing features: efficient boosting procedures for multiclass object detection. In: IEEE CVPR, vol. 2, pp. II–II (2004)
37. Tzeng, E., Hoffman, J., Saenko, K., Darrell, T.: Adversarial discriminative domain adaptation. In: IEEE CVPR, vol. 1, p. 4 (2017)
38. Zhang, L., Zhang, D.: Robust visual knowledge transfer via extreme learning machine-based domain adaptation. IEEE TIP 25(10), 4959–4973 (2016)
39. Zhu, J.Y., Park, T., Isola, P., Efros, A.A.: Unpaired image-to-image translation using cycle-consistent adversarial networks. In: IEEE CVPR, pp. 2223–2232 (2017)

A Deep Learning Framework Approach for Urban Area Classification Using Remote Sensing Data

Rahul Nijhawan, Radhika Jindal, Himanshu Sharma,
Balasubramanian Raman and Josodhir Das

Abstract The main aim of this study is to propose a Deep Learning framework approach for Urban area classification. The research proposes a multilevel Deep Learning architecture to detect the Urban/Non-Urban Area. The support models/parameters of the structure are Support Vector Machine (SVM), convolution of (Neural Networks) NN, high resolution sentinel 2 data, and several texture parameters. The experiments were conducted for the study region Lucknow which is a fast-growing metropolis of India, using Sentinel 2 satellite data of spatial resolution 10-m. The performance observed by the proposed ensembles of CNNs outperformed those of current state of art machine algorithms viz; SVM, Random Forest (RF) and Artificial Neural Network (ANN). It was observed that our Proposed Approach (PA) furnished the maximum classification accuracy of 96.24%, contrasted to SVM (65%), ANN (84%) and RF (88%). Several statistical parameters namely accuracy, specificity, sensitivity, precision and AUC, have been evaluated for examining performance during training and validation phase of the models.

R. Nijhawan · J. Das
Department of Computer Science and Engineering, Graphic Era University, Dehradun, India
e-mail: Rahulnijhawan2010@gmail.com

J. Das
e-mail: jddasfeq@iitr.ac.in

J. Das
Department of Earthquake Engineering, Indian Institute of Technology Roorkee, Roorkee, India

R. Jindal
Jaypee Institute of Information Technology, Noida, India
e-mail: jindal.radhika96@gmail.com

H. Sharma (✉)
National Institute of Technology Hamirpur, Hamirpur, India
e-mail: sharmah70@gmail.com

B. Raman
Department of Computer Science and Engineering, Indian Institute of Technology Roorkee, Roorkee, India
e-mail: balarfma@iitr.ac.in

© Springer Nature Singapore Pte Ltd. 2020
B. B. Chaudhuri et al. (eds.), *Proceedings of 3rd International Conference on Computer Vision and Image Processing*, Advances in Intelligent Systems and Computing 1022,
https://doi.org/10.1007/978-981-32-9088-4_37

Keywords Deep learning · Convolution neural network · Remote sensing ·
Support vector machine · Urban area classification

1 Introduction

An Urban Area is a human settlement with high population density and infrastructure.
The world's urban population has been increasing since 1950, and is further predicted
to grow to 6.4 billion by 2050. Urbanization is subject to various disciplines such
as geography, sociology, economics and public health. The extent of urbanization
helps us keep track of the population density of an area, the rapid changing social and
cultural change, urban planning, human impact on environment, etc. [1]. Monitoring
urban changes can also help support decision making in resource management and
pollution control. These important beneficial factors make it necessary to study the
urban areas and their expansion periodically and with great precision.

Remote sensing is one such cost-effective methodology. The critical capability of
Landsat [2–4], Sentinel, Landsat Thematic Mapper [5–8] and Enhanced Thematic
Mapper plus [8] to provide repetitive and synoptic observations of the Earth, makes
them the most widely used source to extract satellite images. Moderate Resolution Imaging Spectrometer (MODIS) [2, 6], high-resolution SPOT satellite imagery
[9, 10] and LiDAR sensors [11] have also formed the basis of many studies. GIS
based land change analysis is one methodology for strategically analyzing the images
through GIS-based operations [3, 5].

Studies have shown that application of object-based and pixel-based SVM and
DT classifiers [9, 12] deliver satisfactory results for urban area classification. Post-
Classification Comparison method [3, 8] and blend of Post-Classification Compari-
son method with Pre-Classification Comparison method [12] is also favored by many
scientists in their studies. Landscape Metrics helps in Land Use and Land Cover
(LULC) change analysis by calculating the size, pattern and nature of languages [5].

Deep Learning is a new area of which originally started with a vision of bringing
machine learning closer to Artificial Intelligence. Simple ANNs usually contain zero
or one hidden layers, at the same time, deep learning has more than one hidden layers.
It comes from the family of machine learning techniques. It focuses on learning
algorithms via training sets and is then made to run upon test cases. Current boom in
AI is actually the boom in deep learning. This boom is due to the many applications
deep learning is able to do such as natural language generation, automatic speech
recognition, image recognition, etc.

In this study, we have proposed a new framework of ensembles of Convolutional
Neural Network (CNNs) (AlexNet Architecture) for computing various specific features and SVM classifier as a model to classify them. The research was conducted in
the city of Lucknow, India employing high resolution Sentinel 2 satellite data of spatial resolution 10-m. Different supportive parameters have been employed by each
individual CNN to form an ensemble. Further, we performed accuracy assessment
for our proposed CNNs ensemble framework and compared the results with those of

current state of art algorithms viz., Support Vector Machine (SVM), Random Forest (RF) and Artificial Neural Network (ANN), using several statistic parameters, viz., accuracy, kappa coefficient, specificity, sensitivity, precision and AUC.

This paper is organized as follows: Introduction of the research (Sect. 1), description of the study area and dataset used (Sect. 2), details of various key terms (Sect. 3), proposed approach and its architecture (Sect. 4) discussion of the results and conclusion (Sect. 5).

2 Study Area

This study focuses on the city of Lucknow, Uttar Pradesh (Lat: 26.8467°, Long: 80.9462°), India which is home to over 3.3 million people and servers as the capital city for Uttar Pradesh. The city stands at approximately 123 m (404 ft.) above sea level. The area occupied by the city is 2,528 km^2 (976 sq. mi). The city is surrounded by rural towns and villages, rendering a diverse social environment to it. The Gomti River is Lucknow's principal topographical feature, dividing the city into: The Trans-Gomti and Cis-Gomti regions. When it comes to climate, Lucknow encounters all the seasons around the year leaving a humid climate in the city.

2.1 Dataset

In this study, we used the high resolution 10-m sentinel dataset for the city of Lucknow, India. We formed our own dataset of 300 urban and 450 non-urban images. We grouped the bands of the Sentinel 2 satellite into three classes as following: (i) B2 (490 nm), B3 (560 nm), B4 (665 nm), B8 (842 nm) and B8A (865 nm) (ii) B11 (1610 nm) and B12 (2190 nm) (iii) Texture Features (Variance, Entropy, and Dissimilarity) evaluated in ENVI, Version-4.7.

3 Methodology

3.1 Support Vector Machine (SVM)

Vapnik and Chervonenkis designed SVM algorithm in 1963 while Corinna Cortes and Vapnik proposed the current version in 1993 [13]. SVM are a group of supervised learning methods that can be applied to classification and regression. SVM is one of the well-known and widely used binary classifiers [13, 14]. SVM Regression is to create a model to predict future values by implying the known inputs [15]. SVMs can efficiently perform a non-linear classification using the kernel trick [16].

3.2   Random Forest (RF)

The random decision forest was first proposed by Ho in 1995 and was then extended by Leo Breiman and Adele Cutler [17]. Random Forests are a collaborative learning technique for classification [17], regression [18], that function by raising an assembly of decision trees at training time and generating the mode of the class i.e. classification and mean prediction of the individual trees.

3.3   Artificial Neural Network (ANN)

A computational and operational model for neural networks was developed by Warren McCulloch and Walter Pitts (1943). Mathematics and algorithms formed the roots of the model [19]. It created the way for neural network exploration into two approaches. Biological processes in the brain and application of neural networks are the two concepts on which ANN is based. Such systems learn to perform tasks by considering training samples [19], usually without task-specific programming. An ANN is composed of artificial neurons that are collectively connected together and organized in layers. Natural language processing, speech recognition, machine translation, social network filtering are some spheres where neural networks are implemented.

4   Proposed Approach (PA)

Figure 2 shows the working of our architecture in a diagrammatic format. As shown, Sentinel 2 satellite data was used for acquisition of training and testing samples for urban and nonurban area. These large satellite images were subsets to 300 images of urban and 450 images of nonurban regions forming a dataset. Three CNN architectures were employed for generating different series of feature vectors (F1, F2 and F3). The inputs to three CNN were given as following: (i) B2 (490 nm), B3 (560 nm), B4 (665 nm), B8 (842 nm) and B8A (865 nm) (ii) B11 (1610 nm) and B12 (2190 nm) (iii) Texture Features (Variance, Entropy, Dissimilarity). Individual feature vectors combined together furnished a final feature vector, passed through SVM classifier. After the classification, output was generated which was then comparatively analyzed with the results generated by current state of the art algorithms such as SVM, RF and ANN.

The overall architecture of our ensemble of CNNs is depicted in Fig. 1 [20–22]. The architecture of individual CNN is as follows: The network consists of eight layers with weights; in total, there are five convolutional and three fully connected layers. Our network makes the best use of the multinomial logistic regression objective,

Fig. 1 LANDSAT 8 satellite image of study area

equivalent to capitalize the mean across training samples of the log-probability of the correct label under the prediction distribution.

All kernel maps in the second layer are connected to the kernels of third convolutional layer. The kernel maps in the previous layer which exist on the same Graphics Processing Unit (GPU) are the only kernels which are connected to the following second, third and fourth convolutional layer. The neurons present in the three fully connected layers are connected to all the neurons in the preceding layer. The first and second convolutional layers precede the response-normalization layers. Max-pooling layers, are present at two positions, after response-normalization layers and after the fifth convolutional layer. The output delivered by every convolutional and fully-connected layer is passed through the ReLU non-linearity.

The first convolutional layer uses 96 kernels of size $11 \times 11 \times 3$ with a stride of 4 pixels to filter the $224 \times 224 \times 3$ input images. The output of the first convolutional layer goes to the second convolutional layer as input. The second convolutional layer filters it using 256 kernels of size $5 \times 5 \times 48$. The third, fourth, and fifth convolutional layers are directly connected to each another without the interference of any pooling or normalization layers. 384 kernels of size $3 \times 3 \times 256$ are present in the third convolutional layer which are connected to the outputs (gained from the output of first convolutional layer as input) of the second convolutional layer. Along with this, 384 kernels of size $3 \times 3 \times 192$ and 256 kernels of size $3 \times 3 \times 192$ are present in the fourth and fifth convolutional layer respectively. The three fully-connected layers have 4096 neurons each, adding up to 12,288 neurons in total.

5 Results and Discussion

Due to rapid population growth, urbanization has been drastically climbing up the graph. Since the human activities play an extremely crucial role in the environment and society, this growth needs to be constantly monitored to keep check on environment issues and degradation of economical sustainability. Satellite remote sensing provides us with a broad and retrospective view of large regions. This highly enhances our capability to assess the land cover and its changes.

Table 1, demonstrates that the highest classification accuracy (96.24%) with kappa value 0.942 was achieved by our proposed ensemble of CNNs, followed by RF (88%) with kappa value 0.852, and ANN (84%). It was observed that SVM produced the lowest classification accuracy (65%) with kappa value (0.681). For our proposed model, a total of 78 combinations of input parameters were tried as input to the ensemble of CNNs. For a specific set of parameters (Fig. 2), we observed the highest classification accuracy. Further, the results of our PA were compared with those of the current state of art algorithms (SVM, ANN, and RF). A total of 254 combinations of these models were tested on the collected dataset. The results of the highest classification accuracy (Table 1) achieved by these models were compared with our approach. Table 2 represents the obtained statistic measures during the training and the validation phases of the models. It was observed that the highest value of sensitivity (0.88) was observed for our proposed approach, while the lowest (0.55) for the SVM classifier. Whereas the highest value of specificity (0.92) was observed for RF, while the lowest value (0.62) for SVM classifier. In conclusion, Sentinel 2,

Table 1 Accuracy assessment

Classification	Accuracy (%)	Kappa coefficient
SVM	65	0.681
ANN	84	0.823
RF	88	0.852
PA	96.24	0.942

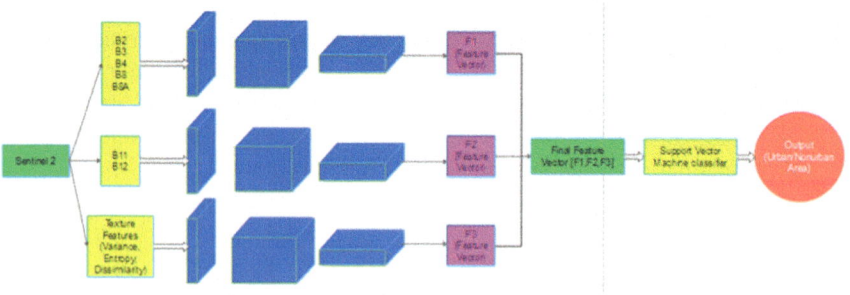

Fig. 2 Flow diagram of Proposed Approach (PA)

Table 2 Statistic measures for the training and validation process for SVM, RF and PA

		Accuracy	Precision	Specificity	AUC	Sensitivity
SVM [13]	Training	0.62	0.59	0.64	0.61	0.55
	Validation	0.65	0.60	0.62	0.63	0.58
ANN [19]	Training	0.83	0.84	0.74	0.80	0.72
	Validation	0.84	0.85	0.71	0.78	0.75
RF [17]	Training	0.89	0.91	0.87	0.83	0.78
	Validation	0.88	0.88	0.92	0.87	0.76
PA	Training	0.95	0.95	0.90	0.92	0.88
	Validation	0.962	0.96	0.91	0.94	0.87

high resolution satellite data, with other parameters is when given to ensembles of CNNs produce optimum classification results for urban and non-urban area.

The proposed accuracy can play quite a crucial role in resource management and urban planning at a time when urbanization is rapidly growing all over the world and possesses the potential to cause harm to environment and society.

References

1. Elmikaty, M., Stathaki, T.: Car detection in high-resolution urban scenes using multiple image descriptors. Pattern Recognit. Remote. Sens. **90**, 36–48 (2014)
2. Li, X., Gong, P., Liang, L.: A 30-year (1984–2013) record of annual urban dynamics of Beijing City derived from Landsat data. Remote Sens. Environ. **166**, 78–90 (2015)
3. Sidiqui, P., Huete, A., Devadas, R.: Spatio-temporal mapping and monitoring of Urban Heat Island patterns over Sydney, Australia using MODIS and Landsat-8. In: 2016 4th International Workshop on Earth Observation and Remote Sensing Applications (EORSA). IEEE (2016)
4. Soliman, O.S., Mahmoud, A.S.: A classification system for remote sensing satellite images using support vector machine with non-linear kernel functions. In: 2012 8th International Conference on Informatics and Systems (INFOS). IEEE (2012)
5. Huang, X., Lu, Q., Zhang, L.: A multi-index learning approach for classification of high-resolution remotely sensed images over urban areas. ISPRS J. Photogramm. Remote. Sens. **90**, 36–48 (2014)
6. Bouhennache, R., Bouden, T.: Using landsat images for urban change detection, a case study in Algiers town. In: 2014 Global Summit on Computer & Information Technology (GSCIT). IEEE (2014)
7. Huang, X., Lu, Q., Zhang, L.: A multi-index learning approach for classification of high-resolution remotely sensed images over urban areas. In: 2014 22nd International Conference on ISPRS Journal of Photogrammetry and (ICPR). IEEE (2014)
8. Gupta, S., et al.: An efficient use of random forest technique for SAR data classification. In: 2015 IEEE International Geoscience and Remote Sensing Symposium (IGARSS). IEEE (2015)
9. Mertes, C.M., et al.: Detecting change in urban areas at continental scales with MODISdata. Remote. Sens. Environ. **158**, 331–347 (2015)
10. Jean, N., et al.: Combining satellite imagery and machine learning to predict poverty. Science **353**(6301), 790–794 (2016)
11. Aydin, B., et al.: Automatic personality prediction from audiovisual data using random forest regression. In: 2016 23rd International Conference on Pattern Recognition (ICPR). IEEE (2016)

12. Kutucu, H., Almryad, A.: An application of artificial neural networks to assessment of the wind energy potential in Libya. In: 2016 7th International Conference on Sciences of Electronics, Technologies of Information and Telecommunications (SETIT). IEEE (2016)
13. Wang, L., et al.: Monitoring urban expansion of the Greater Toronto area from 1985 to 2013 using Landsat images. In: 2014 IEEE International Geoscience and Remote Sensing Symposium (IGARSS). IEEE (2014)
14. Yan, W.Y., Shaker, A., El-Ashmawy, N.: Urban land cover classification using airborne LiDAR data: a review. Remote. Sens. Environ. **158**, 295–310 (2015)
15. Zhou, Y., et al.: A cluster-based method to map urban area from DMSP/OLS nightlights. Remote. Sens. Environ. **147**, 173–185 (2014)
16. Ren, D., et al.: 3-D functional brain network classification using Convolutional Neural Networks. In: 2017 IEEE 14th International Symposium on Biomedical Imaging (ISBI 2017). IEEE (2017)
17. Sun, C., et al.: Quantifying different types of urban growth and the change dynamic in Guangzhou using multi-temporal remote sensing data. Int. J. Appl. Earth Obs. Geoinformation **21**, 409–417 (2013)
18. Jebur, M.N., et al.: Per-pixel and object-oriented classification methods for mapping urban land cover extraction using SPOT 5 imagery. Geocarto Int. **29**(7), 792–806
19. Kemper, T., et al.: Towards an automated monitoring of human settlements in South Africa using high resolution SPOT satellite imagery. Int. Arch. Photogramm., Remote. Sens. Spat. Inf. Sci. **40**(7), 1389 (2015)
20. Rahul, N., Sharma, H., Sahni, H., Batra, A.: A deep learning hybrid CNN framework approach for vegetation cover mapping using deep features. In: 2017 13th International Conference on Signal-Image Technology & Internet-Based Systems (SITIS), pp. 192–196. IEEE (2017)
21. Nijhawan, R., Joshi, D., Narang, N., Mittal, A., Mittal, A.: A futuristic deep learning framework approach for land use-land cover classification using remote sensing imagery. In: Mandal, J., Bhattacharyya, D., Auluck, N. (eds.) Advanced Computing and Communication Technologies. Advances in Intelligent Systems and Computing, vol. 702. Springer, Singapore (2019)
22. Nijhawan, R., Das, J., Balasubramanian, R.: J. Indian Soc. Remote Sens. **46**, 981 (2018). https://doi.org/10.1007/s12524-018-0750-x

Image-Based Facial Expression Recognition Using Local Neighborhood Difference Binary Pattern

Sumeet Saurav, Sanjay Singh, Madhulika Yadav and Ravi Saini

Abstract Automatic facial expression recognition (FER) has gained enormous interest among the computer vision researchers in recent years because of its potential deployment in many industrial, consumer, automobile, and societal applications. There are a number of techniques available in the literature for FER; among them, many appearance-based methods such as local binary pattern (LBP), local directional pattern (LDP), local ternary pattern (LTP), gradient local ternary pattern (GLTP), and improved local ternary pattern (IGLTP) have been shown to be very efficient and accurate. In this paper, we propose a new descriptor called local neighborhood difference binary pattern (LNDBP). This new descriptor is motivated by the recent success of local neighborhood difference pattern (LNDP) which has been proven to be very effective in image retrieval. The basic characteristic of LNDP as compared with the traditional LBP is that it generates binary patterns based on a mutual relationship of all neighboring pixels. Therefore, in order to use the benefit of both LNDP and LBP, we have proposed LNDBP descriptor. Moreover, since the extracted LNDBP features are of higher dimension, therefore a dimensionality reduction technique has been used to reduce the dimension of the LNDBP features. The reduced features are then classified using the kernel extreme learning machine (K-ELM) classifier. In order to, validate the performance of the proposed method, experiments have been conducted on two different FER datasets. The performance has been observed using well-known evaluation measures, such as accuracy, precision, recall, and F1-score. The proposed method has been compared with some of the state-of-the-art works available in the literature and found to be very effective and accurate.

Keywords Facial expression recognition (FER) · Local neighborhood difference pattern (LNDP) · Principal component analysis (PCA) · Kernel extreme learning machine (K-ELM)

S. Saurav (✉) · S. Singh · R. Saini
Academy of Scientific & Innovative Research (AcSIR), Chennai, Chennai, India
e-mail: sumeetssaurav@gmail.com

CSIR-Central Electronics Engineering Research Institute, Pilani, Pilani, India

M. Yadav
Department of Electronics, Banasthali Vidyapith, Vanasthali, Rajasthan, India

© Springer Nature Singapore Pte Ltd. 2020
B. B. Chaudhuri et al. (eds.), *Proceedings of 3rd International Conference on Computer Vision and Image Processing*, Advances in Intelligent Systems and Computing 1022,
https://doi.org/10.1007/978-981-32-9088-4_38

1 Introduction

Recently, automatic facial expression recognition (FER) has gained enormous interest among the computer vision researchers because of its potential deployment in many industrial, consumer, automobile, and societal applications. For example, such a technology could be embedded inside a robot for providing home services like talking to children and taking care of elderly people. In addition, FER-based technology can be deployed in a car to identify the fatigue level of driver and produce warning alarm in case the fatigue level exceeds the threshold limits, which will definitely avoid many accidents and save the lives of people. Automatic FER can also be used by companies to determine the worth of their products before their actual launch. Facial expression provides an important cue which reveals the actual intention and state of the mind of a person. Therefore, researchers are trying to develop systems for automatic FER which will have the capability of reading the facial expression of a person just like us and take necessary actions in response.

The techniques available in the literature for automatic facial expression recognition can be broadly classified into two main categories: geometric-based methods and appearance-based methods, whose details can be found in [1, 2]. As in this work, appearance-based method for facial feature extraction and, more specifically, texture-based approach has been used; therefore, a brief overview of different works related to our approach has been discussed below.

Well-known techniques which come under the category of appearance-based feature extraction methods consist of local binary patterns (LBP), local ternary pattern (LTP), local derivative pattern (LDP), local directional number pattern (LNDP), local directional texture pattern, local directional ternary pattern, and so on. The first successful demonstration of LBP for the purpose of FER has been reported in [3]. In this work, the authors have reported a comprehensive study on the role of LBP for FER. Although LBP is very effective and computationally efficient feature descriptor, it has been observed that the descriptor fails in the presence of non-monotonic illumination variation and random noise. This is because under such conditions a small change in gray-level values can easily change the LBP code [4]. To overcome this limitation, different new techniques have been developed, as well as different modifications to the original LBP have been done over the time. One such modification is Sobel-LBP [5]. The performance of the operator has been investigated on a facial recognition application and found to outperform the traditional LBP operator in terms of recognition accuracy. However, this operator also fails in uniform and near-uniform regions where it generates inconsistent patterns just like LBP as it also uses only two discrimination levels. To overcome this, LDP [6] was developed wherein a different texture coding scheme is used. In contrast to the LBP which uses gray-level intensity values, the LDP descriptor makes use of directional edge response values. Although LDP has been proved to be superior to LBP but it also face issues similar to Sobel-LBP as it also produces inconsistent patterns in uniform and near-uniform regions. In order to overcome the limitations of LDP and Sobel-LBP, LTP was developed. LTP adds an extra discrimination level and uses the ternary code as opposed to binary codes in LBP. More recently, a technique called gradient local ternary pattern (GLTP) [7] has been developed for the purpose of FER which combines Sobel operator with LTP operator. GLTP uses a three-level discrimination ternary coding scheme like LTP of gradient magnitudes obtained after Sobel operation to encode the texture of an image. As expected, GLTP has proved to be more effective for FER task compared to the earlier discussed operators. Another feature descriptor which was developed to overcome the limitations of the LBP is the Weber local descriptor (WLD) [8]. An important property of WLD is that it is less sensitive to noise and illumination changes and has been adopted for the purpose of FER in [9]. A more recent face descriptor called local directional ternary pattern (LDTP) has been developed for FER [2]. In LDTP emotion-related features are extracted using directional information and ternary coding scheme is used to overcome the weaknesses of other texture patterns in smooth regions of the image. Another recent method for FER which has been motivated by GLTP is improved gradient local ternary patterns (IGLTP) proposed by the authors in [10]. The improvements over GLTP include the following: use of a pre-processing step which improves the quality of the input image, use of a better and more accurate edge detection scheme using Scharr gradient operator, use of principal component analysis (PCA) to reduce the dimensions of the features, and finally making use of facial component extraction for discriminative feature extraction. A very recent work for FER has been proposed that makes use of multi-gradient features and five-level encoding called Elongated Quinary Pattern (EQP) [11] and the operator has been proved to be very effective for the purpose of FER.

The remainder of the paper is organized as follows: In Sect. 2 we have provided a brief description of the proposed methodology used in our work, which has been followed by experimental results and discussion in Sect. 3. Finally, Sect. 4 concludes the paper.

2 Proposed Methodology

The algorithmic pipeline used for the implementation of proposed facial expression recognition (FER) system has been shown in Fig. 1, wherein the blue arrowed line indicates steps used during training and the red is used during the testing phase of the proposed FER pipeline.

As shown in the figure, the first step detects a human face in the image under investigation and the detected face is then registered. The face registration step is very much essential as it allows the system to get facial features for different expressions at similar facial locations which give discriminative features. The vector of features extracted is then passed to dimensionality reduction to reduce the dimensions. The reduced feature is finally used for classification of the facial emotion using K-ELM classifier. A brief description of the different sequence of steps has been discussed in the further sub-sections.

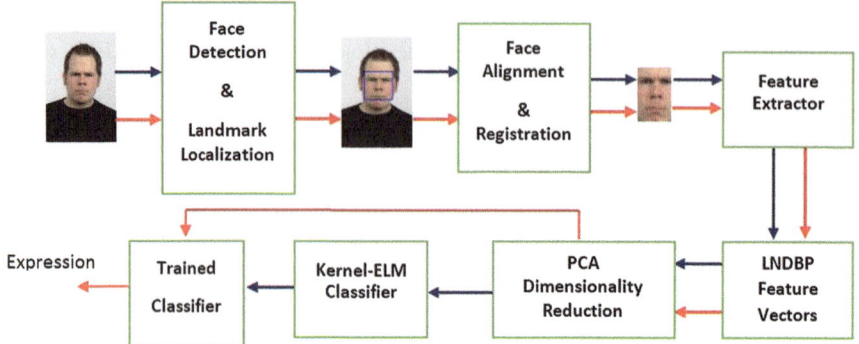

Fig. 1 Algorithmic pipeline of the proposed facial expression recognition system

2.1   Face Detection and Registration

In the first step, human faces are detected using the Viola and Jones frontal face
detector [12, 13], which is then passed to the facial landmark detection unit [14] that
marks the location of different landmarks on the face. Using coordinates of different
landmarks for the left and right eyes, the positions of the eye's center are calculated.
Based on the location of the eye's center, the image is rotated and in the subsequent
step, the area of interest is cropped and scaled to the specified size of (147 × 108)
in order to obtain the registered facial image.

2.2   Facial Feature Extraction and Dimensionality Reduction

The procedure used for feature extraction and dimensionality reduction has been
shown in Fig. 2.

As shown in the figure, a 3 × 3 pixel block taken from the input image is first
coded into its corresponding LBP and LNDP patterns (Fig. 2b, c). An OR operation
is then performed on these patterns to generate the LNDBP pattern, which is then
multiplied by a set of fixed weights assigned to each neighborhood pixels to generate
the LNDBP coded decimal value, as shown in Fig. 2e. Doing so retains the benefits
of both the LBP [15] and LNDP [16] operator. The LNDBP-coded image is then
divided into different cells (Fig. 2f) and an L2-normalized histogram of these cells
is concatenated which encodes the complete facial feature information of the face.
Since the size of the LNDBP feature is too large, therefore, dimensionality reduction
using PCA has been used to reduce the dimension of the features. An important point
to note here is that the algorithm does not encode the input image into corresponding
LBP and LNDP-coded images. These images have been shown in Fig. 2d just to
have a visual representation of these two encoded images (LBP and LNDP). A brief
description of LBP and LNDP descriptors has been mentioned in the following.

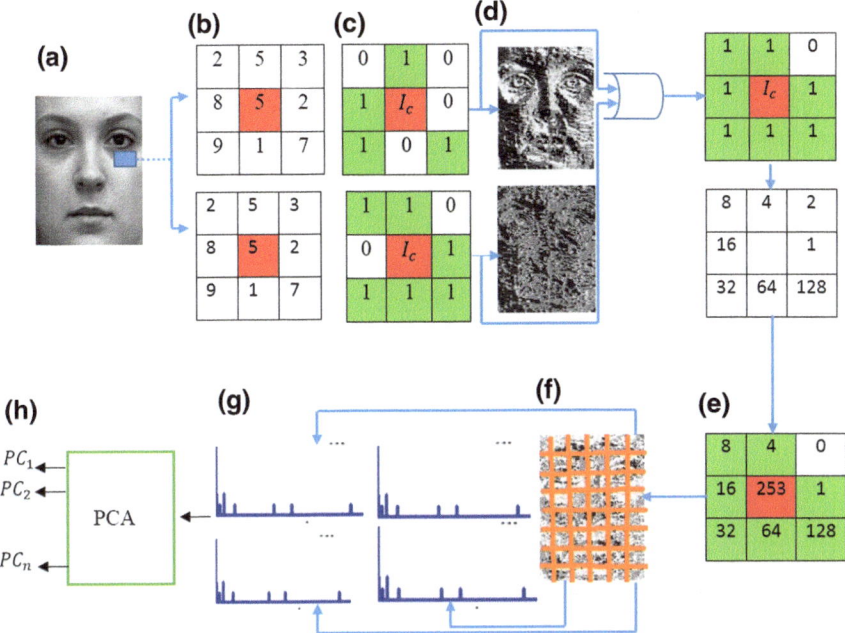

Fig. 2 Representation of feature extraction and dimensionality reduction technique

Local binary pattern (LBP) is a very popular texture descriptor which first computes a corresponding LBP pattern using a 3×3 window by comparing the neighboring pixels with the center pixel. These patterns are then multiplied by some weights and summed to generate the LBP-encoded pixel value for the corresponding center pixel value. Local binary pattern for a center pixel I_c and neighboring pixels I_n ($n = 1, 2, \ldots, 8$) can be computed as in (1) and (2), respectively.

$$LBP_{P,R}(x, y) = \sum_{n=0}^{P-1} 2^n \times F_1(I_n - I_c) \tag{1}$$

$$F_1(I) = \begin{cases} 1 & I \geq 0 \\ 0 & else \end{cases} \tag{2}$$

In (1), R is the radius of neighboring pixels, P is the number of neighboring pixels and (x, y) are the coordinates of center pixel. The histogram of LBP map is calculated using (3) and (4), wherein $m \times n$ refers to the size of image and l is the pattern value.

$$His(l)|_{LBP} = \sum_{x=1}^{m} \sum_{y=1}^{n} F_2(LBP(x, y), l); L \in \left[0, \left(2^P - 1\right)\right] \tag{3}$$

$$F_2(a, b) = \begin{cases} 1 & a = b \\ 0 & else \end{cases} \tag{4}$$

The procedure used for the computation of LNDP feature can be referred from [16], wherein for a 3×3 block of pixels, each neighboring pixel is compared with the two most adjacent and appropriate pixels, which are either vertical or horizontal, and is computed using (5)–(7).

$$k_1^n = I_8 - I_n, k_2^n = I_{n+1} - I_n, \quad for \; n = 1 \tag{5}$$

$$k_1^n = I_{n-1} - I_n, k_2^n = I_{n+1} - I_n, \quad for \; n = 2, 3, \ldots, 7 \tag{6}$$

$$k_1^n = I_{n-1} - I_n, k_2^n = I_1 - I_n, \quad for \; n = 8 \tag{7}$$

In the above equations, I_n ($n = 1, 2, \ldots, 8$) are the neighboring pixels of a center pixel I_c and k_1^n and k_2^n are the difference of each neighborhood pixel with two other neighborhood pixels. A binary number is assigned to each neighboring pixel with the help of (8).

$$F_3\left(k_1^n, k_2^n\right) = \begin{cases} 1, & if \; k_1^n \geq 0 \, and \, k_2^n \geq 0 \\ 1, & if \; k_1^n < 0 \, and \, k_2^n < 0 \\ 0, & if \; k_1^n \geq 0 \, and \, k_2^n < 0 \\ 0, & if \; k_1^n < 0 \, and \, k_2^n \geq 0 \end{cases} \tag{8}$$

Using the above binary values, LNDP value for the center pixel is computed as in (9) and then the histogram is calculated as in (10).

$$LNDP(I_c) = \sum_{n=1}^{8} 2^{n-1} \times F_3\left(k_1^n, k_2^n\right) \tag{9}$$

$$His(l)|_{LNDP} = \sum_{x=1}^{m} \sum_{y=1}^{n} F_2(LNDP(x, y), l); \; L \in [0, (2^8 - 1)] \tag{10}$$

In pattern recognition, it is often found that either having too few features or a very large feature vector often results in classifier failure. Even if we are using the best of classifiers, having too few features results in sub-optimal performance on one hand, and on the other hand, a large feature dimension makes the classification process slow and also causes loss in the classification accuracy. Therefore, in order to overcome the issue of slow classification speed and to increase the classification accuracy of the classifier, a dimensionality reduction technique using PCA [17] has been employed to reduce the dimension of the extracted LNDP features.

2.3 Kernel Extreme Learning Machine (K-ELM) Classifier

An extreme learning machine (ELM) [18] classifier is basically a machine learning algorithm used for fast training a single-layer feed-forward neural network (SLFN). Unlike the traditional back propagation algorithm which is used for training a neural network, the ELM does not involve any iteration; therefore computation of the output weights β does not involve any iteration and have a direct solution. ELM invariably makes use of a feature mapping function $h(x)$ which helps in learning nonlinearity. If the mapping function is not known, kernel technique can be applied into ELM based on Mercers' condition [19]. The output vector $f(x)$ of a kernel ELM can be represented as in (11).

$$f(x) = h(x)\beta = h(x)H^T \left(\frac{I}{C} + HH^T\right)^{-1} T = \begin{bmatrix} \phi(x, x_1) \\ \vdots \\ \phi(x, x_{N_K}) \end{bmatrix} \left(\frac{I}{C} + \Phi\right)^{-1} T$$

(11)

where

$$\Phi = HH^T = \begin{bmatrix} \phi(x_1, x_1) & \cdots & \phi(x, x_{N_k}) \\ \vdots & \ddots & \vdots \\ \phi(x_{N_k}, x_1) & \cdots & \phi(x_{N_k}, x_{N_k}) \end{bmatrix}$$

and N_k is the randomly selected training samples used for training the K-ELM classifier. The kernel ϕ used in this work is the Gaussian function represented as in (12).

$$\phi(x, x_1) = \exp\left(-\frac{\|x_i - x_j\|^2}{\sigma^2}\right)$$

(12)

In the above equation, σ denotes the spread (i.e., standard deviation) of the Gaussian function.

3 Experimental Results and Discussion

In this section, we discuss various experiments which were performed on different FER datasets. All the experiments in this work have been performed using MATLAB 2015a run on a Windows platform.

3.1 Datasets

Two different FER datasets have been used in the experiments. The first one is the extended Cohn-Kanade (CK+) dataset [20] which is an extended version of the CK dataset. In our experimental setup, both six-class and seven-class expression images have been used, which were obtained from 309 labeled sequences selected from 106 subjects. For six-class expression recognition, from each labeled sequence we selected the three most expressive images resulting in 927 images, and for creating the seven-class expression dataset, we simply added the first image of neutral expression from each of the 309 sequence to the six-class dataset, resulting in a total of 1236 images which is similar to that in [7]. The second dataset used in the experiments is the recently introduced Radbound Faces database (RFD) [21]. The dataset contains images of 67 subjects performing eight facial expressions (anger, disgust, fear, happiness, contemptuous, sadness, surprise, and neutral) with three gaze directions. However, in our experiments, the frontal gaze direction images comprising seven expressions (anger, disgust, fear, happy, neutral, sad, and surprise) counted to a total of 469 images were only used for a fair evaluation of results with the work presented in [22].

3.2 Parameter Selection

There are a number of parameters involved in the design of any automatic FER system. The optimal value of these parameters needs to be determined for an accurate and efficient FER, and for this, a number of experiments were performed. First experiment involved determination of the registered facial image size and the size of the cell in which the LNDBP-encoded image was divided. From the recent works [10] and [22], in our work, two different facial image resolution of size 65×59 and 147×108 pixels were adopted with variable cell sizes. The experiments were performed on CK+ 7 expression dataset with ten-fold cross-validation strategy which was repeated ten times. Kernel extreme learning machine (K-ELM) with regularization parameter C and kernel parameter γ value of 100 and 200, respectively, was used. The results of the experiments in terms of average overall accuracy, precision, recall, and F1-score have been tabulated in Tables 1 and 2 for image resolutions of 65×59 and 147×108, respectively, for different cell sizes. On the basis of tables, we find that the facial image with a resolution of 147×108 and cell size of 12×11 performed well compared to all other combinations and therefore, this value of facial image size and cell size was used in all our further experiments.

From the equations of K-ELM, one can find out that there are two parameters involved, which needs to be determined to obtain a better classification accuracy. In order to obtain the optimal value of these parameters, a grid search was performed along with ten-fold cross-validation which was repeated ten times as done in the above experiments. The experiments were performed on both the datasets involved

Table 1 Determination of cell size on 65 × 59 image size

[65, 59]	[5, 4]	**[6, 5]**	[7, 6]	[8, 7]	[9, 8]	[10, 9]	[11, 10]	[12, 11]
Avg. acc. 10 runs	98.8 ± 0.2	**99.3 ± 0.2**	99.0 ± 0.2	98.7 ± 0.2	98.5 ± 0.2	98.2 ± 0.2	96.9 ± 0.2	97.6 ± 0.3
Feature dim.	43,008	**28,160**	20,736	14,336	12,544	9216	6400	6400
Avg. acc.	99.03	**99.44**	99.27	98.94	98.63	98.46	97.24	98.13
Avg. prec.	99.04	**99.52**	99.40	98.87	98.81	98.35	96.58	98.02
Avg. rec.	99.21	**99.53**	99.28	99.37	98.68	98.85	98.03	98.38
Avg. F1-S	99.12	**99.53**	99.34	99.11	98.74	98.59	97.27	98.20

Table 2 Determination of cell size on 147 × 108 image size

[147, 108]	[7, 6]	[8, 7]	[9, 8]	[10, 9]	[11, 10]	**[12, 11]**	[13, 12]	[14, 13]
Avg. acc. 10 runs	98.9 ± 0.3	99.1 ± 0.2	99.3 ± 0.1	99.3 ± 0.1	99.1 ± 0.2	**99.4 ± 0.2**	99.0 ± 0.1	99.1 ± 0.3
Feature dim.	87,040	69,120	53,248	39,424	33,280	**27,648**	22,528	20,480
Avg. acc.	99.27	99.43	99.51	99.51	99.35	**99.51**	99.27	99.43
Avg. prec.	99.55	99.64	99.70	99.64	99.54	**99.70**	99.52	99.64
Avg. rec.	99.21	99.42	99.46	99.52	99.37	**99.52**	99.27	99.42
Avg. F1-S	99.38	99.53	99.58	99.58	99.45	**99.60**	99.39	99.53

in this work and the results have been tabulated in Table 3. The range of values for both C and γ which were taken was [1:10] in logarithmic scale of base 2. In all further experiments, these values of the K-ELM parameters were used for deriving

Table 3 Determination of K-ELM parameter

Performance measure/datasets	CK+ 7 expressions	RFD 7 expression
Avg. accuracy 10 runs	99.5 ± 0.1	97.6 ± 0.2
Kernel parameter (γ)	64	1024
Regularization parameter (C)	32	64

the accuracy of the proposed FER pipeline on the individual FER datasets used in this work.

In the next experiment, the optimal number of principal components (PCs) was determined. To do this, values of all other parameters were fixed to the optimal value determined in the earlier experiments. Here again, ten-fold cross-validation testing protocol was used which was repeated ten times. The experimental result on CK+ dataset has been tabulated in Table 4 and that on RFD in Table 5.

Table 4 Determination of no. of principal component using CK+ 7 expression dataset

No. of PCs	32	64	96	128	160	192	224	256
Avg. acc. 10 runs	98.7 ± 0.2	99.2 ± 0.2	99.3 ± 0.2	99.4 ± 0.1	99.3 ± 0.2	99.3 ± 0.1	**99.5 ± 0.1**	99.4 ± 0.2
Avg. acc.	98.9	99.5	99.5	99.6	99.5	99.4	**99.6**	99.6
Avg. prec.	99.2	99.7	99.7	99.7	99.7	99.6	**99.7**	99.7
Avg. rec.	98.7	99.4	99.4	99.5	99.4	99.3	**99.5**	99.6
Avg. F1-S	98.9	99.5	99.6	99.6	99.6	99.5	**99.6**	99.6
Comp. time (Sec)	0.15	0.153	0.164	0.18	0.20	0.21	**0.27**	0.28

Table 5 Determination of no. of principal components using RFD 7 expression dataset

No. of PCs	32	64	128	160	224	256	288	320
Avg. acc. 10 runs	93.3 ± 0.7	94.7 ± 0.5	96.2 ± 0.3	97.2 ± 0.3	97.3 ± 0.2	97.4 ± 0.3	97.4 ± 0.3	**97.7 ± 0.3**
Avg. acc.	94.2	95.3	97.0	97.6	97.4	97.8	97.9	**98.1**
Avg. prec.	94.2	95.3	97.0	97.6	97.4	97.8	97.9	**98.1**
Avg. recall	94.4	95.4	97.1	97.7	97.5	97.9	97.9	**98.1**
Avg. F1-score	94.2	95.3	96.9	97.6	97.4	97.8	97.9	**98.1**
Conmp. time (Sec)	0.15	0.15	0.18	0.20	0.27	0.28	0.30	**0.33**

3.3 Results on CK+ Dataset

In order to determine the performance of the proposed FER pipeline on CK+ dataset, ten-fold cross-validation using K-ELM was performed, which was repeated ten times with the value of C and γ determined using grid-search. On CK+ 6 expression, the accuracy achieved using both LNDBP and LNDBP+PCA was 100 and on CK+ 7 expression dataset the FER pipeline achieved an accuracy of 99.5 ± 0.1 using both LNDBP and LNDBP+PCA. The performance in terms of different measures has been shown in Tables 6 and 7 corresponding to CK+ 6 and CK+ 7 expressions, respectively.

Table 6 Performance of LNDBP+PCA (CK+ 6 expressions)

Actual/predicted	An	Di	Fe	Ha	Sa	Su	Recall
An	135	0	0	0	0	0	100
Di	0	177	0	0	0	0	100
Fe	0	0	75	0	0	0	100
Ha	0	0	0	207	0	0	100
Sa	0	0	0	0	84	0	100
Su	0	0	0	0	0	249	100
Precision	100	100	100	100	100	100	
F1-score	100	100	100	100	100	100	

Avg. Performance: recall = 100, precision = 100, accuracy = 100, F1-score = 100

Table 7 Performance of LNDBP+PCA (CK+ 7 expressions)

Actual/predicted	An	Di	Fe	Ha	Ne	Sa	Su	Recall
An	135	0	0	0	0	0	0	100
Di	0	177	0	0	0	0	0	100
Fe	0	0	75	0	0	0	0	100
Ha	0	0	0	207	0	0	0	100
Ne	2	0	0	1	305	1	0	98.7
Sa	0	0	0	0	0	84	0	100
Su	0	0	0	0	1	0	248	99.6
Precision	98.5	100	100	99.5	99.7	98.8	100	
F1-score	99.3	100	100	99.7	99.2	99.4	99.8	

Avg. performance: recall = 99.7, precision = 99.5, accuracy = 99.6, F1-score = 99.6

3.4 Results on RFD Dataset

Performance of the proposed FER pipeline on RFD dataset in terms of avg. precision, avg. accuracy, avg. recall, and avg. F1-score corresponding to the best ten-fold cross-validation run out of the 10 runs has been mentioned in Table 8. The overall average accuracy of the 10 runs obtained after ten-fold cross-validation using LNDBP and LNDBP+PCA is 97.7 ± 0.2 and 97.7 ± 0.3, respectively. Here, also the optimal value of the parameters determined in the earlier experiments was used.

In order to test the performance of the proposed FER pipeline, cross-dataset performance evaluation using LNDBP+PCA was also done, wherein one of the datasets was kept as the training data and the other one as the test dataset. The results of the experiment have been tabulated in Table 9 in terms of different performance measures.

Comparison result of the proposed FER framework with other state-of-the-art approaches has been shown in Table 10. As shown in the table, the proposed approach

Table 8 Performance of LNDBP+PCA (RFD 7 expressions)

Actual/predicted	An	Di	Fe	Ha	Ne	Sa	Su	Recall
An	67	0	0	0	0	0	0	100
Di	0	67	0	0	0	0	0	100
Fe	0	0	62	0	1	1	3	92.5
Ha	0	0	0	67	0	0	0	100
Ne	0	0	0	0	67	0	0	100
Sa	1	0	1	0	1	64	0	95.5
Su	0	0	0	0	1	0	66	98.5
Precision	98.5	100	98.4	100	95.7	98.5	95.6	
F1-score	99.3	100	95.4	100	97.8	96.9	97.1	

Avg. performance: recall = 98.1, precision = 98.1, accuracy = 98.1, F1-score = 98.1

Table 9 Performance of LNDBP+PCA (cross-dataset)

Database/perf. measure	Avg. precision	Avg. recall	Avg. F1-score	Testing time (s)	Testing accuracy (%)	C	γ
Train: CK+ 7 expressions Test: RFD 7 expressions	86.8	81.0	78.71	0.02	81.0	32	64
Train: RFD 7 expressions Test: CK+ 7 expressions	81.7	79.9	79.3	0.02	81.1	64	1024

Table 10 Comparison of recognition accuracy (%) on CK+ and RFD dataset

Method	CK+ 6 expressions	CK+ 7 expressions	RFD 7 expressions
LTP [23]	93.6	88.9	–
GLTP [7]	97.2	91.7	–
Improved GLTP [10]	99.3	97.6	–
HOG [22]	95.8	94.1	94.9
LNDBP proposed	**100**	**99.5**	**97.7**
LNDBP+PCA proposed	100	**99.5**	**97.7**

has superior performance compared to the other approaches available in the literature.

4 Conclusion

In the presented paper, a new facial feature descriptor has been proposed and named as local neighborhood difference binary patterns (LNDBP). The proposed feature extractor incorporates the benefits of LBP which computes the relationship of neighboring pixels with center pixel and LNDP which extracts the relationship among neighboring pixels by comparing them mutually. LNDBP is computed by using a simple OR operation on the local binary pattern and local neighborhood difference pattern. The proposed descriptor has proved to be very effective in extracting discriminate facial attributes. Dimensionality reduction technique using PCA was also used to reduce the dimensions of the LNDBP features. K-ELM classifier was used to classify the facial attributes obtained from different facial images into their corresponding labels. Two different testing protocols were used in the experiments: the first one using ten-fold cross-validation which was repeated ten times and the second one using cross-dataset evaluation wherein one dataset was used as the training set and the other as the test set. The experiments were performed on two FER datasets, namely CK+ and RFD and performance has been observed using precision, recall, accuracy, and F1-score. Performance of the proposed approach was also compared with some of the state-of-the-art works available in the literature and the employed performance measure clearly indicated that the proposed method outperforms other methods in terms of recognition accuracy and efficiency.

References

1. Rivera, A.R., Castillo, J.R., Chae, O.O.: Local directional number pattern for face analysis: face and expression recognition. IEEE Trans. Image Process. **22**(5), 1740–1752 (2013)
2. Ryu, B., Rivera, A.R., Kim, J., Chae, O.: Local directional ternary pattern for facial expression recognition. IEEE Trans. Image Process. **26**(12), 6006–6018 (2017)
3. Shan, C., Gong, S., McOwan, P.W.: Facial expression recognition based on local binary patterns: a comprehensive study. Image Vis. Comput. **27**(6), 803–816 (2009)
4. Zhou, H., Wang, R., Wang, C.: A novel extended local-binary-pattern operator for texture analysis. Inf. Sci. **178**(22), 4314–4325 (2008)
5. Zhao, S., Gao, Y., Zhang, B.: Sobel-lbp. In: 15th IEEE International Conference on Image Processing, pp. 2144–2147 (2008)
6. Jabid, T., Kabir, M.H., Chae, O.: Facial expression recognition using local directional pattern (LDP). In: 17th IEEE International Conference on Image Processing, pp. 1605–1608 (2010)
7. Ahmed, F., Hossain, E.: Automated facial expression recognition using gradient-based ternary texture patterns. Chin. J. Eng. (2013)
8. Chen, J., Shan, S., He, C., Zhao, G., Pietikainen, M., Chen, X., Gao, W.: WLD: a robust local image descriptor. IEEE Trans. Pattern Anal. Mach. Intell. **32**(9), 1705–1720 (2010)
9. Alhussein, M.: Automatic facial emotion recognition using weber local descriptor for e-Healthcare system. Clust. Comput. **19**(1), 99–108 (2016)
10. Holder, R.P., Tapamo, J.R.: Improved gradient local ternary patterns for facial expression recognition. EURASIP J. Image Video Process. (1), 42 (2017)
11. Al-Sumaidaee, S.A.M., Abdullah, M.A.M., Al-Nima, R.R.O., Dlay, S.S., Chambers, J.A.: Multi-gradient features and elongated quinary pattern encoding for image-based facial expression recognition. Pattern Recogn. **71**, 249–263 (2017)
12. Viola, P., Jones, M.J.: Robust real-time face detection. Int. J. Comput. Vision **57**(2), 137–154 (2004)
13. Martin, K.: Efficient metric learning for real-world face recognition. http://lrs.icg.tugraz.at/pubs/koestinger_phd_13.pdf
14. Xiong, X., De la Torre, F.: Supervised descent method and its applications to face alignment. In: IEEE Conference on Computer Vision and Pattern Recognition (CVPR), pp. 532–539 (2013)
15. Ojala, T., Pietikäinen, M., Harwood, D.: A comparative study of texture measures with classification based on featured distributions. Pattern Recogn. **29**(1), 51–59 (1996)
16. Verma, M., Raman, B.: Local neighborhood difference pattern: a new feature descriptor for natural and texture image retrieval. Multimed. Tools Appl., 1–24 (2017)
17. Jolliffe, I.: Principal component analysis. In: International encyclopedia of statistical science, pp. 1094–1096. Springer, Berlin, Heidelberg (2011)
18. Huang, G.B., Zhou, H., Ding, X., Zhang, R.: Extreme learning machine for regression and multiclass classification. IEEE Trans. Syst., Man, Cybern., Part B (Cybernetics), **42**(2), 513–529 (2012)
19. Huang, Z., Yu, Y., Gu, J., Liu, H.: An efficient method for traffic sign recognition based on extreme learning machine. IEEE Trans. Cybern. **47**(4), 920–933 (2017)
20. Lucey, P., Cohn, J.F., Kanade, T., Saragih, J., Ambadar, Z., Matthews, I.: The extended cohn-kanade dataset (ck+): a complete dataset for action unit and emotion-specified expression. In: IEEE Computer Vision and Pattern Recognition Workshops (CVPRW), pp. 94–101 (2010)
21. Langner, O., Dotsch, R., Bijlstra, G., Wigboldus, D.H., Hawk, S.T., Van Knippenberg, A.D.: Presentation and validation of the Radboud Faces Database. Cogn. Emot. **24**(8), 1377–1388 (2010)
22. Carcagnì, P., Coco, M., Leo, M., Distante, C.: Facial expression recognition and histograms of oriented gradients: a comprehensive study. SpringerPlus **4**(1), 645 (2015)
23. Ahmed, F., Kabir, M. H.: Directional ternary pattern (dtp) for facial expression recognition. In: IEEE International Conference on Consumer Electronics (ICCE), pp. 265–266 (2012)

Author Index

© Springer Nature Singapore Pte Ltd. 2020
B. B. Chaudhuri et al. (eds.), *Proceedings of 3rd International Conference on Computer Vision and Image Processing*, Advances in Intelligent Systems and Computing 1022,
https://doi.org/10.1007/978-981-32-9088-4

Printed by Printforce, the Netherlands